MIRAGE

MIRAGE

Power, Politics, and the
Hidden History of Arabian Oil

Aileen Keating

Prometheus Books

59 John Glenn Drive
Amherst, New York 14228-2197

Published 2005 by Prometheus Books

Inquiries should be addressed to
Prometheus Books
59 John Glenn Drive
Amherst, New York 14228–2197
VOICE: 716–691–0133, ext. 207
FAX: 716–564–2711
WWW.PROMETHEUSBOOKS.COM

09 08 07 06 05   5 4 3 2 1

Library of Congress Cataloging-in-Publication Data

Keating, Aileen.
    Mirage : power, politics, and the hidden history of Arabian oil / Aileen Keating.
    p. cm.
    Includes bibliographical references and index.
ISBN 1-59102-346-7 (hardcover : alk. paper)
    1. Holmes, Frank, 1874-1947. 2. Petroleum industry and trade—Middle East—History. 3. Petroleum industry and trade—Government policy—Middle East. 4. Petroleum—Prospecting—Middle East—History. I. Title.

HD9576.M52K43 2005
338.2'7282'0953—dc22
                                                                    2005013881

Printed in the United States of America on acid-free paper

CONTENTS

6   Contents

INTRODUCTION

The discrepancy between the Western view of the history of Arabia's oil and the view held by the Arabs of the Persian Gulf first caught my attention in 1973. As a journalist, I was on assignment covering political developments and the drafting of a constitution in postindependent Bahrain. The island was alive with discussion of what was termed the "oil weapon," the Arab embargo on oil shipments that was being used for the first time as a political tool. The American and British media were full of reports expressing outrage that the Arabs could consider adopting such a measure "after all we've done for them" alongside supporting stories about American and British "oil pioneers" who had toiled in the heat and dust of the Arabian deserts to discover the oil.

In government offices, and in the homes and coffee shops of Bahrain, I was told emphatically that neither the Americans nor the British discovered the oil of Arabia. Every Bahraini seemed to know that the true discoverer was a New Zealander—*Abu Al Naft*—though few knew that his name was Frank Holmes. After working outside the Middle East for some years, I returned to Kuwait in 1979, where I was again assured that *Abu Al Naft* was indeed as the name implied, "the Father of Oil" in Arabia.

Settling into Bahrain for a long posting in 1980, I began to pursue this mystery. Most of the current publications I read barely mentioned the New Zealander Frank Holmes. Those that did depicted him not as the heroic discoverer of oil but as a concession hunter of somewhat dubious reputation. This was also the picture I was given of Holmes by the American and British managers of the Bahrain Petroleum Company, the Kuwait Oil Company, and the Arabian American Oil Company (Aramco) in Saudi Arabia, at least by those who had even heard of Frank Holmes.

By now I had spoken to Husain bin Ali Yateem, the nephew of Muhammad Yateem, the Bahraini who was Holmes's colleague and personal assistant from 1922 to 1938. Frank Holmes and his wife, Dorothy, had cared for Husain while he attended school in Brighton, England. Back in Bahrain, as a very young man,

7

Husain had been sent by Holmes on a mission to the shaikh of Qatar. Husain bin Ali Yateem knew, without a shred of doubt, that Frank Holmes had discovered the oil of the Arabian Peninsula. The Yateem family's records, however, had been destroyed by fire in their Bahrain offices some years earlier.

I searched for books published closer to the time of discovery. Tantalizing glimpses of Frank Holmes began to emerge in publications such as Ameen Rihani's 1928 *Ibn Sa'oud of Arabia: His People and His Land*. I pursued a first trawl through the India Office Library in London and the Abu Dhabi Documentation Center. I spoke with Dame Violet Dickson in Kuwait. Her husband, Harold Dickson (1881–1959), was a political officer with the Government of India and is featured throughout this story. Although she was at pains to impress on me what a truly engaging personality Frank Holmes was, her memory at that stage did not extend to dates or hard facts.

In London, I met with Archibald Chisholm (1902–1988). Chisholm was joint negotiator with Holmes in 1934 for the Kuwait Oil Company concession and in 1978 published *The First Kuwait Oil Concession Agreement: A Record of the Negotiations, 1911–1934*. While Chisholm also extolled Holmes's personality, he could not agree that it was Holmes who actually pinpointed the Arabian oil fields, although he did acknowledge that he had first identified Kuwait as rich in oil. Chisholm wanted credit for discovering Kuwait's oil to be given to the British geologists of the Anglo-Persian Oil Company (in 1935 renamed the Anglo-Iranian Oil Company and, in 1954, British Petroleum).

The primary source breakthrough came by tracing Holmes's family members, now scattered throughout the world in New Zealand, Australia, England, the United States, and South America. From Holmes's surviving personal papers, and letters of his wife, Dorothy, I was able to construct a time line of what Holmes was doing, where and when. Documents of Holmes's younger brother, Percy, assisted to fill in the gaps and provided a picture of Holmes's personal and professional background before he entered the Arab world. The personal correspondence between Swiss geologist Dr. Arnold Heim and Frank Holmes held at Zurich University was enlightening. The material of Guy Scholefield, editor of the *Dictionary of New Zealand Biography*, held at the Alexander Turnbull Library was useful and contained some additionally informative personal letters.

The publication in 1990 of Angela Clarke's *Bahrain Oil and Development, 1929–1989* to celebrate the Bahrain Petroleum Company's sixtieth anniversary, was a setback with its comment that "opportunist businessman" Frank Holmes was "instrumental in securing oil concessions" and its claim that "Arabia's first oil well was struck by American pioneers." Daniel Yergin's *The Prize: Epic Quest for Oil, Money, and Power* quickly followed in 1991 and, although this was a little kinder to Holmes, he was still described as a concession broker.

Despite these two publications, the Bahrainis, Kuwaitis, Qataris, and Saudis remained adamant. In conversations with me they still insisted that it was Frank

Holmes who was *Abu Al Naft*. They couldn't all be wrong. I set out to demonstrate once and for all whether or not Frank Holmes was, as the Arabs of the Gulf insist, the true discoverer of the oil of Arabia.

I searched the Chevron Oil Archives in San Francisco and the American Heritage Center's International Archive of Economic Geology at the University of Wyoming. A satisfying exercise was the search through the personal correspondence related to the 1965 publication of Thomas E. Ward's *Negotiations for Oil Concessions in Bahrain, El Hasa (Saudi Arabia), the Neutral Zone, Qatar and Kuwait*, and Edgar Wesley Owen's 1975 *Trek of the Oil Finders: A History of Exploration for Petroleum*. The postpublication correspondence on both books was particularly productive.[1]

I needed to find something on the American mining engineer Karl Twitchell (1885–1968). In 1927 he was sent by the American bathroom fitting millionaire and philanthropist Charles Crane to Yemen to report on mineral possibilities. One of the first people he ran into was Frank Holmes's eccentric British colleague, Cdr. C. E. V. Craufurd of the Royal Navy (Retired). Craufurd was passionately obsessed with locating the site of King Solomon's Mines and was convinced he had traced them to the port of Mukalla, not far from Aden. When the two men met, Twitchell was similarly infected by Craufurd's obsession, but he believed King Solomon's Mines were in what is now Saudi Arabia. Charles Crane sent Twitchell from Yemen to Saudi Arabia in an unsuccessful search for water. On his return to New York, Twitchell was hired by Standard Oil of California to assist in obtaining Frank Holmes's lapsed Al Hasa concession. The myth subsequently developed that Twitchell had discovered the oil of Saudi Arabia.

I had a hunch that there must be, somewhere, the papers of Karl Twitchell. I was frustrated for many months as I pursued this possibility. Discovery of Twitchell's papers was an exhilarating experience. I found them, uncataloged in eleven cartons, at the Seeley G. Mudd Manuscript Library (Public Policy Papers and University Archives) at Princeton University.

In the early years the engaging Lebanese-American writer Ameen Rihani was closely associated with Frank Holmes. I was also convinced that somewhere there should be a body of correspondence between Frank Holmes and Ameen Rihani. Discovery of a cache of Rihani-Holmes letters in Lebanon provided me with a treasure trove of primary source material.

I knew that Frank Holmes and the thirty-first president of the United States, Herbert Hoover (1874–1964), had worked together in the great Kalgoorlie goldfields of Western Australia when both were only twenty-two years old. Holmes and Hoover had gone together to China and had remained friends. For some time I had this connection only through secondary sources. To my delight, the discovery of Hoover's letter of 1908 among the personal correspondence held by the Hoover Library, and mention of the relationship between Hoover and Holmes in the Chevron Oil Archives, allowed me to prove this connection in primary sources.

As I worked through the American documentation on the history of Arabia's oil, I frequently thought that the downplaying of the achievements of Frank Holmes in the written record, in favor of American "oil pioneers," might be attributed to the public perception of the American oil industry itself. From its very inception the industry was forced to defend its reputation. The American oil industry, for example, appeared before some twenty congressional hearings beginning with the trust busting actions of 1890–1911 that ended with the breaking up of the original Standard Oil.

From that experience, the American oil companies recognized the value of positive publicity and did not stint on investing in powerful public relations departments. As early as 1919 Standard Oil of California had a public relations practitioner whose experience included stints as a foreign correspondent in Asia, South America, and England—before becoming chief of foreign intelligence and chief of information at the State Department, from where he was hired by Standard Oil of California.[2]

I also wondered why the British had denied Frank Holmes his place in history. A second careful search of the India Office Library, and the personal papers of various Holmes's contemporaries at the Middle East Center at St. Antony's College Oxford, produced a bigger, and even more startling insight that resulted in my widening the scope of this project beyond demonstrating that it was the New Zealander Frank Holmes who discovered Arabia's oil.

I have dubbed the picture that emerged the "Gulf Raj." It is a picture of an Arabian Peninsula ruled not by "the British" as is usually described, but an Arabian Peninsula ruled by the Government of India.

That this factor is not evident in works dealing with the period remains a puzzle to me. Briton Cooper Busch in his esteemed *Britain and the Persian Gulf* (1967) and *Britain, India, and the Arabs* (1971) comments on the Government of India in Mesopotamia (modern Iraq), but does not extend his attention to the Arabian Peninsula. Most writers imply that the presence of the Government of India faded after World War I, giving way to "the British." While this is true for Mesopotamia/Iraq, it is not what happened in the Arabian Peninsula.

It seems that even some Arab scholars are confused by this loose categorization of "the British." In *Beyond Oil: Unity and Development in the Gulf*, the respected Kuwait University academic Muhammad Rumaihi remarks on the discussion among Arab legal analysts of whether or not the shaikhdoms were ever true protectorates. Rumaihi says Arab scholars also disagree "about the nature of the ties between the Amirates and the British authorities, and about whether relations existed with the Bombay government (the British regime in India) or with London directly."[3]

In entering this discussion, I have drawn on the primary source documentation to show that the relations of the Arab shaikhdoms of the Persian Gulf were indeed with what Rumaihi terms "the Bombay government." This relationship, in the form of the authority of the Government of India for all matters except "policy," was reconfirmed by Winston Churchill at the Cairo Conference of 1921 and remained in force until the Foreign Office took over responsibility in 1947 on the independence of India.

"Policy" was not defined although this was usually accepted to mean such matters as international relations but also auditing of accounts, supply of materials, and grant of local allowances. In practice the Government of India retained full control of general policy as well as administration. Until 1947, while the administrators and the political officers were certainly "British," they were members of the Indian military or the Indian civil service. They received instructions from India and reported, not to the British government, but to the Government of India.

This is clear in the new Orders-in-Council enacted following the August 1947 Indian Independence Act. The new orders refer to previous arrangements linking the Arab shaikhdoms to the Government of India, saying that "the Governor General of India exercised a number of powers." It was now necessary, these documents explain, to "transfer elsewhere all powers hitherto exercised by the Governor General of India." Herbert Liebesny, on the staff of the Foundation for Foreign Affairs in Washington, pointed out in the *Middle East Journal* in 1947 that "the Political Residency, and the Political Agencies in the Persian Gulf are subordinate to the Government of India."[4]

How the true dimension of the role of the Government of India—the Raj— in the Arabian Peninsula could have been obscured was a question I needed to address. As I delved deeper, I found frequent advice by the Government of India and British officials, including Winston Churchill himself, to avoid "attracting attention" to the assumption and nature of the rule in the Persian Gulf.

The Government of India regularly instructed its political officers that the rulers themselves should be persuaded to "request" certain actions. Even without direct instructions from above, the political officers themselves were not averse to authoring requests and letters that the shaikhs then signed, or of writing in the shaikhs' names whether or not they signed. Taken at face value, this gave the impression that the shaikhs had more freedom of action than was actually the case.

The Government of India's political officers were also in the habit of speaking for the shaikhs. The political agents, and at times the resident, forwarded reports purporting to quote verbatim the individual shaikhs. It became very noticeable to me how often the shaikhs were quoted as expressing ideas that were remarkably similar to those contained in the agents' own personal correspondence. One political agent, Harold Dickson, who features prominently in this story, was a master of this.

The technique could thrive because the arrangements reached at Churchill's 1921 Cairo Conference restricted the British government to communicating about, or with, the shaikhs solely through the Government of India's political resident in the Persian Gulf. The same restriction applied to the shaikhs in that they could communicate with what they called the High Government in London only through the Government of India's political agents and the resident. Officials in Britain would have had little chance of independently ascertaining what the shaikh might, or might not, have thought, said, or written.

It seems reasonably safe to assume that it was to the benefit of the political officers to report that the shaikhs were in accord with whatever measures were currently being implemented. Perhaps because it was what they wanted to hear, the officials in Britain accepted information received in this fashion. The British government's meetings and reports on matters relating to the Arab shaikhdoms utilized and incorporated the reports of the political officers of the Government of India. In this way the material entered the records and was assumed to be independent observations of the British government.

It must be said that such unquestioning acceptance by British officials of the reports of the Government of India's political officers in the Persian Gulf was in keeping with the times. The belief prevailed that these were men who understood how to "deal with" the Arabs. The Orientalist view of the "Eastern Mind" was alive and well in the offices of Whitehall.

Once having grasped this all-important dimension, I moved on to track the equally important India associations of the Anglo-Persian Oil Company. That the British government held 51 percent of this company from 1914 did exert influence in matters related to the oil of Iraq, but I found it had little bearing on the development of Arabia's oil. The key factors I identified were the company's historic ties to India and the fact that the Anglo-Persian Oil Company was predominantly staffed by former members of the Indian military, or the Indian civil service.

There was active collusion between the Anglo-Persian Oil Company and its personnel with the political officers of the Government of India. Both institutions were trying to protect their dominant positions in the Persian Gulf. The aim of the Anglo-Persian Oil Company was to maintain a de facto commercial monopoly. The goal of the Government of India was to keep away any outside influence that might not be sympathetic to its somewhat unique ideology. In pursuit of their shared goal the two institutions and their personnel combined in actions aimed first at thwarting the award of the Arabian concessions to Frank Holmes and then to prevent their development with American financial backing.

Now the trials and tribulations experienced by Holmes in his effort to develop the Arabian oil fields he had discovered fell into place. Previously it had seemed to me that, at times, the objections to Holmes were instances of using a sledgehammer to crack a nut. But now the reaction to the perceived threat presented in the person and activities of Holmes came into context. Finally, I had the story. I would explore the episode of the Raj in Arabia through following Holmes from discovery of the area's oil to resolution of the granting and development of the Arabian concessions.

The whole episode was a cruel exercise in imperialism. The people of the Arab shaikhdoms were in dire economic straits, most particularly following the Depression and the coincident collapse of their main source of revenue, the market for natural pearls. Yet, for the purpose of maintaining its dominant position and that of the Anglo-Persian Oil Company, the Government of India refused the possibility of economic relief through development of Frank Holmes's oil concessions, even more so when he secured financial backing from the Americans.

It is not surprising that distrust of "foreign" governments was engendered in the general population of the Arab shaikhdoms. Also not surprising is that this distrust was compounded by the realization that, after the concessions were developed, the American and British governments made more money in taxes from Arabia's oil than Arabia did from its extraction and sale.

There was an obvious benefit gained by diminishing Frank Holmes's achievements. Professing to have discovered the oil themselves allowed propagation of American and British claims to possess a "right" to Arabia's oil—claims that steadily expanded to become the West in general possessing a right to Arabia's oil. Arab recognition of the denial of their economic development through oil, and profit-taking when it was developed, generated strong suspicion of the motivation behind such claims.

This historic sense of entitlement on the one side, and deeply embedded wariness on the other, was apparent in the discussions around the final takeover by the Gulf states of full ownership of their respective oil companies. Both attitudes can be discerned today in the approach, and reaction, of each side to the subject of the Gulf states' policies on production and pricing of their oil.

In maintaining their international relations the Gulf states must take the residual suspicion of other governments' motivation toward them felt by the people of the Arabian Peninsula into account as public opinion. Conversely, the Western nations' relations with the Gulf states are often based on a belief that the Americans and British discovered Arabia's oil and so pulled the region out of its abject poverty. Holding this mythology-induced view leads the West to the conclusion that the Arabs owe them something "after all we've done for them."

Some writers appear to have been aware of the inconsistencies, discrepancies, and misinterpretation in the accepted accounts. Preparing a thesis for Princeton University in 1948, Frederick Lee Moore observed: "The impact of the discovery of the Arabian oil resources is recent, and so great, as to overshadow its origin . . . further investigation of the events will have to wait upon the opening of government and oil company files." He noted that Karl Twitchell's account of the discovery of Saudi Arabia's oil did not ring true but commented kindly: "This is not to infer that Twitchell deliberately has not made a true report . . . but only that the record in his book is incomplete." Archibald Chisholm, who went on to become head of public relations at British Petroleum 1945–62, candidly admitted the bias in his book on Kuwait's oil as he said it "was written primarily for the archives of the British Petroleum Company."

In January 1950 Roy Lebkicher felt it necessary to include in the *Handbook for American Employees* in Saudi Arabia the following passage: "In mentioning the lack of interest of the British oil companies in Bahrain and Saudi Arabia . . . no disparagement of their judgement is intended. They had good reasons for their views (that there was no oil in Arabia). The point is made primarily to correct an impression commonly held that Standard Oil of California 'knew' there was oil in Bahrain and Saudi Arabia before beginning its ventures in these countries."

Standard Oil of California had bought Frank Holmes's original concessions in both Bahrain and Saudi Arabia. Lebkicher would have been dealing with the problem that, despite the public relations efforts of the oil companies, the Arabs of the Persian Gulf "knew" there was oil in these concessions, because Holmes said there was, and after he was proven correct, they continued to openly praise him as *Abu Al Naft*—the Father of Oil, the true discoverer of the Arabian oil fields.

Nevertheless, the American "correction" of what was indeed common knowledge at the time has been so effective that in 1999 *Aramco World* could unflinchingly close an article titled "Prelude to Discovery" by stating that four months after the May 1933 resale to Standard Oil of California of Holmes's lapsed 1923 Al Hasa concession, two American geologists arrived and "the search for oil in Saudi Arabia had begun."

In the introduction to his 1982 PhD thesis for the University of London on the negotiations for the Kuwait oil concession, Yossef Bilovich pointed out "the history of the Persian Gulf oil concessions has never been fully outlined nor analysed in any depth." He added: "The accounts provided by former employees of the various oil companies have been inclined to treat their publications merely as an exercise in public relations." In a 1988 article, Bilovich commented that "the Bahrain case is relatively unknown and the existing record quite erroneous."

In 1971 Pulitzer Prize–winning author Wallace Stegner, in his book on Standard Oil of California and Saudi Arabia, observed that the story of that concession "has been told several times." "But," he noted, "it has never been quite fully or quite accurately told, even by those who participated in it." Stegner did not claim to redress this injustice as he commented enigmatically "some day, some historian . . . will relate the episode in detail."[5]

I hope I have succeeded in telling the whole story, fully and quite accurately, of the discovery of Arabia's oil, and of the role played by the Raj in Arabia.

NOTES

1. The Arab oil embargo was imposed in response to the re-arming of Israel by Western countries, particularly United States, in the midst of the 1973 Arab-Israel Conflict; even as the war raged, President Nixon went to Congress to obtain US$2.2 billion in emergency aid for Israel. Archibald H. T. Chisholm, *The First Kuwait Oil Concession*

Agreement: A Record of the Negotiations, 1911–1934 (London: F. Cass, 1975); Angela Clarke, *Bahrain Oil and Development, 1929–1989*, for the sixtieth anniversary of Bahrain Petroleum Company (London: Immel Publishing, 1991), p. 53, "opportunist businessman"; Daniel Yergin, *The Prize: The Epic Quest for Oil, Money, and Power* (New York: Simon & Schuster, 1991); Thomas E. Ward, *Negotiations for Oil Concessions in Bahrain, El Hasa (Saudi Arabia), the Neutral Zone, Qatar and Kuwait* (New York: privately published, 1965); and Wesley Edgar Owen, ed., *Trek of the Oil Finders: A History of Exploration for Petroleum*, Semicentennial Commemorative Volume (Tulsa, OK: American Association of Petroleum Geologists, 1975).

2. Yergin, *The Prize*, pp. 97–113; for break up of Standard Oil, see Gerald T. White, *Formative Years in the Far West: A History of Standard Oil Company of California and Predecessors through 1919* (New York: Appleton-Century-Crofts, 1962), p. 557; the public relations practitioner was the thirty-five-year-old Philip H. Patchin.

3. Briton Cooper Busch, *Britain and the Persian Gulf, 1894–1914* (Berkeley: University of California Press, 1967) and *Britain, India, and the Arabs, 1914–1921* (Berkeley: University of California Press, 1971); Muhammad G. Rumaihi, *Beyond Oil: Unity and Development in the Gulf* (London: Al Saqi Books, 1986), p. 51, "nature of the ties." Rumaihi remarks on the argument they were not protectorates and so did not relinquish legal sovereignty. Rumaihi refers particularly to Hussein Muhammad Al Baharna, *The Modern Gulf States: Their International Relations and the Development of Their Political, Legal and Constitutional Positions* (Beirut, 1973).

4. India Office Library (hereafter cited as IOL)/R/L/PS/11/193, January 31, 1921, "Report of the Interdepartmental Committee appointed by the Prime Minister to make recommendations as to the formation of a new department under the Colonial Office to deal with mandated and other territories in the Middle East," Clauses 9–12 deal with the "Persian Gulf littoral"; and IOL/R/L/PS/10/1273/PG/Sub/18, December 12, 1929, Committee of Imperial Defence, Persian Gulf, Report by a Sub-Committee on Political Control; Appendix "De Facto" position as regards political arrangements in the Persian Gulf; Herbert J. Liebesny, "Documents: British Jurisdiction in the States of the Persian Gulf," *Middle East Journal* 2 (1949): 330–32, "Governor General of India"; Liebesny, "International Relations of Arabia: The Dependent Areas," *Middle East Journal* 1 (1947): 148–68, "subordinate to the Government of India"; Liebesny, "Administration and Legal Development in Arabia: The Persian Gulf Principalities," *Middle East Journal* 10 (1956): 33–42, "only very recently have the Arab principalities of the Persian Gulf been officially put into the category, at least for some purposes, of British-protected states."

5. Frederick Lee Moore Jr., *Origin of American Oil Concessions in Bahrain, Kuwait, and Saudi Arabia* (BA thesis, Department of History, Princeton University, 1948), introduction to bibliography, "overshadow its origins" and p. 82, "is incomplete"; Ward Papers, March 15, 1965, Chisholm to Ward, "primarily for . . . BP"; Roy Lebkicher, "Aramco and World Oil," *Arabian American Oil Company Handbook for American Employees* 1 (New York: Russell F. Moore, 1950): 25, "to correct an impression"; Jane Waldron Grutz, "Prelude to Discovery," *Aramco World* 50, no. 1 (January–February 1999): 31–34, "search had begun"; Yossef Bilovich, "Great Power and Corporate Rivalry in Kuwait, 1912–1934: A Study in International Politics (PhD thesis, University of London, 1982), p. v, "never fully outlined"; Bilovich, "The Quest for Oil in Bahrain, 1923–1930: A Study in British and American Policy," in *The Great Powers in the Middle*

East, 1919–1939, ed. Uriel Dann (New York: Holmes and Meier, 1988), pp. 252–68, "record quite erroneous"; Wallace Stegner, *Discovery! The Search for Arabian Oil*, abridged for *Aramco World* (Beirut, Lebanon: Middle East Export Press, 1971), pp. 17–18, "some day some historian. . . ."

SETTING THE
HISTORICAL STAGE

THE RAJ IN ARABIA

"When there are no horses, saddle the dogs" is a local proverb that surely illustrates the resilient character with which the Arabs of the Gulf endured the domination of the Government of India.

What the Indians called the *angrezi raj*—English rule—extended across the Arabian Peninsula and had begun in December 1600 when England's Queen Elizabeth I granted a charter under the title *Governor and Company of Merchants of London Trading into the East Indies*. Run by a governor (chairman) and twenty-four directors appointed from among its shareholders, what became known as the East India Company penetrated as far as Japan in its early voyages. In 1610 and 1611 its first factories, or trading posts staffed by "factors," otherwise known as agents, were established in Madras and Bombay. King Charles II (1630–1685), an enthusiastic promoter of a British navy, granted the company sovereign rights that allowed it to acquire territory, exercise the various functions of government including legislation and administration of justice, issue currency, negotiate treaties, and wage war, all in its own name. The long rule of India formally began in 1689 when the Indian provinces of Bengal, Madras, and Bombay were declared administrative districts, termed presidencies, under the direct control of the company.

Under Elizabeth's original 1600 charter the East India Company was granted a monopoly of trade in Asia, Africa, and America. There was only one restriction: the company must not attempt to compete with the prior trading rights of "any Christian Prince."

Under this proviso Muslim Persia seemed to be there for the taking. The company lost no time in establishing a presence at the Persian port of Bushire. The 1617 appointment of the first British ambassador to the Persian Court followed. Entirely commercial at first with a monopoly on silk exports from and woolen imports into Persia, freedom from all taxes and a promise that no other European power would be permitted to open a trading station, the East India Company's presence at Bushire soon assumed a political character.

Employing its own military, the East India Company extended its control in India through conquest and "persuading" local rulers to cede concessions and territories. Robert Clive, in his capacity as an official of the East India Company, emerged victorious over battles with the French in 1751, 1757, and 1761, establishing what was still a private commercial company as the dominating power in India. In the process "Clive of India" became an extraordinarily wealthy man.

In Persia by 1763 the trading station at Bushire had transformed into the Residency, or, more correctly, the Political Residency. Some in Britain were becoming uncomfortable with the unfettered power and privilege a private British company was exercising in India. The 1773 Regulating Act was instituted with the intention of giving the British government the right to oversee and regulate the affairs of India. The idea was to establish a governor-generalship of India that would be accountable to the British parliament. It didn't quite work out when the East India Company managed to get its very own governor of Bengal appointed as the first governor-general of India. Another ten years went by before the British parliament's India Act of 1784 passed limited oversight of India affairs to a Board of Control whose president was a member of the British cabinet.

Everybody wanted to be nice to the East India Company, its ruthless military mercenaries, and its British backers. In the early 1800s the Persian government gave a "gift" of land that quickly grew into Sabzabad located some seven miles inland from the port of Bushire. The importance of Persia increased when envious agitation among those members of the British upper class who were not shareholders in the East India Company forced the 1813 abolition of its monopoly of the India trade (in 1833 it also lost its monopoly on the China trade). The East India Company didn't hurt too much. The loss of its monopoly was compensated by its 10.5 percent dividends being converted into a fixed annual charge—to be paid from India's national revenue. The company's administrative authority continued unchecked.

At Sabzabad the ground space and sprawl of buildings soon expanded considerably, particularly when the British Expedition Force used it as their forward base for a war they waged against Persia in 1856–57. In the subsequent Treaty of Paris, Persia surrendered to Britain all claims to Herat and Persian territories in present-day Afghanistan. In the same year that Britain was waging the Anglo-Persian War, the Indian Mutiny of 1857–59 erupted against the harshness of rule by the East India Company. The immediate result was the 1858 Act for the Better Government of India. Now the British Crown took over, assuming administrative responsibilities and incorporating the private twenty-four-thousand-man fighting force into the Anglo-Indian army.

The new act established two sources of executive power. In London, there would be an India Office with a secretary of state for India to give regular reports to the British parliament. In India, a powerful Viceroy of India would be in charge, appointed by the British government. The British monarch would not be declared ruler until 1877 when Queen Victoria took the title Empress of India. (The company had been dissolved three years earlier by the East India Stock Dividend Redemption Act of January 1874.)

Under its exalted viceroy, the new Government of India re-created mirror images of most British government departments, including its own Department of Foreign Affairs. The hierarchy included a bevy of presidency and provincial governors, commissioners, judges, magistrates, and district officers. Known as agents when the East India Company employed them, senior district officers now became political agents." The province of the men of the Government of India extended into the Persian Gulf.

By 1862 the officer in charge of the Political Residency in Persia had become Her Britannic Majesty's political resident in the Persian Gulf, a title that inferred rights on both sides of that waterway whether the land was Persian or Arab. The people of the Gulf referred to the resident as *balyoz*, a term used by the Turks centuries earlier for the powerful representative of the Venetian Republic at the sultan's court. The purpose of the Political Residency and its work was now officially described as "protecting the route to India and developing influence, prestige and commerce."

Britain's assumption of the Persian Gulf was, more or less, internationally claimed in 1903 with the declaration aimed primarily but not solely at an expansionist Russia. The British government declared it "would resist with all the means at our disposal" any attempt "by any other Power" to establish a naval base or a fortified port anywhere in the Persian Gulf. Citing the doctrine laid down by US president James Monroe (1758–1831) that no other power would be allowed to colonize any part of continental America, Lord George Nathaniel Curzon, viceroy of India (1898–1905), decreed he now had "our Monroe doctrine in the Middle East" and immediately set about strengthening the Government of India's grip on both sides of the Persian Gulf.[1]

The Gulf Raj

Curzon scheduled the Arab shaikhdoms on the southern side of the Persian Gulf into his 1903 almost royal tour. He traveled in a fleet including six British men-of-war and two Indian marine vessels—the largest fleet to visit since the founder of the Portuguese empire in the Orient, Afonso de Albuquerque, terrified the entire region from 1510–15. Lord Curzon carried with him a purpose-built platform decked out with gold embroidered carpets and hangings and crowned with his personal gold and silver throne. From this magnificent vantage, Curzon addressed the Arabs assembled and squatting on the bare ground below. Curzon's majestic example would set in place the imperial style in which British officials of the Government of India would treat the Arabs of the Persian Gulf for the coming half century.

Curzon, and the Government of India, had regard for what they thought of as the sophisticated culture of Persia and for the timeless civilization that was Mesopotamia. Occupying the plain between the Tigris and Euphrates rivers, Mesopotamia had been the center of the Sumerian, Babylonian, and Assyrian civilizations. The Turks initially conquered here in 1534 but did not secure true control until the seventeenth century when the three Mesopotamian provinces of Baghdad, Basra, and Mosul were brought under the direct administration of the Ottoman Empire.

In antiquity it had been known as *Al Iraq*, and the Arabs had always used both Al Iraq and Mesopotamia interchangeably. In modern history others knew only Mesopotamia, as used by the Ottomans. The British and the Government of India continued to use Mesopotamia until the anti-British revolt of 1920 when plans were drawn up to institute a "provisional" government for what would now be called the State of Iraq. (The revived name came into official use in August 1921 when Faisal was proclaimed king of Iraq.)

Possibly they were ignorant that the area that would become Saudi Arabia was a thriving civilization mentioned, together with the province that would become Kuwait, by Herodotus in 430 BCE and again by the Greek geographer Strabo in 22 CE. Bahrain, under its ancient name of Dilmun, was celebrated in the Sumerian texts from the third millennium BCE. The hero of the Sumerian poem the *Epic of Gilgamesh* traveled to Dilmun in his quest for the secret of immortality.

Whether it was a lack of knowledge, or a lack of recognizable commercial opportunity, neither Curzon nor the Government of India displayed much respect for the people and prospects on the Arabian side of the Persian Gulf. Lord Curzon described the sultan of Muscat as "pathetic" and the shaikh of Kuwait as "sinister." He was personally offended by the shaikh of Bahrain's "chilly antagonism" to his proposal that Bahrain's main source of income, the customs dues, be given over in order to meet the cost of the Government of India "administering" his country.[2]

At the start of the First World War there were three high-powered representatives of the Government of India in the region. At Sabzabad, the Government of India's appointee now carried the ponderous title of His Britannic Majesty's Political Resident in the Persian Gulf and Consul General for Fars, Khuzistan, and Luristan. From 1873 he received his orders from the Government of India, before that orders had come from the Bombay presidency. Subordinate to this office were the Political Agencies in the individual Arab shaikhdoms, also staffed and administered by the Government of India. The British minister in Tehran was an entirely separate establishment. The British minister in Tehran reported to the

Foreign Office in London while the political resident at Bushire reported to the Government of India; these two officials had no direct contact.

A second Government of India appointee was the political resident in Baghdad, with a consul subordinate to him stationed in Basra. The Baghdad resident, although drawn from the Government of India's Political Department, reported to London through the British ambassador in Constantinople. The political resident officiating at Aden was a third Government of India appointee. After occupation by a British force in 1839, Aden had been annexed to India where it became part of the Bombay presidency. The conquest and absorption of Aden—an Arab land—to India created a precedent. Aden provided a model to follow in the Government of India's later goal of similarly absorbing Iraq into an expanded Indian empire.

The Government of India paid the cost of all Persian Gulf officers, half the cost of the British minister's establishment in Tehran, half the charges subsidizing mail boats to Baghdad, and an annual contribution to London for an Admiralty provision of navy patrols and "special" assignments as requested.

This was jurisdictional confusion writ large.

One British official later commented that "the nature of our political association" in the Persian Gulf was "an association so unusual and so obscure as to be virtually incomprehensible to the layman . . . and baffling even to the official mind." Even scholars became lost in the maze. One expert wrote, "The anomalous position in relation to the internal affairs of Persia and the affairs of the Persian Gulf, respectively British and Indian Government spheres of interest, were never really resolved" until the independence of India in 1947.[3]

In the Persian Gulf, the Government of India's political officers referred to the actual British government as the High Government. To the people of the Arabian Peninsula the High Government was a remote and threatening concept. Until their independence in the 1970s, the status of the Arab shaikhdoms were described by the British as being "in special treaty relations with the British government." This was a unique category, fitting neither colony nor protectorate.

The British employed two basic models of colonization: direct rule and indirect rule. The first imposed administration by British colonial officials without any involvement of the native authorities. Indirect rule was that in which some administrative functions were in the hands of native institutions but under the control of British colonial officials. The status of Aden, for example, moved from being a district within the Bombay presidency at its 1839 occupation and annexation to that of being a separate province of India in 1932. The "Island of Bombay" had itself been "transferred to the British Crown" as part of the dowry of the Infanta of Portugal on her marriage to King Charles II. After Charles died in 1685 the British government "sanctioned" the takeover of Bombay by the East India Company whereupon the company promptly "raised troops" and instituted its own administration. Aden was bounced a third time when the 1935 Government of India Act separated it from India and further legislation transformed it into a British Crown colony, under direct colonial administration.[4]

A British protectorate, even if it closely resembled a colony, was distinguished in that it did not become part of the national territory of the protecting power but remained legally a foreign country. Natives of a protectorate were not British subjects. They did not owe formal allegiance to the British Crown although a duty of "obedience" was taken for granted. The right of the Crown to exercise legislative and administrative functions in protectorates was based on, apart from treaties with the local rulers, the Foreign Jurisdiction Act of 1890.

The Arab shaikhdoms of the Persian Gulf were not colonies, nor would they ever be declared protectorates. An example of how a protectorate was created was the enactment of the Aden Protectorate Order of 1937 that declared the hinterland to be the Western Protectorate and the Eastern Protectorate. The governor of what was now the Crown colony of Aden also became governor of the two protectorates.

The post–World War I mandates awarded by the League of Nations would introduce yet another form of administration to the Middle East. But, as will be seen, this would not be extended to the Arab shaikhdoms either. As late as 1947, one international legal authority could still remark that "the exact status of the states along the coast of the Arabian Peninsula is not clarified."

The dominance of the British position in the Persian Gulf was achieved through the Government of India's men. To underscore their negotiations they frequently called in the troops of the Indian military and gunboats of the Indian marines. By the beginning of the twentieth century the Arab rulers in the Persian Gulf had been persuaded into a series of treaties, including a pledge not to entertain overtures from any other power and not to sell, cede, or lease any part of their territory without first gaining permission from the High Government.

Ostensibly, the Arab shaikhdoms signed away control only of their foreign relations, but by regular use of the men and ships of the Indian marines, and their armaments, the political resident had no trouble in ensuring that his "advise" was followed in all matters. The smaller shaikhdoms were so locked into Government of India control that they entirely lost their individual identities and were lumped together under the dismissive title of the Trucial States (the modern United Arab Emirates).

The political resident in the Persian Gulf was so powerful, his authority so absolute, that while he was viceroy of India a satisfied Lord Curzon claimed: "The Political Resident at Bushire, who has always been a British officer of the Government of India, is really the uncrowned King of the Gulf."

The last official appointed by the Government of India would not retire until 1958, making the Arabian Peninsula actually the last bastion of the Raj. In practice, it would endure a decade beyond India's own independence of 1947.[5]

Indianization

An agreement was signed on July 29, 1913, although never ratified, between the Ottoman Imperial Government and the Government of His Britannic Majesty. The British recognized the shaikhdom of Kuwait as part of the Ottoman Empire, and Turkey renounced all claims to sovereignty over Bahrain and the Qatar Peninsula. The effect of this formal carve-up and recognition of respective masters was felt first in the Bahrain Islands. While standing by the letter of the agreement and not openly annexing Bahrain, even as the agreement was being signed, the last of a Three-Step Indianization program was being completed.

The program's first stage, coincidental with Lord Curzon's 1904 personal appointment of his own protégé, Percy Zachariah Cox, as political resident, was the upgrading in Bahrain of the 1900 post of political assistant-in-charge to the position of political agent, reporting to the resident in Bushire. The political agent in Bahrain quickly gained jurisdiction over "subjects of the British Empire" on the islands. In 1909 the second stage was completed, with the political agent obtaining jurisdiction over all "foreigners" along the lines of the Indian Foreign Jurisdiction Order of 1890. The third and last step of the Indianization was the 1913 establishment of the "Bahrain Order-in-Council" that imposed on Bahrain the civil and criminal laws of India "as if Bahrain were a district in the Presidency of Bombay."

The effect was to treat Bahrain as a possession of the Government of India, with the political resident in the Persian Gulf "having much the same powers over Bahrain as a Colonial Governor." Similar orders-in-council would be applied to Muscat, Kuwait, Qatar, and the Trucial States. The Indian criminal and civil codes and other Indian legislation were imposed, and appeals from the shaikhdoms' courts, presided over by the Government of India's resident and political agents, lay with the High Court in Bombay.[6]

Kuwait was described in the 1913 Anglo-Turkish Agreement as an "autonomous *Qada* of the Ottoman Empire" using a Turkish flag with a Kuwait logo. But in November 1914, in an attempt to gain the shaikh of Kuwait's cooperation in the "liberation" of Mesopotamia, the British would define the status of Kuwait, in writing, as an "independent state under British protection." The resident, Percy Cox, boasted to T. E. Lawrence that he, Cox, had personally "made an Arab flag for Kuwait." With a Government of India political agent, and the ceding of control of its commercial and external relations through a number of agreements and treaties, despite its "Arab flag" Kuwait was definitely in the British orbit.

Modern Saudi Arabia can be traced to the Anglo-Saudi Treaty, signed by Percy Cox and Abdul Aziz bin Abdul Rahman bin Saud on December 26, 1915. Under this treaty the British government recognized the Central Arabian provinces of

Nejd, Al Hasa, Katif, and Jubail as bin Saud's territories, with bin Saud as their ruler and his sons and descendants after him. In return for this recognition, bin Saud, son of the House of Saud as he was called across the Arab world, also agreed to the usual clauses of not entering into any correspondence, agreement, or treaty with any other foreign power. Like the other shaikhs before him, he also agreed not to cede, sell, lease, mortgage, or otherwise dispose of any part of his territories to any other power, or to the subjects of any other power, without the consent of the High Government, whose advice he would follow unreservedly. By November 1916, "a combination of every sort of restrictive treaty concluded by Britain in the Persian Gulf over the previous century," including a Government of India political agent at Doha, would bring Qatar firmly into the fold.

With the gunboats of the Indian marines ever on the horizon and platoons of Indian Sikhs kept in easy striking distance, there was little the Arabs of the Persian Gulf could do to resist Britain's assumption of power through its proxy, the Government of India. Bin Saud, the future king of Saudi Arabia, expressed it very well. In 1922 he would explain: "I will put my seal, if they say I must, but I will strike when I can. What I cede of my rights under force, I will get back . . . when I have sufficient force."[7]

The means by which the Gulf Arabs escaped the harsh embrace of the Government of India was not force; it was development of their oil resources. New Zealander Frank Holmes would discover that oil in 1922–23, then would have to fight the Government of India, and the British Empire, to develop it. He would bring in the Americans and would have to fight all over again on their behalf. The Government of India and her officials in the Persian Gulf would heap objections, obstacles, and obfuscations in his path in their efforts to deter Frank Holmes from his goal of developing Arabia's oil fields.

NO OIL IN ARABIA

The discovery and development of oil in Arabia is a story set in the context of the imperialism of the day and punctuated by the faults and drawbacks of imperial and colonial systems. The big successes, in Persia and in Iraq, were a matter of hijacking cottage industries that already existed in oil extraction—as had been done in Burma. In Arabia, where the great oil fields waited to be discovered rather than merely appropriated, the oil companies failed completely.

For thirty years, from the first expedition mounted by the Geological Survey of India in 1904 until the seventh "investigation" of Kuwait's oil seepages in 1932, the belief that Arabia held no prospects for oil was an article of faith. Geological survey followed geological survey and one after the other each reached the same conclusion—there was no oil in Arabia.

The conclusion owed as much to the organizational structures in which the

oil seekers were employed, an environment of patronage, paternalism, and pro-
tégés, as it did to the scientific standards of the day. Once Arabia had been
decreed devoid of oil by the acknowledged experts, few men of ambition would
disagree with the opinion of their superiors—certainly not when promotion
depended on loyalty, being a "team player," and keeping the job long enough to
achieve seniority. This obedient respect for superiors was actively approved by a
hierarchical management structure that drew its members from a closed network
of like-minded and well-connected servants of the empire. Early twentieth-cen-
tury British corporations, such as the Anglo-Persian Oil Company, seemed no
more inclined toward meritocracy than was the empire's civil service.

Writing the history of British Petroleum, the company that had originally
been the Anglo-Persian Oil Company, R. W. Ferrier attempted to rationalize the
failure to recognize Arabia's oil riches by claiming, "what has been assumed to
be the political sensitivity or indifference of Anglo-Persian has, in fact, been due
to genuine geological differences of opinion." It is doubtful that any difference
of opinion did occur, genuine or otherwise. There may have been one or two hesitant suggestions about possible further examination of Kuwait or Bahrain, but
no evidence exists to show any firmly held opposing view. In fact, all the geological surveys recorded opinions of an "oil dry" Arabia.

After this unanimous opinion was shown to be seriously erroneous, Frank
Holmes's pinpointing of the oil resources of Bahrain, Kuwait, and Saudi Arabia
was a blow to Britain's national prestige, one that reverberated for years. Despite
the fact that throughout the 1930s outraged members continued to raise questions
in the British parliament about the failure to recognize the oil riches of Arabia.

Winston Churchill personally propagated the myth of British discovery. In
his forward to Henry Longhurst's 1959 *Adventure in Oil: The Story of British
Petroleum*, Churchill referred to his own "close association" with the Anglo-
Persian Oil Company and praised "the pioneering of the vast oil industry of the
Middle East; a story of vigour and adventure in the best traditions of the mer-
chant venturers of Britain."[8]

Poor Discovery Record

Yet actual discovery of oil was not one of the skills of the Anglo-Persian Oil
Company. Prior to the arrival of the British, a cottage industry in oil extraction
already existed in Persia. W. K. Loftus, of the Persian Turkish Frontier Commission of 1848, estimated that twelve thousand pounds of liquid naphtha and prepared bitumen were collected annually in the area of the Bakhtiari Mountains. A
takeover of "certain existing seepages, worked privately at Shustar, Qasr-i-Shirin
and Dalaki" was agreed upon in the 1901 arrangements between the British
entrepreneur Knox D'Arcy and the Persian government for the concession that
would become, in 1909, the Anglo-Persian Oil Company. Anglo-Persian's most

productive field at Maidan-i-Naftun had been producing oil since biblical times. The name translates to "area of oil"; this was perhaps too obvious. Company publications took to describing the area as Masjed Soleyman instead, although this was almost as big a giveaway. It means "temple of Suleiman" and was the site of a sacred fire temple, erected around the permanent gas escape from the nearby oil field.

Simply muscling in on an existing operation was nothing new in the imperial experience. The Anglo-Persian Oil Company's partner and operating company, Burma Oil, had already done exactly this, with great success, after the 1885 annexation of Burma to India. Founded by a group of Scottish traders and investors, Burma Oil had begun by appropriating the oil-gathering activities of Burmese village people, an appropriation on which it built a commercial industry with a refinery in Rangoon. The growth of Burma Oil was aided by the monopoly it obtained from the Government of India over the sale and extraction of oil products throughout the Indian empire.

After World War I, the Anglo-Persian Oil Company, through its 50 percent holding in the Turkish Petroleum Company, moved in on the oil of Mesopotamia (Iraq) after the Germans had already completed substantial work in proving and developing the fields. This would not stop the company subsequently claiming full credit for "sensational" oil discoveries there.

Historian Ferrier reluctantly admitted Anglo-Persian's poor discovery record, noting that "no alternative sources of supply to Persia had been discovered by 1923"—some fourteen years after the establishment of Maidan-i-Naftun in Persia—and "concessionary success had certainly not been commensurate with the efforts taken or the money spent."

Anglo-Persian's concessionary activities were criticized as "primarily based on examination of territory which has been brought to our notice through third parties." The company did have a step-by-step procedure to follow in the search for new oil fields. As Ferrier explained, "the procedure . . . was to acquire concessions, examine them, test them if geologically favourable, and then float a working company if the tests were sufficiently promising." He commented: "This was a simple procedure which kept expenditure to a minimum. It did not involve any cash liability, other than the cost of geological examination, until the company was satisfied with the geological evidence. At that point there would be the expense of testing operations."

Because of Britain's power in the Persian Gulf, the Anglo-Persian Oil Company was able to move directly to "examining" Bahrain, Kuwait, bin Saud's territories, and the Trucial States (modern United Arab Emirates) without even bothering about any preliminary costs such as prospecting licenses. The fact that they did this but did not move on to acquire concessions or attempt to implement any of the other steps shows the strength of their conviction that there was no oil in Arabia. The procedure had been followed in other areas. By 1924 the Anglo-

Persian Oil Company had abandoned interest in Cyprus, Russia, Sicily, Brunei, Timor, Angola, Egypt, the Gold Coast, Guinea, Colombia, Ecuador, Peru, Madagascar, and New Brunswick. It is telling that during this period of hectic concessionary exploration no effort was taken, or money spent, right next door on the Arab side of the Persian Gulf—which certainly makes suspect the Anglo-Persian Oil Company's later claims, mythologized after the successes of Holmes's fields, to have known all along that oil existed there.[9]

American Delight

The Americans took some delight in the British discomfort at being shown up geologically and technically by their failure to recognize the oil riches of the Arab side of the Persian Gulf. Critics made sure everybody knew that "because of the British protected status of the shaikhdoms in the area, only the British were allowed to explore the territory for purposes of future exploitation." The 1950 *Arabian American Oil Company Handbook* assured its American employees that the discovery and development of Saudi Arabia's oil was "a story of distinctly American flavour" based on "common effort on the part of typical Americans working together in a tradition of free enterprise."

An article in *American* magazine of 1939 pointedly titled "Is John Bull's Face Red!" is not nearly so kind as it trumpets: "Bahrain is one of the world's great oilfields. It was discovered and built up by the Standard Oil Company of California." The article gloats: "For years the British government and British oil companies, which are tied in one bundle, told each other, told the shaikh and all parties who wanted them to gamble on a test well, that there was no oil there . . . it is difficult for the British to explain why Americans could find oil where the English said there wasn't any."

Despite all the subsequent American bravado, the fact is that before moving in on Frank Holmes's identification and mapping of the oil fields of Bahrain and Saudi Arabia, Standard Oil of California's own record of discovery was as dismal as that of the Anglo-Persian Oil Company.

In 1920 Standard Oil of California had set out to expand its sources beyond its traditional area of the western United States. A division was set up called Foreign Crude Oil Production, and geologists were sent to Central and South America, Mexico, the Philippines, and the Dutch East Indies. From 1920 to 1928 some $50 million was spent on foreign exploration, drilling, and concessions; Standard Oil of California explored and relinquished one concession after the other. By 1929, when it took over Holmes's Bahrain concession, Standard Oil of California had not been able to develop a single barrel of foreign commercial production.

Vice President M. E. Lombardi would later recall that during this period "a foreign legion of Standard Oil of California men probed various places." And in

1953 Lombardi would admit to being no better than Burma Oil, Anglo-Persian, and Turkish Petroleum in the habit of taking over other peoples' discoveries. He revealed that "Sumatra oil fields, from which oil is now shipped to California, were found for us by the Japanese during their occupation of that country."

And Fred Davis, Standard Oil of California's first geologist in Bahrain and later chairman of the Aramco board, would admit in 1974 that the Al Hasa concession in Saudi Arabia, as identified and mapped by Holmes, "covered almost the entire area we had there at the start *and is where all our oil has been found.*"[10]

THE GEOLOGICAL SURVEYS

It may have been their very political power that allowed the British experts to get it so wrong, so often and for so long. Absolute British hegemony over the Arab shaikhdoms of the Persian Gulf ensured that only the narrowest of views were aired. In contrast, before becoming a British mandate following World War I, Mesopotamia had been visited by an international array of natural scientists and travelers, as had Persia. Consequently, while only the British knew Arabia, the oil resources of Iraq and Persia were widely recognized.

An Austrian geologist, Emil Tietze, summarized the oil possibilities of the Middle East in his 1879 review of the accounts of travelers. Tietze wrote: "Claudius Rich is of the opinion that . . . petroleum might be found in various areas (other than Persia) within that entire mountain range. Indeed, it seems that near Kirkuk, Tuz, and Kifri, one of the most important petroleum areas of the old world is waiting for future exploration, once those areas, at present still much too remote, will be made more accessible to European skill."

Tietze's opinion of the adjacent area appears so startlingly ahead of its time that it is difficult to imagine how it was almost totally ignored by the petroleum seekers of the first decades of the twentieth century. Tietze wrote: "it might be of interest to mention the occurrence of bitumen on the Arabian side of the Persian Gulf . . . the possibility of a geologic connection between the south Persian and the Arabian petroleum areas certainly exists in the same manner as between the bitumen seepages on the Caucasian and those on the Turkmenian side of the Caspian sea." With notable clarity, Tietze added: "Any petroleum industry to be established in Persia should realise the possibility of a foreign, but nearby competition, and therefore reckon with the condition on the Arabian side of the Gulf."[11]

Unlike Persia and Mesopotamia, few "foreigners" made it through the stifling embrace of the Government of India's "protection" of the Arabian side of the Persian Gulf. From the late 1800s what was known about these mysterious shaikhdoms came from British scientific expeditions. And in the hierarchical imperial environment of seniority and mentors, a self-protective herd instinct prevailed even in supposedly independent scientific endeavor. In organizations where promotion was awarded by heads of department, few men would go out on a limb by disagreeing with the opinion of a superior.

Men like New Zealander Frank Holmes, who had spent most of his working life in non-British countries, and Scottish engineer George Reynolds, who was credited with "setting the very foundations of the Iranian oil industry," were outsiders—and they were visibly made to pay a price for their nonconformism as the Establishment closed ranks against them. Holmes was obstructed, slandered, and libeled. Reynolds was tagged as not a company man and fired after bringing in Anglo-Persian's first successful well. Perhaps it was this very factor of being outside the dominant corporate cultures, not socialized into company or imperialist patterns of thought and behavior, that generated their independent mentality and risk-taking approach rather than mere acceptance of the prevailing wisdom. Certainly, when tracing through the geological surveys, from Guy Pilgrim in 1904 to Percy Cox in 1932, a culture of conformism clearly emerges and is, perhaps, one explanation for the repeated failure to recognize the oil riches of Arabia.

Boverton Redwood, 1895–1906

At the turn of the twentieth century the most respected British expert on petroleum was Sir Boverton Redwood. His 1895 *Petroleum: A Treatise*, reissued in 1906 and again in 1913, covered all the known theory and data on geographical distribution and geological occurrence, chemical properties, refining, transport and storage of petroleum, and natural gas. His London-based consultancy provided geologists to every corner of the world. At one time Redwood, whose training was in chemistry, was consultant to the Burma Oil Company and to D'Arcy Exploration (forerunner of Anglo-Persian Oil Company). He also directed large-scale exploration in Mexico while simultaneously servicing other clients scattered across the globe. He was adviser on petroleum to the Admiralty, the Home Office, the India Office, and the Colonial Office. Redwood was either a member, or one of the important advisers, on every official petroleum-related committee and was one of the earliest and most consistent advocates of liquid fuel in warships that would later culminate in Winston Churchill committing the British government to a commercial partnership with the Anglo-Persian Oil Company.

Redwood's *Treatise* remained the standard work for two decades. His influence ran strongly through the early years of petroleum exploration in the Persian Gulf. The recommendations of the first official British geological reconnaissance, mounted by the Government of India, were rooted in Redwood's theories. Under the heading Arabia, Redwood recorded, without comment, Tietze's mention of the observations taken by a British captain while mapping the waters of the Gulf. Redwood recorded: "On the eastern side of Arabia, traces of petroleum occur at Benaid el Oar near Koweit on the Persian Gulf, on the waters of which also films of oil often appear, after earthquakes or storms between the islands of

Kubbar and Garu, and again near Farsi Island. Deposits of bitumen are also reported as found on Bahrein Island, and oil is said to rise in the sea off Halal Island eastward of Bahrein peninsula."[12]

Guy E. Pilgrim, 1904–1905

Following his majestic tour of the Persian Gulf in 1903—designed to leave the Arab shaikhs in no doubt as to who was in charge—Lord Curzon, viceroy of India, commissioned the Geological Survey of India to explore and document minerals in the area that could be of economic importance to the Government of India. The task went to the Geological Survey's deputy superintendent, Guy E. Pilgrim. Beginning in Oman in November 1904 Pilgrim trekked along the coast to Dubai and up to Qatar, went on to Bahrain and to the islands of the lower Gulf, and, until June 1905, traveled across southern Persia. Pilgrim did not visit Kuwait. After the 1906 publication of his report *The Geology of the Persian Gulf and the Adjoining Portions of Persia and Arabia* and with the prestige of the Geological Survey of India behind him, Pilgrim became the accepted authority on the Persian Gulf.

Pilgrim would surely have been familiar with Tietze's 1879 review but appears to have ignored it in favor of the work of Redwood, the éminence grise of oil affairs, and British to boot. Pilgrim firmly adhered to the theory expounded in Redwood's *Treatise* and "now universally accepted" that petroleum originated at the same time as the gypsum beds "in which it is stored." Pilgrim merged Redwood's theory with the belief, popular in Europe since the 1840s, that gypsum originated through the action of volcanic gases on limestone. Pilgrim now decreed "the formation of petroleum is directly due to volcanic action."

Putting forward his volcanic theory, Pilgrim took a swipe at the French, commenting: "I cannot share M. de Morgan's opinion expressed in the *Annales des Mines*, 1892, and in *Mission Scientifique en Perse*, that there is a great subterranean source, which feeds the surface reservoir." In this Pilgrim was echoing Redwood's acerbic putdown of "minds of a certain calibre," which presented the "wildest assertions as to underground 'rivers' or 'lakes' of oil."

The volcanic theory fitted neatly the manifestation in the Gulf waters that Pilgrim noted as "apparently bituminous springs exist beneath the sea, as the British captain records his ship having passed through a sea covered with an oily substance emitting a strong smell of naphtha." Pilgrim reached the same conclusion as Redwood. They agreed that what the captain had noted were "mud volcanoes" and that "the occurrence of these volcanoes, and of hot springs generally, were merely the accompaniment of metamorphic change not possessing any significance in regard to the occurrence of petroleum."

Pilgrim also favored the mistaken, but prevailing, belief that productive reservoirs were to be found in the lowest beds of the Fars mountain series, "which in their lithological character would afford as favourable a reservoir as could be desired." As to the low range of hills in Persia, where "petroleum springs" were situated, Pilgrim surmised, "since the greater part of this formation has been laid bare for ages untold, it is also evident that most of the petroleum that existed in it once has now been wasted."

And he predicted the same barrenness for Bahrain: "The strata which overlie the Jebel Durkhan beds are admirably adapted lithogically for storing petroleum. . . . In the present instance, however, as I have already stated, not only the whole of the 'cover' but also the whole of these porous beds have been removed from the anticline by the operation of atmospheric forces, so that any petroleum that ever existed must long ago have drained away."

Redwood had written of "trivial oozings from shallow and temporary crevices." Pilgrim agreed. He reported that conspicuous "shows" did not signify much because "the laws of hydrostatic pressure would naturally force a limited vein or pocket deposit of petroleum to find an outlet." The only deposit of asphalt, or bitumen, that Pilgrim actually examined was that of Bahrain. He dismissed this completely as he concluded: "The deposit need not be seriously considered in the light of its offering any scope for an export trade, although it would be a pity not to make some use of it locally." For Bahrain, he recommended further consideration of sulphur and iron oxide mining and perhaps quarrying of gypsum and building stone.

Pilgrim thought the whole area he spent eight months examining was a pretty poor prospect. He concluded: "Southern Persia and the Gulf region do not present as favourable a field for prospecting operations as Northern Persia seems to. Petroleum alone seems to offer any promise of great financial success and the ultimate issue of the work which is being carried on in regard to that remains at present quite doubtful." Pilgrim's reference was to the failure of the Persian sites, recommended by Sir Boverton Redwood, and now being closed down by Knox D'Arcy's exploration company. Drilling had begun at Redwood's recommended Persian sites in 1902. At the time Pilgrim was in the area, 1904–1905, original shows had shrunk to a trickle, and the early Redwood sites were being abandoned.[13]

George Reynolds, 1902–11

Pilgrim's negative report sponsored by the authoritative Geological Survey of India squashed any geological interest in the Arabian side of the Gulf. Only George Reynolds seems to have disagreed with Pilgrim's conclusions. Reynolds, a graduate of the Royal Indian Engineering College, was in charge of Knox D'Arcy's operations in Persia. As chief engineer, in May 1908 he proved the rich Persian oil field at Masjed Soleyman, which he had spent two years unsuccessfully advocating to D'Arcy's consultant, Sir Boverton Redwood.

Bailed out of financial difficulties by Burma Oil, D'Arcy's First Exploration

Company that he had set up in 1901 for exploring Persia became with its new investment the Anglo-Persian Oil Company in 1909. Along with the finance came a group of men whom Reynolds referred to disparagingly as "Office Wallahs."

They were to manage the Anglo-Persian Oil Company in the same way they ran affairs in India. Charles Greenway, who favored a monocle and spats and had spent his entire career in India with Shaw Wallace, agents for the Burma Oil Company, was appointed chairman and managing director of the new company.

The new chairman decided the rough-and-ready George Reynolds was not a company man. In December 1910, eighteen months after producing the first commercial oil in the Middle East, Reynolds was summoned to London and unceremoniously dismissed by Greenway, despite the fact that both chairman and company had been secured solely on the basis of Reynolds's work. Eight of Reynolds's wells had struck oil and there was no doubt about the oil field he had proven. After ten years in Persia, Reynolds was paid off with one thousand sterling severance.

After his dismissal from the company that owed so much to his skill and tenacity, Reynolds showed interest in Kuwait where, in 1903, he had had time to look around because of a delay arranging passage back to England. In London in 1911, Reynolds was talking to Royal Dutch Shell about Kuwait. News of the discussions reached Anglo-Persian. Greenway (later Sir Charles Greenway) did not believe there was oil in Kuwait but was nevertheless wary of Reynolds's instinct and demonstrated skill.

The chairman of the Anglo-Persian Oil Company turned to the Government of India's powerful political resident in the Persian Gulf, Percy Cox, requesting he secure for the company a concession in Kuwait. Greenway explained that "it would be very prejudicial to have a powerful foreign rival on our heels in the Persian Gulf." While demanding that Cox block Reynolds, Greenway confided: "I do not know what reasons Reynolds has for assuming that there are any oil deposits of value in Kuwait . . . the question of whether or not there are oil deposits of any value in Kuwait is, of course, entirely problematic."

Although Percy Cox replied that conditions in the area were too disturbed for an application to be put forward, no more is heard of Reynolds in the area of the Persian Gulf. Instead, after proving oil in Persia, Reynolds went on to make major discoveries in Venezuela. Had Greenway not blocked him, George Reynolds may have gone on to prove Kuwait's oil fields, twenty-seven years before an Anglo-Persian/Gulf Oil Corporation combine finally drilled successfully in the Holmes-advocated site at Burgan—after two unsuccessful years at other sites, as directed by the company geologists.[14]

Edwin Hall Pascoe, 1913

It was Reynolds's interest that eighteen months later inspired a secret agreement between the Government of India, Percy Cox, the Kuwait political agent, and

Anglo-Persian's Greenway to arrange with the Geological Survey of India to take a closer look at the oil possibilities of Kuwait.

The scheme was for the geological survey to be put to the shaikh of Kuwait under the guise of reporting on the possibility of finding a water supply for Kuwait Town. The sensitive mission went to Edwin Hall Pascoe, assistant superintendent of the Geological Survey of India, who arrived in March 1913.

The political agent warned Pascoe "on no account must such resources or possibilities be discussed with anyone in Kuwait, European or native." A British consulting engineer was hired to conduct a pretty slapdash water survey. Pascoe was again warned by the political agent that "neither this Engineer nor my Head Clerk has any inkling of the possibility of mineral or oil deposits."

Perhaps not surprisingly, Pascoe, waiting his turn in the promotion line at the Geological Survey of India, displayed much deference to the theory and opinion published by his superior, Deputy Superintendent Guy E. Pilgrim. In his report "Prospects of Obtaining Oil Near Kuwait, Persian Gulf" Pascoe appears to have been trying to please both those conspirators who hired him and his superintendent at the Indian Survey.

He carefully hedged his bet, reporting: "The age of the beds is favourable, if it is correct to assume they belong to the Fars, since this is the oil bearing series in Persia. There is no reason to believe the nature of the beds beneath Burgan should be different from that of the Fars in Persia and they may be looked upon therefore as sufficiently porous to retain oil in workable quantities." And yet, Pascoe warned: "This locality is not on the line of strike of the rich oil deposits now being worked in Persia . . . it is in fact over 170 miles to the southwest of this line. This does not necessarily mean that oil in commercial quantities does not occur below Burgan, but it adds a decidedly speculative element to any operations."[15]

The Admiralty Commission, 1913

At this time, Winston Churchill, First Lord of the Admiralty, was converting the ships of the Royal Navy from coal burning to oil burning. To secure oil supplies, Churchill was keen to commit the British government to buy into the Anglo-Persian Oil Company. Several members of Parliament objected to this expensive government venture. Churchill critic Adm. Sir Charles Beresford declared, "the Government should not contract with any company without having their own geological expert to report."

Immediately after Beresford's criticism, Churchill appointed a Commission of Oil Experts, headed by the naval commander-in-chief, East Indies, Adm. Sir Edmond Slade, and advised by the ever-present Sir Boverton Redwood, to visit Anglo-Persian's concession and report to Parliament on the production potential of the Persian oil fields and the operation of the company refinery.

Although the commission's terms of reference were clearly stipulated as being to "investigate the resources of the oilfields comprised in the Concession of the Anglo-Persian Oil Company," three days after their October 23, 1913, arrival in Persia, Percy Cox grabbed the initiative on behalf of Greenway and Anglo-Persian and approached the shaikh of Kuwait, suggesting that the commission could visit his country "to inspect Burgan . . . and see if there is a hope of obtaining oil." In the same conversation, Cox also "suggested" that the shaikh sign a document agreeing to give a concession only to a company, or person, recommended by the High Government in London.

As Pascoe had only six months earlier completed his secret mission to inspect Burgan, and communicated his less than glowing report, Cox's move must be interpreted as using the Admiralty Commission as bait in order to block any other interest, such as Reynolds and Shell, from dealing with Kuwait. The shaikh signed, stating that "we are agreeable to everything which you regard as advantageous" and "if in their view there seems hope of obtaining oil we shall never give a concession in this matter to anyone except a person appointed from the British government."

By itself this document met Anglo-Persian's goal of denying access to any other party, as laid out in Greenway's letter two years earlier to Cox after learning of Reynolds's interest in Kuwait. While the prospect of an inspection by Churchill's Commission of Oil Experts may have enthused the shaikh of Kuwait to sign, it is open to interpretation how seriously the members of the commission considered the opportunity.

Pascoe was seconded by the Government of India to assist the group. He joined them in Persia and now went back into Kuwait, again to Burgan, only six months after completing his secret assignment there. Pascoe was accompanied by S. Lister James, "one of Redwood's roving geologists." They spent a single day "examining the hills and rocks" of Burgan before returning to Kuwait, "having decided that nothing further was to be learned." The shaikh, who was certain his oil potential was as good as Persia's, provided a guide and boats to reach another reported seepage the next day, but "Mr. Pascoe and Mr. James . . . were unable to reach the spot before sunset and . . . returned unsuccessful." Admiral Slade and fellow commission member John Cadman, at the time petroleum adviser to the Home Office and Colonial Office, showed not the slightest interest in Kuwait's oil possibilities. According to the report of the political agent, they spent the whole visit in the relative comfort of the Kuwait Political Agency, attending to "social" matters.[16]

The shaikh of Bahrain may have avoided persuasion to sign a document similar to the one signed in Kuwait through the simple expedient of being out of town. The Slade Commission arrived in Bahrain on November 21, and so did Percy Cox, aboard the Indian navy's *Lawrence*, accompanied by a major general and a colonel. The Bahrain 1913 Administrative Report records: "The shaikh was away on a hawking and shooting expedition and could not get back in time to see the Resident." As Cox and his military entourage stayed two full days, and Bahrain Island measured only ten by thirty miles, this seems curious indeed.

By December, Percy Cox had left the Persian Gulf, promoted to secretary of the Foreign Department of the Government of India. But the shaikh of Bahrain had achieved only a reprieve. A year later, he gave in to pressure from the Bahrain political agent and signed a statement: "I do hereby repeat to you that if there is any prospect of obtaining kerosene oil in my territory, I will not embark on the exploitation of it myself and will not entertain overtures from any quarter regarding it without consulting the Political Agent and without the approval of the High Government."

In its three-month tour, the Admiralty Commission spent four days in Kuwait and four days in Bahrain. As the Anglo-Persian Oil Company did not have a concession in either Kuwait or Bahrain, the Commission of Oil Experts (or Cox and Slade) was acting outside its Terms of Reference to "investigate the resources of the oilfields comprised in the Concession of the Anglo-Persian Oil Company." Tellingly, while producing "a report of such unanimous enthusiasm on the oilfields of Persia that Churchill laid before Parliament a Bill by which the British government was to become the major and controlling shareholder in Anglo-Persian," there is no mention at all of Kuwait or Bahrain in the Admiralty Commission's official Report to Parliament.[17]

Lister James, 1917

A young Lister James, finding himself in the illustrious company of Admiral Slade, John Cadman, and Edwin Pascoe, agreed with Pascoe's report on Kuwait and Bahrain in the same way as Pascoe had earlier agreed with the findings of his senior colleague, Guy Pilgrim. Lister James, who was a Redwood man, had selected two sites in Persia, both of which were now proving unsuccessful. Following his excursion with the Admiralty Commission he recommended to Anglo-Persian a site on Qishm Island, near the Straits of Hormuz. This site was also unsuccessful.

Following Pilgrim and Pascoe, James reported that Eocene rocks, as found in Kuwait and Bahrain, were "not known to be petroliferous." Nevertheless, he halfheartedly suggested the company might want to try shallow structure drilling near Kuwait's bitumen seepages at Burgan and Bahra. As to Bahrain, he wrote, "in view of the definite occurrence of asphalt and the ideal nature of the structure it appears inadvisable to ignore the area before testing with a fairly deep well."

On the heels of a 1916 request for a postwar monopoly of "all areas that may fall under British influence" made by Greenway and now Anglo-Persian director Slade, Lister James, now also employed by Anglo-Persian, was sent back into Kuwait yet again. In January 1917, accompanied by geologist G. W. Halse, James undertook the fourth inspection of Kuwait. The two geologists apparently concluded the sites of Burgan and Bahra had more curiosity value than oil potential. As a matter of scientific endeavor only, they recommended digging pits or trenches "to find out more about the provenance of bitumen" and "for geological information."[18]

Edwin Pascoe, 1918–19

The division of the Arab world among the victors following World War I delivered Mesopotamia to the British, along with some of history's most famous oil seepages. In 1918–19, immediately after the war but before the award of the League of Nations mandates, Edwin Pascoe, now promoted to director of the Geological Survey of India, arrived to spend five months surveying "as many of the important oil indications as possible." Stimulated by favorable reports of oil potential, Mesopotamia had for ten years or more been the object of diplomatic and commercial maneuvering for oil concessions.

Just one extract from Pascoe's survey reveals the extent of the prize that existed in Iraq:

> Oil was collected here by the Turks, or by the Germans who were in control of the place before the war. . . . Seven borings were made by the Germans before the British occupation. Of these, four are producing or capable of producing oil . . . one was still in process of construction and was evidently on the point of entering the oil horizon, as oil has recently commenced to appear at the casing head. . . . The best well has a considerable pressure and would fill an ordinary kerosene tin in five or six seconds with a black oil containing a certain amount of tar but some petrol as well. The wells are all in the same locality and about 300 yards apart. There is a small refinery of five stills, for which crude oil was used as fuel, and two condensers, all in working order though of crude construction. Some of the refined oil was found in drums.

No great scientific skills were needed to conclude of this particular site, as Pascoe did, that "an oil field of importance exists here." Pascoe's report concluding that Mesopotamia "should rival the Persian fields, and outclass Burma" ensured that attention remained focused on the side of the Gulf where "the principal structural and stratigraphic features extend across the two countries (Persia and Mesopotamia)."

Pascoe grouped the oil prospects from A to F in declining order of importance. He categorized Kirkuk, which would in October 1927 prove to be Iraq's major oil field, as "C—scarcely promising enough to warrant deep boring."[19]

Frank Holmes, 1918–24

The story of Frank Holmes's identification and mapping of the oil riches of Bahrain, Saudi Arabia, and Kuwait, and of his struggle to ensure these fields were developed, is the subject of this story and unfolds in the following chapters.

Arnold Heim, April–July 1924

In late 1923 Frank Holmes commissioned the highly respected Swiss geologist Dr. Arnold Heim of the University of Zurich to provide a second opinion on his, Holmes, own positive conclusion and findings regarding the oil possibilities of Bahrain, Kuwait, and Saudi Arabia's Al Hasa province. Heim, who was also a consultant to Royal Dutch Shell and others, arrived with three assistants in April 1924. The story of Heim's survey is told here in later chapters. Unfortunately for Holmes, the report of Dr. Heim reinforced the previous pessimistic opinions. But even more negative reports were to follow, all concluding that there was no oil on the Arabian side of the Persian Gulf.

Hugo de Bockh, 1923–25

In 1923 Anglo-Persian employed University of Budapest professor Hugo de Bockh (he had changed his name from Von Bockh after the war) as geological adviser. De Bockh was a former undersecretary for mines and would go on to become director of the Geological Survey of Hungary. He was brought to Persia to advise on geological work because many of Anglo-Persian's wells were coming up dry. He made two extensive field trips during the winter seasons of 1923–24 and 1924–25. On the second trip, de Bockh was accompanied by Lister James, now promoted to the position of Anglo-Persian chief geologist, and his assistant George M. Lees along with F. D. S. Richardson.

Oman was included in the first season where "valuable information was gathered on Arabian tectonics, but no immediate oil prospects were revealed." Which perhaps explains why a concession the Anglo-Persian Oil Company began negotiating with the sultan of Muscat in 1924 was not finalized until 1937. De Bockh conceived the "lagoon theory," in which he supposed that oil would be found in large quantities "only in the limestone of the strongly folded foothills zone of the Zagros mountains and only where that limestone had been deposited under lagoonal conditions." Without ever leaving the Persian side of the Gulf, de Bockh reviewed the work of Pilgrim, Pascoe, and James, and pronounced the Arabian littoral "unpromising for oil deposits."[20]

George M. Lees, 1925–32

When he was sent to survey, George M. Lees agreed with his immediate superior, Lister James. Either side of de Bockh's two seasonal surveys, Lees conducted an Anglo-Persian-sponsored six-year comprehensive study of South Eastern Arabia. His report agreed with the unfavorable opinion of the Arab lands as put forward by the company's chief geologist Lister James and was agreed upon by the company's prestigious geological adviser, Professor de Bockh.

The influential *Geographical Journal* published Lees's study in 1928, and each member of the survey was invited to lecture at Royal Geographical Society meetings. During the study, Kuwait was the subject of a sixth geological survey, this time by Lees, B. K. N. Wyllie, and A. G. H. Mayhew in 1925–26. In the *Geographical Journal*, Lees reported that "it is very improbable that, in the total absence of any direct geological evidence, a convincing case could ever be made out by this means for exploring the unknown depths of Kuwait territory with the drill."

Lees agreed with de Bockh, James, Pascoe, and Pilgrim that even if oil were found, it would be in such small quantities as to be economically of little value. His advice was set out in a management memo dated January 6, 1931. Lees advised that "in view of the agreement between the local evidence and the general regional geology and in the absence of any alternative hypothesis . . . the opinion is that the company need not take any further interest in Kuwait." The dismal state of Kuwait's oil prospects was relayed to the members of the International Geology Party then combing Iraq. One American geologist reported "a Scotsman, B. K. N. Wyllie, who was the reconnaissance man for Anglo-Persian, reported to the group having just finished a trek across Kuwait without finding any indications of structure."

When Lister James retired, Lees would go on to take his turn as chief geologist at the Anglo-Persian Oil Company. He was thought to be a geologist of "great promise" when he joined the company after serving with the Anglo-Indian occupation force in Mesopotamia and then as the Government of India's political officer in Kurdistan. He followed the damning report on Kuwait with an equally negative report on the oil prospects of Bahrain.

George M. Lees was to forge a place in history as the man who promised he "would drink any commercial oil found in Bahrain." His vehement vow came at the end of a speech in which he summed up the evidence for and against finding oil in commercial quantities in Bahrain. The respected Dr. George M. Lees opted for the negative. Just a few months after his speech, in June 1932, oil in commercial quantities flowed in Bahrain.[21]

Peter T. Cox, 1932

The final "examination" of Kuwait—the seventh—was performed by Peter T. Cox in 1931–32. In later years Cox would claim his was a "restricted assignment" and that because of the Depression he was on a "shoestring." Cox and his team, including a driller, spent five months in Kuwait before being recalled in early April 1932. Following a now familiar pattern, Cox agreed with those who had been before him. His report concluded, "Kuwait's prospects appear to be the possibility, even probability, of finding only very heavy oil accumulations." In his turn, Cox became Anglo-Persian chief geologist and, later, a director of the Kuwait Oil Company. When Bahrain struck oil in June, just weeks after his negative report, Cox noted dryly: "Anglo-Persian set about getting a concession from the Shaikh of Kuwait as a matter of urgency."[22]

NOTES

1. Lawrence James, *The Rise and Fall of the British Empire* (UK: Abacus, 1998), p. 219: "*angrezi raj*"; Sir Rupert Hay, *The Persian Gulf States* (Washington: Middle East Institute, 1959), p. 11, "commercial at first," the first governor general was Warren Hastings; in 1784 the president of the Board of Control was Viscount Castlereagh; p. 13, "*balyoz*"; Daniel Yergin, *The Prize: Epic Quest for Oil, Money, and Power* (New York: Simon and Shuster, 1991), p. 140, "resist with all our means" and "Monroe Doctrine."

2. Herodotus, *The Histories* (c. 430 BCE) and Strabo, *Geography* (c. 22 CE), http://www.al-bab.com/arab/history.htm. Geoffrey Bibby, *Looking for Dilmun* (London: Collins, 1970), p. 10, "the city of the mound-builders of Bahrain dates to 2,000 B.C., the same people (lived) far to the north in Kuwait, (we found) the extension of civilisation in the Arabian Gulf farther and farther back in time, to 3000 then to 4000 B.C. (and) a second civilisation on the Oman coast in Abu Dhabi and the same civilisation again one hundred miles inland at the foot of the Muscat mountains; p. 14, "the Ubaid sites in Saudi Arabia go back to 5000 B.C."; Briton Cooper Busch, *Britain and the Persian Gulf, 1894–1914* (Berkeley: University of California Press, 1967), p. 179, "pathetic"; p. 225, "sinister"; p. 146, "chilly antagonism."

3. Donald Hawley, *The Trucial States* (London: Allen and Unwin, 1970), p. 74, "first British Ambassador to Persia"; the East India Company assumed "Presidencies" in Bombay, Madras, and Calcutta, where its main commercial activities were located; Hawley, comment by Sir William Luce in his introduction: "virtually incomprehensible"; for the twists and turns of authority for the Persian Gulf Residency see Penelope Tuson's *The Records of the British Residency and Agencies in the Persian Gulf* (London: India Office Library and Records, 1979), pp. 1–9, "the anomalous position in relation to internal Persian affairs and the affairs of the Gulf, respectively British and Indian Government spheres of interest."

4. Hawley, *The Trucial States*, p. 77: "the Island of Bombay was transferred to the British Crown as part of the dowry of the infanta of Portugal on her marriage to King Charles II"; for the evolving status of Aden see Herbert J. Liebesny, "Administration and Legal Development in Arabia, Aden Colony and Protectorate," *Middle East Journal* 9 (1955): 335–96.

5. Herbert J. Liebesny, "International Relations of Arabia: The Dependent Areas," *Middle East Journal* 1 (1947): 148–68, "two types of colonial administration" and "Protectorate . . . right of the Crown" and "not clarified." The Arabs of the Persian Gulf still refer to these tactics as Britain's "Gunboat Diplomacy"; Busch, *Britain and the Persian Gulf*, p. 258, "King of the Gulf." See epilogue (*Legacy*), the Residency was moved from Bushire to Bahrain in 1946. From 1947–51, the resident was Cornelius James Pelly and then, until his retirement in 1958, Rupert Hay of the Indian Political Service, ex-Indian army, one of World War I Acting Civil Commissioner Arnold Wilson's handpicked and personally groomed "Wilson's Young Men" ruthlessly imposing Anglo-Indian rule on Mesopotamia. See Hay, *Persian Gulf States*, p. 26, "the last member of the old Indian Political Service . . . in the Gulf . . . retired in 1958"; this was, of course, Hay himself.

6. Abbas Faroughy, *The Bahrein Islands (750–1951); A Contribution to the Study of Power Politics in the Persian Gulf, an Historical, Economic, and Geographical Survey* (New York: Verry, Fisher, 1951), p. 96 gives the relevant provisions including "Government

of His Britannic Majesty declares they have no intention of annexing the Bahrain Island to their territories (article 13)" and p. 95, "three step Indianisation." Muhammad G. Rumaihi, *Bahrain: Social and Political Change Since the First World War* (London: Bowker, 1976), p. 168: "Indian Foreign Jurisdiction Order 1890"; Faroughy, *Bahrein Islands*, p. 101: "Bahrain Order-in-Council"; Rumaihi, *Bahrain: Social and Political Change*, p. 169, gives full detail of the Order-in-Council that appeared in the London *Gazette* August 15, 1913; John Marlowe, *The Persian Gulf in the Twentieth Century* (London: Cresset, 1962), p. 40, "a district in the Presidency of Bombay"; Marlowe, *Persian Gulf*, p. 40, "as a Colonial Governor; Liebesny, "Administration and Legal Development in Arabia, Aden Colony and Protectorate," pp. 33–42, "appeals . . . High Court Bombay."

    7. Philip Graves, *The Life of Sir Percy Cox* (London: Thornton Butterworth, 1939), p. 203, Cox to Lawrence June 1916, "made an Arab flag for Kuwait." Note that *bin* translates to "son of" (female equivalent is *bint*) the term *ibn* was the Turkish equivalent used by the Ottoman Empire. *Al* is the definitive so, for example, the shaikh of Kuwait Ahmad bin Jabir Al Sabah translates to "Ahmad the son of Jabir of the Sabah (tribe)." While Arabs are mostly known by their first name, Abdul Aziz bin Abdul Rahman bin Saud was unusually known as "bin Saud" (or the Ottoman/Turkish "ibn Saud"), meaning, in the colloquial, "son of the family of Saud"; Graves, p. 197, "follow unreservedly" (this 1915 agreement is also referred to as the Treaty of Darin); Busch, *Britain and the Persian Gulf*, p. 346, "combination of"; Ameen Rihani, *Ibn Sa'oud of Arabia: His People and His Land* (London: Constable, 1928), p. 68, citing a personal conversation with bin Saud.

    8. R. W. Ferrier, *The History of the British Petroleum Company: The Developing Years, 1901–1932* (Cambridge: Cambridge University Press, 1982), p. 541, "genuine geological differences"; Knox D'Arcy (see Who's Who section of this book) formed First Exploration Company in 1901, investment from Burma Oil created Anglo-Persian Oil Company in 1909 in which the British government purchased a 51 percent interest in 1914. Anglo-Persian acquired in 1917 the British Petroleum Company, the Petroleum Steamship Company, and the Homelight Oil Company from the British Public Trustee (registered in Britain and Breman, the three companies had been confiscated as enemy property). Anglo-Persian was renamed Anglo-Iranian in 1935 along with Persia's change of name. In May 1951, the Iranian government nationalized the properties of the Anglo-Iranian Oil Company and declared a new company called the National Iranian Oil Company. Anglo-Iranian Oil Company continued as the vehicle to receive compensation payments and as the 40 percent shareholder in the eight-company consortium formed in 1954 as operators of the National Iranian Oil Company. The agreement officially designated the consortium as "an Exploration and Producing Company and a Refining Company known as The Operating Companies." In 1955, Anglo-Iranian Oil Company was relegated to a subsidiary and the name British Petroleum Company Ltd. became the dominant title; Henry Longhurst, *Adventure in Oil: The Story of British Petroleum* (London: Sidgwick and Jackson, 1959), "in the best traditions of the merchant venturers of Britain."

    9. Edgar Wesley Owen, ed., *Trek of the Oil Finders: A History of Exploration for Petroleum*, semicentennial commemorative volume (Tulsa, OK: American Association of Petroleum Geologists), p. 1263, "liquid naptha and prepared bitumen" and p. 1253, "certain existing seepages." For early German involvement see Stephen Helmsley Longrigg, *Oil in the Middle East: Its Discovery and Development* (London: Oxford University Press, 1954), pp. 13–15, 45, 69, "sensational" oil discovery; Ferrier, *History of the British*

Petroleum Company, p. 542, "through third parties," p. 547, "no alternative sources," p. 524, "procedures," p. 544, "had abandoned."

10. Frederick Lee Moore Jr., *Origin of American Oil Concessions in Bahrain, Kuwait and Saudi Arabia* (BA thesis, Department of History, Princeton University, 1948), pp. 11–12, "only British could explore"; Roy Lebkicher, "Aramco and World Oil," preface to *Arabian American Oil Company Handbook for American Employees* 1 (New York: Russell F. Moore, 1950), "distinctly American flavour"; Jerome Beatty, "Is John Bull's Face Red!" *American*, January 1939, p. 111, "why Americans could find oil." For detail of Standard Oil of California 1920–31 see Moore, pp. 60–62, 78, citing Annual Reports; also see Moore, p. 68, "total expenditure on Bahrain to the date of discovery was a mere $650,000" mostly spent on drilling and equipment; Chevron Archives, Box 0120814, December 18, 1953, Re Standard Oil of California Antitrust Case, in-house interview with M. E. Lombardi, "foreign legion" and "found by the Japanese"; Owen Papers, Box 1a, October 11, 1974, Fred A. Davis to G. Gish, "is where all our oil has been found."

11. Owen, *Trek of the Oil Finders*, p. 1282, Tietze's reference is to Claudius James Rich (1787–1821). His widow published in 1836, "Narrative of a residence in Koordistan, and on the site of ancient Ninevah, with journal of a voyage down the Tigris"; Owen, p. 1319, "possibility of a foreign, but nearby competition."

12. Sir Boverton Redwood, *Petroleum: A Treatise*, 2nd ed. (London: C. Griffin, 1906); the observations were recorded by British captain C. G. Constable.

13. See IOL/R/15/5/236, November 26, 1912, Confidential, Percy Cox to Government of India, "Pilgrim . . . did not visit Kuwait . . . only refers to Kuwait incidentally"; Redwood, *Petroleum: A Treatise*, p. 111, "universally accepted"; Guy E. Pilgrim, *The Geology of the Persian Gulf and the Adjoining Portions of Persia and Arabia*, vol. 34, pt. 4, memoir (Australian National University Library, Document Search Service: Geological Survey of India, 1908), p. 146, "due to volcanic action"; Owen, *Trek of the Oil Finders*, p. 1261, "theory of origin of gypsum"; Pilgrim, p. 147, "cannot share Morgan's opinion"; Redwood, p. 117, "wildest assertions"; Pilgrim, p. 143, "apparently bitumous springs"; Redwood, p. 122, "not possessing significance"; Pilgrim, pp. 146–47, "has now been wasted" and p. 154, "have drained away"; Redwood, p. 118, "trivial oozings"; Pilgrim, p. 147, "find an outlet" and p. 154, "not seriously considered"; p. 32 mentions several times D'Arcy unsuccessful drillings, p. 144, "quite doubtful."

14. Ferrier, *History of the British Petroleum Company*, p. 73, "very foundations"; Archibald H. T. Chisholm, *The First Kuwait Oil Concession Agreement: A Record of the Negotiations, 1911–1934* (London: F. Cass, 1975), p. 87, "entirely problematic"; see Ferrier, p. 86, and Yergin, *The Prize*, p. 143, for the May 1908 strike and "eight of Reynold's wells"; Ferrier, p. 139, particularly citing Greenway's comments about Reynolds and saying he wanted "a good office man"; Yergin, p. 142, and Ferrier, p. 66, refer to Reynolds meeting the Indian viceroy, Lord Curzon, in Kuwait on November 28, 1903; Pilgrim, p. 32, mentions meeting Mr. G. B. Reynolds and examining geological specimens he had collected from beds in the Bakhtiari country; Ferrier, p. 139, "major discoveries in Venezuela."

15. For the secret agreement see IOL/R/15/5/236 November 26, 1912, Cox to Government of India and January 14, 1913, Cox to Captain Shakespear, political agent Kuwait and February 11, 1913, "Secret and Confidential" instructions to Pascoe by Shakespear "on no account . . . discussed with anyone in Kuwait, European or native" and "(no) any inkling of the possibility of mineral or oil deposits"; the consulting engineer to prepare the

supposed water report was C. F. Shaw; for Pascoe's report see Chisholm, *First Kuwait Oil Concession Agreement*, p. 88, and p. 89, "decidedly speculative."

16. Ferrier, *History of the British Petroleum Company*, p. 183, citing Hansard, House of Commons, vol. 55, col. 1622, July 17, 1913, "own geological expert"; the Terms of Reference Are in Final Report of the Admiralty Commission on the Persian Oilfields—Explanatory Memorandum; House of Commons Tabled Papers, 1914, vol. 54 (liv) Cd. 7419; Chisholm, *First Kuwait Oil Concession Agreement*, p. 88, "unable to reach the spot before sunset," p. 89, "we are agreeable"; Ferrier, p. 282, "one of Redwood's"; Administrative Report for the Political Agency, Kuwait, for the Year 1913, memorandum on visit of Sir Edmond Slade and Oil Experts' Commission to Kuwait (the political agent was Capt. William Henry Irvine Shakespear), "returned unsuccessful" and "attended to social duties." Both Slade and Cadman were to become members of the Anglo-Persian Oil Company Board; the latter, as chairman 1927–41, would become Sir John and then Lord Cadman.

17. Administrative Report Bahrain, for the Year 1913 (the political agent was Major Arthur Prescott Trevor), "the Shaikh was away"; Abbas Faroughy, *The Bahrein Islands (750–1951); A Contribution to the Study of Power Politics in the Persian Gulf, an Historical, Economic, and Geographical Survey* (New York: Verry, Fisher, 1951), p. 124 reproduces the shaikh of Bahrain letter dated May 14, 1914; the Admiralty Commission was in Kuwait, November 11–15 and Bahrain November 21–26; Longhurst, *Adventure in Oil*, p. 51 "unanimous enthusiasm"; see 1914 House of Commons Tabled Papers, vol. 54 (liv) Cd. 7419, Final Report of the Admiralty Commission on the Persian Oilfields—Explanatory Memorandum and vol. 54 (liv) Cd. 7419, Navy Oil Fuel: Agreement with the Anglo-Persian Oil Company Limited. Presented to Parliament.

18. Ferrier, *History of the British Petroleum Company*, p. 282, "also unsuccessful"; Owen, *Trek of the Oil Finders*, p. 1322, "fairly deep well," p. 1336, "G W Halse" and "geological information."

19. E. H. Pascoe, *Geological Notes on Mesopotamia with Special Reference to Occurrences of Petroleum*, vol. 43, Australian National University Library, Document Search Service (Memoirs of the Geological Survey of India, 1922), Yergin, *The Prize*, p. 185, "manoeuvring for oil"; Pascoe, p. 74, "should rival"; Owen, *Trek of the Oil Finders*, p. 1283, "across the two countries"; Pascoe, p. 74, "scarcely promising."

20. Owen, *Trek of the Oil Finders*, p. 1262, "coming up dry" and "Pascoe, Lees and Richardson"; Longrigg, *Oil in the Middle East*, p. 101, "no immediate oil prospects"; Thomas E. Ward, *Negotiations for Oil Concessions in Bahrain, El Hasa (Saudi Arabia), the Neutral Zone, Qatar and Kuwait* (New York: privately published, 1965), p. 20, "not finalised until 1937"; for de Bockh's lagoon theory see Owen, p. 1263, and Ferrier, *History of the British Petroleum Company*, p. 541, "unpromising for oil."

21. G. M. Lees, "The Physical Geography of South Eastern Arabia," paper presented to the Royal Geographical Society London, January 16, 1928, *Geographical Journal* 71 (January–June 1928): 441–63; Ferrier, *History of the British Petroleum Company*, p. 567, "very improbable"; Chisholm, *First Kuwait Oil Concession Agreement*, p. 118, "need not take any further interest"; Owen, *Trek of the Oil Finders*, p. 1296, "without finding any indications"; Ferrier, p. 428, "geologist of great promise" and p. 567, "unfavourably on Bahrain"; Chisholm, p. 162, "drink any commercial oil."

22. Owen, *Trek of the Oil Finders*, p. 1336, P. T. Cox claiming "restricted assignment" and "of urgency."

CHAPTER 1

"MAJOR" FRANK HOLMES

A month after Britain's August 1914 declaration of war on Germany, a heavily built man with a sunburned face and very blue eyes appeared in Winston Churchill's anteroom. He wore a waistcoat under a tweed suit, a hat at a jaunty angle, and carried a cane. He completely captivated everyone with his lively anecdotes of people and adventures encountered in his work in the gold, silver, and coal mines of Australia, China, Mexico, Russia, and the Far East, all told in the colorful speech and unique accent that identified him as a New Zealander. He had a fund of knowledge about geology, geography, natural history, astronomy, literature, and the Bible. He spoke fluent Spanish and knew something about the Oriental antiques and various artifacts of the British Empire that decorated the anteroom.

Frank Holmes was in Winston Churchill's waiting room in order to enlist in the Royal Naval Division. Personally created by Churchill, then First Lord of the Admiralty, the Royal Naval Division's purpose was "the seizure, fortification or protection of any temporary naval bases, which might be necessary to the employment of the Fleet, or the provisioning of an army in the field." Many Australians and New Zealanders in Europe at the outbreak of war joined the British forces. Among them was Holmes's friend, fellow New Zealander Maj. George Richardson, in England in 1914 attending the Military Staff College. Richardson was appointed chief of staff of the Royal Naval Division. Churchill liked the "cut" of the tall, hardy, and self-sufficient colonials. They fit right in with his vision of the new force as attracting the very cream of the British Empire's bright young men. The recruitment poster had specifically listed a preferred age of twenty-five to thirty-five.

As Holmes strode confidently across the carpeted office toward him, Churchill took one look at the burly man and barked: "How old are you?" Broadening his smile, the cheerful colonial replied: "Just forty, Sir." Churchill snapped:

"That's a bit old isn't it?" Fixing his bright blue eyes on the First Lord of the Admiralty, Frank Holmes retorted smoothly: "It's exactly your age, sir."

It is said that Churchill always knew when he met his match. Not missing a beat, Churchill shot back: "In that case you're in the prime of life. We'll see what can be done." Churchill made an exception. Holmes was in. As senior supply officer in Churchill's own force, "Major" Holmes was taking the first steps along a path that would eventually lead to his life passion—the discovery and development of the oil of Arabia.[1]

*   *   *

Frank Holmes may have inherited sheer grit from his mother. In the same week of November 1874 that Winston Churchill was born into the upper-class comforts of British society, on the other side of the world Mary Ann Holmes was on horseback, alone, riding across miles of rugged New Zealand terrain. She gave birth to Frank within hours of arriving at the remote work camp where her husband was building a bridge.

Frank's Tasmanian-born mother had married James Holmes in 1866 not long after he arrived from the Lakes District of England. The couple set up house in New Zealand where James worked on the roads and bridges opening up that country. They obtained a government grant of virgin land in Otago on which they developed a sheep farm and raised seven children. Frank completed his education at Otago Boys High School and in 1891, at seventeen, followed his two older brothers already apprenticed to their father's brother in South Africa. They learned the hard practicalities of the mining business in the goldfields of the Rand where their uncle was general manager of the Johannesburg-based Jumpers Co., a pioneer in the cyanide process of recovering gold, a knowledge Frank would use to advantage.

Frank Holmes would later advise a young nephew eager to follow in his footsteps:

> At about your age, I found the necessity of technical knowledge if I wanted to secure appointments in the higher grades. I took the International Correspondence School Mining Engineers Course and have found it a very great help during my active mining career. Consulting Engineers place a high value on the ICS pass diploma as it is generally considered that a man who procures such a diploma and has pluck enough to study seriously while working at the same time to obtain practical knowledge of his profession will be most likely to be of use on the mines.[2]

In South Africa he learned more than mining. On December 29, 1895, he was among the five hundred men whom Scotsman Leander Starr Jamieson unsuccessfully led on a thirty-six-hour raid into the Transvaal in rebellion against Kruger's high taxing of the mostly British-owned mining industry. Holmes left

South Africa for Australia soon after, and while he was employed as a metallurgist with the Melbourne Mint he married Nina Isobel (Nancy), the daughter of Dr. James Howard Eccles, children's specialist at the Melbourne Hospital. They moved on to the arid area of Broken Hill, Australia's "Silver City," 1,180 kilometers west of Sydney. This was where Australia's biggest company, Broken Hill Pty. Ltd. (BHP) was incorporated in 1885 to exploit the rich silver, lead, and zinc deposits discovered two years earlier. Frank and Nina's son, Frank Junior, known as Eccles, was born here while Frank was manager of Seabrook, a metallurgical and ore treatment plant. Across in Western Australia, the first claims were being staked out in what would become the fabulously rich Kalgoorlie goldfields.[3]

WITH HERBERT HOOVER

United States president-to-be Herbert Hoover was hired to inspect these remote Western Australian fields. On Hoover's advice his London principals bought the Sons of Gwalia mine initiated by a group of Welsh miners. At the age of twenty-two, Hoover was appointed mine manager, and Frank Holmes, also twenty-two, joined him, initially as metallurgist. From Western Australia Hoover was promoted to a post in China for the same company, and Holmes, who from his training in the Rand was one of few persons experienced in the cyanide process of gold extraction, went with him. Although Hoover was constantly urged by the Chinese to find gold, as it turned out, much of the time was spent seeking and developing the enormous coal deposits that constituted China's true wealth.

Traveling by boat from Basra to Bahrain in 1922 Holmes would describe the way Herbert Hoover worked. "He was not long in bossing a job he undertook. A man of decision, of quick mind, impetuous but not nervous, full of ambition and grit. He seldom hesitated in making up his mind. He would go down a mine, for instance, look it over and say whether it was worth working or not," Holmes recalled. "I once inspected a mine and thought it was good. I asked Hoover to go down and look it over. He did; and when he came up he told me that it was no good. 'I wouldn't spit on it,' he said. Hoover is like that. He wasted no time or energy on what he thought was of no value."[4]

The period spent in the arid areas of Western Australia was an experience important to Holmes's later recognition of the resources hidden in the deserts of the Arabian Peninsula. Certainly, it taught him how to survive in a harsh climate. Herbert Hoover's description of Kalgoorlie could apply equally to Arabia. It was, he said,

> among the hottest and driest and dustiest places on this earth. The temperature was over 100 degrees at midnight for days at a time. The rain was little more than an inch a year and most of it all at once. The country is unbelievably flat and uninteresting . . . some of our mines lay long distances away in the interior

... the principal means of transport was long strings of Afghan camels. We rode them on inspection trips ... we travelled in 20 to 30 mile daily stages, mostly slept on the ground under the cold stars and were awakened by swarms of flies at daybreak.

Hoover also wrote of Western Australia's water problems that Holmes would immediately recognize when he saw them again in Arabia; a recognition that prompted his successful search for artesian water in Bahrain and Kuwait. "The Kalgoorlie mines were unbelievably rich, but presented difficult metallurgical problems, made more difficult by the lack of water ... such water as we had came from shallow wells in salty depressions ... we must of necessity recover all the water we could from our metallurgical processes in order to use it over again," Hoover recorded.[5]

When the work in China, interrupted by the Boxer Rebellion, ended in October 1901, Hoover went back to California, returning to Australia in 1902. Holmes traveled to London where he joined the mining enterprises of the groups chaired by Edmund Davis, who would later form the Eastern & General Syndicate for work in Arabia. Holmes stayed in touch with Hoover, sometimes referring mining persons he considered particularly good; Hoover often employed them on Holmes's recommendation. In 1926 Herbert Hoover would provide a personal reference to the American oil companies attesting to Holmes's professional abilities and character. It may also have been Holmes's association with Hoover that would later influence his choice of the Gulf Oil Corporation USA as a partner in developing the Bahrain oil concession. The founder and majority stockholder in this company was the Mellon bank family. In 1926, when Holmes began his association with Gulf Oil, Andrew Mellon was US Secretary of the Treasury, working with Herbert Hoover as secretary of commerce.[6]

RUSSIA, MALAYA, AUSTRALIA, AND MEXICO

The Edmund Davis group sent Frank Holmes to Russia, where he had his first taste of the oil industry in the Caspian. He went next to Malaya as assistant general manager of the Penang Corporation, whose many branches included running the Tanjong Pagar Dock Company at Singapore. This experience later echoed in Holmes's advice to the future king of Saudi Arabia, Abdul Aziz bin Abdul Rahman bin Saud, that Ras Tanura was the best prospect in his territories for developing a commercial port. By 1905 Holmes was back in Australia, intending to set up his own business as a mining consultant, but the Davis group made him an offer he couldn't refuse and, now thirty years old, he moved to Mexico.[7]

He managed several mining projects simultaneously, mostly around Chihuahua in northern Mexico, while also acting as consulting engineer to the General Luis Terrazas Estate, headquartered in Chihuahua. Spurred by the 1901 discoveries just across the border in Texas, Weetman Pearson (later Lord Cowdray) spent five million sterling until, in 1908, he found the huge oil reserves in Mexico that would make him one of the richest men in Britain. Very quickly the Hearst family was drilling for oil in Chihuahua, and Holmes's associate General Terrazas—dubbed "the largest individual land and cattle owner in the world"—obtained a ten-year regional prospecting license for oil.[8]

Holmes's sympathy for the native labor, frequently employed in deplorable conditions at that time, was not always well regarded by expatriate mining colleagues. In Mexico, the foreign miners spoke disparagingly of the indigenous people as "Apaches." Holmes, sniffed one of his contemporaries, "does not offend them." Rebellion broke out in Mexico in 1909 in the lead-up to the 1910 elections. In the midst of the uprisings in May 1910, Holmes's Australian wife, Nina, died. She had been very ill after the loss of an infant daughter. Within weeks, continuing fierce rebel activity caused the suspension of all mining operations in Chihuahua. Holmes packed up his son and took him to England.[9]

While the boy settled in with the Shropshire family of his father's friend William Pendlebury—Holmes would marry Dorothy, a daughter of this family—over the next four years, Holmes worked again in Russia, then on various tin- and gold-mining projects in West Africa. For the Davis group he negotiated the purchase of a lightering company in Uruguay and also spent time in Montevideo.

Although Nigeria did not develop its oil resources until years later, Holmes was there before the First World War exploring for gold, tin—and oil. In April 1914 he married Dorothy in England and was preparing to move his family back to South America when, in August, Britain entered the war against Germany. Frank Holmes set out along the path that would lead to his discovery of the oil riches of Arabia.[10]

THE ISLANDS OF BAHRAIN

Britain's war against Turkey was launched from Bahrain. On October 23, 1914, on the heels of Holmes's meeting with Churchill, ships of the Royal Indian Marines carrying the five-thousand-strong Indian Expeditionary Force sailed into the Persian Gulf and anchored off the island of Bahrain. Almost equal distance from the entrance to the Gulf and the mouth of the Shatt Al Arab (the Shore of the Arabs), the tiny speck of land measured thirty miles in length and ten miles across at its widest point. About twelve square miles were covered by fields of mysterious prehistoric burial mounds and a further eighteen square miles by date plantations.

The shaikhdom of Bahrain, under Sunni Arab rulers the Al Khalifa family, was actually an archipelago of thirty-three islets and rocks, of which Bahrain was the largest with two-thirds of the total population. The temperature from June to September averaged 107 degrees Fahrenheit, with humidity of 80 percent or above. Rain averaged 3.25 inches, ordinarily falling three to six days a year. The population numbered close to ninety thousand, the majority being Sunni Arab townspeople along with the indigenous Shia Arab villagers known as *Baharna.* Among the population were nearly twelve thousand African Negroes who, as a British intelligence report observed, were so well integrated that it was "almost impossible to distinguish Negroe families from the communities among whom they live." There was also a small number of foreign permanent residents including Persians, Arabs from Basra, Jews, and Hindu merchants.[11]

The pearl fisheries of Bahrain were the most important in the region, utilizing almost one thousand boats and employing eighteen thousand Bahrainis, augmented each May–October season by additional boats and divers from the Arab shaikhdoms, Turkish Arabia, Persia, Aden, Somaliland, and Zanzibar. As the only steam port in the Persian Gulf, Bahrain had a thriving transit trade with some one hundred Bahrain trading vessels regularly running to the Arab ports of the Gulf, the Persian coast, and to India, Yemen, Aden, and Zanzibar. Bahrain was famous for its boatbuilding, sail making, and the weaving of woolen cloth.

The islands were self-sufficient in fruit, vegetables, sheep, and goats, importing only dried goods and cereals. About six hundred permanent fishing boats could supply five hundred pounds of fish per day. The "finest kind of donkey in the world" was the small white animal bred in Bahrain, of which there were about two thousand, along with fifty or so pure Arab horses, one hundred fifty camels, and a local breed of cattle "famous even on the Persian coast for their milking qualities." Almost any currency was acceptable in Bahrain. Turkish lira, a wide variety of foreign coins including some from Europe, Indian coins, the Persian double kran, and the Tawilah of the Al Hasa oasis circulated freely. Pearl divers were paid in gold Marie Theresa dollars or Riyals, imported from Bombay.

An officer of the Political Department of the Government of India, known as the political agent, resided in the main town of Manama. Attached to the Political Agency was the Victoria Memorial Hospital, financed by local contributions but constructed by the Government of India in 1905 in a move aimed at offsetting the influence of the educational and medical mission of the Reformed (Dutch) Church of America, established in 1893. An Indian Post Office was also attached to the agency. A British mercantile firm and two British shipping companies had offices in Bahrain, and a German commercial firm had recently been established. The political agent reported in 1914 that permanent British subjects in Bahrain numbered two British Europeans, four Anglo-Indians, seventy-one Indians, five "British Jews," and, curiously, two "Native Christians."

When Britain declared war with Germany, the British intelligence report noted that there were probably no more than three hundred rifles on the islands, most of them badly affected by the damp climate, and not more than half a dozen men with any idea of how to use a rifle at over three hundred yards. The ruler was thought to own six "rusty old signalling guns." Bahrain's population was, the report concluded, "unwarlike and has a dread of ships' guns. The shaikh and his family are supposed to keep up to 500 armed retainers, but most of these are more familiar with a coffee pot than a rifle."[12]

The ships carrying the Indian Expeditionary Force remained in Bahrain waters for nine days, waiting for orders. None of the five thousand troops were permitted to disembark. The German firm Robert Wonckhaus and Co. was quickly closed down and the remaining employee "made a prisoner of war and sent to India." According to the political agent in Bahrain the attitude of the shaikh and his family was "perfectly loyal," but the "common people" and the merchants were not happy at the prospect of being dragged into someone else's troubles. When hostilities were launched against Turkey from their islands, they reacted with what the political agent called "a certain amount of 'religious' feeling."[13]

Some five thousand of the "common people" would die of plague in the first six months of 1915. Bahrain's main livelihood of gathering and marketing pearls would be set back for years to come. The island's trade would grind to a halt under an export prohibition on essential goods, including food, and the islanders would be quickly reduced to a state of starvation that continued for the next five years. In 1919 the political agent would comment on "the Islanders' famine conditions . . . the starvation in the villages. . . . It is to be hoped Government of India will see its way to removing the restrictions on the export of rice from India at an early date."

The Kuwaitis would similarly suffer the devastating effects of a sea blockade, enforced on the "suspicion" that supplies were reaching the Turks in Damascus from Kuwait via the desert. The ruler of Kuwait was told that he must prevent, and would be held personally responsible for, "all acts in his territory, whether committed by his own subjects, *or by other persons*, that might be against the interests of the British government."[14]

MESOPOTAMIA: INDIA'S WAR

The Indian Expeditionary Force that sailed from Bahrain was the culmination of a goal long cherished by the Government of India. Two years earlier, in 1912, the commander-in-chief, India, the naval commander-in-chief, India, Chief of Staff India, and the political resident, Persian Gulf had drawn up a detailed plan for the occupation of Fao and Basra by the Indian military. First mooted in 1911, this idea for occupation was promoted as resulting in, if successful, the conversion of Britain's "special privileges" in the area into those of rights by conquest, rights that would devolve to the Government of India. Now, however, the Government of India was dazzled by the possibility of securing Mesopotamia outright—as a colony for India—exactly like Aden.[15]

Although the safeguarding of oil from Persia was thought to be the reason why London consented to India's dispatch of this force to Basra, the fact is that protection of Persian oil was not high on the list of priorities of the Government of India. At the time Persian oil was of little importance to Britain.

At its best during World War I, the 51 percent British government–owned Anglo-Persian Oil Company supplied a mere one-sixth of the British navy's fuel; 80 percent of all Allied oil supplies came from the United States. The military secretary to the Government of India quite openly placed oil as the last of five objectives to be gained by the sending of an Indian Expeditionary Force to the Shatt al Arab. He advocated sending Indian troops and landing them in neutral Persia "*ostensibly* to protect the oil installations." Churchill didn't think Persian oil was important either; he stated that European and Egyptian defense had priority and, anyway, Britain could get her oil elsewhere.[16]

Nevertheless, consent did come from London, and so in the 1914 invasion of Mesopotamia, now known as Iraq, it was India that provided the men, the means, and the management of the campaign. Mesopotamia was India's war, India's frontier, and India wanted it as India's reward. The Government of India confidently expected that occupation would result in the colonization of Mesopotamia and the Persian Gulf and their absorption into an extended Indian empire, ruled from Delhi. British officials in India saw no anomaly in the proposal to subordinate Mesopotamia to an already subordinate government— Mesopotamia as a colony for India. Curzon's successor as viceroy of India, Lord Charles Hardinge, apparently could not see any difference between Arabs and Indians as he distributed statements claiming that the Government of India was the logical controller of an occupied Mesopotamia because "the interests concerned, strategic, commercial, political and religious alike are mainly Indian."[17]

The chief political officer with the Indian Expeditionary Force was Lord Curzon's "uncrowned King of the Gulf," the political resident Percy Cox. Supporting Cox's January 1917 move to be appointed high commissioner of Mesopotamia, Lord Curzon would write to him approvingly: "Nobly have you carried out the mission with which I entrusted you 18 years ago—you have made yourself the King of the Gulf. When the war is over we shall consolidate that Kingdom and see that no one snatches away the crown."

Cox's own protégé, his assistant and deputy Lt. Col. Arnold Wilson, reflected the thinking of many in the Government of India when he declared: "I should like to see it announced that Mesopotamia is to be annexed to India as a colony for India and Indians, that the Government of India will administer it, and gradually bring under cultivation its vast unpopulated desert plains, peopling them with martial races from the Punjab." Wilson also believed that the British Legation in Tehran should be made subordinate to India.

And in March 1915, Sir Arthur Hirtzel, undersecretary at the India Office, enthusiastically advocated that an unrestricted transfer of Indians into

Mesopotamia would at once reward India's war effort while achieving the laudable goal of removing any excuse for Indian immigration into "white man's colonies."[18]

The "Indianization" of Ottoman territory proceeded at a clip as occupation allowed for fast-tracking of the old Three-Steps to Indianization program. Seven days after occupying Basra the military police were replaced with an India-modeled civil police. Customs, finance, the municipality and the government press were taken over. The *Basra Times* was published six times a week by the occupying force, in English, Arabic, and Hindi. So-called lunatics and criminals were disposed of by being sent to India. An Indian judicial system was imposed and Indian currency became the legal tender, as it was already in the shaikhdoms of the Gulf.[19]

Noncombatants from India provided the labor force for the army in Mesopotamia, and Indian immigration boomed. Immediately after the war there were ten thousand British and sixty-five thousand Indian troops stationed in Mesopotamia. Some fifty thousand to seventy thousand Indian laborers, including eight thousand convicted habitual criminals promised sentence remission, were also in Mesopotamia. In addition, hundreds of Indian clerks and office workers were employed to administer the "government." Some fifteen hundred private Indian citizens had taken up residence. Presiding over all this was the Indian Expeditionary Force's chief political officer, now Maj. Gen. Percy Cox, who was also, as he declared, "still the Political Resident of the Persian Gulf for all important matters."[20]

Yet India's management of Mesopotamia was dubious from the start. It hit bottom with the international humiliation occasioned by the retreat of the Anglo-Indian army followed by a five-month Turkish siege that ended in April 1916 with the surrender at Kut. Less than four hundred of the fourteen-thousand-strong force made it out; forty-six hundred were dead, wounded, or missing while nine thousand men and officers surrendered to the Turks.

The disaster triggered a parliamentary inquiry into the Government of India's management and resulted in control of military operations in Mesopotamia being taken from the Government of India and transferred to the War Office in London. Parliamentary debate and public comment on the inquiry's findings began in July 1917 when an epic story of sheer incompetence, political intrigue, personal envy, and backstabbing was revealed. "I regret to have to say," wrote Lord Curzon to his colleagues, "that a more shocking exposure of official blundering and incompetence has not, in my opinion, been made . . . at any rate since the Crimean war."[21]

HOLMES IN EGYPT

When the Indian Expeditionary Force became the British Mesopotamian Expeditionary Force, under the orders of the War Office, Maj. Frank Holmes was sent as senior supply officer. From headquarters in the British Embassy at Addis Ababa, he scouted through Abyssinia (modern Ethiopia), the Italian colony of Eritrea, and across the Arabian Peninsula locating and buying food and supplies for the Anglo-Indian forces in Mesopotamia.

After the meeting with Churchill, Holmes had gone with the Royal Naval Division to Antwerp until the division was committed to the ill-fated Gallipoli campaign. On March 8, 1915, Frank Holmes was among the twelve thousand men who marched from Downs Camp in quiet Dorset to the rural railway stations at Blandford and Shillingstone. From there they traveled to Avonmouth where Holmes took charge of boarding one thousand mules onto the SS *Astrian*. They sailed for Egypt where an exotic seventy-five-thousand-man force was assembling to undertake a plan to force the Dardanelles by sea and land and so capture Constantinople from the Turks. In the tent cities that sprang up within sight of the pyramids were the elegantly uniformed French and their Foreign Legionnaires from Africa and the Levant. Sailors of the British and French navies mixed with Scottish, Irish, and Indian troops and thirty-thousand bronzed colonials from Australia and New Zealand. The men from Down Under were the Anzacs, the Australia and New Zealand Army Corps. Most of them were at least six inches taller than the descendants of the British cousins left behind one hundred fifty years earlier when their great-grandparents had sailed for the other side of the world.[22]

Holmes prowled the bazaars of Cairo and Alexandria looking for supplies. He had the knack, considered the height of good manners in the Arab world, of being able to talk about nothing in particular for a very long time. He could sit and yarn for hours, about the weather, about history, about man's place in the scheme of things. When he finally came to the point, he and the *bazaari* were on the same wavelength. If they had anything to sell, Holmes would get it. He disembarked the one thousand mules and took delivery of cargoes of horses as these arrived. On April 14, after reloading the horses and mules, he was onboard the *Astrian* as the headquarters of the division sailed from Suez to join their colleagues at Gallipoli. Their mission was to stage a pretense at landing in order to divert the enemy while a naval attack was mounted elsewhere.

But the great naval attack never got off the ground, and within two hours of their first sight of the enemy the Naval Division was called back. The attack was rescheduled for April 25. Once again, Holmes disembarked the horses and mules and exercised them on the beach of the Island of Lemnos, the mythical home of the god Vulcan.

THE GALLIPOLI CAMPAIGN

On Saturday April 24 the Naval Division sailed again for the Gulf of Saros. The landing, at dawn on the twenty-fifth, was utter folly. The Anzacs landed a mile from their intended destination and were mown down in the thousands as they attempted to scale the sheer cliffs facing them. Farther up the coast British troops were also met by a hail of Turkish fire and totally devastated in rowboats and on the gangways and decks of the massive *River Clyde* in their landing attempt. A further fiasco took place on May 2 when the Naval Division tried to cover the landing of the Otago Battalion—from Holmes's neighborhood in New Zealand. But they couldn't be found and nobody seemed to know where they were. The Otago men were "digging in" a quarter-mile away, where they, too, had been erroneously deposited.

As the Gallipoli campaign settled into the long and hopeless siege recorded in military history, Holmes's mules worked through the night toiling up the maze of goat tracks to deliver ammunition and supplies to the men in the trenches and returning downhill with the wounded. Water had to be brought by sea from the Nile, seven hundred miles away in Egypt, pumped ashore, and carried to the front on muleback. The supply site was the Island of Lemnos, the Naval Division's initial destination. Ships arrived with their cargo unknown, or the cargoes were sent in the wrong vessels to the wrong place. Holmes's ability to claim the most useful of these cargoes, and his forays into the bazaars of Cairo and Alexandria, made him famous as a truly talented supply officer. His personal bravery during the eventual evacuation of December 1915 resulted in his being twice mentioned in dispatches. Despite being over Churchill's preferred age of twenty-five to thirty-five, Frank Holmes may have been "the new type of Officer" described by the Naval Division's historian. These were men, the historian recorded admiringly, "who had knocked about and seen life in the raw, planted tea in Ceylon and daggers in recalcitrant cannibals, built bridges in India and blown them up in South America. This kind of officer was apt to be unduly affronted by the luxury of a sleeping bag or the elaborate fittings of a dug out."[23]

The Dardanelles campaign went awry almost from the moment of its conception in the plush offices of Whitehall. Of the half-million men sent to Gallipoli over the two hundred fifty-nine days of the campaign, more than half ended up dead or wounded. When it was all over, a Royal Commission concluded in 1917 that the campaign had been "a mistake."

Across the other side of the world, the criticism was more blunt; the "tragedy of Gallipoli" assumed mythical proportions in the official histories of Australia and New Zealand. The Anzacs took a long time to forgive what they saw as the incompetence and arrogance of "old school aristocrats" of the British command that they believed led to the decimation of the very flower of the two colonies' young men. The offense was compounded by the fact that the Anzacs were a non-conscripted force who had voluntarily gone to the aid of the British. Frank Holmes would soon meet this type of British official again, in the service of the Government of India. Perhaps from his experience of the mistakes, miscalculations, and mischances that made up the Gallipoli story, he would feel the same lack of respect for them as his countrymen felt for the "pompous and inflated" upper-class military men of the Dardanelles campaign.

HOLMES IN ADDIS ABABA

The British Embassy at Addis Ababa must have seemed a welcome change for Frank Holmes after the mire of France and the hardship of the Dardanelles. Addis Ababa at eight thousand feet was rarefied, and the large stone buildings of the legation, designed in British-India style, were comfortable. In 1868 Sir Robert Napier of India, with an army of sixteen thousand men and elephants to carry the guns, had established British control over what was then called Abyssinia (it became Ethiopia after World War II). When Holmes arrived in early 1916, the British minister was the urbane, six-foot-two-inch-tall Wilfred Patrick Thesiger. Holmes and Thesiger had some experiences in common. Holmes's engaging personality and talent to amuse with his store of anecdotes ensured the two men quickly became friends. A grandson of Lord Chelmsford (the former Lord Chancellor and brother to the current viceroy of India who had inherited the Chelmsford title), Thesiger had fought during the 1901 Boer War in South Africa and had spent 1906 in Russia before taking up the appointment of British minister to Abyssinia in 1909, at the age of thirty-eight. From his brother, who had been governor of Queensland and New South Wales in 1905–13, he also knew something about Australia. On the declaration of war in August 1914, Thesiger was on leave in England. He joined the Intelligence Branch of the army and served in France before returning to his post in Addis Ababa in January 1915.

Also living in the embassy compound was six-year-old Wilfred Thesiger Jr., the first British child born in Addis Ababa. Wilfred the younger, like Holmes, would go on to develop a love of the Arab people and lands. He would become what some called "the last of the great British eccentric explorers" for his talented exploration, writing, and photography across Africa and Arabia. He was widely acclaimed after the 1959 publication of *Arabian Sands*, the record of his many journeys through the great Rub al-Khali, the desert that stretches for nine hundred miles from the frontier of Yemen to the foothills of Oman and for five hundred miles from the southern coast of Arabia to the Persian Gulf. It was known to Europeans as the Empty Quarter.[24]

Holmes became something of an explorer himself as he turned to the job of scouting the entire Red Sea and Arabian Peninsula in his quest to secure food and supplies for the British forces in Mesopotamia. He was interrupted in this duty when, midyear, internal unrest began to bubble throughout Abyssinia, encircled as it was by British, French, and Italian territories. In expectation of needing to shelter an influx of refugees, including Europeans, Indians, and Arabs, Thesiger and Holmes stocked up on flour, grain, cattle, and sheep. In October 1916 the wildly colorful armies of the north and south joined battle on a dusty plain sixty miles north of Addis Ababa. After several days of hand-to-hand battle the victors entered Addis Ababa with a new ruler at their head. They were officially greeted by the British minister. In Abyssinia, it was business as usual.

London's takeover of military control of the Mesopotamian campaign from the Government of India was vindicated with the capture of Baghdad in March 1917. By July, when revelations of the parliamentary inquiry into the Government of India's management of the Mesopotamian Campaign became public, Frank Holmes was on sick leave in England. Perhaps not surprisingly he was laid low by what the Medical Board diagnosed as "the effects of food and water conditions in his area of active service." On recovery, he was posted to Flanders with the Royal Marines, Sixty-third (RN) Division. In January 1918 Winston Churchill personally recalled Holmes from the trenches of France. Churchill's instructions were that Holmes "be ordered to return to England and report to the Admiralty as soon as possible, as he is required for other duties."[25]

Mission for Churchill

The "other duties" were to assist in capturing for Britain the imagined mineral riches of Abyssinia. Over the previous twelve months Holmes's prewar mining associates, the Edmund Davis group, had been talking with Churchill (during 1917 he was minister of munitions), the Board of Trade, and the Department of Agriculture. Edmund Davis, the group's influential chairman (later knighted for service to British industry), was requesting exclusive rights to take over several concessions covering potash, potassium, and Wolfram (tungsten ore), all minerals used in production of fertilizer and, more importantly, the manufacture of explosives.

In backing the Davis group proposal, the Board of Trade commented, "in regard to the Potash Deposits on the Abyssinian-Erythrea Frontier; From the point of view of the future of English agriculture and the English Chemical Industry it is most important that a new source of potash should if possible be obtained otherwise we shall continue at the mercy of the Germans after the war." The British hoped to completely exclude the Germans. "It would appear that a practical and intensive exploitation of these deposits will enable exploiters absolutely to dethrone German influence re Chloride of Potash," they surmised. The Germans had long been the main suppliers of potash and the French and Italians were already in Abyssinia. Britain had been buying these minerals through the Italians, paying, as Churchill pointed out, almost double the price charged to Italian industry.

The ministry of munitions vouched for the individual members of the Davis group lodging this application, saying, "Edmund Janson is a partner in Percy Tarbutt & Co, mining engineers (see Directory of Directors). I have known Janson for years and he has already spoken to me on this subject. I do not think we need investigate Tarbutt & Co's doings, especially as Janson is vouched for. If an English firm can get control of these deposits so much the better." The Department of Overseas Trade declared the members of this particular syndicate "well known to the two West African Departments of this office as extremely acute men of business connected with serious West African mining ventures."[26]

*   *   *

Abyssinia was tipped in 1917 mining circles as an area of unlimited mining potential. Holmes's associates put forward grand plans for developing the Abyssinian concessions, including construction of several railways, and nominated the group's own consulting engineer, New Zealander Frank Holmes, as the ideal man, perhaps the only man, capable of both negotiating and securing the concessions and then moving on to physically and technically develop them.

Churchill, now minister of war, agreed, and the Davis group got their exclusive rights. In tribute to Holmes the group formed in March 1918 for this venture was named the Frank Syndicate. On his way to Abyssinia to take up the new challenge, Holmes reported to his London colleagues "a new trend" for pharmacies in Alexandria, an observation of interest to one of the Davis investors, the manufacturing chemists Allen and Hanburys.[27]

When the war ended in November 1918 Holmes returned to London to collect his wife, Dorothy. In December, he received another letter from the Admiralty informing him he was to receive the "Diploma of the Order of the Crown of Italy (Cavalier) awarded you by His Majesty the King of Italy." Modestly, Holmes told friends it was "only a dog fight medal."

Over the five months it took to get from London to Addis Ababa, Dorothy wrote home from each port of call. From Paris in January 1919, from Rome in February, where she reported of Frank and an accountant, also going to Abyssinia, learning golf on the links between the new and old Appian Way. In March they were in Athens. By April, in Alexandria, where Dorothy was disappointed to learn that "the Thesigers have left Abyssinia in the middle of the month and have sold all their furniture so it looks as if they will not be returning which is sad as it would have been so much jollier." They spent a May evening in Cairo in the garden of Shepheards Hotel listening to an Australian band, which perhaps to Dorothy's surprise "played very well indeed." Finally, on May 31 Dorothy wrote they were in Port Said "on board the *Sphinx*, this boat is going to Djibouti in five days or so."[28]

The British had predicted that "parts of Abyssinia will prove an El Dorado, a new California, for commercial enterprises after the war." The hoped-for Abyssinian riches did not eventuate. But Holmes thought he knew where riches could be found. As he had moved around the Arab shaikhdoms during the war, Frank Holmes became convinced that great wealth was hidden beneath the vast deserts.

He was so sure that in 1918 he wrote his wife in England, "I personally believe that there will be developed an immense oil field running from Kuwait right down the mainland coast."[29]

Holmes's conviction of an oil-rich Arabia was out of step with all current opinion, which held there was no oil in Arabia. To convince the Davis group to back his hunch he needed to find a way to generate income from within the region. His attention turned to the other side of the Red Sea from Abyssinia, and centered on the lapsed oil concession of the Farasan Islands and also on the salt mines concession of the Salif Peninsula. The policy of Holmes's business associates was to set up different groupings of investors to support individual, and financially separate, ventures. Consequently, the syndicate backing the Abyssinia operations remained intact and a new grouping was formed for activities on the Arabian side of the Red Sea.

The new group was the Eastern & General Syndicate Ltd. (E&GSynd). In April 1920 it was visualized as "developing trade on the Arab Coast, but more especially on the Assir and Yemen portion of the Red Sea." By the time of its August 6 official registration in London, Holmes had managed to have the aims extended to include "dealing with concessions in Arabia."[30]

NOTES

1. See Douglas Jerrold, *The Royal Naval Division: With an Introduction by the Right Hon. Winston S. Churchill* (London: Hutchinson, 1923) and Geoffrey Sparrow and J. N. Macbean-Roos, *On Four Fronts with the Royal Naval Division* (London: Hodder and Stoughton, 1918); Scholefield Collection, July 25, 1960, Dorothy Holmes to Guy Scholefield, "Frank knew General and Mrs. Richardson when he was at the Staff College Camberley. I think it was probably on his recommendation that Frank joined the RND. I know he went to see Churchill who said Frank was too old when Frank commented that Churchill was the same age Churchill said in that case you are in the prime of life and he got the commission." See also Wayne Mineau, *The Go Devils* (London: Cassell, 1958). Mineau interviewed Dorothy and recounts this episode on pp. 179–80; also October 26, 1959, and December 1, 1960, Lord Freyberg, Lt. Gov. Windsor Castle to Scholefield, "Richardson and Holmes were known to me about 1914 . . . they certainly were associated with Winston. . . . I saw a great deal of Holmes when he was with the Naval Division and several times after the war." As Gen. Sir George Richardson he became Administrator of Western Samoa.

2. Frank's father, James Holmes, was born in Westmoreland, UK, 1837; his wife, Mary Ann Smith, was born in Hobart, 1836; they married in Otago, New Zealand, 1866. Frank Holmes was born on November 25, 1874. His two older brothers, Robert and Alan, died during the Boer War. His youngest brother, Percy (1880–1968), followed a similar apprenticeship to their uncle in South Africa. Percy Holmes went on to be manager of the Patino Mines, in his time the world's largest tin mine; he founded a branch of the family in Bolivia; Carole Davidson Collection, December 31, 1928, Frank Holmes at the Maude Hotel Baghdad to his nephew in Mexico, "a man who procures such a diploma and has

grit enough to study seriously and work to obtain practical knowledge of his profession at the same time will be most likely to be of use on the mines."

3. Jamieson surrendered on January 2, 1896; the raid was one of the important causes of the Boer War of 1899–1902. Papers held by Philip Davidson and Michael Stephenson show Frank and Nina married in the Holy Trinity Church, Maldon Victoria, October 14, 1897; see Geoffrey Blainey, *The Rise of Broken Hill* (Melbourne: Macmillan of Australia, 1968).

4. Ameen Rihani, *Ibn Sa'oud of Arabia: His People and His Land* (London: Constable, 1928), p. 80, "I once inspected a mine."

5. Herbert Hoover, *Memoirs of Herbert Hoover: Years of Adventure, 1874–1920* (London: Hollis and Carter, 1952), pp. 30–32. From 1897 to closure in 1963 the Gwalia mine produced 2.6 million ounces of gold. For the work in Australia and China see David Burner, *Herbert Hoover, A Public Life* (New York: Knopf, 1979).

6. Herbert Hoover Library, July 11, 1908, Hoover to Holmes: "I am in receipt of your letter regarding a position for a friend of yours. I am afraid that I have no opening at the present moment although I think I shall want a general clerk when we expand our works a little and will be very glad to oblige you. I owe you many thanks for Travis for he is doing splendid work." Chevron Archives, box 120791, memorandum for historical research project, notes of telephone conversation with Mrs. Mary Stearns, October 20, 1955, "Holmes had worked with Herbert Hoover in Australia at some time pre WW1 and gave Hoover's name as reference. . . . Loomis checked on Holmes with Hoover." Andrew Mellon (1855–1937) was Secretary of the Treasury under three successive presidents, Harding, Coolidge, and Hoover; Herbert Hoover was behind the post–WWI push to obtain a share of Iraq's oil for American companies; he insisted the British afford an Open Door policy to Americans. See, for example, Benjamin Shwadran, *The Middle East: Oil and the Great Powers* (London: Atlantic, 1956), p. 209, then secretary of commerce Herbert Hoover "took it upon himself" to ensure American oil companies would have a share in Mesopotamia's oil.

7. See J. D. Henry, *Baku, An Eventful History* (London: A. Constable, 1905); Dickson Papers, box 2a/file 4a, report, September 30 to November 12, 1922, Dickson to Cox: "Holmes said he had already concluded from charts Ujair (as bin Saud wanted) was useless as a port, Ras Tanura was the place."

8. Anthony Sampson, *The Seven Sisters: The Great Oil Companies and the World They Made* (London: Hodder and Stoughton, 1975), pp. 81, 88, 100–101, for detail Lord Cowdray and Mexican Eagle (Aguila) Company. Among Holmes's private papers, held by Peter Mort, UK, is a card recording the inscription on a gold watch, dated October 31, 1908, "presented to Mr. Frank Holmes, General Manager of the Palmarejo and Mexican Gold fields Ltd, of the Oxnam Prospecting Co No.1, Chinipas, Chihuahua as a token of their esteem by the staff of both."

9. Percy Holmes Papers held by Michael and Gloria Stephenson, Bolivia, the 1909–12 journal of Robert Emmerson, p. 175, "(the mine) has a bunch of Apaches . . . Holmes does not offend them . . . the only Indian on the warpath is myself"; the Mexican troubles continued; in 1911 despotic dictator General Porfirio Diaz resigned; he had been president since 1877. Holmes and his son arrived in London July 1910. Eccles was with the Indian army for fifteen years after joining during WWI, he married Cecily Ronide (no children), was financed by his father in mining projects in Canada, and died in 1939.

10. William Pendlebury would progress from schoolmaster to Director of Education for Shropshire. See Guy H. Schofield, ed., *Dictionary of New Zealand Biography* (Alexander Turnbull Library, Wellington, New Zealand), MS 212, folder 22, letters from Dorothy. She was proud of her Austrian mother, whom she told Scholefield was the daughter of "Baron del Monte von Montaucau an aide de camp to the Emperor Franz Joseph." Dorothy was with Frank in Abyssinia 1919 and on four-to-six-month visits taking in Iraq, Kuwait, Bahrain, and Dubai in 1924, 1934, and 1937.

11. Sheila A. Scoville, ed., *Gazetteer of Arabia: A Geographical and Tribal History of the Arabian Peninsula* (Graz, Austria: Akademische Druck Verlagsanstalt, 1979), 1:408–49, "free negros, known as Sidis, are treated as of the tribe or class among whom they dwell."

12. Ibid., 1:439–47, Intelligence Report prepared 1914 on resources of Bahrain that may prove useful to the British war effort, included such headings as "Fighting Men," "Camping Sites," "Food Resources."

13. Administrative Report for Bahrain 1914. "Of the two members of Wonckhaus and Co. one, Herr Trumper, a Lieutenant of the active reserve returned to Germany via Baghdad at the outbreak of War, the second Herr Harling of the Ersatz Reserve, was made a prisoner of war while the Indian Expeditionary Force was anchored off Bahrain, and was sent to India. Their branch was closed when he was made a prisoner of war. Herr Abraham, of Messrs Rubin and Co was in Bahrain for a short time. On the outbreak of war he tried to return to Europe, but being a member of the Landsturm, was made a prisoner of war in Karachi."

14. Administrative Report for Bahrain 1919, "the starving condition"; H. R. P. Dickson, *Kuwait and Her Neighbours* (London: Allen and Unwin, 1956), p. 243, "early in 1918, the new Ruler of Kuwait, Shaikh Salim, was warned."

15. Philip Willard Ireland, *Iraq; A Study in Political Development* (New York: Russell and Russell, 1937), pp. 22, 28. The committee was Admiral Slade, naval commander-in-chief, East Indies; Lt. Gen. Percy Lake, chief of staff, India; Lt. Col. H. McMahon, foreign secretary, India; and Maj. Percy Cox, political resident Persian Gulf.

16. Briton Cooper Bush, *Britain, India, and the Arabs, 1914–1921* (Berkeley: University of California Press, 1971), p. 4 claims the British cabinet decided to send the forces "to occupy Abadan Island to protect the oil tanks, refinery and pipe lines located there . . . *ostensibly, at least.*" Sir Arnold Talbot Wilson, *Mesopotamia, 1914–1917: A Personal and Historical Record* (London: Oxford University Press, 1930), p. 8, also makes this claim however Wilson placed the decision making in the hands of the Government of India. "Brigadier General Delamain of the Indian Army was in command . . . to occupy Abadan Island in order to protect the oil refineries, tanks and pipelines, cover the landing of reinforcements and assure the local Arabs of our support against Turkey." That protection of Britain's oil from Persia was not a decisive factor is concluded by Marian Kent, *Oil and Empire: British Policy and Mesopotamian Oil, 1900–1920* (London: Macmillan, 1976), p. 118, "during the early part of the campaign, Mesopotamia was clearly not a major battleground, and oil, whether Persian or Mesopotamian, was not a major factor in planning military strategy" and by Ireland, *Iraq; A Study*, pp. 24–25, "General Barrow, Military Secretary to the Secretary of State India, September 1914 'ostensibly to protect the oil installations but in reality to notify the Turks we mean business' Barrow placed oil as the last of five objectives"; Elizabeth Monroe, *Britain's Moment in*

the Middle East, 1914–1971 (London: Chatto and Windus, 1981), p. 98, also makes the point that royal assent was given to the bill ratifying the British government's 51 percent purchase of the Anglo-Persian Oil Company just six days before the outbreak of World War I, "because of its date the purchase is often attributed to the fear of war then hanging over Europe . . . Churchill dispelled this notion saying 'Nobody cares in wartime how much they pay for a vital commodity.'" Excellent documentation on this whole episode, including the aims of the Government of India, can be found in the Public Record Office (hereafter cited as PRO), PRO/WO/106/52 precis of correspondence regarding the Mesopotamian Expedition, Its Genesis and Development, a very long, and secretly prepared report, by the Military Department of the India Office; Kent, p. 118, "Churchill . . . noted . . . buy her oil elsewhere"; Sampson, *Seven Sisters*, pp. 77–81, "80% . . . from USA," also Daniel Yergin, *The Prize: Epic Quest for Oil, Money, and Power* (New York: Simon and Schuster, 1991), pp. 174–75, for details of Britain's World War I oil supplies.

17. Busch, *Britain, India, and the Arabs*, p. 23, "mainly Indian."

18. Philip Graves, *The Life of Sir Percy Cox* (London: Thornton Butterworth, 1939), p. 231, "no one snatches away"; Busch, *Britain, India, and the Arabs*, p. 22, "a colony for India and Indians"; John Marlowe, *The Persian Gulf in the Twentieth Century* (London: Cresset, 1962), p. 115, "Tehran Legation should come under India," and p. 41, "white man's colonies." See Peter Mansfield, *The Arabs* (London: Penguin Books, 1982), p. 216, "imperialists in the British Cabinet argued that everything should be done to obtain 'a continuity of territory or of control both in East Africa and between Egypt and India.'"

19. See Marlowe, *Persian Gulf*, pp. 50, 144, 391, 401, for "Indianization" of Mesopotamia; IOL/R/L/PS/11/193, January 31, 1921, "Report of the Interdepartmental Committee . . . to Deal with Mandated and Other Territories in the Middle East"; note in this report "there is a Treasury of the Government of India" at the Political Residency at Bushire.

20. See Wilson, *Mesopotamia, 1917–1920: A Clash of Loyalties* (London: Oxford University Press, 1931), p. 46, convicted Indian prisoners, originally designated "Disciplinary Military Labour Corps" became known as the "Jail Corps"; Graves, *Life of Sir Percy Cox*, p. 209, July 1916 Cox to India Office, "still Political Resident"; see Penelope Tuson, *The Records of the British Residency and Agencies in the Persian Gulf* (London: India Office Library and Records, 1979), p. 184, "Cox was titular Political Resident until October 1920 although absent in Baghdad, on his 1918 appointment to Tehran, Lt Col Arnold Wilson took over as 'absentee Resident' in Baghdad until October 1920."

21. Busch, *Britain, India, and the Arabs*, p. 130, "official blundering and incompetence."

22. Jerrold, *Royal Naval Division*, p. 48, "Senior Supply Officer"; see also Michael Hickey, *Gallipoli* (Melbourne: J. Murray, 1995); Holmes's diary 1914–15 for detail; also Dorothy Holmes Papers; Alan Moorehead, *Gallipoli* (London. H. Hamilton, 1956), p. 132, "It was the physical appearance of the Dominion soldiers, Colonials as they were then called, that captivated everybody who came to Anzac Cove and there is hardly an account of the campaign which does not refer to it with admiration and a kind of awe . . . those glorious young men . . . their tallness and majestic simplicity of line, their rose-brown flesh burned by the sun and purged of all grossness . . . something near to absolute beauty."

23. Dorothy Holmes Papers, November 7, 1915, Dorothy to her sister, "a piece in

The Times refers to Frank's being mentioned in Despatches in the Dardanelles"; also Guy H. Scholefield, "A Tough Patient New Zealander," *Otago Daily Times* (n.d., n.p.), "served throughout to the evacuation and was twice mentioned" and "NZ Man Walked into Confidence of Sheik Oilmen," *Post* (n.d., n.p.), "headquarters at the British Embassy"; Jerrold, *Royal Naval Division*, p. 56, "new type of officer."

24. The British minister at Addis Ababa was Wilfred Patrick Thesiger (1871–1930); See Wilfred Thesiger, *The Life of My Choice* (London: Collins, 1987), pp. 23–57, for description of the Addis Ababa embassy and the revolution of 1916–17 that a decade later would lead to the installation of Haile Selassi as emperor of Ethiopia. The viceroy of India, Wilfred Jr's Uncle Frederick John Napier Thesiger (1868–1933), may have been influenced by his time in the newly federated Australia. On his appointment as viceroy of India he became a strong advocate of Indian rights. He went on to be First Lord of the Admiralty under the Labour government of 1924; Wilfred Thesiger, *Arabian Sands* (London: Longmans Green, 1959).

25. PRO/WO 339/69521 is Frank Holmes's WWI military file, containing Medical Board report, "effects of food and water," January 25, 1918, Admiralty to Secretary War Office request that Holmes be returned to England, February 23, 1918, advises that "Major Frank Holmes, Royal Marines, reported at the Admiralty, on return from France, on the 20th instant."

26. PRO/BT/66/11, October 10, 1917: memo/comment Board of Trade re potash deposits on the Abyssinian Erythrea Frontier, and November 30, 1917, abstract of information re Abyssinian potash deposits, and November 7, 1918, Board of Trade to under secretary of state Foreign Office, "so far as Mr. Churchill is aware . . . the price which (Compagnia Mineraria Colonials of Turin) are charging is 38 sterling per ton fob. And Mr. Churchill has reason to believe although direct information on this point is not forthcoming that they are supplying potash to Italy at a price of 20 sterling"; and October 19, 1917, Edmund Janson of Percy Tarbutt and Co. to R. Sperling, Foreign Office, cc. Ministry of Munitions, "we now wish to present to you a scheme we have in mind for the development of Abyssinia." Directors were Chairman (later Sir) Edmund Davis who was chairman of the Chartered Company and director of many other mining and railway companies; Mr. F. W. Gamble, a director of Allen and Hanburys; Messrs. E. W. Janson and P. C. Tarbutt, both of P. C. Tarbutt and Co. Consulting Engineers and directors of various mining companies; and Mr. J. E. H. Lomas, a mining engineer and director of mining and rubber companies. November 13, 1917, interbranch memo from R. B. Ransford, Potash Branch, Munitions, re Abyssinia; and PRO, CO 727/3, Colonial Office memo/letter from Department of Overseas Trade.

27. Papers held by Philip Davidson and Michael Stephenson, registration of the Frank Syndicate, dated London March 25, 1918, and power of attorney to Frank Holmes, same date, witnessed by Davis's associates listed as Directors, Edmund Janson and William Pocock; Dorothy Holmes Papers held by Stella Pendlebury, May 3, 1918, "Frank writes Alexandria is full of chemist shops and he ought to be able to get to Abyssinia in about 10 days time from here."

28. Dorothy Holmes Papers, January 21, 1919, letter from Frank's sister, Agnes, in New Zealand to Frank's son with Dorothy's family in England, "I am enclosing a clipping about your Dad's mission to Abyssinia, most interesting I think, and apparently successful." Dorothy's letters home February–May 1919; Dorothy thinks the boat tickets at fifty sterling each from Port Said to Djibouti are extremely expensive.

29. PRO/WO 339/69521, Admiralty to Holmes, "Citation from King of Italy . . . published London Gazette 17/1/1919"; Holmes Papers, December 13, 1918, Dorothy to her family, "dog fight medal" and Scholefield Collection, October 27, 1959, Dorothy to Scholefield, "we never could quite fathom what this was in aid of"; PRO/BT66/11, November 30, 1917, re Abyssinian potash deposits, "The British Consul General Cairo, and General Clayton say . . . Eldorado. . . ." This statement "running from Kuwait right down the mainland coast . . ." in Holmes's 1918 letter is verified by Dorothy Holmes, January 18, 1960, in a letter to Guy Scholefield and is mentioned in Mineau, *Go Devils*, p. 192. Angela Clarke, *Bahrain Oil and Development, 1929–1989*, for the sixtieth anniversary of Bahrain Petroleum Company (London: Immel Publishing, 1991), rather cheekily, but quite erroneously, cites this as from Bapco Archives.

30. Prior to WWI the Turkish government had ceded the salt revenues of the empire to the Ottoman Public Debt Administration, but during the war the Salif Mines, with salt deposits estimated at 3 million tonnes, were barely operated except for some export to India; Eastern and General Syndicate Ltd. was incorporated in London on August 6, 1920, with an authorized capital of 50,300 sterling, of which 15,400 had been issued by June 9, 1922, when its first Directors Report and Accounts for the period ending August 31, 1921, was sent to shareholders. The syndicate's principal shareholders were the Chartered and General Exploration and Finance Co. Ltd., Allen and Hanburys Ltd. (Pharmaceutical Chemists), and B. M. Messa (an Aden merchant). Its first directors were Chairman Edmund Davis, Gamble, Janson, Tarbutt, and Lomas. The first Directors Report stated that "Major Frank Holmes, of Charterland, and Lt-Cdr C E V Craufurd proceeded to Arabia to look into several propositions."

CHAPTER 2

THE WAY TO HANDLE NATIVES

After Frank Holmes was repatriated from Mesopotamia suffering "the effects of food and water conditions in his area of active service," possession of Baghdad led to much tension between the bureaucrats of the Government of India and those of Whitehall. The question was, who would have authority for civil administration now that all of Mesopotamia was occupied? The London-constituted Mesopotamia Administration Committee advocated that, as long as the military occupation continued, authority should be taken from the Government of India and given to the India Office in London. After that, the Committee recommended, Basra would be retained as a British possession. Baghdad would be administered as "an Arab state with a local ruler or government, under British Protectorate in everything but name."

The Government of India had been severely damaged by the military and management fiascos revealed through the parliamentary inquiry into Mesopotamian matters. The disclosure of the true horrors resulting from the failings brought about the resignation of the secretary of state for India. The Government of India had to concede that its dream of Mesopotamia as a colony for India was slipping away. But they had no intention of giving up without a fight. India had contributed five hundred thousand fighting men and donated one hundred million sterling to the imperial war chest. India now demanded that if they expected it to continue providing military manpower, financial contributions or any other aid, then they had better give it a quid pro quo and some compensation for sacrifices already made.

In the trade-off, despite the incompetence, malingering, and venality revealed by the inquiry into the management skills of officials of the Raj, the Government of India was allowed to keep its grip on the Arab shaikhdoms and on the southern half of Persia that bordered the Gulf. Power would still be exercised through the Government of India's own man, the resident in the Persian Gulf.

Strongly supported by Lord Curzon, Maj. Gen. Percy Cox was appointed civil commissioner of Mesopotamia. Cox, who had wanted the title high commissioner as used in Egypt, was not entirely happy about this downgrading of his expected kingdom. He blamed the Foreign Office and "rivals" among the British officials administering Egypt for snatching away the crown that Curzon had promised him. Acknowledging that Mesopotamia could not now be claimed as a colony for India, Cox commented sourly that "Cairo has practically forced us into agreeing not to annex these parts, but merely to administer them."[1]

Cox did not suffer this indignity for long. In April 1918 he was posted as minister to Persia. His mission was to fulfill the dream of his own patron and mentor, Lord Curzon, by negotiating under Curzon's supervision the Anglo-Persian Treaty signed in August 1919 that imposed British political and economic dominance on Persia.

Curzon was thrilled with Cox's success, describing it as a "great triumph." Unkindly, he did not credit Cox. "I have done it all alone," Curzon bragged. The Persian people did not share Curzon's sense of triumph. Of the twenty-six newspapers and publications circulating in Tehran, twenty-five agitated against the unequal treaty.

ARNOLD WILSON

In Baghdad, Cox's own assistant and protégé, the archimperialist Arnold Wilson, exercised a deadly stewardship as acting civil commissioner of Mesopotamia. He also took over Percy Cox's other role, that of political resident of the Persian Gulf. Wilson's designation was absentee resident residing in Baghdad.

As soon as he was in charge, Wilson set about pursuing policies that prompted one official to comment, "Wilson's policy would have been splendid . . . half a century ago." Despite the instructions from London that an Arab government was to be created, even if in facade only, Wilson's view was that "the entire population of Iraq is unfit to form a government or even to have a voice in the process."

Many Iraqis felt they were more than fit to form their own government and they were not willing to settle for anything less than independence. In fact a push for independence in Mesopotamia, Kuwait, and the Arab provinces of southern Persia had begun before the war.

A meeting aimed at achieving independence was held in March 1913, four months before the signing of the Anglo-Turkish Agreement in which Britain designated Kuwait as a Turkish *qada*, or subdistrict. The meeting was attended by the ruler of Kuwait, Shaikh Mubarak, and Shaikh Khazaal of Muhammerah on the left bank of the Shatt al Arab (Southern Arabistan, later known as Khuzestan). Although Khazaal was an Arab, he was a Persian subject, and the Anglo-Persian Oil Company's refinery at Abadan was in his territory. Also at the meeting was the highly placed Turkish official Saiyid Talib Pasha bin Saiyid Rajab, one of the outstanding figures of prewar Mesopotamia and, until his enforced removal from Baghdad by Sir Percy Cox in April 1921, a prominent player in the creation of modern Iraq.

The meeting participants chose Mesopotamia as the best place to begin their campaign and agreed that the three leaders would do everything in their power to advance the claim of all Mesopotamia to independence, or at the very least, to self-government. A "Reform Committee" with Saiyid Talib at its head set about actively seeking transfer of power from the Turks. A second meeting was planned for Kuwait in early 1914 at which the founding members were to be joined by Sharif Husain of Mecca and by the leaders of two great tribes, the Amir bin Rashid of the Shammar and Shaikh Ajaimi of the Muntafiq. Abdul Aziz bin Saud of Nejd was cautious. He responded to his invitation by saying that when the time came to rise he would be ready, but the time had not yet come. Unfortunately for them, the time never came. The outbreak of World War I ensured that the second meeting never took place.[2]

"WILSON'S YOUNG MEN"

There were some who were paying attention to the tide of democratic theory that advocated the consent of the governed and self-determination for all people. Sir Arthur Hirtzel, undersecretary of state to India, urged on his Government of India colleagues a softening of the old imperial ways. In mid-1919, in response to Arnold Wilson's obsessive refusal to countenance a local government for Iraq, Hirtzel advised him: "I hoped you would have realised when you were here in London that the idea of Mesopotamia as the model of an efficiently administered British dependency or protectorate is dead. The same thing is dying in India and decomposing in Egypt, and an entirely new order of idea reigns." Several weeks later, Hirtzel reinforced his advice, again writing to Wilson, "We must swim with the new tide which is set towards the education and not towards the government of what used to be called subject peoples."

Hirtzel was advising Wilson to loosen his grip on Mesopotamia. Perhaps he did not know the full extent of Wilson's enforcement of his own ideology, particularly in the Arab shaikhdoms. Wilson was rewarding the finest exponents of the imperial ideas of the Raj for faithful service in Mesopotamia by installing them in positions of power across the water. Before the war, the political agents in the Persian Gulf were recruited from the Political Department of the Government of India. Under Wilson, they were appointed directly from his own Civil Commission in Baghdad.

Political officers in Mesopotamia were personally selected and groomed by Arnold Wilson. An "in" group of these were tagged derisively as "Wilson's

Young Men"; their careers favored on the basis of how finely their ideas were attuned to Wilson's own. One reported that during his interview he was dazzled by Wilson. He said that "with his old style Indian Army tunic buttoning right up to the neck (instead of the new fangled collar and tie) and the belt with the two parallel shoulder straps (instead of the Sam Browne) with his flashing eyes, his beetling eye-brows, his close cropped hair, and his biblical quotations," Wilson reminded him of a hero of the Indian Mutiny.

Complaints were being made about Wilson's autocratic ways. Curzon had received a report showing that "of the 233 officers employed in Wilson's civil administration only four are over 45 years of age." In November 1919 Curzon wrote to Cox in Tehran, telling him: "The present situation in Mesopotamia is causing us considerable anxiety. It is rigid, costly and hampering the development of either civil administration or whatever form of native government is decided in future. It is for the most part in the hands of young officers who are necessarily lacking in experience. . . . We receive very disquieting reports from some of our own officials expressing growing anxiety about the existing trend of Wilson's administration."[3]

Wilson himself had an inflated idea of his own importance and was convinced he "exercised more independent powers than the Viceroy of India and over a wider area than the High Commissioner in Cairo." He claimed that "my show runs from the Indian frontier to Syria, from the Caspian to Central Arabia; it is twice as big as it was when I first took over from Percy Cox."

In a letter to Sir Arthur Hirtzel, Wilson pointed out: "Professionally, I am a rank Imperialist whose trade for the last twelve years has been the acquisition of territory and influence in these parts." He wrote to his parents, saying: "I am not willing to take on the job of High Commissioner myself yet, although people tell me I should."

Wilson had never hesitated to claim credit for his cadre of adoring political officers. He earnestly explained to Hirtzel: "The only way I keep this show together is by maintaining a strong personal bond between myself and my officers. They are practically without exception my nominees. Almost all the friends I have in the world are included in their ranks and I regard Political Officers and Assistant Political Officers as being with the same family as my own brothers."[4]

RULED BY "WILSON'S YOUNG MEN"

In Bahrain, the first of Wilson's Baghdad appointees was Capt. Norman E. Bray, who arrived in November 1918. He immediately set about bringing into force the Bahrain Order-in-Council, which was the third and last step in "Indianization." Imposed in 1913, the Order-in-Council had not yet been acted on because of the war. Modeled on the civil and criminal laws of India, the Bahrain Order-in-

Council, made up of seven parts, twenty-nine articles, and a schedule, installed the political agent, backed by the Government of India, as virtual ruler.

Captain Bray set up a Chief Court, with all the powers of a High Court, over which the political resident presided. The political agent himself presided over a District Court. A third court, the Joint Court, was instituted over which the political agent also presided together with an official appointed by the shaikh. Moreover, the political agent held jurisdiction over all non-Bahrainis, although the controversial issue of who were subjects of the shaikh by dint of being Bahrainis and who were under the jurisdiction of the political agent by virtue of being foreigners was not exactly defined.

In addition, the political agent appropriated the right to appoint half the members of the *Majlis al Urf*, the commercial or mercantile court. For the appointment of judges to the *Sulfa* Court that handled matters related to pearl diving, the shaikh had to reach agreement with the political agent. All these appointments were previously exercised exclusively by the shaikh. The *qadis*, or judges, of the religious, or *Sharia* court, could still be appointed by the shaikh— but only after acceptance by the political agent. Furthermore, the judgment passed by the *qadis* of the Religious Court could not be implemented until ratification by the political agent, who had the power "to revise the findings and the sentence." Over and above these powers, the political agent had the right to deport any person judged—by him—"about to commit" an offense against the Order-in-Council or "endeavouring to excite enmity" between the people of Bahrain and His Majesty's government.[5]

Not surprisingly, the ruler of Bahrain, Shaikh Isa bin Ali Al Khalifa, was only brought around "with difficulty" to accepting "all the terms" of the Order-in-Council. The Bahrainis soon began to grasp the full effects on their daily lives of being policed by the Government of India through the Order-in-Council that came into effect at the beginning of February 1919.

In May, Arnold Wilson in his capacity as acting civil commissioner invited Shaikh Isa of Bahrain and Abdul Aziz bin Saud to each send one of their sons on an official visit to London. How the Arabs interpreted this invitation is clear in a letter written by Jassim bin Mohamed Al Shirawi, a Bahraini who accompanied Shaikh Isa's son Abdulla on the trip. Soon after the party's return, the political agent deported Jassim to India. In his letter to the secretary of foreign affairs of the Government of India requesting that he be returned to Bahrain, Jassim Al Shirawi explained that the son of the shaikh of Bahrain was sent to England for the purpose of "paying homage to His Majesty the King and also to pledge complete loyalty and support to the Crown of England."[6]

"HOSTILE" BAHRAINIS

Bray's paranoid reporting on the hostility and noncooperation of the Bahrainis provoked a response from the deputy political representative at Bushire, a wartime appointee and civilian not attached to either the Government of India's civil service or the Indian military. "Regarding Bray's report from on board the *Lawrence*," he queried, "is it or is it not desirable that we should take a pronounced step further and assume the responsibility for a gradual education of the Gulf population on the lines of Western civilisation?" He reminded Arnold Wilson that "on the Arab side, with the disappearance of the Turks, we have an absolutely free hand and the time has certainly come for a definite decision on policy. If we simply pursue our present policy, the Arab of 2019 will be exactly the same as his forefather of today."

G. H. Bill, deputy political representative, appears to have had some romantic ideas, but not of the noble savage type. He warned Wilson that education of the Arabs was "a very arguable position." After all, Bill said, his personal belief was that "the Arab is, though an animal, a fairly happy animal, and the heady wine of Western civilisation may merely turn him into a discontented decadent."

Bill had his reservations concerning Bray's conspiracy theories and advised Wilson: "Enthusiastic local officers, like Bray, must be informed that if devout Americans like to spend their dollars enlightening the Bahrainis, it is no concern of ours. In any case we are not going to enter into competition with them at government expense." Bill did not consider "the case against the American Missionaries as proved" and thought their activities were more "idiosyncrasies rather than deliberate policy from their headquarters." Anyway, he concluded: "The State Department is not paying any attention to this region. The whole of the Persian Gulf, including even Basra, is designated by them as 'British India' and American subjects undertake not to transgress the laws of 'British India' while residing in Muscat, a state with which the American Government itself has a treaty."[7]

BAHRAIN PROTESTS TO LONDON

In London the son of the shaikh of Bahrain and the son of Abdul Aziz bin Saud "shook hands with the King at Buckingham Palace." The Administrative Report declared that they "thoroughly enjoyed the trip and realised the greatness and magnificence of the Capital of the British Empire. It is doubtful, however, whether Shaikh Abdulla's visit has not done more harm than good. He saw much of an unhealthy and demoralising nature, especially in Paris, and come back in October with many wild political ideas which he had not got before."

Despite the opinion of an earlier agent, Abdulla would be out of favor with the British from then on because of his "wild political ideas." In 1917, political agent Capt. Percy Loch had reported: "Shaikh Abdulla, the third son of Shaikh Isa, often does business on behalf of his father. He is very intelligent and a thoroughly nice person to deal with. Much of the present smoothness of relations between the Agency and the shaikh is due to him."

Abdulla was in trouble because he had tried to carry out his father's orders and deliver a protest to London about the treatment being meted out to Bahrain. "Before leaving Bahrain and unknown to Captain Bray," the Political Agency report noted, "Shaikh Isa instructed his son to approach the King Emperor in London with a petition." In fact, shaking hands with the king at Buckingham Palace did not give Abdulla the chance to personally deliver the protest to the "King Emperor" as his father had instructed. Eventually, he handed it over to Arthur Hirtzel at the India Office. The letter contained a number of requests, including that Shaikh Isa should exercise jurisdiction over all subjects except those of Great Britain and of the other European powers, that he alone should be the judge of who were suitable persons to serve in the *Majlis al Urf* and the *Sulfa* courts and that he should have the right to correspond directly with London.

Capt. Norman Bray did not return to Bahrain. He obtained a six-month sick leave after which he went back to Mesopotamia. Bray's loyalty to Wilson and personal paranoia would again be evident in a report he wrote after the Iraq Revolt of 1920, circulated to the British cabinet, in which he claimed the revolt was the result not of Wilson's mismanagement but of a far-reaching international conspiracy. The conspiracy involved, he alleged, "Nationalist and Pan-Islamic movements deriving their inspiration from Berlin, through Switzerland and Moscow . . . further complicated with Italian, French and Bolshevist intrigues."[8]

HAROLD R. P. DICKSON

While Bahrain waited for an answer to their petition, on November 6, 1919, a new Baghdad appointee arrived. Arnold Wilson had heeded the warnings in Bray's letter written onboard the *Lawrence*. In addition to his post in Bahrain, the newly appointed political agent was also assigned as liaison officer between Arnold Wilson in Baghdad and Abdul Aziz bin Saud in Nejd.

The new agent was Maj. Harold Richard Patrick Dickson, a thirty-seven-year-old Indian army career officer. His father, John Dickson (1847–1908), was born on the Barbary Coast and was British consul general at Beirut 1879–82 and at Damascus, Homs, and Jerusalem 1882–89. His Scottish grandfather was Dr. Edward Dickson (1815–1900), born in Libya. He was employed by the amirs of Western Tripoli (Al Babat) before joining the Ottoman military service as a doctor. He died in Istanbul.

Harold Dickson was born in Beirut and, because of his father and grandfather, possessed an almost mystical belief that he had inherited a special gift for "handling" Arabs. Like Bray, Dickson had served as a political officer in Mesopotamia, was an admirer of Percy Cox, and was one of Arnold Wilson's handpicked "Young Men"—"young officers who are necessarily lacking in experience," according to Lord Curzon.

Dickson had had a wonderful war. From what he described as "wild and untamed" areas, Dickson reported in enthusiastic letters to his mother the progress of "your young son Harold." "I am the only WHITE [*sic*] man who can manage the Muntafiq tribe," he assured her. "I have got a wonderful hold on these fellows." His achievements included, he boasted, the Muntafiq's tribal area, "which I regenerated and rendered a decent country from a howling wilderness of anarchy."

Bragging that he was "becoming a big man now, aren't I mother?" he was sure he was going places. "I expect you will find me at Hail or Boreida in Central Arabia, leading a revolution and marrying the Amir of Nejd's daughter," he babbled. Dickson likened himself to Richard Burton, the great explorer who discovered the source of the Nile and immortalized his 1853 pilgrimage to Mecca and Medina in a classic of Victorian travel, and to explorer, travel writer, and poet C. M. Doughty, who published his *Travels in Arabia Deserta* in 1888. He might need a bigger stage, Dickson told his mother: "I wish I was in Palestine. Palestine is a WHITE [*sic*] man's country whereas here, Mesopotamia, isn't really exactly. . . ."

In September 1918 Dickson reported, "I am to be posted to Kuwait as Political Agent. Then there is a probability of my being sent to Nejd, Central Arabia, after I have put the situation in Kuwait on a proper basis." In October he announced that although "Cox said he didn't have anyone more qualified than myself to do the Kuwait job" his plans had changed. "Your son," he told his mother, "was ordered to stay in Mesopotamia indefinitely." Dickson soon distinguished himself by his skill at press-ganging Arabs as labor to toil on British projects. "As Political Officer I provide the labour for the railway construction. I have 2,000 working now and hope to have 4,000 in a few days . . . those Arabs won't work for anyone but me," he told his mother proudly. He gleefully confided that part of his "wonderful hold on these fellows" was based on an outright lie.

"I am putting around a story about being wet nursed by an Arab woman," he wrote her, "it usually gets them in." With no compunction at the deceit, he expanded: "I far prefer dealing with these wild Arabs than with the Indians. The Arabs are truly delighted to hear this 'yarn.' They believe in the foster mother theory and that the milk affects the blood of the person drinking it. They say my love for the Arabs and the influence I have with them is because of this 'wet nursing.' I often laugh to myself while I pull their legs so beautifully with this yarn."

Over the years, Dickson would elaborate shamelessly on this story. In his 1949 *The Arab of the Desert*, Dickson claims:

I was born in Beirut, Syria in 1881, and as a small child was taken to Damascus where my mother's milk failed early. It so happened that Shaikh Mijwal of the Mazrab section of the Sha'a, the well known sub-tribe of the great Anizah group, was in Damascus at the time and gallantly stepping into the breach, he volunteered to secure for me a wet nurse or foster mother from among his tribes'-women. A Bedouin girl was duly produced and according to my mother's testimony I drank her milk for several weeks. In the eyes of the Bedouin this entitles me to a certain "blood affinity" with the Anizah; for to drink a woman's milk in the desert is to become a child of the foster mother. This fact has been of assistance to me in my dealings with the Bedouin of the high desert around Kuwait.

Unblushingly, Dickson continued: "Shaikh Mijwal al Mazrab, as my readers will doubtless remember, was the desert Chieftain who married Lady Digby (at one time also Lady Ellenborough) in the 1860s and provided her with a suitable residence in Damascus where she always spent the summer months."

"SULLEN" BAHRAINIS

Dickson's status was high, he assured his mother. "I am living in much the same style as we all used to live in Jerusalem. Actually, my house servants are better than we had in Jerusalem." Dickson was very concerned with status. Almost as soon as he arrived on Bahrain he set about upgrading his circumstances. "I have had large numbers of filthy huts belonging to Persian subjects removed from the immediate vicinity of the Agency," he wrote Arnold Wilson in December 1919. "This is an improvement," he commented as he turned the wretched workers out.

Dickson followed Arnold Wilson's instructions, as inspired by Bray's *Lawrence* report. His schedule was, he told Wilson: "Monday morning *Majlis* 10 am–2 pm in my office. Visit notables four days a week 5:30–7 pm to enable me to educate, enlighten and draw out the sullen Bahrainis. There is a deep-seated suspicion of us English among most of the people of Bahrain."

Not to be outdone, Dickson was soon giving Wilson his own opinion "compiled from close personal observation and study of the people of Bahrain during the last eight weeks and also from valuable material left behind by my predecessor Capt. Norman Bray."

When considering the Bahrainis, Dickson said, the analogy should be borne in mind of the "wild animal which depends on the instinct of suspicion" for survival. He said that this would change because it was well known that once an Oriental state established a close relationship with a European power it became more like "a domesticated animal which depends on confidence (in its master) for its comfort and well-being." But until then, "very few Bahrainis can write. Geographical knowledge is appalling. Politics are of the most amazing conception. They cannot understand the simplest measure of administration or reform. They are incapable of clear statement or sound reasoning and are intellectually dull and naturally stupid."

The way to handle the Arabs, Dickson said in an echo of Wilson's own views, was to ensure that British "prestige" was upheld and to avoid anything that might make "British domination and control appear weak kneed or sentimental." Dickson was as good as his word. The island's business activities were crippled because Dickson refused to lift the closure of the island's commercial court imposed by Bray as punishment of the shaikh of Bahrain. Dickson agreed that Bray had been forced to impose his will because of an offense to his personal dignity "owing to the Ruler abusing his privileges and dismissing a member of the *Majlis al Urf* (the Commercial Court) without consulting the Political Agent. As the Ruler refused to admit himself wrong, I, in turn, refused to allow the Court to meet."

In Bahrain, Dickson faithfully copied the methods of his mentor, Arnold Wilson, and devised a scheme to "tax the town of Manama on the principles in force in the various towns of Mesopotamia." The tax raised was to be mainly allocated to meeting the costs of Dickson's Political Agency. The shaikh of Bahrain, Dickson declared, would be "persuaded" to sanction the enforcement of the tax and an Indian would be put in charge to collect. Opposition to Dickson from almost every level of Bahrain society grew steadily. His refusal to allow any authority outside that of his own office generated unrest among Bahrainis and ill feeling toward those designated non-Bahraini and so "protected" by the political agent.[9]

BIN SAUD

Dickson was not concerned. He saw himself as having bigger fish to fry than the tiny islands of Bahrain. Like Burton and Doughty, he was off to Arabia. He told his mother: "I go to meet bin Saud at Hofuf. I hope to go with him to Riyadh. I have volunteered to live there as British Representative for two years at his Court. It would be the greatest joy of my heart if I could go and live in the heart of Nejd for a good long time. Arnold Wilson has promised to do what he can for me."

Dickson was instantly infatuated with the willowy, physically graceful desert leader, Abdul Aziz bin Saud of Nejd, when he met him in February. Surrounded by a cloud of heavy perfume, wearing the traditional dress of billowing camel-hair cloak over long gown, a double-looped rope of gold thread holding in place the red-checked headcloth, he seemed to Dickson the very embodiment of all his Oriental fantasies.

"Bin Saud is a genius and also has the faculty of making men worship him," Dickson gushed to his mother, inadvertently revealing more about his own state of mind than about bin Saud's abilities. Breathlessly, he detailed the gifts he had received: "One mare, two oryx, two complete sets of Arab clothing and Arab riding gear." That his Indian assistant had also been favored put him out a bit. "He gave my assistant sterling pounds 100 cash," he said, but explained, "it is impossible to forbid such gifts as bin Saud is a Sultan rather than just an Arab Amir."

LONDON REBUFFS BAHRAIN

Back in the less glamorous atmosphere of Bahrain, Dickson noted with some satisfaction that the Bahrainis were getting anxious as to why they did not have a reply to the petition Shaikh Abdulla had lodged in London on behalf of his father. "He now knows he should not have done this," Dickson intoned grimly, because it had probably "annoyed" the High Government. In May 1920, nine months after submission to Hirtzel, the answer came. "As was to be expected each of the requests were disallowed by the High Government as unreasonable," Dickson gloated.[10]

This was not exactly true. The High Government had simply passed on the whole issue to the Government of India. Shaikh Isa was informed that the selection of members of the *majlis* was clear in the Bahrain Order-in-Council and the Government of India would not alter this. On the right to correspond directly with London, Shaikh Isa was told that his business would be conducted through the officers of the Government of India who would consider, where they thought necessary, passing on particular correspondence to the High Government.

However, the Government of India had actually agreed that Shaikh Isa could exercise jurisdiction over the subjects of other Arab rulers, subject to the formal concurrence of these rulers. Dickson promptly put paid to this possibility by secretly writing to bin Saud and to Shaikh Abdulla bin Qasim al Thani of Qatar, the two rulers he knew would refuse. He did not write to other rulers who probably would have agreed, including those of Kuwait, Sharjah, Abu Dhabi, and Muscat. After receiving the two replies Dickson rounded out this sabotage by publishing a notice declaring: "All foreign subjects including Persians and subjects of Arab rulers and Chiefs other than those of Bahrain, are while in Bahrain entitled to British protection." The publication of this notice aroused vigorous opposition from Shaikh Isa, his supporters, and other tribes. Even the petty shaikhs outside the towns felt it to be a direct challenge to their traditional powers.[11]

Arnold Wilson actively encouraged Dickson. His assessment of Dickson for the employment records of the Political Department of the Government of India declared Dickson to be "An officer with marked abilities for political work. He has shown marked capacity in dealing with political questions both in Mesopotamia and Bahrain."

The refusal of the British government to consider Shaikh Isa's petition—and Dickson's arrogant actions—was a strong indication to both the Arabs and the British that the Arab shaikhdoms would not soon escape the grip of the Government of India.

POSTWAR CARVE-UP

When the guns fell silent in Europe in November 1918, Britain and France, the allies that had borne the brunt of the fighting, knew what they wanted in the Arab world.

The Turks agreed to surrender all their dependencies that were inhabited by an Arab majority (on this basis they argued to hold on to Mosul and Kurdistan, neither of which had an Arab majority). France wanted control over all of geographical Syria, including Lebanon. Britain wanted Palestine, Mesopotamia, and the Persian Gulf, where the Government of India intended to retain its authority and not let it pass to Baghdad, where they were no longer in charge.

The British had a complication in the form of the October 24, 1915, letter from the British high commissioner for Egypt, Sir Henry McMahon, to Sharif Husain of Mecca. This correspondence is now infamous as the source of Arab bitterness based on the belief that, in return for Husain's assistance with T. E. Lawrence's so-called Arab Revolt, an independent Arab region would be acknowledged, including Palestine.

The McMahon letter promising to recognize and support the independence of the Arabs, except in "portions of Syria lying to the west of the districts of Damascus, Homs, Hama, and Aleppo" had a second exclusion. This second clause denied the possibility of independence to "regions affected by our existing treaties with Arab chiefs"—meaning the shaikhdoms of the Persian Gulf.

Despite the treaties, Britain's postwar position in the Persian Gulf was, in international legal terms, an "assumed" one and therefore valid only insofar as it remained unchallenged by any other power.[12]

An interdepartmental committee under the chairmanship of Lord Curzon was set up to plan how Britain would approach the peace negotiations. Curzon's committee had no trouble in deciding that no change needed to be made for the countries on the Arabian side of the Persian Gulf. The committee was confident in deciding, as a matter of policy, that there was "no question of submitting our longstanding relations with the Arabian Chiefs" to the judgment of the powers assembled at the 1919 Paris Peace Conference.

US president Woodrow Wilson believed Article XXII of the League of Nations Covenant stipulating the wishes of the local people should be "a principal consideration" in any arrangements reached at the peace negotiations. Wilson sent Dr. Henry Churchill King, education activist of Oberlin College, and Charles R. Crane, philanthropist heir to the Crane bathroom fittings fortune, to the Middle East to ascertain the wishes of the people for the form their postwar governments should take. King and Crane reported that the scheme devised by Britain and France for the future of the Middle East could not be approved, most particularly so in Palestine.

The two-man mission considered visiting Iraq, but at no time was it even suggested they cross the water to visit the Arabs of the Gulf. Not that it would have mattered. The King-Crane Report simply vanished at Versailles; it was never presented for discussion.[13]

After scenes that President Wilson described as "the whole disgusting scramble" for the Middle East, the technical extension of the Paris Peace Conference was the meeting held in early 1920 at San Remo for the assignment of the mandates for former Turkish territories—Syria and Lebanon to France, Mesopotamia and Palestine to Britain. France was to inherit the 25 percent share, previously held by Germany, of the old Turkish Petroleum Company of Mesopotamia, reconstituted by Britain as a legal ploy through which to exploit Mesopotamia's oil.

At San Remo the policy was maintained by the Foreign Office of not seeking allied recognition of Britain's position in regard to the Arab shaikhdoms. When negotiating the Treaty of Sevres signed with Turkey in August 1920, Lord Curzon again deliberately avoided any discussion of the Arab shaikhdoms. In this way the Arab states of the Persian Gulf were left totally isolated and without any form of international protection after World War I.[14]

TRANSJORDAN

There was a slight hiccup when the second son of Sharif Husain arrived in Amman with a fighting force ready to enter Syria and drive out the French. The fiery Abdulla had been in charge of the Arab army during the Arab Revolt against the Turks that featured T. E. Lawrence, later raised to celebrity status as Lawrence of Arabia.

Abdulla was seeking to avenge his older brother Faisal who, with British encouragement, had set up a provisional Arab government in Damascus; the Arabs had dubbed him "King of Syria." The French hadn't taken too kindly to that. The French army had marched eastward from Beirut, easily defeating the Arabs, capturing Damascus, and driving Faisal out. It was inconceivable that the British would respond by declaring war on their ally, France, particularly while final stages of the carve-up were still going on at San Remo and Sevres.

Winston Churchill personally met Amir Abdulla in Jerusalem and persuaded him to halt his march and instead accept—temporarily, until the French left Syria—the rulership of an amirate of Transjordan with Amman as its capital. The British would guarantee to preserve it against the ambitions of the Zionists who wanted the entire area of Palestine and both sides of the Jordan River open to Jewish settlement and eventual statehood. (Faisal would become king of Iraq.)

Transjordan had been an independent state in antiquity. During the Crusades it was known as *Outre-Jourdaine*; its capital was the Crusader's castle of Kerak. Defended on all sides by wild hills and forbidding cliffs, its famed "rose red city" of Petra had been a wealthy emporium for the ancient caravan trades and had resisted both Greeks and Romans until the end of the first century, when it was absorbed into the Roman Empire.

The severing of a reconstituted Transjordan from Syria had some logic; it bordered the Hashemite's other territory of the Hijaz, ruled by Faisal and Abdulla's father.

The House of Hashem had held the title of amir, or grand sharif, of Mecca since at least 1073, controlling the holy cities of Mecca and Medina and most often, give or take a tribal battle or two, the entire Red Sea coastal area known as the Hijaz. (Direct descendants of the Prophet, through the Hashemite line, are entitled to use the honorific "sharif." The Prophet's full name was Muhammad bin Abdulla bin Abdul Muttalib Al Hashem.)

At the end of all the negotiations and meetings, the sum result of the efforts of the victorious allies was the creation in the Arab world of four mandates and one amirate.

The Arab shaikhdoms missed out in every way. They were not granted independence. They were not covered by a mandate that at least would have ensured international oversight through the League of Nations. They were not even declared official British protectorates, a legal status that would have, for one, guaranteed them defense against territorial predators.

Britain would later attempt to explain the status of the shaikhdoms. In a superb display of diplomatic doublespeak Britain would define the status of Bahrain, for example, as: "The principality is an independent Arab State . . . under the protection of His Majesty's government . . . but not a British Protectorate." Instead of independence, a mandate, or a protectorate, the Arabian Peninsula was given over by Britain to the Government of India as compensation for failure to achieve its World War I goal of annexation of Iraq as a colony of India.

Although they may well have been what T. E. Lawrence scathingly described as "a form of colonial administration that can benefit nobody but its administrators," the mandates were a step forward from the excesses previously possible under rampant imperialism. Under the mandates an operative document was drawn up defining the obligations to be undertaken on behalf of the League of Nations. The mandatory powers were supervised, and each year had to face a detailed review conducted by the international members of the Permanent Mandates Commission of the League of Nations. Implicit within the mandates was the understanding that they would end sometime.[15]

THE IRAQ REVOLT

While the Paris Peace Conference and the follow-up at San Remo were concerned with parceling out the Arab world (and Mesopotamian oil), Arnold Wilson was pursuing his own version of what he believed the postwar situation should be.

Wilson was a fervent advocate that the way to handle "native" populations was to show no mercy. He strongly resented what he saw as interference from London in his handling of Mesopotamia. Not that this mattered too much to Wilson. He simply ignored any instructions with which he did not agree.

In January 1920 he wrote to Percy Cox in Tehran: "The Home Government interferes unjustifiably in local internal matters . . . e.g., they wired out forbidding flogging of any kind at all for anybody . . . regardless of our careful regulations on the subject and the conditions of the country. I need hardly say that I have not accepted their orders. But we should not be put in that position."

In the mandated areas of Syria, Palestine, and Mesopotamia the people had taken to political protest by displaying the red, green, and black colors of the Hashemite house of the amir of Mecca. Sharif Husain was now a local hero because of his conduct during the Arab Revolt and his dogged demand for an independent Arabia, for which he had sent his son Faisal to the Paris Peace Conference. Sharif Husain and the Hashemites had a wide following of ardent admirers from Damascus to Aden.

Arnold Wilson was not prepared to brook even a silent display of protest. He allowed no show of Hashemite colors; no mention in the only (government) newspaper of Arab activists including Sharif Husain; no communication with Syria where Arab nationalism was rife; no return of Iraqi military men who had fought in Syria, Palestine, or with the Hashemites; and no association of groups to discuss the political situation. Wilson also forbade all ranks in his own military any talk on the future of the country or listening to anyone else doing so in their presence.

Wilson complained to Percy Cox that London had "wired the other day saying that I was not to forbid Mohammedan children going to government schools in Mosul wearing Hashemite colours. Here again I need hardly say that I did not take action on their instruction." Wilson was sure his tactics had the complete confidence of Cox and that Cox shared his view that the people of the entire region were "unfit" for self-government. "If you keep me on as Deputy High Commissioner," he assured Cox, "you can tour in Mesopotamia and the Persian Gulf, which we should endeavour to keep along with Mesopotamia, at all events on the Arab side."[16]

Wilson's no-holds-barred approach appears to have had a fan in Winston Churchill. While some in Whitehall were attempting to make Arnold Wilson curb his propensity for flogging the natives, Churchill, at the time minister for war and minister for air, was approaching the pioneer of air warfare, Sir Hugh Trenchard, asking him to deal with the growing dissent among the Iraqi tribes. Churchill recommended that this should entail "the provision of some kind of asphyxiating bombs calculated to cause disablement of some kind but not death . . . for use in preliminary operations against turbulent tribes."

When the British cabinet balked at the indiscriminate dropping of poison gas bombs as the means of colonial control in Mesopotamia, Churchill grumbled: "I do not understand this squeamishness about the use of gas. I am strongly in favour of using poison gas against uncivilised tribes." He complained that the use of gas, "a scientific expedient," should not be prevented "by the prejudices of those who do not think clearly."

By June 1920, armed resistance had spread in areas opposed to the prospect of a British-imposed government, even if it was to have an Arab facade. A desire for independence carried no weight with Arnold Wilson.

He wired the India Office in London, claiming that "it is held by many here, both British and native, that I made a mistake in not dealing drastically with prominent agitators here before the movement had reached its present dimensions." The country had reached the point, he said, where "a constitutional movement becomes so seditious as to demand or justify repression." He had no truck with "insufficient severity" and dealt unmercifully with anti-British feeling. Wilson was soon at the center of reports of hangings and bombings of defenseless villages accompanied by waves of arrests and deportations. By August the whole area of the Euphrates was in flames and a holy war, or jihad, against the British had been proclaimed in the Shia religious center of Karbala.

By the time the Iraq Revolt was put down at the end of September, the cost to Britain of containing it was some forty million sterling. Gas was used against the Iraqi tribes with "excellent moral effect." Churchill's idea of dropping gas bombs from aircraft was not adopted, but only because of practical difficulties involving the unstable winds. It was hardly required considering the unequal strength of the parties.

One participant was Wing-Cdr. Sir Arthur Harris, later known as Bomber Harris, head of Britain's Bomber Command and forever associated with the phosphorus firebombing of Dresden that immolated more than 150,000 at the very end of the Second World War. Of his part in putting down the Iraq Revolt, Harris remarked happily: "The Arab and Kurd now know what real bombing means in casualties and damage. Within forty-five minutes a full-size village can be practically wiped out and a third of its inhabitants killed or injured."

The final tally was four hundred fifty British and ten thousand Iraqi deaths and thousands of casualties on both sides. For the Iraqis the result was a disaster. As Lawrence commented wryly, while the Turks were in control of Mesopotamia they "only managed to kill" an average of about 200 people each year, "but Britain has managed to eliminate 10,000 in one summer. At this rate the population will soon be no more trouble."[17]

In the subsequent inquiry, Arnold Wilson listed thirteen factors as having caused the Iraq Revolt. He ignored the wholesale hangings, village bombing, arrests, and deportations that took place under his management prior to the uprising. Instead, his list included what he called "congenital Iraqi anarchy and fanaticism," the liberal ideas of US president Wilson, intrigue by Syrians, the role of Turkey's Mustafa Kemal Atatürk, and conspiracies among the American Standard Oil companies.

When detail of Wilson's draconian policies became known, the British government removed him from Mesopotamia. In October 1920, Percy Cox was transferred back from Persia to again take charge. He held out until his demands for increased authority and the title high commissioner were met. An additional incentive was the knighthood he was awarded in London before his return.[18]

DICKSON BACKS BIN SAUD

While the final phase of the tragic Iraq Revolt was being played out, Dickson was attempting to ingratiate himself with Abdul Aziz bin Saud. "I have recommended Saud to Sir Percy Cox for a Government of India honour or the title of Sultan (this is confidential)," he wrote home.

Dickson's mother unwisely expressed sympathy for Faisal, eldest son of the sharif of Mecca, saying he had been treated very badly at the Paris Peace Negotiations and she didn't think it was right that he had then been turfed out of Damascus by the French in their anxiety to entrench their mandate over Syria.

Dickson replied sharply to his mother. He shared Arnold Wilson's intense dislike of the Hashemites and their unrelenting push for Arab independence, a view also expounded by Dickson's hero and the Hashemite's rival Abdul Aziz bin Saud. In a most ungracious retort, Dickson told his mother: "You ask if I don't think Faisal has been badly treated. I am afraid I don't. He has brought the whole trouble on his own head. I believe bin Saud's view of him, which is that Faisal is a double faced rascal."

Dickson was soon reporting home that Percy Cox had "finally invested Shaikh Sir Abdul Aziz bin Saud, Ruler of Nejd, with an Indian empire Knighthood (again at my insistence) and a gift of 5,000 sterling. Bin Saud is, I think, on the high road to the top of the ladder. HMG has at last decided to back my horse, Bin Saud. I hope before long that the name 'Sultan of Central Arabia' will be given to him and that useful financial help etc will be forthcoming." Dickson was now almost dizzy at the prospect and assured his mother: "Saud is as superior to Faisal, late of Damascus, as a thoroughbred horse is to a mule. I say this although I have never seen Faisal."

"Sir" bin Saud was soon inciting his tribes to attack the Hashemite's territory in Transjordan and the Hijaz. As Percy Cox passed through on his way to persuade bin Saud to pull back his tribesmen, the shaikh of Bahrain grabbed the opportunity to try again for redress. He appealed to a higher authority than Dickson in a formal complaint addressed to Percy Cox and signed by Shaikh Isa and a large number of his citizens.

Dickson dismissed this second petition as "prepared at the instigation of certain disaffected persons, including Shaikh Abdulla, that purported to complain against me for so called high handed actions." Tellingly, he wrote his mother: "I have had enough of Bahrain despite my nice house. Cox has promised to get me back to Mesopotamia shortly."[19]

CLIVE DALY

There would be no reprieve for Bahrain. Dickson's replacement in January 1921 would be Maj. Clive Daly, another of "Wilson's Young Men" of already formidable reputation for keeping "native" populations in check.

Major Daly had been a lieutenant colonel but was demoted because of his activities during the Iraq Revolt of 1920. He had been political governor of Diwaniya in Mesopotamia and it was in his region that the uprising began. Some said the fact the tribes "hated" Daly had actually precipitated the uprising. He was a "tough single-minded soldier" who, like Dickson, vigorously pursued Arnold Wilson–style policies.

Daly would isolate the ruler and disrupt the traditional structures. He would create acrimonious division in Bahrain's society by designating more and more individuals and groups as "foreigners" under his protection. He would deport anyone who opposed him, whether "foreigners" or Bahrainis. He would recruit Persians to build up what had previously been a small group of market guards into a "law keeping" force, answerable only to him. He would convict and deport to India the shaikh of Bahrain's personal secretary for alleged conspiracy against the Order-in-Council and follow this a few weeks later by deporting, on the same charge, the progressive headmaster of Bahrain's first modern school.

Under Daly, Arab-Persian and Shia-Sunni riots would break out and civil unrest become commonplace. The entire tribe of Al Dawasir would leave their home in Bahrain and move to Dhamman in Saudi Arabia. Bahrain would be subjected to a personal rule by Major Daly that Dickson himself later criticized, perhaps enviously, as "worse than a Persian Potentate."[20]

CHURCHILL'S "FORTY THIEVES"

When he took over as colonial secretary in early 1921, Winston Churchill arranged the Cairo Conference at which the method by which Britain was to control her territories in the Middle East was decided. The Cairo Conference of March 1921 gathered some forty participants—Churchill's "Forty Thieves," quipped the British press.

Although, according to T. E. Lawrence, the details were worked out beforehand "at the Ship restaurant in Whitehall," the upshot of the conference was that the Colonial Office would have responsibility for both policy and administration in Aden, Iraq, Palestine, and Transjordan. The Foreign Office would be responsible for relations with Sharif Husain of Mecca in the Hashemite territory of the Hijaz.

The Cairo Conference gave in to Sir Percy Cox's demand and a special arrangement was agreed upon whereby, in recognition of his long relationship with bin Saud, and with Kuwait, built up over his years as the political resident in the Persian Gulf, he would retain command of these two areas until his retirement. As the Government of India was vehemently opposed to any interference from Baghdad, the arrangement was to remain strictly on a personal basis and not officially attached to either the position of high commissioner for Iraq or the Government of Iraq per se.

With the temporary exceptions of Kuwait and bin Saud's domain, the Arab shaikhdoms of the Persian Gulf and the southern provinces of Persia were to remain under the authority of the Government of India.

The formal document stated that the Colonial Office would be responsible for all Arab areas of the Middle East. However, in the case of the Arab shaikhdoms of the Persian Gulf, the Colonial Office would be interested only in matters of policy. "Policy" was not defined. In practice the Government of India was to retain full control of general policy as well as administration.

Churchill's "Forty Thieves" agreed to the recommendation that "the Gulf is dealt with as a single administrative unit . . . the currency of the Persian Gulf is almost exclusively Indian . . . almost all trade is with India, the Gulf lives almost wholly on cereals imported from India. Indian merchants handle the bulk of the trade."

The recommendation, lodged to the Cairo Conference, said there were "large Indian interests" in the Persian Gulf and that "only the Political Resident" was sufficiently "in touch with the political, strategic and commercial interests involved . . . to take responsibility . . . in times of stress in the Persian Gulf." While the administrators and the political officers would certainly be "British," as before, they would be members of the Indian military or the Indian civil service. Their instructions would come from India and they would report, not to the British government, but to the Government of India.

The autonomy of the Government of India in the Persian Gulf was cemented when the Colonial Office promised it would never attempt to communicate directly with the various shaikhs. The Colonial Office agreed to the arrangement that all contact both in and out of the Persian Gulf would take place only through the Government of India's own man, the political resident in the Persian Gulf.[21]

ARNOLD WILSON JOINS ANGLO-PERSIAN

In keeping with the Raj tradition of looking after its own, at almost the very moment the British were removing him from Mesopotamia, the Government of India was rewarding Arnold Wilson with a knighthood. As Sir Arnold Wilson he was soon in charge at the Anglo-Persian Oil Company's headquarters at Abadan with the promise, soon fulfilled, of a place on the board of directors. His first official duty was to attend, as one of Churchill's "Forty Thieves," the Cairo Conference.

Curiously, Wilson appears to have been in discussion with the Anglo-Persian Oil Company right throughout the period the Iraq Revolt was building. In June 1920, as armed resistance was breaking out across Mesopotamia, chairman Charles Greenway was proposing to a meeting of the Anglo-Persian Oil Company Board that "Arnold Wilson be engaged at a salary of about 5,000 sterling." This was more than one thousand sterling above Wilson's earnings as acting civil commissioner of Mesopotamia. After his unceremonious removal, Greenway pushed through the appointment of Wilson against strong opposition by some directors. Terms and conditions were agreed upon in January 1921, and by February Greenway was promising Arnold Wilson that "you may rest assured that we shall, when the time comes for your retirement to this country, recommend your election to the Board."[22]

This was not so surprising in view of the fact that the notion of the Anglo-Persian Oil Company as an icon of India was firmly fixed. In 1909, when the original Persian concession holder, D'Arcy Exploration Company, needed an injection of funds, it was the Burma Oil Company of India that came to the rescue. The Burma Oil Company had thrived through the Government of India's blanket concession, given in 1889, that protected the company from all foreign competition.

In 1914, when the British government purchased 51 percent of the Anglo-Persian Oil Company, Burma Oil retained a third of the voting stock. Their own man, Charles Greenway, remained as chairman. Strick, Scott & Co., long associated with shipping and trade to and from India, were Anglo-Persian's managing operators.

The Indian military was associated with the Anglo-Persian Oil Company and its forerunner almost since the beginning. In 1907 a young Arnold Wilson, then with the Indian army, was posted to Ahwaz. His job involved being the British officer in charge of Indian army troops protecting the oil company site—against the native Persians.

The connection with India was so strong that when oil was struck in 1908 in Persia, Arnold Wilson wired enthusiastically, "I hope it will turn out well as it will establish British influence, or rather Indian influence, more strongly in these parts than anything else could do."[23]

Wilson's appointment to the Anglo-Persian Oil Company stirred public controversy. Questions were asked in the British parliament about the danger of corruption in the public service "arising out of the situation that a civil servant's power may cover the area to which he subsequently receives, on resignation, a more lucrative appointment."

Newspaper comment ran hot. One letter in the *Times* of March 3, 1921, criticized the appointment: "For two and a half years he was the principle adviser on the spot to the Government of India and to the Home Government. . . . Apparently he is about to return . . . as a private man of business as the representative of British oil interests. . . . We wish Sir Arnold had gone into the oil business in Borneo or the South Pole rather than in the country over which he has so recently been in political control."

On April 20, the *Times* published a damning letter from G. T. Garratt, who had been personal assistant to Wilson when he was deputy chief political officer in Basra. Garratt charged:

> Surely, the chief argument against allowing Sir Arnold to accept this post is unanswerable. Sir Arnold, supported by a few Anglo-Indians, organised an autocratic form of government which, rightly or wrongly, cost England some fifty million sterling pounds a year and led inevitably to rebellion. . . . For some years he deposed and sometimes exiled or imprisoned paramount shaikhs who opposed his rule while his agents supplied him with an immense amount of secret information which he has no right to use in the service of a private trading company. There is something incongruous and undignified in a man appearing one day as the Chief Representative of Government and the next as a private trader.[24]

Nevertheless, Wilson's appointment stood and so it would be that Cox and Wilson continued to dominate the Persian Gulf, as they had done during the war. Both these men were, as Wilson's biographer observed, "raised in Victorianism and trained in Indian paternalism" and both were "the product of Indian attitudes and systems of government." Both men would continue to exert these influences on the last bastion of the Raj—the Arab shaikhdoms of the Persian Gulf—one from a powerful government position and the other from the Anglo-Persian Oil Company.[25]

NOTES

1. Briton Cooper Busch, *Britain, India, and the Arabs, 1914–1921* (Berkeley: University of California Press, 1971), p. 145, "India . . . must have a *quid pro quo*" and p. 150, Cox wanted the title of high commissioner as in Egypt; Chamberlain compromised with civil commissioner; Elizabeth Monroe, *Philby of Arabia* (London: Faber and Faber, 1973), p. 52, "forced us into agreeing not to annex."

2. David Fromkin, *A Peace to End All Peace: Creating the Modern Middle East, 1914–1922* (London: A. Deutsch, 1989), pp. 455–62, "a great triumph"; Busch, *Britain, India, and the Arabs*, p. 356, "splendid . . . half a century ago" and p. 399, "unfit to form a government." For details of these early independence meetings and nationalist movements in Iraq and the surrounding areas see Philip Willard Ireland, *Iraq; A Study in Political Development* (New York: Russell and Russell, 1970), pp. 222–39; for Sayid Talib Pasha, see Who's Who section of this book.

3. John Marlowe, *Late Victorian: The Life of Sir Arnold Talbot Wilson* (London: Cresset, 1962), pp. 165–66, "of what used to be called subject peoples." Curiously, on his return from London Wilson had criticized Hirtzel as "too anti-Turk and too anti-Muslim," p. 105, "Wilson had a particular appeal to young officers, to many of whom he appeared as the authentic hero of the adventure stories of their schooldays, and as the very epitome of all that they would themselves like to become. Most of those whom he recruited, who were known sometimes derisively as 'Wilson's Young Men,' became extremely attached to him, and he to them," and p. 58, Major Young, ex-Mesopotamia, now Foreign Office, himself well under forty-five, wrote to Lord Curzon "of the 233 officers . . . only four are over 45 years of age." Curzon to Cox in Tehran in November 1919, "hampering the development of either civil administration or whatever form of native government is decided in future."

4. Marlowe, *Late Victorian*, p. 168, October 1919, Wilson writing to his parents, "more independent powers than the Viceroy" and "not willing to take on the job of High Commissioner myself yet . . . although people tell me I should," and p. 183, March 1920, Wilson writing to Sir Arthur Hirtzel, "they are practically without exception my nominees. . . ."

5. Muhammad G. Rumaihi, *Bahrain: Social and Political Change Since the First World War*, p. 169, gives full details of the Order-in-Council, "The Order also introduced various rules and regulations unlike any the Bahrainis had experienced before. It gave the Political Agents great power and at the same time deprived the traditional local courts of their authority . . . the British not only controlled the Judiciary with respect to cases between Bahrainis and foreigners but also interfered in the traditional religious law courts by which the community set great store . . . a British subject could be prohibited from being within the limits of this Order . . . for a period not exceeding two years." Besides these provisions, which were unacceptable to a number of people in Bahrain, the Order raised the controversial issue of who were foreigners and who Bahrainis. Article 8 (2) defined foreigners as those "with respect to whom the Shaikh of Bahrain has agreed with His Majesty for, or consented to, the exercise of jurisdiction by His Majesty," a vague phrase that did not define who was or was not subject to British rule of law.

6. Administration Report Bahrain for the Year 1919 (Dickson), "The Bahrain Order-in-Council was brought into force from noon on February 3rd 1919. Shaikh Isa was brought round with difficulty to accept all the terms mentioned therein." Rumaihi, *Baharain: Social and Political Change*, p. 185, Wilson to Shaikh Isa bin Ali dated May 13, 1919. The same invitation was sent to Abdul Aziz bin Saud. Isa sent his third son Abdulla and Abdulla's own son Muhammad. Abdul Aziz sent his second son Faisal. PRO/FO/371/8941/E5661, p. 149, statement of Jassim bin Mohamed Al Shirawi dated May 1, 1923, addressed to the foreign secretary, Government of India, "Son of the Shaikh of Bahrain was deputed to proceed to England to pay homage to His Majesty the King and also to pledge complete loyalty and support to the Crown of England." Jassim Shirawi was deported from Bahrain on November 19, 1921, but the political resident had not bothered reporting this to Government of India's secretary of foreign affairs until April 17, 1922—and then only after a lawyer for Shirawi had written to the secretary for foreign affairs.

7. Dickson Papers, box 2a, files 4a and 4b, January 30, 1920, Wilson to secretary to Government of India in the Foreign Political Department enclosing a May 27, 1919, letter written to Wilson from onboard the *Lawrence* by Capt. Norman Bray; Dickson Papers, box 2a, files 4a and 4b, June 17, 1919, G. H. Bill, deputy political representative Bushire to Arnold Wilson in Baghdad, "the Arab is, though an animal, a fairly happy animal."

8. Administrative Report for Bahrain 1920 (Bray and Dickson), "realised the greatness and magnificence of the Capital of the British Empire"; Administrative Report Bahrain 1917 (Loch), "He is very intelligent and a thoroughly nice person to deal with"; Rumaihi, *Bahrain: Social and Political Change*, p. 170, for details of the ruler of Bahrain's letter of protest delivered to Sir Arthur Hirtzel; Administrative Report Bahrain 1919 (Dickson), "probably Bray's illness was aggravated by the fierce (Bahrain) weather of May and June"; Fromkin, *A Peace to End All Peace*, p. 467, "Bray urged the government to track down the secret 'comparatively small central organisation' at the centre of the far reaching international conspiracy."

9. Dickson Papers, box 2, file 2, August–December 1918, Dickson letters to his mother from Mesopotamia. H. R. P. Dickson, *The Arab of the Desert* (London, Allen and Unwin 1949), "A Bedouin girl was duly produced . . . I drank her milk for several weeks. In the eyes of the Bedouin this entitles me to a certain 'blood affinity' with the Anizah"; Dickson Papers, box 2a, files 4a and 4b, Dickson letters to Arnold Wilson from Bahrain, December 1919, "huts removed," January 1920, "intellectually dull and naturally stupid"; Administrative Report for Bahrain for the Year 1919 (Dickson), "refused to allow the court to meet." Shaikh Isa held out from April 1919 to January 1920 but eventually had to give in as transactions ground to a halt; Busch, *Britain, India, and the Arabs*, p. 410, in Iraq, "the Turks had collected only 2.5 million sterling per annum in taxes, the British were extracting 6 million"; for opposition to Dickson see Rumaihi, *Bahrain: Social and Political Change*, p. 172.

10. Dickson Papers, box 2, file 3, December 10, 1919, and February 7, 1920, Dickson letters to his mother from Bahrain, "bin Saud . . . has the faculty of making men worship him"; in his Administrative Report of 1919, Dickson commented on the High Government's reply to the petition that "Shaikh Isa and his sons were extremely disappointed and showed their disappointment and irritation in a number of petty ways, chiefly by trying to persuade some of the ignorant foreigners to take their cases and complaints to Shaikh Isa and his officials . . . till at last the Shaikh was reminded of the decision that his jurisdiction is limited to natives of Bahrain . . . there has been no more attempt of this kind on the part of Shaikh Isa or his sons."

11. Rumaihi, *Bahrain: Social and Political Change*, p. 171, "Naturally enough, both rulers gave a negative answer. Ibn Saud was at that time an adherent of the British and was well aware of the Bahrain internal situation and bin Thani, with his longstanding animosity for the Al Khalifa was not prepared to grant the request either . . . (this) was probably the breaking point in Shaikh Isa's relations with the British." Arnold Wilson encouraged Dickson's way of dealing with the Bahrainis, see Dickson Papers, box 2, file 3, Wilson's assessment of Dickson, "An officer with marked abilities for political work. Has shown marked capacity in dealing with political questions both in Mesopotamia and Bahrain."

12. Elizabeth Monroe, *Britain's Moment in the Middle East, 1914–1971* (London: Chatto and Windus, 1981), p. 32, for text of the McMahon letter; see also Busch, *Britain, India, and the Arabs*, p. 83. Note that the Arab shaikhdoms were also excluded from the considerations of the Sykes-Picot Agreement of December 1915 that divided the Arab world into areas of influence between Britain and France. Prior to the discussions Mark Sykes defined Britain's ideal outcome as "an external protectorate over the Arabian littoral from Kuwait to Hodeida in the southern Red Sea and an internal and external protectorate over part of Mesopotamia with an area of French influence to the north." See also Arthur

Goldschmidt Jr., *A Concise History of the Middle East*, 4th ed. (Boulder: Westview Press, 1988), pp. 194–96, "Britain wanted Husain to start a revolt against Ottoman rule in the Hijaz . . . (Britain) knew Husain could not rally other Arabs to their cause unless the Arabs were assured that they would get their independence in the lands in which they predominated, Arabia, Iraq and Syria including Palestine and Lebanon. . . . In my judgement, given that Britain cared more in 1915 about keeping its French connection than about reserving Palestine for the Jews, the area excluded from Arab rule in the McMahon-Husain correspondence must have been (Christian) Lebanon. Only later would Jewish claims to Palestine become the main issue."

    13. Monroe, *Britain's Moment*, p. 52, for this committee chaired by Curzon; Busch, *Britain, India, and the Arabs*, p. 436, "concerned to see to it that Iraq did not control the Persian Gulf" and p. 285, "no question of submitting" and pp. 320–41, for the King-Crane Commission.

    14. Monroe, *Britain's Moment*, pp. 304–308, nor were the Gulf shaikhdoms included in the discussions held prior to the San Remo meeting to thrash out terms between Britain and France. These became known as the Long-Berenger Terms of April 1919 and decreed that 25 percent of Mesopotamian oil would go to France in return for France giving up its claim to Mosul; see Busch, *Britain, India, and the Arabs*, p. 388, for detail San Remo and Sevres and "not to suggest Allied recognition." See also Monroe, p. 54, for Britain's "plans" for the Middle East; Busch, p. 263, somewhat unkindly comments that perhaps Britain's policy in Arabia could be termed a success at the end of WWI because "Britain had taken an undeveloped peninsula of warring tribes and left it an undeveloped peninsula of warring tribes."

    15. Monroe, *Britain's Moment*, p. 66, "disgusting scramble"; Goldschmidt, *Concise History*, pp. 199–202, "Winston Churchill met Abdulla in Jerusalem and persuaded him to accept—temporarily, until the French should leave Syria"; Mahdi Abdalla Al Tajir, *Bahrain 1920–1945: Britain, the Shaikh, and the Administration* (London: Croom Helm, 1987), p. 7, Britain's submission to the League of Nations re Persia's claim to sovereignty of Bahrain "an independent Arab state . . . under the protection of HMG . . . but not a British protectorate"; Monroe, p. 68, "form of colonial administration"; on p. 72, Monroe contends "the Permanent Mandates Commission of the League of Nations was the best body of its kind ever so far constituted, particularly in its experience, impartiality and effectiveness."

    16. Philby Papers, box XV11–9, January 2, 1920, Wilson to Cox, "forbidding flogging of any kind" and "need hardly say I did not take action on their instruction"; Busch, *Britain, India, and the Arabs*, p. 382, for detail of Wilson's political repression.

    17. See Geoff Simons, *Iraq: From Sumer to Saddam* (New York: St. Martin's Press, 1994), pp. 179–81, for Churchill, poison gas, and the 1920 Iraq Revolt. Churchill wrote to Sir Hugh Trenchard in February 1920. Simons notes, "Iraq and Kurdistan were useful laboratories for new weapons; devices specifically developed by the Air Ministry for use against tribal villages. The ministry drew up a list of possible weapons, some of them the forerunners of napalm and air-to-ground missiles: Phosphorus bombs, war rockets, metal crowsfeet [to maim livestock] man-killing shrapnel, liquid fire, delay-action bombs. Many of these weapons were first used" in Iraq in 1920. The bombing of Dresden took place on February 13, 1945; Marlowe, *Late Victorian*, p. 214, "not dealing with prominent agitators . . . drastically"; Busch, *Britain, India, and the Arabs*, p. 415, Wilson memo of August 1920, "British weakness through insufficient severity . . . and statement on the point at

which a constitutional movement becomes so seditious as to demand or justify repression," and pp. 395, 405, 408 for reports of hangings, bombing, deportations, etc., and p. 408 for final tally, p. 416, quoting T. E. Lawrence interviews with the *Observer*, the *Daily Herald*, and the *Sunday Times*, "at this rate the population would soon be no more trouble."

18. See Arnold Talbot Wilson Sr., *Mesopotamia, 1914–1917: A Personal and Historical Record* (London: Oxford University Press, 1930) and *Mesopotamia, 1917–1920: A Clash of Loyalties* (London: Oxford University Press, 1931). For further detail of Wilson's policies and actions see Busch, *Britain, India, and the Arabs*, pp. 395–416; Fromkin, *A Peace to End All Peace*, p. 453, "anarchy and fanaticism" and p. 535, "British officials suspected American oil interests were behind the anti-British insurrection in Iraq and the Kemalist movement in Turkey. Allegedly an arrested insurrection leader had in his possession a letter from one of the Standard Oil companies showing American funds were dispensed by the American consul in Baghdad to the rebels." See Philip Graves, *The Life of Sir Percy Cox* (London: Thornton Butterworth, 1939), p. 264, "On August 11, 1920, Cox was invested by the King with the KCMG."

19. Dickson Papers, box 2, file 3, September 14, 1920, and October 4, 1920, Dickson letters to his mother from Bahrain; Administrative Report Bahrain 1920 (Dickson), "so called high handed actions"; Rumaihi, *Bahrain: Social and Political Change*, p. 172, notes that Dickson considered the majority of Persians in Bahrain were not against him.

20. Rumaihi, *Bahrain: Social and Political Change*, p. 186, "Daly's region"; Rumaihi cites Ameen Rihani's Arabic edition of *Kings of Arabia* (Muluk al-'Arab) (Lebanon, 1924), p. 773, stating Daly was demoted from lieutenant colonel to major for his role in the Iraq Revolt; see also H. St. J. B. Philby, *Arabian Days, An Autobiography* (London: R. Hale, 1948), p. 193, "Daly was one of Wilson's toughest men." Philby claims he "insisted" Daly be transferred out of Iraq; see also Violet Dickson, *Forty Years in Kuwait* (London: Allen and Unwin, 1971), p. 158, citing Gertrude Bell, "the tribal rising is due mainly to hatred of individual Political Officers . . . the fact that they hated Daly precipitated matters; it was in Diwaniya that it began"; Philby Papers, box XVII–9, July 31, 1923, Dickson to Philby, "I was shocked greatly to see Daly's methods at close hand when my wife and I stayed in Bahrain. Never have I seen such a boor or anything so opposite to the word gentleman in his dealings with the local Arabs . . . Daly's methods are worse than those of a Persian Potentate"; Rumaihi, *Bahrain: Social and Political Change*, p. 173, "single minded" and deportations, p. 178, for decampment of Dawasir; also see IOL/R/15/2/88, Dawasir Decamp to Nejd (complete file August 1923–February 1924) and IOL/R/15/1/341/C25/19/166 (complete file 10/5/23–21/7/23), Disturbance at Bahrein between Nejdis and Persians.

21. IOL/R/L/PS/11/193, January 31, 1921, report of the Interdepartmental Committee appointed by the prime minister to make recommendations as to the formation of a new department under the Colonial Office to deal with mandated and other territories in the Middle East. Clauses 9–12 deal with the Persian Gulf littoral "only the Political Resident" is sufficiently "in touch with the political, strategic and commercial interests involved . . . to take responsibility . . . in times of stress in the Gulf." And IOL/R/L/PS/10/1273/PG/Sub/18, December 12, 1929, Committee of Imperial Defense, Persian Gulf, Report by a Sub-Committee on Political Control; Appendix "De Facto" position as regards political arrangements in the Persian Gulf.

22. Marlowe, *Late Victorian*, p. 194, for Wilson's knighthood from the Government of India. For detail of Anglo-Persian's hiring of Wilson at a very generous salary, and promise of a directorship, see R. W. Ferrier, *The History of the British Petroleum Company: the Developing Years, 1901–1932* (Cambridge: Cambridge University Press, 1982), p. 309.

23. Graves, *Life of Sir Percy Cox*, p. 48, "from the native Persians" and p. 52, "rather Indian influence."

24. Marlow, *Late Victorian*, p. 237, claims (erroneously) that Anglo-Persian first approached Wilson after his return to London and that Churchill introduced Wilson to Greenway in London on February 17. See also pp. 242–43 for detail of the letters to the *Times* and Questions in Parliament.

25. Ibid., p. 485, "Indian paternalism" and p. 421, "Indian attitudes"; Busch, *Britain, India, and the Arabs*, p. 420, "Wilson himself was the product of Indian attitudes and systems of government—India's legacy as it were to postwar Middle East policy."

CHAPTER 3

AN IMMENSE ARABIAN OIL FIELD

In November 1920, as Sir Percy Cox entrenched his High Commission in Iraq, Arnold Wilson discussed terms of employment with the Anglo-Persian Oil Company, Harold Dickson departed a hostile Bahrain, and Frank Holmes made his way to the ancient Red Sea coastal towns of Jaizan and Hodeida in the province of Assir, bordered by Yemen on one side and the Hijaz on the other.

Hodeida, mentioned as a port in the old Arab histories, had been a garrison for the Turks and served as the main outlet for Assir, the richest province in Southern Arabia. Palm groves and gardens lay to the north and sandy roads, suitable for camels, fanned out from the town for some thirty miles. Bedouin raiders were deterred by barbed wire fencing around the town. Three gates with customs stations allowed entry from the caravan routes. During the war British ships had bombarded Hodeida, but its bazaars were active again and traders had returned. Holmes was to meet with the ruler of Assir, Saiyid Muhammad bin Ali bin Muhammad bin Ahmad bin Idrissi, a Sufi of the Ahmadiyah cult. He claimed descent from the Prophet and was a man of huge physical proportions—and the only black ruler on the entire Arabian littoral.

Located across the Red Sea from Abyssinia, Jaizan was dominated by a cannon-bedecked citadel of pure Yemeni design incorporating deep walls, small windows, and hundreds of arrow slits. Its six thousand people lived in circular reed huts with pointed straw roofs or masonry houses styled like small Egyptian temples. Both a religious center and a thriving trade hub, Jaizan was serviced by sea, and also by some five hundred camels coming in caravan each month, bringing grain and food and leaving with the product of the rock-salt mines and the goods shipped from Egypt and North Africa, especially Morocco. Jaizan and Hodeida were the two centers from which the Idrissi ruled.

Born in 1876 in Assir and educated at Al Azhar University of Cairo, Muhammad Al Idrissi was a pure Negro; one of his grandmothers had been a Sudanese slave. The Arabs called his territories the "Country of the Sid." *Sid* is the diminutive of *Saiyid*, the title used by the Idrissi and descendants of the Prophet through his grandson, Hussein. The Assir province numbered some one million persons and included various Sunni and Shia Muslim sects, Christians, Jews, Zoroastrians, Parsees, and Hindus.

The Sufi cult of Ahmadiyah that Muhammad led was derived from the teachings of his grandfather, Ahmad bin Idrissi, born in 1785 in a small village on the North African coast. Ahmad first studied and taught in the Persian city of Fars (Persia and North Africa are the two sources of Sufism in the Muslim world) before moving to Mecca. At the age of seventy-two, he left Mecca on a missionary tour, finally settling not far from Jaizan, at Sabya, where his Sufism was "sealed" and where the Ahmadiyah cult that bears his name was first established.

Muhammad Al Idrissi, the current ruler of the "Country of the Sid," had confirmed his heritage by marrying a dark-skinned Sudanese woman, a daughter of the leader of the Ahmadiyah sect in that country. During his rule, Muhammad had skillfully dealt with the Turks, the British, the Egyptians, and the Italians, and the rising tide of militant Wahhabi Islam across his border in the territory of Abdul Aziz bin Saud. There were many who considered the Idrissi a political genius.[1]

AL YEMEN

Until the mid-nineteenth century all of Southern Arabia was known as Al Yemen; the Romans called it *Arabia Felix*, meaning "Fortunate Arabia." Arab legend held that Shem, the son of Noah, was the ancestor of the Arabs and that the Arabian Peninsula was divided in two when one of his sons, Adnan, settled "Arabia," or the northern portion. Shem's other son, Qahtan, is said to have settled "Al Yemen," or the southern portion. It was first the British and then the Ottomans who subdivided Al Yemen by instituting an artificial frontier separating the north of Yemen from its south; the British appropriated Aden as they did so.

Ancient Al Yemen was the home of the queen of Sheba and the source of frankincense, myrhh, ambergris, and other incense vital to the religious rituals of Babylon and Pharonic Egypt. The world's first skyscrapers were built in the queen of Sheba's city of Saba, seventy-five miles east of Sana'a, some time in the tenth century. The fabled Palace Ghamdan was said to be twenty stories high. The towering buildings were so wondrous that, with the passage of time, it came to be believed that they had been built by the jinn (genie). A verse in *The Book of 1001 Nights* tells this legend: "Twenty of the Jinn of King Solomon were sent, bottled and sealed, to the Queen of Sheba to build for her the loftiest palace in the world." (This is the origin of what the Western world thinks of as the genie in the bottle.) The story goes that the jinn so liked Al Yemen that they elected to remain and built a whole city of tall buildings with, almost as marvelous, windows of fine opaque alabaster.

The European discovery of the coffee of Yemen, which they named "mocha" after the Red Sea port from where it was shipped, attracted the British of the East India Company in the early 1600s, the Dutch East India Company in the early 1700s, the French, and finally the first American ship in 1798. Most of the coffee trade eventually passed into American hands.

Throughout Al Yemen, and in the forbidding mountains of the interior, both powerful and petty chiefs established dynasties interrupted through regular incursions from the Romans, the Persians, the Abyssinians, the Egyptians, and the Dutch, as well as French, Portuguese, German, Italian, Ottoman, and British imperialism. The city's proud Islamic history included the Prophet Muhammad's sending of his own cousin and son-in-law, Ali bin Abu Talib, on a personal mission to Yemen in 631. (Ali was married to the Prophet's daughter, Fatima.) The Yemenis loved Ali for his bravery, evenhanded justice, generous nature, poetic talent, and eloquence, and they were passionately supportive of him.

When the Prophet died in 632, he left no instruction on how the succession should be handled. Muslims were divided. Some believed that succession should only go in the Prophet's direct line through Fatima and Ali and their descendants; this group became known as the Shia (literally, the "followers"). The other group, the Sunnis, believed the method of succession could be discerned in the Sunnas (verses) of the Koran, from which they took their name (*Sunna* means "the path"). Three caliphs followed the Prophet, Abu Bakr, Omar, and Othman, until Ali assumed the rule in 656. He ruled successfully until his savage murder in 661 by an opponent to the principle of hereditary succession.

The consequent divisions became bitter. Ali's younger son Hussein was called to Iraq to take over the leadership but was cut down by opposition troops. The date of Hussein's death is still observed as a solemn day of mourning throughout the Shia world. Important to Yemen was the 739 bid by Hussein's grandson, Zaid, to assert his claim to leadership on the grounds of hereditary succession. Zaid was killed in an unequal battle with the caliph's forces. His followers fled to the East African coast and to the mountains of Yemen, a natural home for what became the Zaidi sect of Shia Islam.

THE ZAIDI IMAM

When Frank Holmes arrived in late 1920, "Yemen" referred only to the province on the Red Sea coast between Aden and Assir. This Yemen, about half the size of France, was ruled from its mountainous interior by the Zaidi Imam, Yahya bin Hamid al Din. As his title implied, he was both the religious leader—the Imam—and the ruler. Turkish advisers were on his staff in the capital city of Sana'a where multistoried buildings, descendants of those said to be built by the jinn, still stood. He styled his rule on Western lines and was assisted by a prime minister and a cabinet, among whom were five of his sixteen sons. His army, which was still instructed in Turkish, wore Western-style uniforms and was adept at drills and ceremonies.

Often called "the old man of the mountains," the Imam Yahya had never been photographed (he would not permit it), had never left the interior of his country, had never seen the sea, but was respected for his piety, religious knowledge, and uncompromising stand for Arab independence, a goal he shared with Sharif Husain of Mecca. The Imam's Yemen hosted a population of three million or more, including Zaidis, Isma'ilis, Sunnis, and fifteen thousand Jews.

During the war the British had swung between support for the Zaidi leader, the Imam Yahya of Yemen, and support for the Sufi leader, the Idrissi of Assir. In 1919 the Government of India clearly laid out its policy: "What we want is not a united Southern Arabia but a weak and disunited Arabia split into little principalities as far as possible under our suzerainty—but incapable of coordinated action against us." Despite setting them at odds through their application of the divide and rule principle, the British were now endeavoring to bring about a rapprochement between the two rulers. Neither the Idrissi nor the Imam was proving amenable. They and their followers remained, quite literally, at daggers drawn.[2]

CRAUFURD

After departing Abyssinia for London, Frank Holmes had enthused Edmund Janson, son-in-law of the Eastern & General Syndicate's chairman Edmund Davis. Janson was convinced by Holmes's plan that the search for the "immense Arabian oil field" that Holmes was convinced lay hidden below the deserts of Arabia could be financed through developing resources and trade in Aden and Southern Arabia along the Red Sea. Edmund Janson was, and would remain, Frank Holmes's most ardent supporter in the newly formed syndicate. But the more cautious of the syndicate's investors had to be catered to and so, inspired by Holmes's 1918 observation of the boom in pharmacies in Alexandria, E&GSynd decided their first business would be to open, "under efficient European management," a chemist and druggist business. This decision particularly pleased one of the syndicate's investors, the pharmaceutical chemists Allen & Hanburys. Holmes had been delayed in setting off to Jaizan to visit the ruler of Assir by the need to find a suitable location for the shop. The business he opened at Aden's Steamer Point was imaginatively called the "English Pharmacy."

Locating to Aden with Holmes was Cdr. C. E. V. Craufurd, Royal Navy (Retired). Employed personally by Sir Edmund Davis, the syndicates' own chairman and chief investor, Craufurd was to be E&GSynd's local adviser. Craufurd had patrolled the Red Sea for sixteen years; during the war he was in command of HMS *Minto*. He professed to know every spot on the Arab coast suitable for anchorage "so that natives can note the ship's guns and realise that an unfriendly reception is likely to meet with prompt reprisal." Taking him at his own word, Edmund Davis believed Craufurd had "considerable experience of dealing with Arabs and is *persona grata* with the Idrissi."

Craufurd seems to have oversold himself in London. Although he did have good contacts among British officials in Aden, he did not know many Arabs and certainly did not know the one person with whom Holmes would now deal, the Idrissi, ruler of Assir. Unfazed, Craufurd had repaired this small omission on arrival by having the Aden resident write him a letter of introduction.

In fact Craufurd was, if not at the time then certainly later, a genuine eccentric. He was obsessed with locating King Solomon's Mines. He pored over the Bible and minutely studied Richard Burton's 1878 *The Gold-Mines of Midian* and 1879 *The Land of Midian*. He believed he possessed the "curious power" of being able to divine gold. "If I am in good form," he once told the Royal Central Asian Society, "my twig trembles as I pass over water, but if I pass over gold, my twig instead of dipping gives a throw." He claimed to the scholarly assembly that, using his twig, he had uncovered "the Imam Yahya's secret mountain caves where all the wealth of Yemen has been stored for hundreds of years."

"All the wealth of Yemen" was a local legend. The folktale says that the jinn constructed below their city of lofty palaces a web of cellars and tunnels where the fabulous wealth of the queen of Sheba was stored. The entrance was said to be through a secret mountain cave. This legend is the source of the story of *Al A'Din*, which the Western world knows as "Aladdin's Cave."

Craufurd eventually became convinced he had traced King Solomon's Mines to the port of Mukalla, on the Arabian Sea coast of the Hadramawt between Aden and Oman. As late as 1933 he was still trying to convince the learned of London by delivering speeches with titles like "Treasure of the Lost Lands of Ophir." When one-half of President Wilson's 1919 King-Crane Commission, the American philanthropist Charles Crane, sent American engineer Karl Twitchell on a mission to Yemen, Craufurd would infect him with a similar obsession for locating King Solomon's Mines. Twitchell, who knew Frank Holmes when both were in Abyssinia, would come to believe the mines were in Saudi Arabia. His search for the ancient gold mines would intersect dramatically with Holmes's goal of developing the Arabian oil fields. Twitchell would lay specious claim to having discovered the oil of Al Hasa in Saudi Arabia, and he would play a prominent role in Standard Oil of California's 1933 purchase of Holmes's lapsed Al Hasa concession.[3]

THE FARASAN ISLANDS

Holmes chose to begin in Assir because he was aware that the Farasan Islands off this coast were thought to hold rich oil possibilities. In 1912 British geologist,

Prof. Arthur Wade, had predicted "commercial success" for oil there. A concession for Farasan had lapsed, as had the Salt Mines concession of the Salif Peninsula. Prior to WWI, financial difficulties had forced the Turkish government to cede the empire's entire salt revenues to the Ottoman Public Debt Administration set up by the combined European powers. During the war the Salif mines, with salt deposits estimated at three million tons, had been barely operated except for a little export to India. Salt was an important, and profitable, commodity. In 1924 Mahatma Gandhi would focus on salt in his Home Rule campaign of civil disobedience against the British rule in India. Thousands of ordinary Indians would follow Gandhi on foot from Ahmadabad to the Arabian Sea in order to manufacture tax-free salt by evaporating seawater.

In an area where the summer heat regularly reached 140 Fahrenheit, Frank Holmes made his base in the Idrissi's "guesthouse," a rambling fortlike palace in the shadow of the Jaizan citadel. He regularly waded through the stretch of knee-deep black mud to climb aboard a six-oar native *sanbook* and set off, thirty miles into the Red Sea, to examine the reported oil shows on the cluster of tiny islands known as the Farasans.

But the Farasans had had their moments. In 1901 the Germans wanted them for a coaling station. The Ottomans refused. A Constantinople coffee and sugar merchant trading with Yemen, who might have known something, obtained a concession for oil exploration. Never having begun any activity, he was agreeable to pass it on when approached in 1910 by a mining engineer, also from Constantinople, representing a group that included British investors. They formed the Farasan Island Oil Company, and it was they who commissioned the geological survey from Professor Wade that reported: "I think indications foreshadow success from a commercial point of view."

Before anyone could get started, the Idrissi took advantage of the Italian-Turkish War to seize the islands in March 1912, only to lose them back to the Turks in October 1913. A wrangle developed over transfer of the concession between the British government, the Turkish government, the British investors, and the two original title holders from Constantinople.

The argument was still unresolved when, in January 1915, while the British were making plans to seize the islands in order to prevent the Italians doing so, the Idrissi preempted everybody by himself evicting the Turks and once again laying claim. In April 1915 his occupation was formalised in the terms of the Anglo-Idrissi Treaty in which Britain, in return for agreement not to surrender the islands to any other power, undertook to "safeguard" the Idrissi's territories from "all attack on the seaboard." Yet the British still feared an Italian grab. In December, as a warning that the islands were under its protection, British troops stormed ashore and the Union Jack fluttered significantly atop the tiny Farasans.

By 1917 a supplementary agreement had been signed in which the Idrissi agreed to the usual clauses of not ceding, mortgaging, or surrendering any of his seaboard territory, including the islands, to any other power. The British confirmed that, by dint of capture, the islands "have become part and parcel of the Idrissi's domains." The troops and Union Jack remained. When the war ended, despite protests from the Germans, Italians, Turks, and the original holders of the concession, the Farasans were a preserve solely for British companies.

Throughout the war a source of income for the Idrissi was the rock-salt mine just outside of Jaizan. About twenty feet deep and forty feet wide, the salt was blasted, broken into small pieces, packed into eighty-pound bags, and loaded onto camels for sale in the Arabian interior. But Holmes's interest was in the huge salt deposits located on the Salif Peninsula, along the Red Sea coast about 100 miles south of Jaizan.

Exporting to India, the Salif Salt Mines employed several Europeans and a British manager. In 1914 the European staff, along with the British and French consuls at Hodeida, were interned by the Turks. For their part, the Government of India and the Ministry of Blockade confiscated "until the end of the war" funds of the Salif Salt Mines and "invested" them in British War Loans.

Just off the Salif Peninsula was Kamaran Island where, in 1513, Alfonso de Albuquerque had quartered his seventeen hundred Portuguese troops and twenty ships before giving up on his bid to conquer Aden and the lands on both sides of the Red Sea. Albuquerque also had a somewhat ill-conceived plan that included diverting the water of the Nile into the Red Sea in order to leave Cairo stranded. He had also planned to force the Muslims to surrender Jerusalem by kidnapping the body of the Prophet from the mosque in Medina. Needless to say, neither scheme came to fruition.

HOLMES'S GENEROUS TERMS

After Holmes's hot, dusty, and muddy explorations of Farasan and Salif, his advice was that efficient exploitation of these two resources alone could put the Idrissi in a good financial position and provide a base on which to develop other commercial possibilities. At the end of May 1921 he gave his terms. Holmes's offer that included participation in overall profits through shareholding was revolutionary. His habit of giving similar terms to the Arabian shaikhs for oil concessions would later arouse strong opposition and much animosity from European and even American companies dealing in the Arab world.

British concessions, known as the Colonial Office Model, were agreements with a sovereign under which the company had the right to explore for, own, and produce oil in that ruler's territory. Oil was not viewed as an asset belonging to its owner—that is, the country in which it was found, but belonging by right to those who extracted it. (This principle would later reverberate in the American claim to have earned a "right" to Arabia's oil by dint of having brought it out of the ground.) The concept of the developer "owning" the resource beneath the land they leased would not be overturned until the Organization of Petroleum Exporting Countries (OPEC), formed in 1960, succeeded in gaining international acceptance of the idea of national ownership of natural resources.[4]

Under the terms Holmes offered in 1921 the Idrissi would have a 15 percent shareholding in the company formed to operate the mines, a royalty of one shilling per ton of salt and two hundred fifty sterling per month to provide "protection" to the operation. His offer for Farasan, Holmes said, would be along similar lines.

The Idrissi was immediately enthusiastic and urged Holmes to get started as soon as possible. On behalf of E&GSynd, Holmes applied to the Colonial Office to be awarded a concession to develop and operate the Salif salt deposit and for "permission to enter into arrangements with Muhammad Al Idrissi to develop the trading and other resources of the Assir and Yemen territories." The syndicate assured London they were "not in any way out to exploit the Arab or his country" and that, as the Foreign Office would confirm, Holmes and Craufurd had "acted through British Official circles only in all their dealings with the Arab rulers." The Foreign Office, and the Board of Trade, were all for it and wrote to Churchill on E&GSynd's behalf.[5]

While waiting for the reply from Whitehall, Holmes drew up a draft agreement for the Idrissi. With the British officials in Aden he agreed to buy out the local salt-processing plant now belonging to the Ottoman Public Debt Administration. More tricky was the Aden resident's idea that the Idrissi himself should repay part of the Ottoman Public Debt through future income from the Salif Salt Mines. Cautiously, the directors of E&GSynd agreed that they would "consider" the suggestion that the syndicate become responsible for an amount, to be decided in the terms of the forthcoming peace treaty with Turkey, and deduct this from the royalty and rent due to the Idrissi. The ruler of Assir absolutely refused to hear of it. Stating that the Turkish occupation of Assir had brought no benefits, from his great height he thundered balefully, "See if the Arabs are justified in paying any portion of this debt."

CHURCHILL HESITATES

With this matter still unresolved, E&GSynd applied again to the Colonial Office in October 1921, this time for permission to negotiate for the oil exploration rights for "the whole of the territory over which the Idrissi rules." The application stated that the syndicate's intention was to "carefully explore and prospect all areas likely to be oil bearing" and later to indicate definite areas for exploitation on terms agreeable to both sides.[6]

But at the Colonial Office, Churchill was having second thoughts about the very validity of the British position in the Assir and Yemen. After receiving Holmes's letter, Churchill wrote to Lord Curzon, the foreign secretary. Churchill said he was worried about the implications if His Majesty's government recommended the E&GSynd application to the Idrissi, "over whom we exercise a somewhat imperfect center."

Churchill pointed out to the foreign secretary that "the exact status of the Idrissi and of His Majesty's government's position vis-a-vis that ruler remains undefined." In reality, Churchill told Lord Curzon, "the extension of His Majesty's government influence to this portion of the Red Sea littoral has never been officially recognised by the Powers."

While he fully appreciated the "considerable benefits," both to the British Empire and to the Idrissi, likely to result from the exploitation of oil and mineral resources of that territory by an all-British firm "of substance and good standing" such as E&GSynd, Churchill considered the political risk too high. It was Britain's policy to avoid any action that could result in scrutiny of the "special position" it had awarded itself in the Arabian Peninsula and across large parts of the former Turkish empire in Southern Arabia. Winston Churchill decided, he informed Lord Curzon, to reluctantly tell E&GSynd that His Majesty's government "does not, at present, see their way to accord to them the desired permission."[7]

Realizing that not a lot could be achieved for the time being, at least not until Britain could feel secure there would be no challenge to its "assumption" of power in the region, Holmes returned to London in early 1922 for discussions with his syndicate colleagues and to plan the focus for the coming year.

He would return to Aden, Assir, and Yemen many times in the coming years, sometimes on secret assignment for Churchill and the Colonial Office, sometimes pursuing the oil concession of the Farasans and the salt of Salif. Holmes and his associates were committed to the Arab lands on the Red Sea and, even more so, along the Persian Gulf.

E&GSynd's interest in this region was known in the small circle of venture capitalists operating in the city of London. Midyear, a Dr. Alexander Mann came to speak with the syndicate's chairman, Sir Edmund Davis. [8]

ALEXANDER MANN

Indirectly, Dr. Alexander Mann owed his new position of personal commercial representative in London of the sultan of Nejd to Sir Percy Cox. The political motivation of medical personnel at the American Mission in Bahrain was the subject of intense suspicion by the Government of India's men in the Gulf. In 1921, when Abdul Aziz bin Saud called for medical assistance, Cox thought he could stall the American missionaries by sending Dr. Mann, then nearing retirement from his job on Cox's own staff in Iraq.

As the political officers had observed, it was true that bin Saud, tending toward hypochondria as he did, had a particular weakness for medical doctors. In a very short time Mann was hired on bin Saud's personal staff. With a four-year contract, Mann's brief included acting as Abdul Aziz's personal representative and agent as well as "striving to cement the existing friendly relations between Great Britain and the Nejd." He was also expected to carry on his profession of "Physician and Surgeon."[9]

Dr. Mann knew of Edmund Davis through the London Jewish community, to which both men belonged. His Jewishness had proved to be immaterial to bin Saud and now, in his new capacity as agent, Mann wanted to discuss with Davis and his partners the possibility of financing development projects in Nejd.

HOLMES'S TARGET: ARABIA'S DESERT

Mann was correct in assuming bin Saud's territory was of interest to E&GSynd. Eastern Arabia was one of the areas that Holmes had been researching since his arrival back in London. As already noted, he had been convinced of the existence of great oil reserves on the Arabian Peninsula since his first contact with the area during the war and had written his wife in 1918 of his belief in an "immense oil-field running from Kuwait right down the mainland coast." After the war, exploration of the Arabian Peninsula was the task he set himself, a task that would result, eventually, in the development of Arabia's great oil fields.

Holmes had drawn a map detailing a string of possibilities he wanted to look into, running from Kuwait at the head of the Persian Gulf to Qatar at its foot. Holmes told his London colleagues that he certainly wanted to undertake a geological examination of as much of Eastern Arabia as possible. He said that, as Dr. Mann appeared to already have the confidence of Abdul Aziz bin Saud, a visit in the company of Dr. Mann should be worthwhile.

Apart from Holmes's drive to search for oil, the Davis groups identified a number of possibilities including harbors and inlets for the feasibility of port development along with an electricity supply for Riyadh and Al Hasa and the erection of a wireless station at Hofuf.[10]

Now chairman Edmund Davis made an error of judgment that almost cost Holmes and E&GSynd all future prospects in the Arab shaikhdoms. Davis decided it was not necessary to run the Eastern Arabia visit through the Colonial Office before setting out. As Holmes would later explain to an offended Percy Cox, Dr. Mann had assured the syndicate in London that to approach the Colonial Office would be an affront to Abdul Aziz. His Highness the Sultan of Nejd, Mann asserted, was a completely independent sovereign "in a position to do as he pleased in such matters."

From his own experience, Holmes had some misgivings about the wisdom of not first consulting His Majesty's government he would tell Cox, but the directors decided "in view of what Dr. Mann told them" from his position as bin Saud's representative, that there was no need to do so.[11]

The various Davis groups, and Holmes, had long dealt with the Colonial Office and the Foreign Office. As far as Abyssinia, Aden, and Yemen were concerned, Frank Holmes was virtually a Colonial Office pet. But he had not come up against the Government of India before. His success with bin Saud, Bahrain, and Kuwait would plunge him, the Davis groups—and the Colonial Office—into a conflict of gothic dimension with the Government of India and its political officers in the Persian Gulf. It would be a massive power play that would continue unabated until Kuwait, the last of Holmes's concessions, was signed in December 1934.

DICKSON'S GREAT EXPECTATIONS

At first it seemed that Edmund Davis was correct. There didn't appear to be a problem as the arrangements made by bin Saud to receive Mann and Holmes did not arouse the attention of the British. They traveled via Bombay and Basra, arriving in Bahrain on September 28, 1922.

Guests of Abdul Aziz bin Saud were always welcomed by his Bahrain representative, Abdul Aziz Al Gosaibi, and accommodated in the rambling two-story Gosaibi house on the seafront at Manama, the main town on Bahrain Island. Along this waterfront were palaces, trading houses, and bazaars dealing in goods coming from Iraq, Persia, India, and the Far East. A regular bevy of small boats carried merchandise across to bin Saud's Al Hasa coast, from where the camel caravans trekked into the interior.

When Holmes and Mann arrived, Gosaibi was already hosting Maj. Harold Dickson and his wife. After his zealous twelve months as political agent in Bahrain, Dickson had transferred back to Iraq where his anti-Faisal fervor and imperialist ideas soon ran afoul of the post–Arnold Wilson British officials and members of the fledgling Arab government.

At the time Holmes arrived in Bahrain, Dickson was out of a job. He was, however, convinced that he was about to fulfill his dream and be appointed the British representative to bin Saud in Riyadh.

Dickson would have to forgo the other fantasy he had confided to his mother, the one about "marrying the Amir of Nejd's daughter," because on his way back from leave in England, he had walked into the Cox Bank in Marseilles and cashed a check with a secretary called Violet. To her astonishment, a week later she received a cable from Dickson in Port Said asking her to marry him. Although Violet was already engaged to a British air force officer on the Northwest Frontier, she agreed and sailed for India and on to Iraq with her new husband. Dickson informed his mother, saying: "I suppose you have now gotten over your surprise at the news that your son had got married. I shall send you a photo of my little wife. She is really such a sweet girl. . . . Vi is simply gorgeous, a dear sweet wife and companion—I did not even remotely know how happy a man could be married."[12]

With his remarkable talent for self-delusion and self-aggrandizement, Dickson had convinced himself, and Violet, that Percy Cox was personally going to "fix me up" in an important job with the sultan of Nejd. Maj. Clive Daly, the political agent in Bahrain who had taken over that job from Dickson, was on leave in India. The Dicksons had expected to use the official residence, but Daly left instructions prohibiting them doing so. They had now been in Bahrain since July and without Gosaibi's help, they might have been in real trouble.[13]

From his own days in Palestine in the 1880s, Dr. Alexander Mann remembered Dickson "in sailor suits" when Dickson's father was British consul in Jerusalem. They met again when both were in Mesopotamia. Although he would later claim that he had identified oil sites in Bahrain and on the mainland during his 1919 posting as Bahrain political agent, and that he knew immediately that bin Saud had sent for Holmes because of oil, the fact is that meeting him for the first time Dickson had no suspicion that Frank Holmes's interest in Arabia was in anything other than civil engineering projects.

In his weekly letter to Percy Cox, Dickson described Holmes as "a Civil Engineer" sent to examine various development possibilities on the mainland. In her letters, Violet Dickson described Holmes as "an engineering expert, out from London to examine the possibility of making Qatif a harbour for bin Saud." When he returned from vacation in mid-October, Daly reported that Dr. Mann's companion, a civil engineer, was "no longer in the army" [*sic*]. He added that, "Dr. Mann is believed to have brought him out on his own initiative to induce him to embark on development schemes."[14]

HOLMES MEETS BIN SAUD

After one night in Bahrain, Holmes and Mann carefully stowed their gifts for bin Saud onboard two large motor-driven dhows provided by Abdul Aziz Al Gosaibi. They had done their shopping in London and Bombay, and the baggage included forty large boxes of medicines, fifty suitcases, an array of leather bags filled with gadgets and canned food delicacies, and a battery of small arms. The dhows would take them the forty miles from Manama to the small-craft harbor of Ujair, the gateway to Abdul Aziz bin Saud's territory. Traditionally a boat and its passengers changed territory when the coast of Bahrain disappeared and that of Al Hasa appeared.[15]

When Holmes met Abdul Aziz bin Saud, both men were forty-nine years old. Abdul Aziz was tall, of excellent physique, and used to command. In 1900, at the age of twenty-four, after a childhood spent in exile in Kuwait, he had led a small band of followers to the town of Riyadh in Al Hasa to recapture his traditional family seat from the rival House of bin Rasheed. The Ottoman Turks had first entered Al Hasa in 1520, at which time they considered it part of Al Yemen. In 1557 they lost both Al Yemen and Al Hasa. In 1870 they reconquered and this time they declared all of Al Hasa to be the *vilayet* (province) of Nejd. When bin Saud retook Riyadh and established authority in the immediately surrounding area, the Turks recognized the inevitable and appointed him their *wali* (governor) over Nejd, under their jurisdiction. He adopted the Turkish title sultan of Nejd.

The charismatic bin Saud skillfully resuscitated the strict Sunni reformist teachings of Muhammad bin Abdul Al Wahab (died 1792), whose creed urged a return to the pure Islam of the Prophet and the four so-called rightly guided caliphs who ruled prior to the schism that developed between Sunni and Shia on the death of Hussain. Bin Saud's ancestor, Muhammad bin Saud (died 1765), then a petty chief in Central Arabia, had allied to Wahab, cementing the relationship through marriage to Wahab's daughter. The combination of the religious zeal of Wahab and the sword of Muhammad bin Saud succeeded in dominating Central and Eastern Arabia.

In their zest to rid Islam of innovation, including the cult of saints, the early Wahhabis sacked the Mesopotamian Shia holy city of Karbala in 1801 and captured Mecca in 1803 and Medina in 1804, destroying venerated tombs and anything they believed smacked of idolatry. The next year they invaded Syria and soon extended their domain from Palmyra in Syria to Oman in the Persian Gulf. The Turkish Porte finally ordered a military campaign that ended in 1818 with the destruction of Wahhabi power.

THE IKHWAN

After his capture of Riyadh, bin Saud had been accepted as the Imam (religious leader) and also the paramount chief of the Bedouin tribes of Nejd. Following the recipe for success employed so well by Muhammad bin Abdul Al Wahab and Muhammad bin Saud in the eighteenth century, Abdul Aziz bin Saud galvanized these tribesmen into missionary-warriors. Known as bin Saud's "Wahhabi Warriors," they were the backbone of his fighting forces.

Bin Saud gathered clusters of Bedouin tribesmen and their families into what were optimistically called *hijar*, or agricultural colonies, where the call for unity and reform would flourish. The first colony was established in 1912; there were sixty within a decade. These cooperatives were run by the Ikhwan, a name that literally translates to "Brotherhood" and most commonly refers to a group of brothers come together to achieve a religious or political goal. The colonies doubled as military cantonments. Bin Saud supplied arms and ammunition, agricultural requirements, money, teachers, and mosques—on the condition that the men were always at his call for military campaigns.

From the agricultural point of view the settlements never really took off, in part because breaking for prayers five times a day and other religious duties left little time for farming. Besides, the Ikhwan far preferred to be on the active duty of rooting out any decadent popular Islam they could find in the surrounding areas. They soon attracted angry criticism for their habit of forcing fellow Muslims to convert to Wahhabism. In battle they were preceded by their reputation for taking no prisoners.

In the early days bin Saud was fairly indulgent of what he termed the Ikhwan's "exaggerated zeal." But, after the 1924 capture of Mecca and Jedda the Ikhwan would turn against him. Successive waves of Ikhwan insurrection would only be quelled with British military assistance. In November 1921 bin Saud and his Wahhabi Warriors had captured Hail, the home of the bin Rasheeds. When Frank Holmes met bin Saud for the first time, he held sway over the entire coast from Qatar to Kuwait and inland to Riyadh and Hail.

Dickson and his "little wife" whiled away the weeks that Holmes and Mann were in Eastern Arabia. They went horseback riding and played cards. As the weeks of his exile in Bahrain became months, Dickson took to barraging Sir Percy Cox in Baghdad with letters and voluminous "reports."

One such report was a marathon sixty-five-page handwritten effort in which Dickson enthusiastically outlined a grandiose proposal for setting up "a British Consulate General" in bin Saud's territory to which a subordinate consul and Political Agency, located at Bahrain and Kuwait, would report. Dickson's not-to-be missed implication was that he should be appointed as this powerful consul general. The only snag, as Dickson saw it, was that the Government of India would have to give up its hold over Bahrain, Kuwait, and Nejd in favor of the Colonial Office because "bin Saud would not dream of coming under the Government of India's influence again" when Cox retired. Cox tried to bring Dickson back to earth. As soon as he received the sixty-five-page slab, he immediately cabled Dickson: "As regards yourself please remember that I have done no more than express pious personal hope my forthcoming meeting with bin Saud may result in some congenial opening for you."[16]

HOLMES MAPS THE AL HASA FIELD

The return of Frank Holmes from the territory of the sultan of Nejd gave Dickson some fresh material at last. He wrote to Cox that Holmes emphasized the need to locate fresh water supplies in Nejd. "Holmes says he will bet anything that there is an enormous volume of sweet water existing over the whole Qatif hinterland," Dickson told Cox.

On the matter of oil, Holmes wisely told Dickson nothing at all, merely remarking that the possibilities of the Al Hasa area from a geological point of view were "full of interest." Distracted by his dreams of fame and glory in Arabia, Dickson failed to notice that Holmes had spent more time on geological matters than on potential engineering prospects. "Holmes hopes, from data already obtained by him, to lead a paper before the Society of Engineers when he gets home," Dickson told Cox.[17]

Yet, on that four-week expedition, Frank Holmes had identified precisely where the first oil would be found in what was to become the fabulous oil riches of Saudi Arabia. As it came out later, he had gone directly into the hinterland, beyond Al Hasa's palm-belted Qatif Oasis.

Single-handedly he had prospected in the dust and sand until he was absolutely certain he had indeed located a rich oil field. When he emerged from his lonely weeks in the desert to join bin Saud in Hofuf, the old double-walled city inland from the coast of Al Hasa, Holmes had with him a number of geological samples, including samples of "earth strangely impregnated with oil" that he had collected. Holmes was so sure he had identified a rich field that, on the spot, he made a generous offer to Abdul Aziz bin Saud for a concession over the area as he had defined on his map.[18]

The geological samples didn't mean much to bin Saud. Until the first well flowed in Al Hasa in 1938, in the area first identified and mapped by Holmes, bin Saud never really believed he had oil. But Holmes was so certain of his find that he immediately laid out a very attractive offer and unhesitatingly agreed to any amendments bin Saud requested. His enthusiasm was contagious.

Like the Arabs themselves, Holmes wasn't fazed by the desert. He had lived and worked in the great desert barrenness of Western Australia, opening up the goldfields there with US president-to-be Herbert Hoover. From his time in other countries, particularly among the indigenous workers in the mines of Mexico, he had developed empathy for other cultures and peoples. He understood their pride and sympathized with their aspirations. He didn't talk down to them. The Nejdis already loved his spontaneity, friendliness, frankness, and tales of adventure.

Abdul Aziz bin Saud would not refuse him. If Frank Holmes wanted to dig in his desert, and pay him handsomely for the privilege, that would be okay by bin Saud.

Bin Saud sent his own secretary to keep Holmes company on the journey to Basra, where Holmes wanted the tentative agreement he and bin Saud had drawn up at Hofuf transformed into a legal document. Bin Saud's secretary and scribe of the *Diwan*, Saiyed Hashem, was a prominent Kuwaiti who had moved to Nejd to work in bin Saud's Amiri *diwan* after the death of his much-loved wife.

The garrulous Hashem and outgoing Holmes became firm friends as they covered the distance from Hofuf to Basra by camel, foot, and boat. In Basra, Holmes dropped his Al Hasa geological samples into a laboratory for analysis.

Within hours Basra was abuzz with the news that Holmes had been with the sultan of Nejd and was now visiting a Basra legal office in the company of the secretary to Abdul Aziz bin Saud.

Dickson heard this in Bahrain and, on November 4, cabled Percy Cox in Baghdad, saying, "Major Holmes who apparently represents large provincial interests in London is believed to have attempted to get bin Saud to sign an agreement permitting exploitation of oil etc in his dominions." An urgent cable followed a week later in which Dickson told Cox: "I have reliable information that an oil field was definitely located by Frank Holmes."[19]

Lt. Col. Arthur Prescott Trevor, the political resident, was alerted. So, too, was Sir Arnold Wilson at the Anglo-Persian Oil Company's office in Abadan. A barage of cables began to fly up and down the Persian Gulf.

NOTES

1. In Mecca, Ahmad Idrissi was teacher to Saiyed Ahmad Al Senoussi, founder of the Senoussi Sufi cult that subsequently spread across North Africa and is based on Ahmad Al Idrissi's teachings.

2. See Ameen Rihani, *Arabian Peak and Desert: Travels in Al Yaman* (London: Constable, 1930) and *Around the Coasts of Arabia* (London: Constable, 1930). See also William Harold Ingrams, *The Yemen, Imams, Rulers and Revolutions* (London: John Murray, 1963) and Robin Leonard Bidwell, *The Two Yemens* (Harlow, Essex: Longman, 1983).

3. Holmes got "The English Pharmacy" up and running and the British pharmacist/manager reported to him. It was eventually sold in 1925. PRO/CO/727/3, June 2, 1921, Department of Overseas Trade (Development and Intelligence) and PRO/R20/A2A/96/10/1 contains a letter of introduction on behalf of Craufurd to the Idrissi from the resident in Aden; Cdr. C. E. V. Craufurd, "The Dhofar District," *Geographical Journal* 53 (1919): 97–105, "Lost Lands of Ophir," *Journal of the Central Asian Society* 14, pt. 3 (1927): 227–37, *Treasures of Ophir* (London: Skeffingtton and Son, 1929), and "Yemen and Assir: El Dorado?" *Journal of the Royal Central Asian Society* 20, pt. 4 (October 1933): 568–77; Twitchell Papers (uncataloged), Seeley Mudd Library, Princeton University, personal diary for 1927 records Craufurd frequently calling on Twitchell; they spent Christmas Day 1927 together.

4. OPEC was formed in 1960 with twelve members, Algeria, Gabon, Indonesia, Iran, Iraq, Kuwait, Libya, Nigeria, Qatar, Saudi Arabia, United Arab Emirates, and Venezuela (Ecuador joined in 1973 and left in 1992); United Nations resolutions concerning permanent sovereignty over natural resources are mainly contained in General Assembly resolutions 3201 (S-VI) and 3202 (S-VI) of May 1, 1974, containing the Declaration and the Programme of Action on the Establishment of a New International Economic Order and 3281 (XXIX) of December 12, 1974, containing the Charter of Economic Rights and Duties of States. The Resolutions agreed that countries possessed undisputed rights to resources within their own borders.

5. PRO/CO/727/3, June 2, 1921, E&GSynd to Colonial Office, "worked entirely with the full knowledge of the Resident at Aden and his Officers," also June 2, 1921, from Foreign Office and Board of Trade. Department of Overseas Trade (Development and

Intelligence) to Colonial Office, "for the information of Mr. Secretary Churchill . . . this Department would welcome the granting of a concession to this Syndicate."

6. PRO/CO/727/3, September 3, 1921, E&GSynd to Colonial Office and PRO/L/PS/10/563 and PRO/CO/727/3, October 26, 1921, E&GSynd to Colonial Office.

7. PRO/CO 727/3, November 6, 1921, Colonial Office to Foreign Office; Churchill also expressed concern that "the Italian Government also has aspirations in the region."

8. Arriving at the exact genesis of Mann's association with E&GSynd is a riddle. Before switching their attention to Holmes, the British seem to have thought it was Mann negotiating an oil concession with bin Saud. Holmes has added to the confusion by himself giving several versions. PRO/CO/730/26, December 20, 1922, Sir Percy Cox to duke of Devonshire, "Holmes (said) he was an engineer by profession and for some years had been Consulting Engineer to the group of financiers which controlled the E&GSynd among other companies. The leader of the Group is a well-known Jewish financier named Davis (or Davies) operating in the City. Dr. Mann had been brought to him by a certain Captain Cheney . . . as a person in a position to obtain an oil concession from the Sultan of Nejd, whose Agent he was." PRO/FO/371/8944, August 2, 1923, Farrer, Foreign Office, Dept. of Overseas Trade, personal and confidential to Mallett at the Foreign Office, "Holmes (said) last summer Sir Charles Addis of the Hong Kong and Shanghai Bank mentioned through a mutual friend that Holmes might be interested to meet a certain Dr. Mann who was in the service of Ibn Saud. A meeting took place at a house in Lancaster Gate . . . Holmes introduced Dr. Mann to the Eastern and General Syndicate." IOL/L/PS/10, vol. 989, March 20, 1924, attached to letter of March 28, resident (Trevor) to Colonial Office, "Holmes said he had been working around the Red Sea and Dr. Mann had been sent to him (in London I think) and told him that bin Saud had heard of Holmes and had asked Mann to invite Holmes to come to Nejd. Major Holmes added that he had come as a result and had seen bin Saud and secured a concession."

9. PRO/CO/730/26, December 20, 1922, Cox to duke of Devonshire, secretary of state for India, enclosing agreement between Mann and bin Saud, February 23, 1922 (effective April 8, for four years), this appointment was personally approved by Cox.

10. At Ujair Holmes updated this map to delineate the Neutral Zones just created by Sir Percy Cox. Ameen Rihani, *Ibn Sa'oud of Arabia: His People and His Land* (London: Constable, 1928), p. 85, reprints this map with acknowledgment to Holmes. Frederick Lee Moore Jr., *Origin of American Oil Concessions in Bahrain, Kuwait and Saudi Arabia* (BA thesis, Department of History, Princeton University, 1948), p. 21, mentions sighting the original map, with Holmes's signature, attached to the Neutral Zone/Hasa Option documents in the Chevron Archives. I did not find it in my May 1999 search of the Chevron Archives. Dickson Papers, box 2a, file 4a, September 30, 1922, Dickson to Cox, "identified a number of possibilities."

11. Dickson Papers, box 2a, file 4a, September 30, 1922, Dickson's diary and PRO/CO/730/26, December 20, 1922, Sir Percy Cox to duke of Devonshire, note of interview (Cox and Holmes) at Ujair on December 1, 1922.

12. Dickson Papers, box 2, file 5, August 13, 1922, Violet Dickson to her sister, "Colonel Cornwallis (Adviser to the Iraqi Minister of Interior) has got to the stage where he believes the word of an Arab official against Harold . . . what can you do? Cox is his friend—I am sure all the local Shaikhs will rush to Baghdad to protest at Harold being removed." Although they only had the use of two rooms in Gosaibi's huge house, in

H. R. P. Dickson, *Kuwait and Her Neighbours* (London: Allen and Unwin, 1956), p. 269, Dickson claims in his brazenly self-promoting fashion that "Gosaibi came to me . . . said he would be obliged . . . if I could put up (Holmes and Mann) while they were in Bahrain." Violet Dickson, *Forty Years in Kuwait* (London: Allen and Unwin, 1971), pp. 12–13, "he spent a short time talking to me over the counter." Dickson Papers, box 8, contains a typed manuscript headed "Violet Dickson: Notes for Harold Dickson Biography"; her description of first meeting Harold is different in this to the version that appears in her book; Dickson Papers, box 2, file 6, January 12, 1920, "my little wife" and January 24, 1921, "such a sweet girl"; Dickson in Iraq, to his mother.

13. Dickson Papers, box 2, file 5, March 28, 1923, Dickson to his brother, "I had hoped to get a permanent job in Hasa or Nejd with bin Saud." August 13, 1922, Violet Dickson to her sister-in-law, "Sir Percy sent for Harold and said . . . he would try and get Harold taken on by bin Saud as his private agent, either to live inland or on the Coast as Harold would like. Cox told him he could sell all his furniture at once and be ready to go at any moment. So we came up here last Friday week. Major Longrigg is to take over Hillah. We sold him all our furniture, very cheap." October 1, 1922, Violet to her sister-in-law, "we stayed with 'Mespers' agent here for the first week but now have two rooms in a most palacial [*sic*] house on the shore lent to us by Gosaibi, bin Saud's agent here. Since we arrived here a fortnight ago today Harold has received no instructions whatsoever from Sir Percy Cox. We were told to come here and wait for his instructions. We hope (every) day that he may come down to visit bin Saud . . . and also fix up Harold as special agent for bin Saud. (Major Daly) is in India but is due back in ten days. He informed (the Resident) when we asked if we might put up in the Agency for a few days 'that it was all closed until October and after that date it certainly would not be available!!' We unpack nothing as we may move any day."

14. In *Kuwait and Her Neighbours*, Dickson draws a very different picture of this first meeting with Holmes, and of immediately subsequent events, than appears in the letters, diaries, and documents of the time. On p. 268 Dickson claims that two years earlier he, Dickson, had identified oil sites on Bahrain and the mainland. He claims that he had made known a mysterious Turkish report of oil seepages in Qatif and that he concluded because bin Saud had been told of this that Holmes had been sent for to investigate. On p. 270 he also tells a fantastic story of Holmes posing as a butterfly collector searching for the "Black Admiral of Qatif," which Dickson immediately recognized as a pun on oil. No evidence exists to support this story, the existence of the mysterious Turkish report, or of Dickson ever making "several visits to the mainland in search of oil seepages." In fact the personal letters and diaries of both Dickson and his wife, and the official records, show that Dickson had no thoughts of oil on the mainland—or that Holmes may have any interest in, or connection with, oil—until first alerted to the possibility some six weeks later by cable from Sir Percy Cox. Unfortunately, most authors have relied on Dickson's substantially fabricated, and certainly erroneous, version of events from July 1922 to January 1923 (pp. 267–80); PRO/R/15/2/109-5/17, October 16, 1922, political agent Bahrain (Daly) to resident (Trevor), "no longer in the army."

15. Dickson, *Kuwait and Her Neighbours*, p. 269, "fifty suitcases"; sadly, Rihani, *Ibn Sa'oud of Arabia*, p. 158 (after the British had removed Dr. Mann) mentions seeing these boxes of medicines "open, half opened and all in a state of woeful neglect" and comments there was no one in Riyadh "who understood the use of these drugs."

16. Dickson Papers, box 2a, file 4a, Dickson's sixty-five-page handwritten "report" from September 30 to November 12, 1922, "bin Saud would not dream of coming under the Government of India's influence again" and October 12, 1922, "private" telegram, Cox to Dickson, "pious personal hope."

17. Dickson Papers, box 2a, file 4a, November 1, 1922, Dickson to Cox, "Mann and Holmes returned evening of 30th, both left for Bombay 31st October, former via Bushire latter via Basra." October 12, 1922, "private," Sir Percy Cox to Dickson and November 3, 1922, Dickson to Cox.

18. Dickson Papers, box 2a, file 4a, November 12, 1922, Dickson reporting to Cox detail of Holmes's visit to bin Saud in Hofuf as told to him by Sa'ad Al Gosaibi and Shaikh Jassim bin Abdul Wahab of Qatif "directly into the hinterland" and "strangely impregnated."

19. Dickson Papers, box 2, file 5, November 12, 1922, cable, Dickson to Cox, Dickson's informant seems to be Dr. Mann as the cable continues, "bin Saud drew up document setting forth his rights with proposals for developing the oil and sent a confidential messenger to Basra to have the document translated. Messenger and Holmes return Bahrain tomorrow. Mann attempted to get copy of the Holmes's document from bin Saud but was curtly refused . . . I believe bin Saud intends to show you document. Undoubtedly Holmes (perhaps with Mann) trying to get a concession, cannot say how far he has been successful."

CHAPTER 4

THE INTRUDER

O ver the year that Frank Holmes had spent in the "Country of the Sid" and in the great desert spaces of Al Hasa, Sir Arnold Wilson had consolidated his new appointment as resident manager of the Anglo-Persian Oil Company. His first job had been to attend, as one of Churchill's "Forty Thieves," the March 1921 Cairo Conference. His report that, officially, even if only on paper, the Colonial Office might now have a say in the running of the Persian Gulf had alarmed the company.

Immediately, Anglo-Persian had begun agitating in order to ensure the de facto monopoly it held under the Government of India. By May of that year Charles Greenway, Anglo-Persian's managing director in London, had put forward the company's "representations" to the Colonial Office. He drew attention to a 1916 letter in which Anglo-Persian had recorded its expectation that it would be given monopoly rights "over any portion of the Turkish Empire which may come under British influence."

He failed to point out that, at the time, the company was aiming for the oil rights of Mesopotamia, not the Persian Gulf. The 1916 letter referred specifically to an area "100 miles inland from the Shatt al Arab lying between Kuwait and Ur of the Chaldees . . . which shows there are possibilities of finding oil in the area surveyed."

The great Sumerian city of Ur on the banks of the Euphrates was already an ancient city thirty-seven hundred years ago when, it is said, Abraham lived there before setting out for the land of Canaan. Clearly, Anglo-Persian's interest was not in the Arab shaikhdoms, which they were convinced were devoid of oil, but in Mesopotamia, where a thriving local industry already existed in oil extraction. "British influence" was never in doubt in Bahrain, Kuwait, and with bin Saud, and the shaikhs had signed the foreign exclusion agreement back in 1913. Yet, because Anglo-Persian believed the Arabian Peninsula to be oil dry, the company had made no effort to move in there, either before or after the war.[1]

Nevertheless, a monopoly was a monopoly, and prudence dictated that all means should be applied to ensure that the company's dominance of the entire area remained unchallenged. Especially as, after the war, the Americans had begun to show interest in the oil of Iraq. Following the Cairo Conference the Anglo-Persian Oil Company demanded that the political resident be "instructed" to personally "apply to the Sultan of Muscat, the Trucial Chiefs, the Shaikh of Bahrain, the Amir of Nejd and the Shaikh of Kuwait, for an exclusive prospecting licence in favour of this company in the territories of these Rulers for oil and its allied products viz bitumen, pitch, ozckerite, etc."

Anglo-Persian strongly emphasized that "no payment would become due to the Chiefs in respect of these exclusive prospecting licences." On "general political grounds," they declared, "it appears very desirable that all oil concessions in these territories should be in the hands of a single British company." By which they, of course, meant themselves. Greenway warned the Government of India, the Colonial Office, and the Foreign Office that prompt action should be taken "in this direction" because of "the activity" shown by foreign oil companies in obtaining a foothold in "undeveloped lands." His warning was a thinly veiled reference to the Americans who were already criticizing the way in which the British and French appeared intent on taking the oil of Iraq for themselves.

Arnold Wilson told Percy Cox that he would "open negotiations with the shaikhs of Bahrain and Kuwait" and repeated the board's emphasis that no payment would be made in return for exclusive prospecting rights. Wilson said a free monopoly was entirely justifiable on the grounds that they probably were not going to find anything anyway because of "the somewhat unpromising prospects of finding oil at either Bahrain or Kuwait."

Wilson added that, in the Arab shaikhdoms, he would conduct negotiations in the name of the D'Arcy Exploration Company, the subsidiary totally controlled by the Anglo-Persian Oil Company. Probably wisely, the company judged that the Arabs might not take too kindly to negotiating with an entity called "Persian," particularly in light of Persia's continuing claims to sovereignty in parts of the Arabian Peninsula.

Wilson told Cox he would be happy if the political agents stepped aside and let him get on with the job of "persuading" the shaikhs but Cox should instruct the political resident to give him "all necessary diplomatic support and assistance." Arnold Wilson had not dropped his habit of expecting the Government of India's political officers to jump to his command as they had done while he was acting civil commissioner of Iraq. He told Cox that the political resident, Lt. Col. Arthur Trevor, "should be authorised by you to accompany me to Kuwait in the *Lawrence* to assist me, on your behalf, in these negotiations."

Perhaps Wilson saw that the shaikhs might not be enthusiastic about replacing the previous exclusion agreement reluctantly signed with the British government by one with the Anglo-Persian Oil Company. At least the original agreement brought with it the faint possibility of military protection from territorial predators. The Anglo-Persian agreement offered nothing at all, not even a token payment. Wilson referred to these original exclusion "conventions" signed in 1913 and surmised to Cox that "without you or the Political Resident present the Shaikh of Kuwait might, in view of his engagements on the subject with the Government, not feel justified in negotiating with the Anglo-Persian Oil Company."

For the purely commercial purpose of traveling to Kuwait and Bahrain in order to persuade the two shaikhs, Wilson had no hesitation in demanding the use of the political resident's flagship, the Indian marine's *Lawrence*—and the personal services of the political resident himself. Wilson also expected Trevor to personally assist the oil company by "obtaining" from the Trucial Chiefs agreements similar to the 1913 exclusion "covenants" that locked out all non-British companies. Wilson was not disappointed.

SIGNING UP THE SHAIKHS

Trevor's swift reaction was to offer, in his official capacity as political resident, to get the new agreements signed *on behalf of the company*. Actually, Trevor was keen to do it all by himself. He advised the Government of India that "the most convenient plan" would be for him to "arrange both" on the grounds that "the Shaikh of Bahrain is very old and obstinate and will require a lot of talking to before he will agree to sign any such agreement." The shaikh of Kuwait, he said, would certainly "not sign such an agreement straight off."

Trevor thought the political agents could also be counted on to use their official positions on behalf of the Anglo-Persian Oil Company. His plan was, he told the Government of India, "when the Political Agent thinks the Shaikh of Bahrain has adopted a reasonable attitude, and shows signs of being willing to sign, a representative of the company will go over with a draft agreement for final discussion and signature. . . . I think it will be best for the Political Agents in both Bahrain and Kuwait to do the preliminary negotiations . . . the most convenient plan would seem to be for me to arrange both."

Instructing the political agents in Kuwait and Bahrain to begin the softening up of their respective shaikhs, Trevor set off around the Gulf. He was soon jubilantly reporting to India on his success in getting the various rulers' signatures. In March he dispatched "two undertakings regarding oil which I obtained during my recent tour on the Trucial Coast. These were given to me PERSONALLY [*sic*] by the two shaikhs. . . . I hope to obtain similar undertakings from the other shaikhs on my next visit." And in May he reported, "I fancy the Shaikh of Umm el Quwain is in a CONTRITE [*sic*] frame of mind at present . . . and will sign an undertaking." In February 1922 both the ruler of Ras Al Khaimah and the ruler of Sharjah signed. By May, Trevor had netted signatures from the ruler of Dubai, the Ruler of Abu Dhabi, the ruler of Ajman, and the ruler of Umm Al Quwain. In January 1923 he would report: "The Sultan of Muscat has signed!!!" [*sic*].[2]

Arnold Wilson was sure he could count on the Government of India's political officers. After all, as he had told Sir Arthur Hirtzel, they were "practically without exception my nominees." When Wilson took over at Abadan, he appointed a staff consisting almost entirely of British ex-Indian military officers. Mostly, they were more of "Wilson's Young Men" from Mesopotamia, now barely transformed from occupying military officers to oilmen. His attempt to re-create with Anglo-Persian the fief he had had in Baghdad, complete with absolute authority, was attractive to fellow Britons in the Indian Political Service and Indian military. These officials viewed the Anglo-Persian Oil Company as an extension of the Government of India in the Persian Gulf. Some were cracking their necks to join. All were seeking to preserve the strength of the oil company as a sort of parallel power and, maybe, a bolthole should they ever need a job. Wilson could call on these officials anytime.[3]

Into this cozy arrangement walked Frank Holmes. And, if the rumors were true, right under the noses of the Government of India's political officers and the men of the Anglo-Persian Oil Company, he had clinched an oil concession with Abdul Aziz bin Saud.

AMEEN RIHANI

In an indignant letter to the Colonial Office, Anglo-Persian said it had just received a cable from Arnold Wilson informing them that Dr. Alexander Mann "accompanied by a Major Holmes" has submitted to bin Saud "on behalf of the Eastern & General Syndicate" a draft concession covering mines and oil wells in Nejd. Anglo-Persian claimed it had been prevented from applying for "a prospecting licence there" because it had been told "political conditions" in bin Saud's territory made it "impracticable." They huffed: "It would appear that the present political situation does not prevent applications being made by others."

Moreover, Anglo-Persian had "heard" that bin Saud was being visited "at frequent intervals" by a US citizen "who may be connected with American oil interests." There was no need for Anglo-Persian to tell the Colonial Office, the letter warned in reference to the shadowy US citizen, the "embarrassment" that might result from the grant of a concession to "such interests."

The mysterious American whom Anglo-Persian gravely warned was a danger to the British Empire was the diminutive, urbane, and very affable Lebanese-American writer Ameen Rihani. Born in the Lebanese mountains, the eldest son of a Maronite Christian silk manufacturer, he had migrated in 1888 at the age of twelve with an uncle to New York. After being stranded with a touring theatrical group in Kansas City, he had returned to Lebanon in his early twenties to teach English and study Arabic. Rihani had already published *The Book of Khalid*, illustrated by fellow Lebanese American Khalil Gibran. In 1926 *The Prophet*, written and illustrated by Gibran, would be a runaway bestseller. Rihani was in the area gathering material for a three-volume series aimed at introducing the general reader to the kings, rulers, amirs, shaikhs, people, and culture of Arabia, the Yemen, and the Persian Gulf. The first title was to be *Ibn Sa'oud of Arabia: His People and His Land.*

Rihani had been intensely irritated, and not a little offended, by the tortuous route he had had to follow to reach bin Saud. In Aden, despite his assurance that bin Saud was ready to welcome him, the Government of India's political resident, Aden, had refused him permission to travel to Arabia. He had gone to Bombay where British officials had granted him, as "a special privilege," permission to go to Baghdad.

In Baghdad he was grilled by the formidable Gertrude Bell, Oriental Secretary to Sir Percy Cox. The slightly built Rihani was "terrified" of the tall, silver-haired dynamo that was Gertrude Bell. Striding up and down her office while Rihani huddled into the sofa, the chain-smoking Bell loudly announced, in her machine-gun-speaking style, that she had called him there in order to satisfy herself as to whether or not he was a political agitator, or associated with American oil companies.[4]

The next day Rihani was interrogated by Sir Percy Cox himself. When Cox commented he would soon be meeting with bin Saud, the now completely demoralized Rihani begged: "Take me with you and I'll serve you in any way I can . . . without any charge to you or the British government."

While waiting to see whether or not these two luminaries would bestow upon him the gift of permission to enter the Arab shaikhdoms, Rihani beat a hasty retreat out of Baghdad. He moved down to Basra where he ran into Saiyid Hashem, who was, as Rihani discovered, "on a mission" for bin Saud. Rihani had no way of knowing that Hashem's "mission" was to accompany Frank Holmes.

Just as Ameen Rihani was on the verge of giving up on the whole idea of a three-volume series, Gertrude Bell cabled he could go "and wait for Cox either in Bahrain or in Ujair." Without wasting a minute, Rihani sped off to the Basra port and boarded a steamer heading for Kuwait and on to Bahrain. To his pleased surprise, "as chance would have it," Saiyid Hashem was on the same boat "returning to bin Saud through Bahrain." Also onboard was Frank Holmes.

BIN SAUD AND THE BRITISH

"We remind you of the application made by us on July 11th 1914, with the approval of the Foreign Office, to the Turkish authorities at Basra, for a *permis de recherche* for the area including Hofuf," Anglo-Persian wrote to the Colonial

Office. Ignoring the fact that there had been a war in between, and that bin Saud no longer had any relations with Turkey, nor Turkey with the area in question, the Anglo-Persian Oil Company dubiously asserted: "Any application made now would therefore be merely a renewal, or an expansion, of an application made before the war." The company demanded the Colonial Office immediately cable Cox, "instructing him to support an application by the D'Arcy Exploration Company for an exclusive prospecting license over the whole of bin Saud's territory and to press for the prior claims of this company to be recognised."[5]

Cox cabled Dickson, saying Holmes should be "instructed" that any approach by him for a concession on the mainland "is premature and cannot be supported by His Majesty's government who, in any case, have received prior requests from other parties." Dickson was happy to have an assignment. When Holmes arrived in Bahrain from Basra, Dickson replied to Cox on November 14: "Holmes denies having prospected for oil, or found oil. He admits having asked bin Saud for a general mining concession on behalf of Eastern & General Syndicate." Holmes was "very reasonable," Dickson said, and, as he was to be in Bahrain for ten days, would personally speak to Cox when he arrived on the way to the meeting that Cox would be attending in Ujair. "Holmes," Dickson said, "states that he entirely understood from Dr. Mann that bin Saud is an independent sovereign so could be approached without permission or prejudice to Government's interest."[6]

This was indeed the nub of the matter. The extent of bin Saud's independence, if indeed he was independent—or whether it was the British government or the Government of India who had responsibility for him if he was not independent—was a contentious issue. Bin Saud had raised himself to the status of a ruler through his 1913 capture from the Turks of the Al Hasa district between Qatar and Kuwait. In November 1914 as the Government of India launched from Bahrain Britain's war against Turkey, Percy Cox had written to bin Saud, promising recognition of independence and guarantees against reprisals in return for aid against the Turks.

Subsequently, in December 1915, bin Saud had signed with Cox the Anglo-Saudi Treaty and in early 1916 Britain had supplied him with twenty thousand sterling plus one thousand rifles. The cash had gone for debts and the rifles to his own guards. A monthly ten thousand sterling subsidy, plus extras, was originally intended to cover the cost of bin Saud mounting offensives to echo those being played out in the Hijaz with the sharif of Mecca and his sons in cooperation with Colonel T. E. Lawrence—Lawrence of Arabia.

In November 1917 Cox had sent from his own office his star tax collector charged with persuading bin Saud to do more to support the British campaign against the Turks. On arrival in Riyadh, Cox's tax collector, Harry Philby of the Indian civil service, took one look at the muscular, six-foot-four-inch-tall Abdul Aziz in his flowing desert robes and golden *ighal*—every schoolboy's dream of a noble desert shaikh straight out of *Boy's Own* adventure stories—and was immediately smitten with a bad case of bin Saud hero worship. The same thing would later happen to Dickson. And, like Dickson, Philby was struck by a vision in which he saw himself as a great Arabist and desert explorer.

Unfortunately for Cox, the mission he sent to bin Saud in 1917 recorded no advantage for the British. Philby, however, grabbed the opportunity to further his own personal glory by talking bin Saud into furnishing him with an armed Bedouin escort, which led him, unauthorized, from Riyadh to Taif—where his appearance added further friction to the distrust and tension between Sharif Husain of Mecca and bin Saud.

Although Abdul Aziz bin Saud did nothing to help their war effort that the British could point to, they considered it expedient to continue the subsidy in order to stop him from doing what he wanted to do—which was to attack his rival, Sharif Husain of Mecca, Britain's ally in the Arab Revolt involving Lawrence. Philby would continue to be driven by his vision and, after a number of false starts, would convert to Islam and attach himself firmly to Abdul Aziz bin Saud. He would plot, scheme, intrigue, and double-cross in his efforts to get hold of the Al Hasa oil field that Frank Holmes had discovered.

As the fuss about Holmes and his possible dealings with bin Saud made the rounds of the Gulf, the resident Lt. Col. Trevor, fresh from his triumphs extracting signatures to the exclusion "conventions" from the various rulers along the Trucial Coast, volunteered to do the same with bin Saud. He cabled the Government of India, saying: "In the circumstances, the time has now arrived for me to open negotiations with bin Saud." Reluctantly, the Government of India had to recognize the temporary compromise reached at Cairo under which Cox was given responsibility for Kuwait and bin Saud until his retirement. The reply to Trevor was curt, stating: "As bin Saud's affairs are managed by the High Commissioner for Iraq please communicate your suggestions to him." Arnold Wilson made the next try. He cabled Cox: "Can meet bin Saud with you if considered desirable."[7]

HOLMES MAPS BAHRAIN

Still cooling his heels in Bahrain, Dickson was out of the loop and unaware of the turmoil. He may have been somewhat distracted by his vision of the glorious Oriental future that he believed Cox would deliver him. In Dickson's view, Holmes's visit to bin Saud was an excellent event inspiring him to confide to Cox that "the arrival of Major Holmes is distinctly optimistic and gives me hope that bin Saud is seriously contemplating the question of opening up Al Hasa." Dickson had learned from the local bank manager that Holmes was associated with "a trust syndicate of at least 14 companies in London interested in oil and other development schemes."

Back from Basra, and having no idea of the fracas he was causing, Holmes had ten days before he was due to meet bin Saud again. Unknown to Dickson, he was examining Bahrain's seepages and asphalt shows previously written off as worthless by respected geologists like Pilgrim, Pascoe, and James.

Dickson passed on the message that Cox wanted to speak to Holmes and reported back that the "very reasonable" Holmes would be available for a meeting in Bahrain when Cox arrived on the way to the Ujair meeting being held to settle border disputes between bin Saud, Kuwait, and Iraq. On November 21 he cabled Cox that Holmes would see him in Ujair because "Mr. Holmes, the engineer mentioned in previous reports, left hurriedly for Hofuf yesterday having been summoned there by bin Saud."[8]

DICKSON'S DREAMS

Although Dickson had been with the Indian army his entire career, his current ambition left no room for the Government of India nor its personnel. Dickson was sure the visit of Frank Holmes and Dr. Mann to bin Saud was an omen portending wonderful things about to happen in which he, Dickson, would play a pivotal role. He explained to Cox, "The object of Major Holmes visit is as already reported. He is a widely travelled man and apparently intimately known in other parts of the world like the Red Sea littoral, Abyssinia, Somaliland etc. Major Holmes informed me he had already concluded from charts etc that Ujair was useless as a port and he recommended Ras Tanura as the place worth examining, also Darin anchorage."

Dickson said Dr. Mann had shown him "a new passport form that he hopes to get introduced in Nejd. Considerable revenue should result from the thousands of Nejdi merchants who annually leave for foreign countries, and the idea is to make every pilgrim proceeding to the Haj take on a passport." Enthusiastically, Dickson continued, explaining to Cox that Dr. Mann "is also arranging for sample postage stamps and hopes bin Saud will shortly open a proper post office in his coast towns. I suggested as suitable stamps something in the nature of a dhow, an oryx, and a camel with two palm trees."

Dickson subtly reminded Cox that he was more than ready to dump his employers, the Government of India, and throw in his lot with Eastern Arabia. He pointed out to Cox that "times have changed and bin Saud is getting more and more a power to be reckoned with. I know he would not dream of coming under the Indian Government's influence again."

Dickson's personal position at this point was precarious. He had had no posting since his removal from Hillah by the new broom movement sweeping out the old Government of India types from Iraq. Since then he had been completely dependent on Cox's goodwill. Dickson and his wife had been stranded almost five months in Bahrain and now they had another problem. Dickson had to tell Cox that Gosaibi was unable to accommodate them much longer. "Al Gosaibi warns me that I must soon vacate the house as Rosenthal Brothers, the French pearl merchants, have a lien and will be arriving soon for the pearl season," he wrote and pleaded, "I trust I shall be away in Eastern Arabia before then?"[9]

THREE INTRUDERS

The political officers of the Government of India and the Anglo-Persian Oil Company had three intruders in their sights. They were Dr. Alexander Mann, the US citizen Ameen Rihani, and the New Zealander Frank Holmes, all of whom had brazenly walked into their territory and talked to bin Saud about an oil concession.

Perhaps wisely, Dr. Mann returned very quickly from bin Saud and traveled immediately, directly to Bombay. But this would not save him. Dickson had done him no favors by revealing Mann's proposals for improving bin Saud's revenues. Dr. Mann quickly became a victim of the Government of India's political officers. Heated exchanges between these officers, the Government of India, and the Colonial Office would eventually result in Cox dictating a dismissal letter, which bin Saud duly signed, and the Colonial Office forwarded to Mann in January 1923.[10]

Ameen Rihani would be cleared of the charge of being associated with American oil interests. Cox dismissed his importance, telling the Colonial Office: "Anglo-Persian's reference to an American citizen no doubt means Ameen Rihani . . . a guest of bin Saud while I was at Ujair . . . he showed no indications of being specifically interested in oil . . . bin Saud told me he regards him as a well meaning idealist with a very great admiration for America and the American people . . . he made no attempt to disguise his Christianity." The fact that Rihani was vouched for by Cox would not relieve the political officers' suspicion of him. They would continue to watch Rihani and intercept his mail as long as he remained interested in the Arabs of the Gulf.[11]

That left only Frank Holmes.

Unaware that he was arousing intense animosity, Holmes arrived from Bahrain on his way to the ancient Carmathian city of Hofuf. At Ujair he found bin Saud's people busy setting up camp in readiness for the arrival of Percy Cox and his official party.

Well versed in the requirements of British officials, the Gosaibis had shipped in, to what was no more than a windswept sand dune, Indian cooks, Perrier water, and crates of crockery, cutlery, and glassware to be laid out on white linen-draped dining tables. On bin Saud's instruction that the British should not lack anything to which they were accustomed, the supplies included bottles of Johnny Walker and boxes of cigars. The British were to be accommodated in tents fully furnished with proper European beds and mattresses, dressers and washstands, writing tables and overstuffed armchairs. To go under the European beds, the Gosaibis thoughtfully supplied porcelain chamber pots.

Hofuf was a full day's ride from Ujair, and as Holmes took a break, walking around the campsite, he met again Ameen Rihani, the man whom Anglo-Persian was whipping up fear about with their stories of the "American citizen" visiting bin Saud.

RIHANI AND HOLMES

Ameen Rihani and Frank Holmes had taken part in an almost surreal scene onboard the steamer *Bajora* as it made its way from Basra to Kuwait and Bahrain. The Arab travelers spread themselves and their belongings out along the deck, laying out their carpets and cushions and ensuring the comfort of their prized saluki hunting dogs and majestic falcons. All along the deck, between boxes, baggage, and bales, groups of Arabs set up their charcoal stoves to make the aromatic coffee that would accompany the long days and nights of pleasant conversation with which they would fill the voyage.

Against this colorful backdrop, the Indian chief steward rounded up his only two salon passengers, Frank Holmes and Ameen Rihani, and insisted on serving "English afternoon tea." Like characters in a Rudyard Kipling story, the two men sat, amid the Oriental chaos of the deck, at a spotless white-clothed table while the Indian steward hovered attentively, elegantly pouring tea and proffering dainties.

Immediately Rihani picked Holmes as "too effusive for an Englishman." He discovered his companion had seen more of America than he had. Then he learned that Holmes knew Herbert Hoover—"a second surprise," that he had worked with Hoover in Australia and China, and had lived in America, the Far East, South Africa, Russia, Latin America, and Mexico.

Holmes told Rihani he was traveling in Arabia "for his health." Attempting to regain his poise by showing himself as an expert on the area, Rihani spoke knowledgeably about the weather in Iraq. Holmes replied: "It's worse in Assir because of the dampness." A bit nettled, Rihani asked with a hint of sarcasm: "And have you traveled in Assir?" Rihani's nose was right out of joint when Holmes replied that he had indeed, and had stayed with the Idrissi in the guest-house in Jaizan, and also lived in "that big house" in Hodeida. "And do you intend to visit bin Saud?" Rihani asked in a tone verging on malice. "I've already seen him," said Holmes cheerfully, "in Al Hasa."

Now Rihani was truly annoyed. After all he had been through with the British just to get this far, he was genuinely piqued to discover that Frank Holmes, with his far-from-learned Arabic, was on intimately friendly terms with all the Arab rulers that Rihani was now trying to meet, including Abdul Aziz bin Saud. Although he got along well enough to host dinners (and bridge nights) with solely

Arabic-speaking guests—and once at a Gulf ruler's request to elucidate some details held a complex business conversation in Arabic with the ruler's wife— Holmes wisely played dumb around Europeans and never admitted to speaking Arabic. Holmes, who was also fluent in Spanish, put Rihani off the scent by telling him he "had thrown his Arabic language primer overboard" on an early voyage from England. That night Rihani confided to his diary: "so the Major is travelling for his health?? Indeed!!" They had gone their separate ways at Bahrain. [12]

When they ran into each other again amid the bustle of the preparations at Ujair, Rihani was quite touched by Holmes's politeness toward the people he was visiting. For the benefit of the less sophisticated men erecting the camp, Holmes had donned an Arab cloak, an *aba*, on top of his European suit. Over his cork topee, he had draped a *gutra*, the red-and-white-checked headcloth of Eastern Arabia. Keeping it in place was the double-looped headband, the *ighal*. In his gesture toward "good Arab form," Rihani commented, Holmes "certainly looked funny" and "his head looked colossal."

Rihani discovered that health had nothing to do with it when Saiyed Hashem passed a copy of the twenty-page draft concession he and Holmes had brought back from Basra. Hashem said bin Saud would like Rihani to check the translation and give his opinion on the content. While not thinking much of the legal translation, Rihani was genuinely impressed with both Holmes's geological opinion and the generous terms of his offer. Rihani's written report "strongly recommended" the contract.

At heart Rihani was fervently anticolonial. His wish was for an independent Arabia, a Pan-Arabia unified from coast to coast, supporting its independence through economic viability. He was critical of the Anglo-Persian Oil Company, pointing out it was "practically the British government," and advised bin Saud that, in his opinion, "the less a company applying for a concession had to do with politics," the better it would be for the Arabs.

Accompanied by his longtime Somali servant and an interpreter, Holmes had left the camp earlier in the evening for the journey to the scheduled Hofuf meeting with bin Saud. Rihani had watched him go. Now aware of Holmes's true purpose in Arabia, Rihani sensed a soulmate. In his diary that night he recorded poetically:

Bent on the accomplishment of his mission,
he trails his caravan under the stars,
in a country as strange to him, on the surface, as its people and their language.

But he knows what's in the bosom of the land, this man;
can see the invisible streams of water that flow from the Persian mountains
    under the Gulf, through the veins of Al Hasa's soil;
can track the bubbling oil and the sparkling minerals to their depths and
    beyond;
has the modern Argus eye of science and finance.

And for his knowledge, his energy, and his pluck,
which are after all the compensations of our modern civilisation,
all honour to him,
and success.[13]

THE UJAIR CONFERENCE

What became known as the Ujair Conference lasted five days, from November 28 to December 2, 1922. As Holmes was ill in Hofuf, bin Saud had gone ahead. When Holmes joined the group on November 30, he found bin Saud had wasted no time in raising the issue of the oil concession with Percy Cox. Despite all the earlier fuss, Arnold Wilson of the Anglo-Persian Oil Company decided "after consulting" with Cox, not to "accompany" him to the meeting with bin Saud. He did write to bin Saud's agent in Baghdad attending the conference, whom he knew from his time in Mesopotamia, saying he would be coming "soon" to see bin Saud and "maybe we can strike a deal about oil." Responsibility for quashing bin Saud's dealings with Frank Holmes and meeting Anglo-Persian's demand that its somewhat dubious "prior claims be recognised" was left to Sir Percy.[14]

Cox began by explaining to bin Saud "his position and obligations in the matter" and reminded him of the clause in the Anglo-Saudi Treaty of December 1915 by which he had agreed not to "cede, sell, lease, mortgage or otherwise dispose of any part of his territories to any other foreign Power, or to the subjects of any such power, without the consent of the British government whose advice he would follow unreservedly." Cox told bin Saud that, while Holmes and his investors did not appear to be foreign, he personally did not know the syndicate and, anyway, the important detail here was the wording "whose advice he would follow unreservedly."

Bin Saud was aghast. He protested that he had the right to enter into negotiations with Holmes. He had been a recognized ruler since his 1913 capture of Al Hasa from the Turks, and in 1914 Cox had personally promised him recognition of independence, which was why he had signed the 1915 treaty. Nevertheless, he agreed that he would not "conclude or sign anything until he had obtained the advice" of the High Government. Bin Saud gave Cox a copy of the document prepared in Basra, adding that there were some minor alterations to be made that he would discuss with Holmes. And bin Saud proceeded to do exactly that.

Holmes pitched his tent closer to those of the Arabs than those of the British. He moved between the two groups, sometimes visiting Rihani's tent where the clerical staff and boundary experts gathered to chat and indulge in a secret cigarette (bin Saud's puritan Wahhabi followers frowned on smoking). Rihani was kept busy interpreting for both sides. He was moved to comment: "The English of bin Saud's interpreter is as bad as the Arabic of Cox's secretary." Cox's secretary for this occasion was Harold Dickson, who was confidently expecting he would come away from the conference "fixed up" with a "really important job" with Abdul Aziz bin Saud.

Sometimes Holmes dropped by Cox's tent, which was set up like a military campaign headquarters, and sometimes he visited bin Saud's tent, which was richly decorated with Persian carpets and flanked by a guard of huge Negro slaves resplendent in their scarlet robes, richly sheathed swords, and dangerous-looking daggers. Sometimes he ate with the Arabs and sometimes with the British. He rarely joined the British at night, however, because Holmes, who was sartorially challenged at the best of times, had not come prepared. Every member of the British party had brought their evening clothes and there, in the middle of the desert, they dressed for dinner.[15]

As Cox later reported to the Colonial Office, he did not want to "humiliate" bin Saud by ordering him not to speak with Holmes. Yet, he could see that "when not doing political business with me, bin Saud was discussing the concession with Major Holmes." Cox tried another tack. He called Frank Holmes in for a man-to-man. He invited him to consider the reality that the "Arab potentates in this region, e.g., Nejd, Bahrain, Kuwait," while they may be thought of as "quasi locally independent rulers," were, in fact, "the creation of Great Britain" and His Majesty's government was "bound to mother them as unsophisticated Arabs."

Therefore, Cox said, there was an obligation to safeguard the interests, both political and commercial, of the unsophisticated Arabs against all comers whether British or foreign. He declared it was because bin Saud accepted "the spirit" of this mothering that he had not actually signed the concession with Holmes before "consulting" with Sir Percy. In fact, Cox told Holmes, His Majesty's government had been approached some time back by "other parties" who had been told no "serious project" could be negotiated in bin Saud's territory because the High Government had not yet determined either "his frontiers, or his relations with his neighbours." The other parties, Cox said, had "contemplated" obtaining an exploring or prospecting license with a "prior right" to negotiate for a concession, should the results of "expert examination" ever warrant this.

As the other parties had only "contemplated" a no-commitment low cost—or, in the case of Anglo-Persian, no cost—prospecting license, Cox was curious indeed to learn what was inspiring Holmes to immediately lay down a very generous offer for a direct concession, without even the safeguard of first obtaining a prospecting license and option. Holmes knew enough not to give anything away. He downplayed his month of geological work in Al Hasa as "a quite superficial examination." Carefully, he told Cox that while he had found "no specific signs of oil," he thought "on the whole his principals would be justified in risking a few thousands to get a concession and make experiments." Disarmingly, he told Cox that in his opinion there might even be potash of commercial value.

COX PITCHES FOR ANGLO-PERSIAN

Sir Percy Cox, high commissioner of Iraq, now made his pitch to Holmes on behalf of the Anglo-Persian Oil Company. He said he didn't know who the "other parties" were but he did know Anglo-Persian had made an approach. The best idea now, he said, was for Holmes to take all his geological findings and "the result of his discussion" with bin Saud along to Arnold Wilson. He suggested Holmes should make an effort with Wilson to come up with a "joint proposal" for bin Saud's territory. Cox clinched his argument by saying that, anyway, he would not "allow" bin Saud "to sign anything" until the whole matter had been submitted back to the High Government.[16]

But Holmes had not come this far to be so easily intimidated. He had not trekked up and down during the war and carefully researched old Admiralty maps and old reports in London to give up now. He had spent an exhausting month alone in bin Saud's desert not far from the area of Qatif where, until the mid-1800s, minor offenders were banished in the sure knowledge they would meet a slow death from marsh fever. He had not discovered and mapped his Al Hasa oil find to pass it all on to the Anglo-Persian Oil Company because Cox said so.

Holmes knew that his syndicate chairman, Edmund Davis, had influence in Whitehall. He was confident of his own good relations with the Colonial Office. He put it in writing for Cox. He felt sure, he wrote Cox, that his principals in London "would be pleased to come to an arrangement" with the Anglo-Persian Oil Company that "would remove Your Excellency's and the Home Government's objections." This being the case, Holmes wrote, "I respectfully submit that I have continued my negotiations with His Highness the Sultan and have now arranged with him that he is to grant my syndicate a concession, subject to the approval of Your Excellency and the Home Government." He closed the letter, saying: "Trusting this will meet your approval."

This did not at all meet with Cox's approval. Just hours before the conference broke up, Cox sent the text of a letter to bin Saud with the demand that he sign it and have it delivered to Holmes. Although bin Saud refused three times, eventually he gave in to Cox and signed. The letter, written by Cox, informed Holmes that "before final discussion of and before entry upon a project of such magnitude and importance" bin Saud felt himself "called upon" to "profit by the friendly advice of my friends the British government." The letter promised that as soon as "their views," together with the "expression of their opinion in respect of the details," reached bin Saud, he would "be ready forthwith to enter upon a transaction on this basis." Confidently, Sir Percy reported to London that Holmes would now "take the course indicated," which was to give all his geological information and negotiated documentation to Sir Arnold Wilson at the Anglo-Persian Oil Company.[17]

While he had succumbed to Cox's pressure and signed the letter, bin Saud had slipped out to speak with Holmes before the conference ended. He assured Holmes he had not changed his mind, that, as far as he was concerned, it was Holmes who had identified and mapped his oil field, if indeed there was an oil field, and Holmes who had asked him for a direct concession and made him a generous offer. He would not be giving his concession to anyone but Frank Holmes.

When he said farewell to Rihani, Holmes remarked ruefully that it appeared as if "my own government is against me." But both Holmes and Rihani knew that Abdul Aziz was "well disposed." Rihani assured Holmes that he was confident bin Saud was leaning "towards you and your company." Personally distressed by the display of colonial power inherent in Cox's forcing bin Saud to sign his dictated letter, Rihani would become one of Holmes's strongest supporters and would actively assist him as he continued to travel in the Arab world researching and writing his trilogy.

COX REDRAWS THE BORDERS

But Cox had much more on his mind at Ujair than preserving the monopoly of the Anglo-Persian Oil Company. He had come to settle border disputes between bin Saud, Kuwait, and Iraq. He was ideally qualified to tackle this subject. His authority was absolute over all the parties involved as he held the position of high commissioner of Iraq and, since Churchill's Cairo Conference, also had authority for bin Saud's territories and for Kuwait. Sir Percy Cox could have done the whole thing by himself, which in essence he did, merely offering a facade of consultation.

He had not even bothered inviting the shaikh of Kuwait although his border was under discussion and, as it turned out, Kuwait would be the biggest loser. When the British had a use for Kuwait, before WWI—if nothing else to prevent its being used as a Russian coaling station or the terminus of a German-financed Constantinople-Persian Gulf railway—they had gone to some trouble to secure its territory. They had delineated its boundaries in the unratified Anglo-Turkish Convention of July 29, 1913, by drawing, from the central point of Kuwait Town, two semicircles with a twenty- and forty-mile radius respectively. In the inner or red circle the ruling Al Sabah family had total autonomy. Although control was more vague in the outer or green circle, the shaikh did collect tribute from the tribes there.

But, in 1917, when Shaikh Salim took over as ruler of Kuwait and set about extending his authority into the outer green circle, Abdul Aziz bin Saud challenged him and claimed all territory right up to the walls of Kuwait Town. Salim requested British assistance on the grounds that the area to which he was extending his authority was within the limits set by the British themselves in the Anglo-Turkish Convention. The British told him that the Anglo-Turkish Convention was now superseded by the Anglo-Saudi Treaty of 1915. This was a matter of some surprise to the shaikh of Kuwait as, until this moment, he had no idea it affected him. He had not been asked to attend the discussions nor had he been given any opportunity for input to that 1915 treaty negotiated between the British and bin Saud.

By the time the Ujair Conference came up in December 1922, Britain was concentrating its efforts on Iraq and had little interest in Kuwait. They calculated that by passing some of Kuwait's territory to bin Saud, peace might be maintained in areas of more direct concern, namely Iraq, Transjordan, and the Hijaz ruled by Sharif Husain of Mecca.

With breathtaking imperiousness, at Ujair, Sir Percy Cox singlehandedly removed from Kuwait most of its traditional territory in the outer green circle and awarded it to bin Saud. The ruler of Kuwait wasn't even there to witness this. Maj. James More, the Kuwait political agent, signed the document. Cox had no regrets. "Existing facts were recognised at Ujair," he stated. It is hard not to read the "facts" as being the difficulties caused to Great Britain by the aggressive expansionism of bin Saud.[18]

Just the year before, coveting the customs dues earned by Kuwait (and Bahrain and Iraq) on goods passing into his territory, bin Saud had imposed a damaging trade boycott against Kuwait.

His goal was to gain these tariffs by forcing the people to use his own ports, which, he believed, would then develop to rival the long-established entrepôt trade of Kuwait and Bahrain. During Frank Holmes's first visit to Al Hasa, Abdul Aziz had specifically requested him to examine the opportunities for expanding his ports in the Hasa region.

Bin Saud's enforcement of the boycott against Kuwait was ferocious. In February, only months before the Ujair Conference, the political agent recorded: "bin Saud issued fresh orders to his subjects that nothing was to be imported into Nejd through Kuwait." A group returning to their tribe was stopped by bin Saud's men and asked what they had been doing in Kuwait. "They replied that they had not been there for the purpose of trade but to collect some debts. They were told there was no excuse. Bin Saud's orders were that Nejd tribesmen were forbidden to enter the town for any purpose whatsoever. Their money and camels were confiscated and the men returned to Kuwait on foot," he wrote.[19]

Percy Cox did not require bin Saud to lift his boycott of Kuwait. Yet, when he departed Ujair on December 3, 1922, Cox had stripped Kuwait of two-thirds of its territory, redistributed from Nejd to Iraq swathes of the Muntafiq and Shamiya Desert—along with the tribal inhabitants, swiped a couple of areas from Transjordan and passed these over to bin Saud and created two neutral zones, one between Kuwait and Nejd and the other between Iraq and Nejd.

Cox had also put paid to "the apparent inclination of bin Saud to absorb the Qatar principality." Cox had spotted that "bin Saud had apparently included the Qatar Peninsula within the tract of country for which he was prepared to negotiate a concession" with Frank Holmes. The high commissioner "at once took bin

Saud to task" by reminding him that under the terms of his 1915 treaty with Britain, he had "nothing to do with Qatar except to respect it." While bin Saud accepted this reprimand "without argument," Cox informed London, he was "fully prepared to return to the attack if there should be any fresh evidence of disposition on the part of the Sultan to encroach upon Qatar."

Of major importance to bin Saud at this conference, beside the borders, was the withdrawal of the annual subsidy of the sixty thousand sterling Britain had been paying him. Prudently, Cox left this matter until after he had achieved the borders he wanted. Cox added insult to the injury of withdrawing bin Saud's subsidy by preventing him closing the concession with Holmes, for which there was an immediate cash payment, and instead pushing him toward an agreement with the Anglo-Persian Oil Company, which was not offering any money at all.[20]

All in all, for an imperialist, Sir Percy Cox had had a pretty good conference.

COLLATERAL DAMAGE

In January, Dr. Alexander Mann received an official letter from the Colonial Office that said: "We are directed to inform you that we have received a dispatch from the High Commissioner for Iraq transmitting a message from His Highness the Sultan of Nejd regarding your appointment as His Highness' Personal Agent. The Sultan requests that an intimation may be conveyed to you to the effect that you must consider your appointment to have terminated as His Highness has intimated that he has no further need for your services." Cox, being the person who had first sent Mann to bin Saud, and approved his subsequent appointment, came in for some ridicule. Many enjoyed seeing Sir Percy Cox being taken down a peg or two. The Foreign Office, for example, noted: "Although this is primarily a matter for the Colonial Office it is quite interesting. Sir Percy Cox evidently chose the wrong man when he let bin Saud engage Dr. Mann."[21]

The biggest disappointment went to Harold Dickson. In letters to his family, Dickson explained that the Ujair Conference "was a great triumph for Cox. My show, however, fell through. Bin Saud did not require my services, at any rate, not just then." Dickson left his pregnant wife in Basra while he went on to Baghdad to try to get a job "but there were no vacancies." Instead, he was peremptorily "ordered" to go on leave. "We got off in three days of getting that order. We left forty-two cases and six boxes in Basra," Dickson wrote.

He told his brother that Percy Cox had given him letters "to important people such as Sir Charles Greenway, Head of the Anglo-Persian Oil Company," but nothing had come of these; "all drew a blank." Dickson was reduced to "protesting vigorously at being axed from the Government of India." Finally, the India Office arranged to take Dickson back into the Indian army "but over to the Infantry, there being no room in Cavalry . . . better than being pensioned off."

Dickson gave a different picture in a letter to Harry Philby, his colleague from Mesopotamian days. After his initial posting in Bahrain, Dickson had returned to Iraq and taken took over the Hilla post from Philby. Dickson and his new "little wife" had stayed with Philby and his wife during the transition. "Cox let me down," he wrote, "I went to Bahrain and kicked my heels. No honour. No reward. After all I've done!"[22]

The version Dickson gave Philby was not that bin Saud didn't want a British representative (or perhaps didn't want Dickson) but that "after Sir Percy arrived and we fixed up the Iraq-Nejd-Kuwait boundary, I was quietly told that the Foreign Office did not approve of my going to Nejd, or even living at Hasa or Qatif, as I offered to do, as they did not wish to multiply their officials in out-of-the-way places. This, I may mention, happened after Dr. Mann had been got rid of." By July 1923, Dickson would be back in India where he spent the next five years pitifully writing in search of work in the Persian Gulf.

He tried with Philby. After being knocked back on applications to join the financial mission to Persia or the staff of Sir Herbert Samuel, high commissioner of Palestine, Philby had been dismissed in July 1921 from his job with the Iraq Ministry of the Interior. His open and active hostility toward Faisal had not gone down at all well. Amazingly, by October that year he had maneuvered himself into the job of chief British representative to King Faisal's brother, Abdulla bin Husain al Hashem, amir of the reconstituted Transjordan bordered by the Hijaz, Palestine, Iraq, and Syria. The job was in line with Philby's skills. His duties were to collect and control revenue and set up an administration. (He would resign in January 1924, just days ahead of his scheduled sacking.) Philby answered Dickson's pleas, warning: "At present there is no job going on my staff . . . don't get your hopes up."

Lamenting that in India he was "unknown and certainly not respected for having been in the Mesopotamian Political Department," Dickson wrote to Arnold Wilson. "Always willing to help," Wilson replied, "but there is nothing doing here at the Anglo-Persian Oil Company."[23]

Dickson's persistence would pay off when, in May 1928, an old friend from Mesopotamian days would get him the job of secretary to the political resident in Bushire. From there, in mid-1929, he would be appointed political agent in Kuwait. He never really forgave Frank Holmes for the ease with which he and Abdul Aziz bin Saud became close friends while Dickson—so admiring of bin Saud and desperate to enter his inner circle—was relegated to the sidelines. In the position of political agent in Kuwait, Dickson, with the connivance of officers of the Government of India, would harass Frank Holmes unmercifully.[24]

At the close of 1922 the loss of Churchill as colonial secretary was a new factor the effects of which could not be predicted. He had lost his parliamentary seat in Prime Minister Lloyd George's disastrous snap election of November 15, 1922. Middle Eastern issues had figured strongly in the election campaign (though, as usual, not including the Arab shaikhdoms of the Persian Gulf), and the British public had been treated to the extraordinary spectacle of a slanging match between Colonial Secretary Churchill and Foreign Secretary Lord Curzon over matters related to the Middle Eastern mandates. Victor Cavendish, duke of Devonshire, took over as colonial secretary.

NOTES

1. PRO/FO/371/2721 February 25, 1916, confidential, Charles Greenway, Anglo-Persian Oil Company (APOC) chairman, to Foreign Office, "between Kuwait and Ur."

2. IOL/R/15/1/618/ F52, December 22, 1921, Wilson to resident, "I solicit your personal assistance in the matter" and January 10, 1922, resident (Trevor) to Denis Bray, foreign secretary to the Government of India, "best . . . for me to arrange both" and PRO/R/15/1/618, vol. F2, pp. 39–51, reports of signatures, and p. 99, dated October, 1, 1923, "the Sultan of Muscat has signed."

3. Note that APOC staff included, but was not limited to, J. B. Mackie formerly in Mesopotamia; Sampson formerly in the Sudan Civil Service; Brig. Stephen Longrigg, later negotiator for Iraq Petroleum Company, who moved to APOC immediately after leaving the Indian army in Mesopotamia; and C. Stuart Morgan who also moved from Mesopotamia to APOC before joining Standard Oil of New Jersey as manager for Near East Development. See also John Marlowe, *Late Victorian: The Life of Sir Arnold Talbot Wilson* (London: Cresset, 1962), p. 60, in 1909, "Wilson . . . was visited by E. B. Soane, an English expert on Persia and the Kurds. Soane had just been fired by the Imperial Bank of Persia over a matter involving a local woman. Wilson got him a job with APOC . . . and later hired him as an official in the Mesopotamia Administration." Also p. 236, "Wilson . . . employed two former confidential assistants with whom he had formed personal relationships, one worked with him at Abadan and the other in London and Paris."

4. Gertrude Lowthian Bell (1868–1926), Oxford-educated, a competent archaeologist and travel writer, the first woman to obtain a first-class degree in Modern History. Joined the Arab Bureau in Cairo in 1914 with the job of gathering intelligence on the Bedouin tribes and shaikhs of Northern Arabia. In 1916 she became Oriental Secretary, in 1917 moved to Baghdad as adviser to Percy Cox as civil commissioner, stayed on during the reign of Arnold Wilson, and served Cox again when he became high commissioner. See Bell, ed., *The Letters of Gertrude Bell, Selected and Edited by Lady Bell* (London: Ernest Benn, 1927).

5. PRO/FO/371/8944, November 15, 1922, H. E. Nichols, general manager, APOC to undersecretary of state Colonial Office, "a renewal or an expansion" and "American citizen" and "instruct him to support." It must be noted that exactly this legal ruse was being used at the time to reconstitute the Turkish Petroleum Company in order to take over Iraq's oil. And November 24, 1922, duke of Devonshire to Cox, "I should be glad to learn what ground there is if any for the reports as to the alleged frequent visits of an American citizen to bin Saud (referred to by) APOC"; ILO/R/15/1/618/F.2, p. 91, memo, November 27, 1922, Petroleum Dept./Board of Trade to undersecretary Colonial Office, "if there is any danger of American competition for these rights (Bahrain, Kuwait, Muscat etc) as is stated to be the case, it may be desirable to allow APOC to secure a definite license."

6. PRO/R/15/1/618/F.52, Dickson's Bahrain diary, no. 44 for week ending April, 11, 1922; Dickson Papers, box 2a, 4a, November 11, 1922, Cox to Dickson: "Holmes is reported to have left Basra November 9 for Bahrain," and November 14, 1922, cable Dickson to Cox, "Saud is an independent sovereign."

7. PRO/R/15/2/109/5/17, November 10, 1922, resident Norman Bray to foreign secretary Government of India, "time has now arrived"; PRO/FO/371/7716, December 4, 1922, Bray to resident, "Saud's affairs are managed"; IOL/R/15/1/618/F.52, November 13, 1922, cable, Wilson to Cox, "can meet bin Saud."

8. Dickson Papers, box 2a, file 4a, sixty-five-page handwritten "report" from September 30 to November 12, 1922, "gives me hope"; November 14, 1922, cable, Dickson to Cox, "very reasonable"; IOL/R/15/1/618/F.52, November 21, 1922, Dickson to Cox, "at least 14 companies" and "summoned by bin Saud"

9. Dickson Papers, box 2a, file 4a, September 30 to November 12, 1922, Dickson "Reports."

10. Dickson Papers, box 2a, file 4a, September 30–November 12, 1922, "report," Dickson to Cox, "I suggested as suitable stamps . . ."; PRO/CO/730/26, January 12, 1923, duke of Devonshire to Dr. Mann, "consider your appointment to have terminated"; Philby Papers, box XVII/9, May 1, 1923, Dickson to Philby, "after Mann had been got rid of. . . ."

11. PRO/FO/371/8944, December 24, 1922, Cox to Colonial Office, "idealist"; IOL/R/15/5/237, vol. 11, February 22, 1924, resident (Trevor) to political agent, Kuwait (More), "extract from intercepted letter dated 20, January 1924, from Abdur Rahman al Naqib, Kuwait, to Prof. Ameen Rihani, Beyrouth, received from Baghdad" and February 28, 1924, political agent Kuwait to resident, "I was very much interested in the intercepted letter to Ameen Rihani, extract of which reached me yesterday afternoon."

12. PRO/FO/371/8944, December 24, 1922, Cox to duke of Devonshire, "APOC reference to American Citizen no doubt means Ameen Rihani . . . a guest of Ibn Saud during my visit at Ujair. He showed no indication of being specifically interested in oil . . . he intends accompanying Ibn Saud to Riyadh and then proceeding overland to Lebanon. He made no attempt to disguise his Christianity."

13. Ameen Rihani, *Ibn Sa'oud of Arabia: His People and His Land* (London: Constable, 1928; Delmar, NY: Caravan Books, 1983), pp. 69–90. There are some inaccuracies in Rihani's account, e.g., "the problem was where to lodge Holmes." Rihani obviously did not know that Holmes was merely transiting Ujair on the way to a prearranged meeting with bin Saud in Hofuf. Holmes "incorporated himself into the Ujair Conference"; again Rihani does not seem to know this was a prearranged meeting between Holmes and Cox. Unfortunately, subsequent writers have repeated, and creatively built on, Rihani's errors.

14. PRO/FO/371/8944, August 2, 1923, personal and confidential, Cecil C. Farrer, Department of Overseas Trade to V. A. L. Mallett at the Foreign Office, "Major Holmes contracted cystitis (in Egypt) which remained with him for most of the trip"; PRO/FO/371/8944, December 23, 1922, Cox to Colonial Office, "Wilson decided not to accompany."

15. Criticizing other people's Arabic was almost a parlor game among Europeans; in *Forty Years in Kuwait* (London: Allen and Unwin, 1949), p. 89, Violet Dickson says that Dickson spoke the "colloquial" language and never learned to read or write Arabic. In *Kuwait and Her Neighbours* (London: Allen and Unwin, 1956), p. 274, H. R. P. Dickson claims, "Sir Percy's Arabic was not too good, so I did the translating." At one of Cdr. Craufurd's 1927 London lectures, Percy Cox got into a loud and prolonged argument with the

learned members of the Central Asian Society about correct Arabic grammar and pronunciation. Disdainfully he closed his argument, saying, "that is a point only those of us with knowledge of Arabic would appreciate." Bin Saud's advisers and officials lampooned Philby's Arabic as "rough" and "uncultured." Rihani castigated everybody's Arabic.

16. PRO/CO/730/26, December 20, 1922, secret, Cox to Victor Cavendish, duke of Devonshire, "rights and obligations," "follow unreservedly," "not conclude or sign," "when not doing political business with me bin Saud was discussing the document with Holmes," "creation of Great Britain," "other parties," and "joint proposal."

17. IOL/R/15/1/618/F.52, December 2, 1922, Holmes in Ujair to Cox, "my principals"; Rihani, *Ibn Sa'oud of Arabia*, p. 85, "will Sultan please write letter in above terms to Major Holmes and send me (Cox) a copy of it"; PRO/CO/730/26, December 20, 1922, Cox to duke of Devonshire, "the course indicated."

18. Christine Moss Helms, *The Cohesion of Saudi Arabia: Evolution of Political Identity* (London: Croom Helm, 1981), pp. 206–43, "that is Abdul Aziz was potentially more troublesome to Great Britain than was the Shaikh of Kuwait." Helms gives an excellent scholarly examination of the effect on Kuwait of the rise of Wahhabism, territorial conflict, and the Ikhwan rebellion. See also Ralph Hewins, *A Golden Dream: The Miracle of Kuwait* (London: W. H. Allen, 1963), pp. 201–205; see Dickson, *Kuwait and Her Neighbours*, p. 276, "Major More (attended the conference) on behalf of the Shaikh of Kuwait, throughout the talks More, who was supposed to be watching the interests of the Shaikh of Kuwait, said nothing."

19. Administrative Report for Kuwait for the Year 1923: "the question of the re-opening of trade between Kuwait and Nejd, and to collect Nejd customs in Kuwait had been discussed at the Ujair Conference in December 1922. . . . Subsequently Cox persuaded Shaikh Ahmad to meet a Nejdi envoy and discuss this . . . (in April 1923) the Nejdi envoy took the line that there were only two ways in which a Customs Agreement was possible, either Abdul Aziz would keep a customs official in Kuwait to look after his interests or the Shaikh of Kuwait should pay him a fixed annual sum."

20. IOL/R/15/5/242, vol. VII, January 19, 1923, Cox to Government of India, "bin Saud's apparent inclination"; see Gary Troeller, *The Birth of Saudi Arabia: Britain and the Rise of the House of Sa'ud* (London: F. Cass, 1976), pp. 164–67, for breakdown of all subsidies to the various shaikhs. Troeller says by March 1920, bin Saud had received 265,000 sterling to which must be added three more years, making a grand total of 283,000 at the time of cancellation. See also Rihani, *Ibn Sa'oud of Arabia*, p. 78, "I'm afraid he also fears losing his annuity" and p. 76, "Five months later I met in Baghdad bin Saud's agent who had come to the Residency to urge the payment of the sum of money agreed upon in cancellation of the yearly stipend of 60,000 Stirling which the British Government was paying to bin Saud."

21. PRO/CO 730/26, January 12, 1923, secretary of state, Colonial Office to Dr. Mann and FO 371/8944, August 2, 1923, Cecil C. Farrer, Dept. of Overseas Trade to V. A. L. Mallett at the Foreign Office; Mallett's comment is on the Department Minute Sheet.

22. Dickson Papers, box 2, file 5, March 28, 1923, Dickson to his brother, "my show fell through"; Dickson Papers, box 2, file 6, January 12, 1920, Dickson in Iraq to his mother, "Vi and I are staying at present with Mr. & Mrs. Philby . . . I am taking over this district from Philby"; and Philby Papers, box XVII-9, May 1, 1923, Dickson in London to Philby, "After all I've done!!"

23. Dickson Papers, box 2, file 6, March 7, 1923, Dickson wrote to Sir Henry Dobbs of the Indian civil service who was due to take over in April from Cox as high commissioner, Iraq. Dobbs replied: "You did good work in Iraq but you climbed up the wrong rope, the call now is for commercial and technical experts or professions such as judges and doctors"; July 14, 1923, Philby in Amman to Dickson, "no job going on my staff"; December 8, 1923, Arnold Wilson in Abadan to Dickson in Quetta, "nothing doing here," December 27, 1923, Dickson in Quetta to Wilson, "I do not like it here . . . I would give anything for a job in Iraq once more in spite of the sickening way some of us have been let down there. . . . Do you know of any opening in Persia now? The Shaikh of Muhammerah couldn't fit me in somewhere?" May 21, 1924, Dickson to Norman Bray, Government of India, reminding him that he had visited him in India with a letter from Cox and that he "is still wanting work in Iraq or the Gulf or Persia."

24. In *Kuwait and Her Neighbours* H. R. P. Dickson gives a largely fictitious account of this period of his life together with a seriously erroneous, and quite malicious, version of Frank Holmes at the Ujair Conference together with a highly inflated picture of his (Dickson's) involvement in that conference.

CHAPTER 5

CONFLICT, ANGER, AND ABDICATION

It didn't take long for Frank Holmes to discover that, far from being "against him," as the behavior of Sir Percy Cox at Ujair had led him to conclude, "his own Government" was seeking his help in a sensitive undertaking.

Active conflict had broken out between the Sufi Muhammad Al Idrissi of Assir and the Zaiyidi Imam Yahya of Yemen. The British did not want either party to gain the upper hand. What they did want was to maintain the status quo.

Obviously, they could not actually be seen to supply armaments to either side, but the Idrissi was outmanned and outgunned. In keeping with his reputation as a "political genius," the Idrissi himself came up with the solution. Send him a cartridge-making machine with which he could manufacture his own bullets, he said, and he, and the British, would present this practical gift as being part payment on the Salif salt concession.

The resident in Aden wrote the Colonial Office, advocating this as "a suitable alternative to the British government itself supplying rifles or Le Gras ammunition to the Idrissi of Assir" for use in fighting the Imam Yahya of Yemen. The resident included a confidential dispatch from the military adviser in Aden. Conveniently, the Colonial Office judged that an undertaking from the Idrissi "though brief" could be regarded as "satisfactory" in terms of the 1922 Arms Convention established at Saint-Germain-en-Laye. Therefore, the delivery of a cartridge-making machine could be just the ticket.

In the first week of January 1923 the new colonial secretary, Victor Cavendish, Duke of Devonshire, wrote personally to the foreign secretary, Lord Curzon, suggesting the mission should be "effected through Major Frank Holmes, privately, and not through the Eastern & General Syndicate." Lord Curzon agreed absolutely that Holmes was the ideal "intermediary." Curzon sent letters to the India Office and the War Office advising them of Holmes's delicate task.

Frank Holmes successfully dodged the skirmishing parties and completed his task. His secret missions for the Colonial Office and his negotiations for the Farasan Islands and Salif salt concessions would continue for several years against a background of the Idrissi on the coast of Assir and the Imam Yahya in the interior of Yemen struggling for power in Southern Arabia—and the British maneuvering for control.[1]

BIN SAUD KEEPS HIS WORD

Holmes was soon back in Baghdad getting to grips with the oil he was certain existed on the Arabian Peninsula. In mid-February, Abdul Aziz bin Saud wrote Sir Percy Cox, saying he had received from Frank Holmes the agreement "which is to be made between me and the Eastern & General Syndicate of London." True to his word, he made his intention absolutely clear, telling Cox: "My desire in granting the concession in question has considerably increased because I have already promised the company that I shall agree to grant the concession. Therefore I request that Major Frank Holmes may be approached for starting with the work urgently."

Cox thought he had put paid to this deal. Irritably, he reminded bin Saud that the Anglo-Persian Oil Company had "previously contemplated entering into negotiations with you" and they were a "stronger and more expert organisation." He advised bin Saud to give Anglo-Persian a year "in which to select a section of your territory so arranged that the whole seaboard is not taken up." After that, he said, "negotiations with E&GSynd could then be resumed by you." (Presumably, for whatever might be left over.) Cox assured bin Saud that Anglo-Persian had "expressed their willingness" to give terms "at least as favourable . . . in many aspects" as those offered by Frank Holmes. Without explaining how, Cox told bin Saud a royalty, as paid in Persia, was better than a share of the overall profits that Holmes was offering. He added that Anglo-Persian's representatives had now been "instructed" to open negotiations directly with bin Saud.[2]

Cox's "instructions" did not move the company along. What Anglo-Persian wanted was for Holmes to be kept out of what they considered their territory. They did not believe there was any oil and had little desire to spend anything at all to obtain what they considered a useless concession, as Cox was now suggesting they do.

A full month passed before Arnold Wilson dispatched "(J. B.) Mackie, formerly in the Mesopotamia Administration" by the slow mail boat to Bahrain en route to visit bin Saud on behalf of the Anglo-Persian Oil Company. The company's previous interest in the area, "contemplated" or otherwise, had been so minimal that they did not even have a map. Wilson requested the political resident, Lieutenant Colonel Trevor, to "furnish Mackie with one copy of Philby's recent map of Nejd." The recent map was one Harry Philby had drawn during his 1917 Bedouin-escorted dash from Riyadh to Taif.[3]

Trevor, previously so enthusiastically pro-Wilson and pro-Anglo-Persian, was beginning to harbor doubt about the oil company's intentions toward the shaikhdoms. He was miffed that Arnold Wilson had not bothered to attend the Ujair Conference despite knowing an oil concession was to be discussed. After all the fuss about its "prior claims" Anglo-Persian had waited more than four weeks to make a move *after* Cox arranged direct negotiations with bin Saud. They hadn't even made the effort to get a map. Trevor had no reply for two weeks to a cable he sent Wilson on the subject of the political agents in Bahrain and Kuwait still trying to talk up the new Anglo-Persian agreement to replace the 1913 conventions.

Their instructions were to continue talking "until they detected signs of the shaikhs being willing to sign," at which time Anglo-Persian would appear with a contract "for final discussion and signature."

Perhaps Trevor perceived that the officers of the Government of India in the Persian Gulf were more keen to sign up Bahrain, Kuwait, and bin Saud than was the Anglo-Persian Oil Company. Perhaps he perceived that he might have been too hasty in his rush to "persuade" the Trucial shaikhs to sign exclusion agreements that, it was beginning to seem, would be of little benefit to them. Trevor was due to take a six-month leave from the end of April and hand over to Lt. Col. Stuart George Knox.

Due to retire, Knox was an old hand at pursuing the Government of India's interests in the Arab shaikhdoms. As a captain, he was the first political agent personally installed by Lord Curzon in Kuwait where he remained from 1903 to 1909. Promoted to major, he was political agent in Bahrain from 1910 to 1911. Rising to lieutenant colonel, he was moved around the Gulf from 1914 to 1916 on special duties, including two previous stints as acting resident. Knox was no diplomat. He would go on to chair the abortive, farcical even, Kuwait Conference the Colonial Office set up at the end of 1923 to resolve the difficulties Abdul Aziz bin Saud was fomenting with his Hashemite neighbors in the Hijaz, Transjordan, and Iraq. After a number of adjournments, Knox would lose patience and irritably dissolve the conference altogether. The task assigned to Lt. Col. Stuart George Knox while Trevor was on vacation was to remove the ruler of Bahrain.

Before departing, Trevor cautiously cabled the Kuwait agent: "You should not discuss Wilson's agreement with your shaikh until you are certain that Wilson is determined to push this arrangement through . . . despite our doubts." Trevor said he suspected Arnold Wilson would attempt to claim he had "the approval of His Majesty's government." If the agent noticed this, Trevor advised, he must "protest" vigorously. He explained further: "The main objection is that Wilson's agreement is a lease and not an agreement to lease and therefore gives neither party an opportunity to resile or reconsider detail. An agreement to lease is all that HMG has authorised. A lease can be prepared at leisure, when oil has been found and conditions can be more properly determined. Then the matter can receive full consideration of HMG before signatures are appended."

Trevor figured Wilson and Anglo-Persian were remarkably skilled at pursuing their own interests. He warned the agent: "Our attitude is that the Anglo-Persian Oil Company can be trusted to look after itself while neither Political Kuwait nor Resident Bushire nor Shaikh of Kuwait are competent to safeguard the shaikhs' interests, for which safeguard they must rely on independent expert advice from England."

When Wilson's reply finally came, Trevor's suspicion that Anglo-Persian was concerned to keep Holmes out but did not actually want any Arab concessions appeared to be confirmed. What did seem certain was that they did not want them if they were going to have to pay for them.

About Kuwait, Wilson's reply to Trevor stressed "the highly speculative nature of the undertaking. The traces of oil are meagre in the extreme. There is an almost total absence of visible geological structure. No previous exploration has been attempted. Kuwait territory is not an easy place to operate." Wilson's assertion of "no previous exploration" was untrue. By 1923 the British had examined Kuwait four times including by Winston Churchill's 1913 Commission of Oil Experts, two surveys by Edwin Pascoe, and another by Anglo-Persian's own Lister James. All agreed there was little prospect of oil in Kuwait.[4]

BIN SAUD ERUPTS

In Riyadh, bin Saud was getting testy. It was more than eight weeks since he had flagged his "considerably increased desire" to grant his concession to Holmes and five months since he had been prevented from doing so at the Ujair Conference.

On learning that Mackie was on the way to visit him, bin Saud fired off a letter addressed to Percy Cox in Iraq, the political resident in Bushire, and the political agent in Bahrain. He said: "I have already informed the High Commissioner that it would be inconvenient for me to break my word given to the Eastern & General Syndicate, unless the syndicate is not English, in which case I should not accept on any account." He had been, he said, "formerly approached on the matter by the Anglo-Persian Oil Company" but had found it "difficult to come to any lasting agreement with them" and therefore would "find it difficult to break my promise to Holmes."

Knowing bin Saud's weakness for medical doctors, Cox arranged to send from his Baghdad office Dr. Abdullah Damlouji, who frequently acted as bin Saud's physician. He wrote: "Just as Dr. Abdullah is on the point of leaving to proceed to Bahrain, I have received your letter." Cox began by lecturing the sultan of Nejd:

> Your Highness will I am sure recognise that I only know what took place at Ujair
> in which Your Highness made it quite clear both to me and to Major Holmes that

you would come to no decision until you received the views of His Majesty's government. There was no understanding that this decision of yours depended only upon whether the company proved to be English or not. I am sending you here a copy of your Ujair letter to Major Holmes with which in truth Your Highness' present communication does not seem to me to be consistent.

He switched to the indulgent father tone that he frequently employed when dealing with stroppy shaikhs:

However, since your letter in question was written you have probably seen Mr. Mackie who I understand proceeded to Bahrain by last mail and bore a note of introduction from me. For this reason and also because Dr. Abdullah will be with you in a few days I will not take any special action upon Your Highness' communication until I receive a further communication from you, after you have seen Dr. Abdullah, in case your conversations with Mr. Mackie, and with the Doctor, should have prompted you to change your mind.[5]

But Mackie wasn't doing at all well. Bin Saud was very clear indeed on the terms that he wanted in any agreement and these were not the same as those in Mackie's instructions. After receiving bin Saud's demands in writing, Mackie cabled these to Arnold Wilson suggesting that Wilson should come and personally negotiate with bin Saud on the basis of the points listed. Far from being willing to give terms "at least as favourable" as those offered by Frank Holmes as Cox had promised, Wilson called the whole thing off.

Mackie could do little but write to bin Saud on May 1: "It is not hid from Your Honour that following the receipt of your esteemed letter (April 22nd 1923) I despatched my assistant to Bahrain in order to cable a message to His Excellency Sir Arnold Wilson. I communicated to him a summary of our correspondence and discussions and asked him to come here should he be willing to open negotiations on the basis of the terms offered by Your Honour."

He was left squirming when he had to tell bin Saud: "Today I received a reply from His Excellency Sir Arnold Wilson informing me that as Your Honour does not agree to open negotiations on the basis of the terms offered by our company he can see no use in proceeding in person to meet you, and that I should not prolong my stay here." He added that he would depart as soon as possible. "I ask your leave to proceed tomorrow afternoon to Ujair. I must now thank Your Honour for the kindness you extended to me during my stay among you. I very much regret that we have not been able to arrive at an agreement with regard to the question of the concession."[6]

Mackie was wise to beat a hasty retreat. Bin Saud's temper was legendary. Ameen Rihani could hardly believe he was "in the presence of the same man" when he first witnessed bin Saud in a tantrum. "His dark brown eyes which throw a soft light when he is in a good mood inflame his words when he is roused to anger," Rihani observed. "His red rose-leaf mouth becomes like iron. The lips shrunk and taut, white and trembling, the colour disappearing from them, suggest a blade of vibrating steel." Rihani was almost struck speechless and described the experience as: "bin Saud in anger changes suddenly and completely. All the charm of his features gives way to a savage expression. Even the light in his smile becomes a white flame. He is then terrible." Rihani was relieved that bin Saud's anger subsided as quickly as it blazed and then he would become "kind, amusing and completely disarming."

On this occasion bin Saud did not become either kind or amusing. He was deeply offended, and very, very angry. The Anglo-Persian Oil Company had humiliated him by preventing his freedom to act in his own territories in that he had been unable to grant Frank Holmes a concession over the oil field Holmes had identified. Because of interference from the Anglo-Persian Oil Company, Cox had spelled out the previously unspoken limits of his independence inherent in the 1915 Anglo-Saudi Treaty. Cox's insistence that bin Saud sign his dictated letter, even though he refused three times, had been witnessed by many at Ujair, including several of bin Saud's own followers. And now, after months of delay on the concession, Arnold Wilson had delivered a public snub by arrogantly declaring there was "no use" in even visiting the sultan of Nejd unless he first bowed to Anglo-Persian's terms. On this occasion bin Saud did not quickly calm down.

AL HASA CONCESSION SIGNED

Bin Saud sent to Baghdad for Frank Holmes and received him in the audience hall of Hofuf. Reminiscent of old Baghdad, the large vaulted room was lined with windows decorated above and below with plaster frieze of lattice, geometric patterns, and swirling arabesques. Three pointed arches spanned the distance from the stone floor to the high ceiling. Low stone benches, suitable for an august Arab assembly, sat against the walls. Here, on May 6, 1923, Abdul Aziz bin Saud awarded Frank Holmes the concession over the Al Hasa oil field that he had identified and mapped eight months earlier.

Sir Percy Cox seems to have been particularly poor in his judgment of the character of medical men. He would have been surprised to learn, as Holmes wrote to Rihani, "our good mutual friend Dr. Abdullah Damlouji was of great assistance." And Harold Dickson might have been very miserable to know of the gifts bin Saud bestowed on Holmes, including a pair of magnificent Arab breeding horses. Indulgent of Holmes's fairly obvious disregard for fashion, bin Saud gave him a full set of embroidered silk Arab clothing—for his wife—and a richly decorated and colorfully sheathed "Sword of Honour" for Holmes. Disappointed that he hadn't yet completely convinced him, Holmes confided to Rihani: "I know bin Saud is doubtful about there being oil in Al Hasa. I do not agree. I think it a very promising area."

The sultan of Nejd may have had his doubt about there being oil in his territory, but he was willing to take a punt. Besides, with no costs to him and genuine participation in any profit, he couldn't lose.

Holmes explained the terms to Rihani. The up-front payment was six thousand sterling, to be forwarded within sixty days of the grant of the concession. After activities began, three thousand sterling would be paid each six months for "security protection" provided by the sultan for the men and equipment working in the leased area.

Furthermore, as Holmes explained, "bin Saud receives free 20% of the capital of any company formed to work the Al Hasa Concession and has the right to take up for cash an additional 20% of such company. Therefore if he exercises his right of purchasing 20% of the capital, he will hold 40% of the company."[7]

Bin Saud wrote to Shaikh Ahmad bin Jabir bin Mubarak Al Sabah, the ruler of Kuwait, suggesting that Ahmad join him in granting to Holmes a combined concession for the Nejd-Kuwait Neutral Zone. Holmes cabled his London office, asking them to immediately credit five hundred sterling to bin Saud's account at the Eastern Bank Baghdad, even though no payment was due for sixty days. Such a polite gesture of goodwill would, he told his colleagues, go a long way in assisting him to secure "a very strong hold over areas other than Al Hasa in the territory of Nejd and also in Kuwait and Bahrain."[8]

Signing the concession to Holmes and personally recommending him to the shaikh of Kuwait was not the end of bin Saud's response to Arnold Wilson's rude rebuff. He made absolutely certain Wilson and the Anglo-Persian Oil Company knew the full extent of his wrath at the cavalier way in which he had been treated. Before finalizing the agreement, he insisted Holmes sign a separate undertaking. It read: "The Syndicate shall not sell to the Anglo-Persian Oil Company, either as to the whole or part thereof, any oil or mineral concession or concessions that may be granted by Your Highness to Eastern & General Syndicate."

ANGLO-PERSIAN INSULTED

As bin Saud shrewdly calculated, when they became aware of the agreement with Holmes, the officers of the Government of India and the officials of the Anglo-Persian Oil Company viewed as a direct insult the clause specifically, by name, excluding transfer to or joint operation by the Anglo-Persian Oil Company. They demanded redress, claiming that as the majority shareholder, the insult extended also to the British government. Cox was relieved of the need to deal with this particular contretemps. He had officially retired at the end of April. The new high commissioner of Iraq was Sir Henry Dobbs, formerly of the Indian civil service and the Mesopotamia administration and great admirer of the Anglo-Persian Oil Company. He promptly addressed a "please explain" to bin

Saud. He also enclosed a copy of the letter written by Cox, signed by bin Saud, and delivered to Holmes at the Ujair Conference.[9]

Bin Saud was now thoroughly sick of seeing this letter. He responded in high dudgeon. He referred to "the zeal we have noticed in British officials for the protection of our interests" and said:

> We addressed that word to Major Holmes because Sir Percy Cox, as he confided to us verbally, said he was not acquainted with the company of E&GSynd. Nothing else was sought except this point; although the wording of our letter was not clear as regards the point, that the question depended solely upon whether the syndicate was English or not. I cannot find in the contents of that letter anything, explicit or implied, that constitutes a cause for the suspension, or giving up of, the negotiations between us and E&GSynd.

Even so, he wrote "We have acquiesced" in the request to consider Anglo-Persian's terms before making a final decision. Scornfully, his letter to Dobbs continued:

> Accordingly we have received the representative of that company and considered their terms, which it was impossible for us to accept owing to the inferiority of the terms offered in comparison with E&GSynd. Lastly, when we laid our own terms before that company—which terms did not differ from those accepted by the E&GSynd—Anglo-Persian were unable to accept them and they gave up the matter in a letter written to us by their agent Mr. Mackie. I am able to state that we have acted and behaved in strict accordance with the spirit of our letter to Major Holmes at Ujair.[10]

In London, Holmes's colleagues were also feeling the heat. They were called into the Colonial Office and raked over the coals. Holmes had not taken his Al Hasa "findings" and the "result of his discussion" with bin Saud to Arnold Wilson as confidently anticipated by Percy Cox. Instead, to appear to meet Cox's demands the London office of E&GSynd had made an approach to the London office of Anglo-Persian with, it must be said, little serious intention of pursuing Cox's vision of a "joint venture." Now the offended parties, Anglo-Persian and the Officers of the Government of India, were suggesting that Frank Holmes himself had instigated bin Saud's Anglo-Persian excluding clause. Furthermore, perhaps E&GSynd had a secret agenda to sell the Al Hasa concession to a "foreign" company. E&GSynd vehemently rejected both allegations. They emphasized "our right as a purely English company." They vowed the syndicate "has no intention of selling to any foreign company." They declared they would indeed be prepared to "work jointly" with the Anglo-Persian Oil Company "if the undertaking given to the Sultan is cancelled."

But, E&GSynd would not agree under any circumstances with the Colonial Office request that they themselves bring about a retraction from bin Saud. "It is probable the Sultan would regard our initiative in such a matter as a breach of faith," they pronounced gravely. In the end they agreed they would not object to the cancellation of Holmes's undertaking, "provided it is made clear in any instructions sent to the Political Resident that the request for such cancellation does not emanate from this syndicate."[11]

With the officers of the Government of India and the officials of the Anglo-Persian Oil Company breathing down his neck, the duke of Devonshire sent instruction to the resident on how bin Saud should be handled. While the duke met E&GSynd's condition his proposal would not have pleased them. He advised that a letter should be sent, reading:

> It must not be hid from you that E&GSynd, although a reputable and solid organisation, have not had as much experience in working oil as has the Anglo-Persian Oil Company. It was not unnatural that they should have sought, as soon as their representative approached Your Highness with a view to obtaining a concession in Nejd, to make an arrangement with Anglo-Persian for cooperation and the joint working of the concession. These negotiations had already proceeded some way when the Anglo-Persian Oil Company learned that Major Holmes had undertaken that no part of the concession should be sold to that company.[12]

It would take a whole lot more than careful wording by the duke of Devonshire to bring bin Saud around. He was offended by Arnold Wilson's snub, humiliated by Anglo-Persian's interference in his freedom to grant the concession, and angry with the British government for not recognizing his rights as a ruler.

GOVERNMENT OF INDIA REIMPOSED ON BIN SAUD

When Sir Percy Cox retired as high commissioner of Iraq, bin Saud requested that he now transact his affairs directly with the Foreign Office. Strident objections from the Government of India and the political resident in the Persian Gulf saw to it that this request was denied. He then asked to be allowed to cable directly to London and for permission to station a representative there. This, too, was denied. He was told emphatically that his channel of communication was the Colonial Office—but never directly. Like every other shaikh, his contact would be permitted only to the Government of India's political resident in the Persian Gulf who would, at his discretion, pass on messages to the High Government.[13]

Without Cox as an intermediary, the machinations of the officers of the Government of India in the Gulf and the Anglo-Persian Oil Company intensified. They were aimed at the repeal of the offending Anglo-Persian exclusion clause, or better still, total annulment of the Al Hasa concession given to Holmes.

Arnold Wilson updated his former colleague in Mesopotamia, Harry Philby, on the situation. "Things are in a pretty mess with bin Saud," he wrote.

He has, I understand, been told that his subsidy is to stop and he has been given 12 months subsidy in advance. But it appears from what Acting Resident Knox tells me that Cox's letter making this announcement was so vague and diplomatic that it is doubtful whether bin Saud realises there is to be no more subsidy after this final payment.

Bin Saud has given Holmes a concession for 70 years with a clause in it, inserted by bin Saud, prohibiting transfer to Anglo-Persian of whom he is suspicious owing to its governmental connections. I sent Mackie to see him but bin Saud was in no way prepared to do any business with us at all, and after a certain amount of fencing told us that unless we were prepared to adopt Holmes system (i.e., payment of a percentage of profits instead of our system of royalty on oil exported) it was no good our discussing the matter.

Arnold Wilson may have been hoping the pointed remarks in his letter would be passed on in the correspondence Philby had initiated with bin Saud. He continued:

Frank Holmes represents a man called Edmund Davis, a Jew with many connections in the gold mining industry, but with no experience in oil. Edmund Davis is trying to sell the concession to us. But I doubt whether we should want it at the price Davis is paying for it as there is, as far as I know, no trace of oil in Al Hasa, nor is there any indication of a favourable geological formation. It seems improbable that Davis will be able to raise the capital and all the necessary technical knowledge to work as a private venture and the whole thing is likely to peter out. Bin Saud has not yet received his money. 6,000 down and 3,000 a year after was the price Davis was to pay.

Apart from getting the figures wrong, Wilson's description of Edmund Davis as "a Jew" would have cut no ice with bin Saud, who hadn't thought twice about hiring Dr. Alexander Mann as his personal representative. Wilson's view that there was "no trace of oil in Al Hasa" corresponded to the opinion of Anglo-Persian's expert geologists who were unanimously convinced there was no oil anywhere on the Arabian Peninsula. Wilson told Philby, "I personally cannot believe that oil will be found in bin Saud's territory. As far as I know there are no superficial oil shows and the geological formation does not appear particularly favourable." Wilson confirmed the suspicion that Anglo-Persian had no intention of even attempting to explore in Arabia when he told Philby: "In any case no company can afford to put down wells into a formation in these parts unless there is some superficial indication of oil."[14]

MUHAMMAD YATEEM

Holmes was absolutely convinced the Arabian Peninsula did have oil. He pointed out to the Arabs that tapping this resource would quickly begin to alleviate the shaikhdoms' severe economic stress. His positive attitude was in sharp contrast to the negative approach of Wilson and the Anglo-Persian Oil Company. Holmes's enthusiasm was so contagious that it drew people to his side.

Ameen Rihani became a Holmes devotee, even though Holmes had originally told him over their "traditional" English afternoon tea on the deck of the Basra-Bahrain steamer that he was "travelling in Arabia for his health." A Pan-Arabist, Rihani was ideologically opposed to the stifling "protection" by the Government of India of the Arabs of the Gulf. He believed in Holmes's vision of the potential oil wealth of the shaikhdoms, so much so that he became an active participant. In August 1923 Holmes would put their understanding on a businesslike, legally documented footing. "Your interests are safeguarded and you have done right in sending to your London bankers the letters I gave you," Holmes would inform Rihani. "All that is in order and you will have your interests in this new company formed to explore and exploit the Al Hasa Concession. I hope we do well. I feel certain we will."

After he left bin Saud in Hofuf, brimming over with ideas of how he would go about developing the Al Hasa concession, Holmes's sheer optimism won over bin Saud's representatives in Bahrain, the Gosaibi brothers. They introduced him to the Yateem family, influential Bahraini merchants with connections throughout the Gulf and in Bombay, Beirut, London, and Paris. Muhammad Yateem of Bahrain joined Holmes as his personal assistant and would stay with him through all the ups and downs of the years to come.

Before opening negotiations with the Bahrain shaikhs for a concession over the area he had examined while waiting to travel to Ujair, Holmes followed up the letter bin Saud had sent recommending him to the shaikh of Kuwait. He cabled to Ahmad, saying: "I have most important letters from Ameen Rihani (who has made inquiries concerning myself and Company) advising Your Excellency not to grant oil concessions to any other company without first seeing the terms offered by my company."

Holmes's new colleague, Muhammad Yateem, added his weight by cabling his close friend, Mullah Saleh, personal secretary to the shaikh of Kuwait and secretary of Kuwait's Council of State. Saleh had also held this influential combined position under the two previous Kuwait rulers. Saying he would come immediately and Holmes would follow within the week, Yateem "strongly urged" Saleh to advise Shaikh Ahmad of Kuwait to see the "liberal terms offered," which "has been successful" with bin Saud. Yateem added that, personally, he considered the question of a concession "most important and vital for your country." Bin Saud's Kuwaiti secretary, Hashem bin Ahmad, who accompanied Holmes to Basra, was also close to Mullah Saleh. He, too, wrote recommending Holmes.

The departure of Muhammad Yateem was reported to acting resident Knox in Bushire, who immediately cabled to warn Major More, the political agent in

Kuwait: "Major Holmes may be endeavouring to get an oil concession out of Shaikh Ahmad. His emissary Muhammad Yateem will arrive at Kuwait almost immediately." If More thought there was any likelihood of the shaikh "being tempted" he was to remind him of "his grandfather's 1913 convention which we consider binds him and his successors." Shaikh Ahmad must be told, Knox instructed, "Major Holmes is in no way a person appointed from the High Government."

Maj. James Carmichael More was supremely confident of the total authority of the Government of India in Kuwait and replied, "I do not think there is any fear of Shaikh Ahmad giving a concession without, or contrary to, our advise." But perhaps he was mindful of the warning sent by Lieutenant Colonel Trevor before going on leave. He added: "At the same time it is known here that bin Saud has turned down the Anglo-Persian offer and is dealing with Holmes. It is presumed that Sir Percy told bin Saud that Holmes's syndicate was sound as it is known bin Saud asked him to find out. If I am asked why we should favour the Anglo-Persian Oil Company, over another sound British firm, what reply should I give?"[15]

THE SHAIKHS OF BAHRAIN

Having arranged his business in Kuwait, Holmes turned his focus on Bahrain where the political situation had been volatile for some twelve months past. It would climax within days with the forced abdication of the ruler of Bahrain. In the period leading up to this humiliation the elderly Shaikh Isa (he was elected ruler in 1870) had attempted to avoid pressure from the political agent Maj. Clive Daly by retiring to his winter palace on Muharraq Island. Behind him, in Manama Town, both the exercise of authority and the people's support had veered between two of his sons, Abdulla and Hamad. It was Abdulla whom the political officers had branded as having developed "wild political ideas" from seeing "much of an unhealthy and demoralising nature, especially in Paris" during his trip abroad to shake hands with the king emperor at Buckingham Palace.

Gossip concerning oil negotiations with bin Saud had been circulating in Bahrain since Holmes's first visit to Al Hasa in October. Isolated in his court on Muharraq, Shaikh Isa was nevertheless aware of his people talking of possible oil wealth on his islands. Previously he had been adamant in his refusal to even discuss with Arnold Wilson, or the resident, the new, no-fee exclusive prospecting license to Anglo-Persian that was represented as updating the earlier "convention" he had signed in 1914. Although the political officers carefully spoke of the D'Arcy Exploration Company rather than the Anglo-Persian Oil Company, all the Arabs of the Gulf knew they were really one and the same. Just three months before, in January, Daly had reported to the resident that fulfilling his instructions to persuade Shaikh Isa was a tough assignment. He was absolutely certain that Shaikh Isa of Bahrain would not under any circumstances grant a concession to the D'Arcy Exploration Company: "he will never agree . . . except under pressure."

But Shaikh Isa made a totally unexpected move. Perhaps, closeted in Muharraq, he was confused and thought it was Anglo-Persian, rather than Frank Holmes, who was the instigating party to the talks with bin Saud about Al Hasa. Perhaps he hoped to reestablish some authority by personally initiating a development project; much of the political officers' criticism of his rule centered on charges that he was antidevelopment. Whatever may have prompted him, Shaikh Isa contacted J. B. Mackie as he passed through Bahrain on his way to the negotiations with bin Saud that would be abruptly abandoned on the orders of Arnold Wilson.

After meeting with Shaikh Isa, Mackie cabled to Wilson: "The main object in his desire to see me was to inform me that he has wells in Bahrain from which bitumen and sulphur are obtained and he is sure there must be oil there." Mackie said the shaikh wanted the Anglo-Persian Oil Company to "consider the question of developing the oil resources of the islands and make an arrangement with him in the way that we are now trying to make one with bin Saud." He had given the shaikh of Bahrain "a very non committal reply," Mackie told Wilson.

Arnold Wilson failed completely to take advantage of Shaikh Isa's overture. He contacted Knox, saying he would come to see him about this matter in a fortnight or so, sometime in mid-May.

In his usual fashion, Wilson proceeded to issue orders as to how he wanted Anglo-Persian's business handled on his behalf. He said he might tell Mackie to contact Shaikh Isa again on his return from Al Hasa, but it was best that any action should be "deferred" until "you (Knox) and I have discussed the question together." Telling Knox that he was "quite prepared" to negotiate with Shaikh Isa without the assistance of Daly and "in some respects it is probably preferable to do so," Wilson requested Knox to instruct his political agent in Bahrain "not to impede the course of my negotiations." Any offer he might make to Shaikh Isa, Wilson informed the acting resident, would be on the basis of a draft agreement recently prepared by Anglo-Persian for negotiating with Kuwait except that, for Bahrain, if ever oil was found, "the guarantee of minimum royalties will be omitted."[16]

BAHRAIN CONCESSION SIGNED

Three days after signing the Al Hasa concession and full of energy, Frank Holmes made his way from the town of Manama on Bahrain Island to Muharraq Island, five miles to the east. The short trip was made in small local boats. To reach them it was necessary to trek a quarter mile through the mire and water of the receding tide, or hire one of Bahrain's famous white donkeys.

The trip from Manama to Muharraq had a festive air as village Bahrainis— men, women, and children—hiked their clothing to waist height and waded through, laughing and splashing each other as they went. This was a startling scene for some European visitors as the modestly veiled Bahrain women raised high their skirts without a shred of self-consciousness.

When Rihani completed his visit to bin Saud in Nejd and came to Bahrain he commented delightedly: "No bathing scene in Europe can surpass in its nudity and merriment this of Bahrain. Here there is no cause for a whispering mind—nothing that betrays a suggestive thought, a prurient fancy." An edict ordering everybody to wear underwear, issued by the straight-laced political agent, Major Daly, would later cause some bewilderment among Bahrainis.[17]

Holmes made his way through the Muharraq bazaar with its powerful perfume of ambergris and musk, past the shaikh's retainers who sat at the palace gate with the hunting falcons perched on brightly cushioned poles, and into a vast open court where stately sworded and daggered Arabs rested on stone benches around the walls. Shaikh Isa Al Khalifa held court in the center of this light-filled square. The Muharraq Palace was famed for the glamour of its coffee ritual. Shaikh Isa's coffee bearer was the most impressive in the Gulf. A tall and broad-shouldered Arab, he wore an orange robe with a saffron sash and a multicolored, gold-flecked *ighal* to hold in place the traditional Bahraini *ghutra*, an embroidered cashmere headscarf. A boy, dressed in blue and gold, assisted him by carrying the stack of delicate coffee cups. Bahrain's coffeepots were of polished and engraved brass; they were about two feet tall and shaped like an hourglass with a high arched spout and domed lid. Shaikh Isa's private reception room was reached through a tangle of internal lanes, hallways, adjoining rooms, and a flight of stone stairs to the second floor.

Holmes had begun his Bahrain discussions with Shaikh Isa's son Abdulla, and had continued with both brothers as Ahmad brought in Hamad. When they took him to see their father, Holmes explained that he had just signed a concession with Abdul Aziz bin Saud. No fool, Shaikh Isa asked to see it. It may have been then that Isa realized he had been mistaken in thinking it was the Anglo-Persian Oil Company that was proposing to develop oil in Al Hasa. Besides, he hadn't heard from Anglo-Persian since his approach to Mackie. The Bahrainis didn't hesitate. They were infected by Holmes's conviction, backed by technical detail from his personal examination of the island, that there was oil in their land. On May 12, just six days after signing the Al Hasa concession, the ruler of Bahrain granted Frank Holmes an exclusive concession over oil in his territory. The concession was signed by Shaikh Isa, Shaikh Abdulla, and Shaikh Hamad.

Holmes wrote to tell Rihani and generously acknowledged his assistance: "I negotiated with your friend Shaikh Abdulla . . . after I showed the old gentleman (Shaikh Isa) the agreement with bin Saud I also showed him your two letters, one you gave me for the Shaikh of Kuwait and also the one to the Najib of Baghdad." The Najib of Baghdad was the custodian of the tomb of the Shia saint Abdul Qadir Gilani from whom the Najibs were descended. This religious position belied the extraordinary political influence the Najib exerted right across Iraq. Holmes complimented Rihani, saying: "These letters did more good and gave more help than any other single thing."[18]

While Wilson was proceeding in his stately fashion, Holmes had been in and out of Al Hasa and had finalized the Bahrain concession. He was about to depart for Kuwait to join Muhammad Yateem, who was already explaining to the Kuwaitis the benefits of the revolutionary terms offered by Holmes, when two British gunships, *Throad* and *Crocus*, and the resident's own flagship, *Lawrence*, sailed into Bahrain.

RULER OF BAHRAIN DEPOSED

Spectacular in full dress uniform Lieutenant Colonel Knox and his retinue, escorted by a Sikh platoon, came ashore on May 17, 1923, to carry out the instructions of the Government of India to depose Shaikh Isa, the elected ruler of Bahrain and install his reluctant son, Hamad.

Major Daly recorded the event in bureaucratese in the annual Administrative Report. "In view of the Ruler's advanced age and manifest inability to rule, he was told that His Majesty's government desired he should hand over full powers, including control of the Revenues, to Shaikh Hamad retaining as a matter of courtesy the title of Ruler." The report continued: "Shaikh Isa, as has been his custom for years past, proved quite intractable and resolutely refused. He was finally informed that we would go ahead anyway and transfer the active conduct of Bahrain affairs, together with control of the Revenues, to Shaikh Hamad."

The Government of India had recognized Hamad, one of Isa's four sons, as his father's designated successor in 1897. In return, Isa had to accept a political agent in Bahrain. It was noted at the time that this would be an excellent move "to tighten our hold on the place" and that while the Government of India was in a treaty position with "independent" Bahrain, "we must however mean independent of all Governments except our own." Hamad was less than enthusiastic about being viewed as condoning the deposition of his father. But, as Major Daly reported, he was persuaded to take up the rulership "in spite of the opposition of all his family and others." No doubt this had something to do with acting political resident Knox threatening that if Hamad refused, the whole family would be disinherited. "We will take the rule out of the hands of your family, the Al Khalifa," Hamad was told.

Knox also found time to ignominiously deport Abdulla Gosaibi, one of the brothers acting as bin Saud's representative in Bahrain (the eldest brother, Abdul Aziz, was on business in India). Before sailing to Bahrain, Knox had advised the Government of India: "I propose to take the following action; arrest Abdulla Gosaibi and take him to Kuwait . . . and have him returned to Nejd for his

'Master' to deal with . . . I may have to get rid of a Persian or two as well." When bin Saud objected and specifically criticized Daly as being responsible for Abdulla Gosaibi's arrest, Knox replied that Daly was a trusted officer "who discharges his duties with unflagging zeal."[19]

In response to the removal of their ruler the Bahrainis formed the "Bahrain National Congress" aimed at working toward gaining participation in their own government. Knox would invite them to meet with him one evening at the Political Agency, ostensibly for the purpose of discussing their demands. When they were assembled, he and Daly would arrest the leaders and force them at gunpoint onto an Indian marines' launch moored alongside the building. Within minutes, under cover of darkness, they would be on the way to deportation in India, accused of being "agitators."

The Bahrainis would send envoys and messengers to all the Arab shaikhdoms asking for support against the removal of their ruler and subsequent actions. The Government of India's political officers would react by intimidating each of the Gulf shaikhs by warning them "not to receive any member of the Al Khalifa family and to ignore their overtures." The sole protest to escape the wall of intimidation would come from Sharif Husain of Mecca who would plead the Bahrain case to Sir Herbert Samuel, high commissioner in Palestine. The British would simply ignore him. There would be no international reaction to the forced abdication of the elected ruler of Bahrain. With no international body responsible for oversight of the Arabian Peninsula, unlike the areas under mandate, no questions would be raised about Britain's assumption of control of the Arab shaikhdoms, nor of its passing this control to the Government of India.

The Government of India was well aware of its political officers' actions. Some officials wondered whether they had "gone too far" in Bahrain. Privately the opinion was expressed that "the Political Resident and Political Agent have shown a tendency to treat the island too much like a native state of India." The "impropriety" of the political agent, Major Daly, running Bahrain to all intents and purposes as if he were the ruler would be criticized. But, lacking the impetus of world opinion, nothing would be done to correct the situation.

By 1927 the position would be so entrenched that the foreign secretary to the Government of India would remark to the political resident: "Our involvement in Bahrain is more than is desirable. A British Financial Adviser, British Police Superintendent and British Customs Manager, this is more British than Katal, which is an Indian Border state." On another occasion the Government of India's foreign secretary would note: "Our present interference in Bahrain doubtless goes beyond what flows inevitably from our Treaties. But no Treaty could adequately cover it unless it were a Treaty extinguishing the shaikh's sovereignty to a degree less than that possessed by an Indian Chief."[20]

While he was in Bahrain orchestrating Shaikh Isa's abdication, acting resident Knox, although warned that Frank Holmes was about to target Kuwait, was not aware that the shaikhs of Bahrain had already signed their concession to Holmes. Perhaps Knox did not investigate more carefully because he thought Arnold Wilson had the matter well in hand. After all, Wilson had declared that he would personally manage the Bahrainis, in his own time, and there was no need for the officers of the Government of India to do anything. It was after Knox departed that Daly learned the concession had been granted.

He called in Shaikh Abdulla who told him that "since Sir Percy Cox had apparently approved of the Al Hasa concession, and Bahrain's was similar, they had not expected there to be any objection." Daly informed Knox that Abdulla and Hamad had brought the concession along to a meeting intending to show it but "they were taken aback when you said that Government strongly disapproved of Holmes's company and were frightened and dropped the matter."

Daly said the Bahrain concession contained a clause that "it should not take effect if officially disapproved by Government, in accordance to Shaikh Isa's 1914 undertaking." He said that "after much hesitation" Abdulla had told him, in the strictest confidence, that Bahrain would not deal with Anglo-Persian "unless compelled" because everybody regarded it as a government concern with Government of India officials "who brought political pressure to bear." He had "disabused" him, Daly said, but to no avail. Shaikh Abdulla vowed that Bahrain "would prefer no concession to the Anglo-Persian Oil Company."

He summoned Holmes and interrogated him as to "why he had concealed from the Government of India the fact that he had got the concession." Holmes replied that he had intended to apply to Knox for approval while he was in Bahrain, but before he could do so Knox had told him in no uncertain terms that he was personally "all out against me." Daly reported that Holmes had said "seeing you (Knox) were not neutral, as he expected," he preferred to first report to his directors who were approaching His Majesty's government in London. Holmes had assured Daly there was no clause in the Bahrain concession excluding the Anglo-Persian Oil Company as in the Al Hasa concession.

Daly informed Knox that Holmes had "asked me whether I officially disapproved of him or his syndicate, and, if so, on what grounds and said that his directors would get satisfaction in London. I said I had no authority to reply but would report to you." Adding that Holmes had "complained bitterly of our support of the Anglo-Persian Oil Company," the political agent confirmed Holmes's suspicions by closing this cable to Knox by saying: "Mackie has been informed."[21]

NOTES

1. See John Baldry, "The Powers and Mineral Concessions in the Idrissi Imamate of Assir 1910–1929," *Arabian Studies* 11 (Cambridge: Middle East Center, University of Cambridge, 1975), p. 85; while Baldry confuses some dates, most particularly whether certain events occurred in 1922 or 1923, his "Powers and Mineral Concessions in the

Idrissi Imamate of Assir 1910–1929" is a valuable contribution on a mostly unfamiliar subject; PRO/FO/371/8945, January 8, 1923, duke of Devonshire to Lord Curzon and January 16, 1923, Lord Curzon to duke of Devonshire.

2. PRO/CO/730/39, February 25, 1923, bin Saud to Cox, "starting the work urgently" and March 12, 1923, Cox to bin Saud, "terms at least as favourable."

3. IOL/R/15/1/618/F.52, April 9, 1923, telegram, Wilson to resident, "recent map of Nejd . . . and also to arrange, in the circumstances, that quarantine regulations be waived." Mackie was accompanied by "Sampson, formerly in the Sudan Civil Service."

4. IOL/R/15/1/618/F.52, April 15, 1923, telegram, resident (Trevor) to political agent Kuwait (More), "no reply," "Anglo-Persian . . . can look after itself"; April 28, 1923, Wilson to resident (Trevor) Wilson's claim of "no previous exploration attempted" was entirely untrue.

5. IOL/R/15/1/618/F.52, April 15, 1923, bin Saud to Cox via political agent Bahrain and repeat to political resident, "inconvenient to break my word," "difficult to break my promise"; PRO/CO/730/39, April 19, 1923, Cox to bin Saud, "change your mind."

6. PRO/L/PS/10, vol. 989, May 1, 1923, Mackie of Anglo-Persian to bin Saud, "no use proceeding."

7. Rihani Papers, August 1, 1923, Holmes in London to Rihani, "a very promising area" and November 22, 1923, Holmes in London to Rihani, "would hold 40% of the company"; the original concession document is among Frank Holmes Papers in Peter Mort Collection, UK; it is written between "Abdul Aziz bin Abdul Rahman bin Faisal bin Saud, Sultan of Nejd and its Dependencies and Major Frank Holmes of 20 Cecil Street SW London in his capacity as the true and lawful attorney of the E&GSynd."

8. Archibald H. T. Chisholm, *The First Kuwait Oil Concession Agreement: A Record of the Negotiations, 1911–1934* (London: F. Cass, 1975), p. 98; May 13, 1923, Holmes in Bahrain to E&GSynd, "to secure a very strong hold."

9. Cox's last official act was the signing of the Anglo-Iraq Treaty on April 20, 1923, although Sir Henry Dobbs had already been in Baghdad for several weeks and had taken on much of the work. Sir Percy Cox departed Baghdad on May 1, 1923.

10. PRO/L/PS/10, vol. 989, May 21, 1923, Abdul Aziz bin Saud to Sir Henry Dobbs, high commissioner for Iraq, "acted and behaved in strict accordance."

11. PRO/CO/730/54, May 24, 1923, E&GSynd to undersecretary Colonial Office, "re Mr. Janson's interview (your office) today . . . as a breach of faith . . . does not emanate from the Syndicate"; curiously, Rihani Personal Papers (June 3, 1924), Holmes to Rihani contains an intriguing reference speaking of matters in Assir, "I would much like to see Abed with you *He could be told that we have not only American money but American directors on the board of the Eastern & General*" (my italics).

12. PRO/L/PS/10, vol. 989, October 1923, colonial secretary, duke of Devonshire to acting political resident (Knox), "have not had as much experience."

13. Gary Troeller, *The Birth of Saudi Arabia: Britain and the Rise of the House of Sa'ud* (London: F. Cass, 1976), pp. 196–98, "the Colonial Office acceded to the importunities of India . . . bin Saud's request to be dealt with through the Foreign Office was rejected."

14. Philby Papers, box XVII-2, June 5, 1923, Wilson to Philby, "unless we were prepared to adopt Holmes's system" and August 18, 1923, Wilson to Philby, "a Jew with many connections in gold mining."

15. Rihani Papers, August 1, 1923, Holmes to Rihani, "Your interests are safeguarded . . . you will have your interests in this new company . . . I hope we do well, I feel certain we will"; Chisholm, *First Kuwait Oil Concession Agreement*, p. 5, cable, May 9, 1923, Holmes to Shaikh Ahmad of Kuwait, "without first seeing the terms"; IOL/R/15/2/96, vol. 1, May 22, 1923; cable, political agent Bahrain (Daly) to acting political resident (Knox), "endeavouring to get"; IOL/R/15/5/237, vol. 11, May 13, 1923, cable, acting resident (Knox) to political agent Kuwait (More), "we consider binds him and his successor"; and Kuwait agent reply of same date, "favour Anglo-Persian."

16. Muhammad G. Rumaihi, *Bahrain: Social and Political Change Since the First World War* (London: Bowker, 1976), p. 177, citing January 1923 letters from political agent to resident, "except under pressure" and resident to Government of India; IOL/R/15/1/618-F52, December 22, 1921, Wilson to Cox, resident, political agents, "negotiations (with the Arab Shaikhs) should be conducted on behalf of the D'Arcy Exploration Company"; IOL/R/15/1/618/ F.52, April 16, 1923, J. B. Mackie in Bahrain to Arnold Wilson, "[Please tell me] by wire what attitude I should adopt in dealing with this subject. I understand that Gaskin sent samples of the bitumen and the surrounding rock some years ago and it may be you have other information which will give you an idea whether there is any likelihood of there being any oil in these islands." And April 30, 1923, confidential, Wilson to resident (Knox), "guarantee . . . will be omitted."

17. Ameen Rihani, *Around the Coasts of Arabia* (London: Constable, 1930; Delmar, NY: Caravan Books, 1930), p. 266, gives a wonderful description: "most of them (locals) men, women and children wade merrily across to the boats and then from the boats to the shore lifting their garments above their legs, even above their bosoms and wading they laugh and play splashing each other with water. The naturalness of the women, however, is spoiled by a tradition: they will wade like the men, but they will not uncover their faces" and "My Bahraini companion, seeing me hesitate with the camera said, 'it is something familiar, there is no harm in taking the picture.'"

18. IOL/R/15/2/96, vol. 1, May 22, 1923, cable, political agent Bahrain (Daly) to acting political resident (Knox), "Shaikh Abdulla informs me that Isa with his and Hamad's approval gave Holmes an oil concession on May 12th." Rihani Papers, August 1, 1923, Holmes to Rihani, "signed by Isa, Abdulla and Hamad" and "Isa asked to see bin Saud's concession"; Rihani Papers, August 1, 1923, Holmes to Rihani, "your two letters."

19. Administrative Report for 1923 (Daly), "manifest inability to rule" and "in spite of the opposition"; PRO/L/PS/10, vol. 989; memorandum, March 17, 1924, political agent Bahrain (Daly) to political resident (Trevor), "take the rule out of." See also Rumaihi, *Bahrain: Social and Political Change*, p. 178. For this incident see also Rumaihi, p. 177, citing letters from political agent to resident and resident to Government of India; Briton Cooper Busch, *Britain and the Persian Gulf, 1894–1921* (Berkeley: University of California Press, 1967), p. 139, "to tighten our hold on the place" and "independent of all Governments except our own"; IOL/R/15/1/341, May 11, 1923, Knox to Government of India, "arrest Abdulla Gosaibi . . . get rid of a Persian or two as well . . ."; Administrative Report Bahrain for the Year 1923 (Daly), "necessity" of deporting bin Saud's representative, "which bin Saud resented"; IOL/R/15/1/341, June 14, 1923, resident to bin Saud, "unflagging zeal."

20. Rumaihi, *Bahrain: Social and Political Change*, pp. 179–83, for detail of abdication and subsequent events. In the 1923 Administrative Report Daly described the Bahrain response as a "calumnious campaign ostensibly aimed at what was termed British

interference"; see also Administration Report of the Bahrain Political Agency for the Year 1922, "A campaign has been conducted in the Persian Press (by) a few malcontents in Bahrain who designate themselves the 'Nationalist Party.'" Rumaihi, *Bahrain: Social and Political Change*, pp. 179–83, "native state" and "impropriety"; p. 182, cites a December 1923 telegram, Government of India to resident, querying whether they had "gone too far"; Mahdi Abdalla Al Tajir, *Bahrain, 1920–1945: Britain, the Shaikh, and the Administration* (London: Croom Helm, 1987), p. 5, "extinguishing the Shaikh's sovereignty."

21. IOL/R/15/2/96, vol. 1, May 22, 1923, cable, political agent Bahrain (Daly) to acting political resident (Knox), "frightened and dropped the matter," "you were all out against him," "get satisfaction in London."

CHAPTER 6

RESOLUTE RULERS

I n Kuwait, despite Knox's best endeavors to stop him, Holmes was outlining for Shaikh Ahmad bin Jabir bin Mubarak Al Sabah his terms for a concession. He assured Ahmad he knew where the oil was hidden in Kuwait, had known since his first visit during the war, and would waste no time in developing it. Holmes was so certain of the viability of his project, so certain there was oil, so sure there would be excellent returns from the rich oil fields he visualized that he felt fully justified in making substantial investments in terms and conditions. His offer to the shaikh of Kuwait was generous. Holmes's terms were far and away more attractive than those of Anglo-Persian. Major More had let slip Anglo-Persian's terms in his concerted efforts to convince Ahmad to sign an agreement with that company in replacement of the 1913 convention.[1]

THE SHAIKHS OF KUWAIT

The Kuwait to which Holmes had now returned had become a walled city. In 1920, following a surprise April raid on Kuwait's tribes by Ikhwan Bedouin from across the border in Abdul Aziz bin Saud's territory, the then ruler Salim anticipated an attack on Kuwait Town and its eighty thousand population. Although bin Saud denied any involvement—he said the raid was carried out by an unruly group of zealous Ikhwan—Salim suspected bin Saud of coveting the caravan trade with Syria developed during the war and Kuwait's excellent port. Men, women, and children of Kuwait responded to Salim's order and constructed a wall of limestone, rock, and mud around Kuwait Town, completed in just eight weeks. The wall was a five-mile circle with its two ends on the water's edge of the Gulf. It was five feet thick and ten feet high with turrets, three gates, and a running path for riflemen.

A concerted Ikhwan attack did come in October. It was met at the town of Jahra by Salim with some of his tribal forces and armed citizens. They suffered catastrophic losses and retreated to the nearby Red Fort where a siege ensued. The British finally intervened. Nevertheless, Salim's strategy had succeeded, and the Ikhwan failed to breach the wall and invade the town of Kuwait. Ahmad, accompanied by the son of the shaikh of Muhammara, traveled to meet Abdul Aziz bin Saud in Riyadh in March 1921 to ask him to rein in his tribesmen. They had just arrived when news reached them that Salim, Ahmad's uncle, had passed away. Ahmad and Abdul Aziz agreed the matters they were intending to discuss were a nonissue now that Salim was dead. With bin Saud's best wishes, Ahmad returned home and was chosen by the members of his family, the Al Sabah, as the tenth ruler of Kuwait.

Ahmad bin Jabir Al Sabah was thirty-eight years old when Frank Holmes met him. He was strikingly handsome, of medium height, soft-spoken, elegant in his gestures, and meticulous in his dress. Following the tradition begun by his grandfather, Mubarak, Shaikh Ahmad conducted most of his affairs in public and held an open *majlis* in the town square every day. Any person could approach him, on any matter, as he sat surrounded by his Council of State, officials, and relatives. It was here that he received visitors, including foreigners, in this way endorsing the visitor's status as an honored guest and ensuring the traveler free movement throughout his domain.

The democratic attitude of the Kuwaitis extended to their bazaar where the Bedouin, mainly from inland Arabia, were the majority customers. Each summer they would purchase food and clothing at the Kuwait market and return the year after to pay for what they had bought. Occasionally they paid in cash, more often in kind by giving sheep or camels. Some were unable to settle their accounts. The Kuwait merchants did not press. They waited two, three, even more years. They were rarely let down. The Kuwaitis spoke of instances when children or relatives came to pay a debt, perhaps years after the death of a purchaser. This homespun form of credit ensured that the nomadic tribes favored the Kuwait market over those of Basra, Baghdad, or Bahrain.

As Ahmad's uncle had suspected, bin Saud did indeed covet the customs charges earned by Kuwait on goods that passed into his territory. Bin Saud's goal was to gain these tariffs by forcing the people to use his own ports, which, he believed, would then develop to rival the long-established entrepôt trade of Kuwait and Bahrain. Despite his best wishes expressed to Ahmad as the new ruler of Kuwait, bin Saud imposed a fierce trade boycott against Kuwait enforced, though never admittedly, by bin Saud's Ikhwan tribesmen. The Ujair Conference followed almost on the heels of bin Saud's boycott. When Holmes met Ahmad, Kuwait was struggling under a combination of bin Saud's boycott and the loss of 70 percent of its revenue in the form of taxes traditionally collected from the tribes in the territory now excised by Percy Cox.[2]

After Shaikh Ahmad listened to Holmes's proposal and discussed it with his advisers, Holmes was assured of Ahmad's intention to sign the Kuwait concession as proposed. Holmes was confident that, once outside the Gulf, he would not have difficulties dealing with British officials. As he told Ahmad he was sure it was only the officers of the Government of India who were blocking anybody, other than the Anglo-Persian Oil Company, operating in their area of authority.

Holmes departed for London via Baghdad. He was anxious to organize a geological team to check his findings on oil fields in Al Hasa, Bahrain, and Kuwait, where he now held, or almost held, concessions. He also needed to be in London to complete the registration formalities for the new companies that would be formed to carry out exploration and development in Al Hasa, Bahrain, and, he expected, Kuwait. Muhammad Yateem remained with Holmes's agreement in readiness for Shaikh Ahmad to sign as soon as he was officially permitted to do so.

KNOX AND WILSON COLLUDE

Arnold Wilson moved a lot faster this time, spurred on by the agitation of the officers of the Government of India and the Anglo-Persian Oil Company. He arrived in Kuwait on May 31, only days after Holmes's departure. He was joined by acting political resident Knox who, "in the face of Major Holmes's active methods," saw himself on a mission for which he was determined to "interpret as literally as possible my obligation to assist the Anglo-Persian Oil Company."

To help persuade the shaikh of Kuwait, Wilson brought with him the partner of Ahmad's own legal adviser, a Basra-based lawyer who handled Anglo-Persian affairs. He also brought along Shaikh Khazaal of Muhammerah. Khazaal had moved a long way from the days when he had joined Mubarak of Kuwait and Saiyid Talib of Baghdad in establishing the Reform Committee aimed at achieving independence in Mesopotamia and the Arabian Peninsula. Since the war, Khazaal had taken to praising the Anglo-Persian refinery constructed in his territory at Abadan. In doing so, he had lost the respect of his fellow Arabs who now described him as "a tool in Anglo-Persian's hands."

On June 2, joined by the Kuwait political agent, Major More, the powerful group of five called on Shaikh Ahmad who had only three months before become the ruler of Kuwait following the untimely death of Shaikh Salim.[3]

Despite his long career in the area, acting resident Knox was peculiarly insensitive to the atmosphere prevailing in the Arab shaikhdoms. He was oblivious to Kuwait's bitter reaction to Cox's excision of two-thirds of its territory at the Ujair Conference. He had no concept of the depth of bin Saud's anger at being forced back under the authority of the Government of India after Cox's retirement, nor of the humiliation at the arbitrary deportation of his agent, Abdulla Gosaibi, from

Bahrain. Nor did Knox appreciate the seething resentment in Bahrain at the removal of their ruler, in which he personally played the prominent role. While he and every other political officer viewed Anglo-Persian as an extension of the power of the Government of India, he could not recognize that the Arabs viewed it in exactly the same way—but to them, this was a good reason for avoiding it.

Knox was totally unable to comprehend the far-reaching effects on the Arabs of actions and policies of the Government of India. Neither he nor his fellow officers saw that the shaikhs' uncooperative spirit—disobedience, as some political officers termed it—had its roots in their own conduct. In apparent sincerity, Knox advised the Government of India: "Hostility to Anglo-Persian has only manifested itself seriously since Holmes arrived on the spot and it is difficult to resist the conclusion that it is chiefly due to him and his agent Muhammad Yateem."

As Lieutenant Colonel Trevor had pointed out to the Kuwait political agent, and as Knox well knew, Arnold Wilson was authorized to obtain in Kuwait "an agreement to sign, ultimately, a lease." Knox dismissed this as sheer nonsense. Incredibly, he saw the powerful Anglo-Persian Oil Company as the victim. He believed his role was to protect the company from the shaikhs, not the other way around.

"The only purpose of an 'agreement to sign,'" must be in order to "safeguard Anglo-Persian in the matter of profits that the shaikh might endeavour to force up later in the event of oil being found in abundance," he declared. Knox and Wilson were in agreement about the "damaging nature of Major Holmes activities." Knox told the Government of India: "The arrival of Frank Holmes and his E&GSynd as an energetic and open rival to Anglo-Persian in Arabia" had resulted in the "general undermining of Anglo-Persian's position."

As it was not the shaikh of Kuwait but Anglo-Persian that needed protection, Knox reported, "Sir Arnold Wilson's desire to get the position of his company defined once and for all and without more delay seemed to me entirely reasonable." Because of this, he wrote the duke of Devonshire, he had "literally interpreted" his instructions and joined with Arnold Wilson in submitting to Shaikh Ahmad a document that "differs considerably" from that which he was "originally directed" to help Anglo-Persian negotiate. Trevor had anticipated this move by Wilson and had advised Major More to "protest vigorously." But More remained silent as Wilson and Knox presented what they said was an "authorised agreement" to the shaikh of Kuwait that combined "exploration license, prospecting license, and a definite lease for sixty years."[4]

Shaikh Ahmad of Kuwait did not commit himself during the meeting with the powerful group of five, nor would he sign anything. Nevertheless, Knox believed he had extracted a verbal promise "in satisfactorily clear terms" that contained "the required assurance that the shaikh would not enter into a concession with any company other than Anglo-Persian." Although he personally would have preferred the shaikh to sign immediately, he said, Wilson had declared himself satisfied. Acting resident Knox was certain the meeting with

Shaikh Ahmad would bring the desired result of "the disappearance of Major Holmes from the scene."

Almost in passing, he reported: "For their part Anglo-Persian are in no hurry to exploit the shaikh's territory; their chief anxiety is to be certain that no other party will be able to obtain a footing there." Triumphantly, Knox declared, "in this we have succeeded!!!" (The copy of this report circulated at the Colonial Office has a handwritten, unsigned note in the margin: "This may be very unfair to the Shaikh, who presumably wants revenue?")

KNOX REPRIMANDS HAMAD

The Government of India was thrilled with Knox's report. They told him to hammer the point to the shaikh of Kuwait that "any oil concession granted to Holmes will not be confirmed" as his syndicate "is not a firm approved by Government." He was instructed to follow through on his success by ignoring the contract signed between Holmes and the shaikhs of Bahrain and "make every effort to induce Shaikh Hamad of Bahrain to conclude an agreement with Anglo-Persian."

And while he was on a winning streak, the Government of India told Knox, he should "similarly endeavour to persuade the Sultan of Nejd" to withdraw from his contract with Holmes. In the very least he should obtain the immediate cancellation of the offending Anglo-Persian exclusion clause.

Less than a week after Wilson revealed Anglo-Persian's true intentions—or nonintentions—in Kuwait, Knox was again clearly told that Anglo-Persian had no interest whatsoever in sinking any effort, or money, into the Arab shaikhdoms, which they considered geologically worthless. Major Daly, the political agent in Bahrain, sent him a personal cable: "Mackie yesterday told me, in confidence, that Anglo-Persian have no desire to work a concession in Bahrain . . . their object would be secured if they could block other companies from getting a concession here." At least momentarily, Daly appears to have had some sympathy for the Bahrainis as he added, "This seems to be a pity from the point of view of Bahrain development and the shaikh's interests."

Knox was singularly untouched and replied irritably: "Please take action on instructions. At the same time inform Shaikh Hamad that Anglo-Persian are a firm approved by High Government and give Mackie such assistance in obtaining the concession as he may properly ask from you by using your influence with Shaikh Hamad in Bahrain."[5]

Knox penned a letter designed to put Hamad firmly in his place. He commanded: "Inform that gentleman (Holmes) in writing that you already consider that agreement void and of no effect, since you have received an official intimation from the representative of the High Government in these waters that it is not approved." He told Hamad that "even if Holmes has concluded an agreement with the Sultan of Nejd . . . what may suit the High Government in regard to bin Saud may not suit them in regard to Bahrain." Knox ordered Hamad to furnish him with "a copy of your written communication to Major Holmes." Pompously, Knox warned the new ruler of Bahrain: "I trust this letter will impose caution on yourself for the future."

Unexpectedly for Knox, Hamad held his ground. He answered that the terms and conditions offered by Anglo-Persian were not "considered satisfactory by me or by our subjects" and "we have done nothing opposed to our agreement with Government." The Bahrain political agent reported Hamad as vowing, "if Government cancelled the present concession, he preferred to give none at all, unless he were allowed to deal with some company other than Anglo-Persian." Daly was now getting anxious about his own position in Bahrain. He told Knox: "Hamad appealed to my friendship to ask Government not to force him to sign with Anglo-Persian. It is evident that they will not yield short of compulsion. Such action would be unfortunate for my relations with the shaikhs that are now very satisfactory. I have done all I can to persuade them."

Major Daly appeared to be wilting. Knox took up the cudgels and again wrote to Hamad, this time adopting the high moral ground. He said he was "surprised to notice in yourself and others that while you are scrupulously mindful to observe your promises to a man like Major Holmes you appear to be very unmindful and neglectful of the promise made by your respected father to the High Government."

His reference was to the 1913 convention and Shaikh Isa's letter stating: "I do hereby repeat to you that if there is any prospect of obtaining kerosene oil in my territory, I will not embark on the exploitation of it myself and will not entertain overtures from any quarter regarding it without consulting the Political Agent in Bahrain and without the approval of the High Government."

Knox charged that because Hamad and his brother Abdulla had participated in the negotiations they were "equally guilty" along with their father of "breaking the promise." He ordered the ruler of Bahrain to "make the only reparation in your power" and inform Holmes that "owing to a previous promise made to Government, which for the moment you forgot, you must ask him to consider all his negotiations null and void."[6]

KUWAIT LETTER OF INTENT

For his part, Shaikh Ahmad waited three weeks after the visit of Knox and Wilson and then called a meeting of his Council of State and put both offers before it.

For a seventy-year concession Holmes's initial up-front payment was two thousand sterling against Anglo-Persian's seven hundred fifty. Under the somewhat loose category of "protection" Holmes's minimum annual payment was three thousand sterling against Anglo-Persian's twenty-three hundred. Moreover, although this amount was designated as "protection," under Holmes's terms salaries of all men provided by the shaikh were to be paid by E&GSynd, including guards, and the shaikh would not be held responsible for raids, forays, or attacks.

There would be a 1 percent export duty paid on all oil leaving the country and customs duty paid on everything E&GSynd brought into the country, except machinery. From Holmes the shaikh would receive 20 percent of the net profits and the right to subscribe to a further 20 percent of the capital, the same as in bin Saud's agreement. Anglo-Persian would not countenance paying either customs duty or tax, and certainly would not consider profit sharing of any description. Anglo-Persian payment would be strictly on the basis of royalty, as paid to Persia.

It did not take Ahmad and his council long to decide in favor of Holmes and his Eastern & General Syndicate. Even Knox later conceded that Holmes's terms were "so much more favourable" than anything ever put forward by Anglo-Persian. The next day Shaikh Ahmad wrote to the Kuwait political agent:

> I beg to inform you that Major Frank Holmes, the accredited agent of E&GSynd, has come to Kuwait and has submitted to me a draft concession for mineral oil in Kuwait territory which, on careful consideration, I find to be beneficent and profitable.
>
> What I particularly like about it is that it is a British company and that it undertakes not to sell the concession to any but British companies. In addition to this, I propose inserting a clause to the effect that this concession will only come into force if approved by His Majesty's government and that, if Government do not approve, it becomes null and void.
>
> I feel confident that if I grant this concession to the company in question, His Majesty's government will raise no objection as my friend Shaikh Abdul Aziz, Sultan of Nejd, has granted them a similar concession and His Excellency Sir Percy Cox, who was present at the time, raised no objection. My friend the Ruler of Bahrain has given them a similar concession and there was no objection to this either. I hope therefore from the justice of Government that this will also be approved.

Major More forwarded this letter to Knox with the comment: "Holmes is said to have assured the shaikh there will be no question whatever of the High Government not approving the proposed agreement, as this is a question which will be settled in London." He reported that Holmes had told Shaikh Ahmad: "It is only the Political Officers of the Government of India who are trying to 'boom' Anglo-Persian for reasons of personal friendship etc and they will go back on what they have said when taxed with it from London as Sir Percy Cox did in the case of bin Saud." More felt sure the selection of Holmes's offer was "not looked upon with favour by the majority of the leading merchants and other influential men in the town." Furthermore, he reassured Knox: "Shaikh Ahmad is a man of no strength of character and I do not think it at all likely that he will try and take his own way in opposition to both public opinion and your advice, if, as I presume will be the case, you still advise him to deal only with the D'Arcy Exploration Co."[7]

KNOX REPRIMANDS AHMAD

When he received the ruler of Kuwait's letter, with More's comments, Knox immediately addressed a "stiff remonstrance" to Shaikh Ahmad, who noted in reply: "I of course recognise Government as the foundation of my welfare and that of my country to whom obedience is incumbent on me and I indeed strive to render it that I may enjoy lasting favours." Nevertheless, like Hamad in Bahrain, Shaikh Ahmad stood by his decision and declared that he regarded his intended acceptance of the Holmes's agreement "in no way contrary" to the 1913 convention signed by his grandfather.

In the thick of it now, Knox reported to the Government of India that Shaikh Hamad's attitude was "hostile" and that of Shaikh Ahmad not much better. He was, he said, "awaiting a visit from Arnold Wilson before taking any further definite action" on Bahrain.

Obviously concerned about the wobbling Major Daly, Knox said he would be instructed to "stand aside." The matter of Kuwait, he advised, "will be taken up with Sir Arnold Wilson." Knox added: "I hesitate to approach the Shaikh of Kuwait before consulting Sir Arnold Wilson as it is highly probable that the Shaikh of Muhammerah may at this moment be persuading the Shaikh of Kuwait to come to an agreement with the D'Arcy Exploration Company."[8]

KNOX REPRIMANDS BIN SAUD

While waiting for Wilson, Knox turned his attention to the third recalcitrant, Abdul Aziz bin Saud of Nejd. He drafted a long paternalistic letter and forwarded this to the Colonial Office for authorization. He began by sharply reminding bin Saud that he was obliged to conduct all business through the Government of India's political resident "in accordance with the arrangement lately come to by my government."

Next he told bin Saud: "Your Highness, of course stands on an entirely different plane from the Shaikhs of Bahrain and Kuwait over whom, for long years, in their interests, His Majesty's government have established a protectorate." In a contorted display of bureaucratese considering the reminder just given, he continued, "Naturally we do not deal in the same way with Your Highness, who is an independent sovereign, as we do with these two shaikhs." Knox declared himself "sorry to say" that both these shaikhs had "undoubtedly broken their promise and have entered into negotiations with Major Holmes who is not in any way a person approved by the High Government."

Knox was verging on the hysterical, if not the libelous, in this draft, as he continued:

> If in the future, Holmes or his company should deal treacherously, dishonestly or oppressively with either of the shaikhs concerned, Government have no hold on either the man himself or on his company and can in no way bring pressure to bear on him to act honestly and rightly. It is therefore not in the least surprising that both Your Highness and the Sheikhs of Bahrain and Kuwait should have found Major Holmes much more amenable and complaisant and desirous to promise all that Your Highness and these two shaikhs may require of him.
>
> But it is not necessary to remind Your Highness, who is thoroughly well versed in the affairs of the world, that it is a very long way from promise to performance. The Eastern & General Syndicate have no experience of working oil and it is quite likely that they will find it very unprofitable to attempt to look for oil and exploit oil, if found in Your Highness' dominions . . . oil exploration and exploitation in the field, even of Eastern Arabia, is a restricted one and there is no room for two British companies to work there.
>
> The probabilities are that the weaker, the Eastern & General Syndicate, will go to the wall in a conflict between the two companies. . . . Their only hope of working successfully is to sell a portion of the concession to the more powerful company, Anglo-Persian, and get them on their side.

Sending this draft to Victor Cavendish, Duke of Devonshire, at the Colonial Office, Knox said he intended dispatching it "after consulting Sir Arnold Wilson in one week's time." He had a plan, Knox confided to the duke. "I think I can force bin Saud into the open," he said, and "get him, in a fit of pique, to cancel the concession. Kuwait and Bahrain will at once desert E&GSynd and come into line." The two shaikhs were also on the receiving end of a strong line from Knox as part of this "Plan B," he said. Knox explained his alternative as "if we can secure Bahrain and Kuwait, Nejd in the end may follow."

The committed Knox may have had his private doubts about Anglo-Persian. He confided to Major Daly, "Wilson may, or may not, be correct when he says that it is quite impossible for the E&GSynd to work the concession they have obtained from bin Saud without the aid of Anglo-Persian."

Nevertheless, Knox said, Holmes and his principals could have "no reasonable ground for complaint whatever if they find that His Majesty's Servants are up against them" because they had not first obtained permission to approach the shaikhs. Knox was unable to explain "the conduct of Sir Percy Cox. He is big enough to defend himself and it seems obvious that he must have had sources of information which are not available to me. But then, on this supposition, I am entirely at a loss to explain why the Secretary of State for the Colonies, the Duke of Devonshire, has let Cox down."

A DANGEROUS PRECEDENT

Knox thought it a "most dangerous" proposal to allow the shaikhs to deal with whom they wished, even if the companies were British. He could "conceive of no course of conduct that could be more distasteful to the British government." Besides, he said, even if there was oil, it "will not run away and the concession is quite likely to be a more valuable matter to the shaikhs five years hence, say, than it is now. Why hurry?"

Even though he was well aware that Anglo-Persian did not want the Arab concessions, and even though he admitted that Holmes's terms were "so much more favourable" than anything ever put forward by Anglo-Persian, Knox had no hesitation in castigating Holmes, who, he said, "has acted unscrupulously and deserves no consideration." Sanctimoniously, Knox declared that, presumably unlike the men of the Government of India or the personnel of Anglo-Persian, Holmes was furthering "his own selfish interests" and in doing so had "done a great deal of harm here." Nevertheless, he reassured his superiors, "it is very unlikely that, in view of our opposition, the agreement between Nejd and Holmes will come to anything."

Replying to Knox's request for prompt instructions, the duke of Devonshire thought it prudent not to mention the fact that the British government was the majority shareholder in the D'Arcy Exploration Company through its holding in Anglo-Persian, "unless specifically referred to by bin Saud." If this was raised, he said, it should be made clear that "HMG do not interfere in the policy of the company and that company has no political significance whatever."

The Duke considered it perfectly acceptable to imply that bin Saud may have been coerced by Holmes into excluding Anglo-Persian, but not to state, as Knox had, that in doing so Holmes had "exceeded his instructions." Advising Knox to again stress that "unlike D'Arcy Exploration, the E&GSynd have no experience of oil development," the duke of Devonshire gave Knox virtual carte blanche. He gave him permission to take "whatever steps you consider advisable and justifiable to secure cancellation of Nejd concession, or failing that, of condition excluding Anglo-Persian, and to induce Kuwait and Bahrain to come into line."[9]

But before Knox could launch either his plan A or B, Arnold Wilson succumbed to cold feet. The matter of bin Saud now involved not just the Government of India but also the Colonial Office. Knowing Anglo-Persian's true lack of interest in the area, Wilson appears to have hesitated when the duke of Devonshire himself became involved. Acting resident Knox had to back down and again write to the duke saying he had "discussed the whole matter today with Sir Arnold Wilson." He had been informed, Knox said, "that Wilson is not very anxious that we should approach bin Saud with any haste as he is not quite certain what the attitude of his company will be in the event of bin Saud suddenly cancelling Holmes's lease altogether."

Knox could no longer ignore the fact that, no matter how much the political officers wanted it, even if Holmes's concession were cancelled Anglo-Persian did not seem ready to step in and conclude its own arrangements with bin Saud. But Knox was not giving up altogether. "In the meantime," he suggested to the duke, he would send the letter anyway.

For Kuwait—"on an entirely different plane" from Nejd—Knox and Wilson simply ignored the letters written by Shaikh Ahmad. They instructed Major More to speak to the shaikh of Kuwait. More, like his counterpart in Bahrain was also beginning to worry about his own position vis-à-vis the ruler. He told Shaikh Ahmad he had just received a cable from Wilson asking when Ahmad could be expected "in Abadan to discuss the agreement further." Shaikh Ahmad, far from being a man of "no strength of character," as More had described him, was getting very annoyed indeed. He replied firmly to the advice that he should conclude an agreement with Anglo-Persian by saying he had already written to "my friend" Arnold Wilson telling him it was impossible to conclude an agreement on "those" lines and "definitely rejecting his agreement."

To the surprised embarrassment of More, Shaikh Ahmad produced a copy of his letter written to Wilson three weeks previously, on June 23, 1923, the day after he and his Council of State had met to consider both offers. It read: "In answer to your letter of June 3rd sent me on your departure from Kuwait, I have the honour to state that the oil concession which you offered me has been, after deliberations, found unsuitable to our private and public interests and I regret to inform you of our rejecting it."[10]

LONDON, AUGUST 1923

When Holmes arrived in London at the end of July 1923, he had a concession signed from bin Saud and another signed by the Bahrainis together with the Kuwaitis' firm intention to conclude an agreement for their own territory and to share with bin Saud in the Neutral Zone. On arrival he found an encouraging letter from Rihani in Beirut: "The news of your success has been coming to me from all quarters—from Baghdad, Kuwait, Bahrain. . . . The Kuwait concession will yet be signed . . . Shaikh Ahmad wrote me about it."

Holmes set to work in London. He reported to Rihani, "We begin work in November of this year at the Al Hasa Concession. We will begin at the same time in Bahrain." He explained that a new company, with a capital of three hundred thousand sterling, had been formed to explore the Al Hasa concession. "This amount is ample money" for the Al Hasa exploration work, Holmes said. To "explore and bore" the Bahrain concession, the capital of E&GSynd had been increased from fifty thousand sterling to two hundred fifty thousand. "We have more than 200,000 sterling now available in E&GSynd" with which to begin work on the Bahrain concession, Holmes told Rihani. He planned to leave London in mid-September, taking with him "an expert staff of geologists" to undertake the required preliminary work on the concessions, prior to beginning actual operations in November.

As Holmes expected to be in the Gulf for quite some time managing the development of the concessions, his wife, Dorothy, would set up a home for them in Bahrain. The mood was so celebratory that they were planning to make it a family occasion. "Captain J.F Eccles Holmes 3-16th Punjab Regiment, has applied for permission to accompany his father, Major Holmes, to Al Hasa to visit oil concessions. He states that both he and his father have been invited by bin Saud. Proposed duration of visit two months. What reply should be given?" the viceroy of India cabled the secretary of state for India. For a lowly captain in the Indian army, Holmes's son was arousing a lot of high-level attention. From the viceroy of India to the secretary of state to India, Eccles's request was passed to the Colonial Office. They quickly dismissed the matter as of no consequence, informing the Government of India that "no objection is seen to the visit of Mr. Eccles Holmes."[11]

Holmes was still planning to "see about the Neutral Zone Concession and get this signed by both Sultan bin Saud and the Shaikh of Kuwait." He was also attending to matters related to his negotiations in Assir and Yemen including the Salif salt concession. Holmes was pleased with his visit to the Colonial Office, where he discovered the officials "rather amused" to find that "Wilson did not have it all his own way." He told Rihani, "Here in England they did not think anyone could upset Anglo-Persian as they were told the company was much liked out there."

Anglo-Persian's London management may have been thinking along similar lines as Arnold Wilson was forced to defend himself in a letter to his managing director. Referring to himself in the third person—and adopting the royal "We"—Wilson wrote: "We venture to point out that there seems little reason to think that the personality and past career of your General Manager (Wilson himself) or the Government shareholding have been factors of importance in deciding the attitude of the Shaikh of Kuwait."

Wilson claimed he had seen Ahmad in 1922 when the shaikh had "expressed anxiety to come to an agreement as soon as possible." He laid blame on the London management, saying at that time he had been told "it was not desired that we should take active steps in the matter of the Kuwait agreement." He also criticized acting resident Knox, saying his opinion had been to "sit tight and do nothing for some time," presumably until Holmes had been forced out of the region as Knox expected. In Arnold Wilson's view the fault was entirely Frank Holmes's. The shaikhs were refusing to cooperate with Anglo-Persian, Wilson told his London headquarters, because "the market has been spoilt by Holmes's proposals."[12]

CORRECT CHANNELS

Holmes did not expect to have any problems complying with the Colonial Office request that "for political reasons" the syndicate should now apply "as they wish us to do (i.e., through official channels)." This was a mere formality, Holmes told Rihani, because "the Government have told us they will support us if we make an application through them." The process of requesting approval for the Bahrain concession was already under way. The Colonial Office had told him that, under pressure from Anglo-Persian, they had recently written to Kuwait stating that the British government "do not approve of the E&GSynd *application* and that they will not ratify any agreement" concluded on the basis of this particular application.

Holmes said the Colonial Office stressed "they had made it clear that it is *the application and not E&GSynd* that is not approved" [*sic*]. Percy Cox was also supportive of Holmes. "He talks most openly with me. I wish I could send you one or two of his letters to me so that you could gauge our relationship," Holmes wrote Rihani. Cox had been suspicious of Anglo-Persian's sincerity since Arnold Wilson had embarrassed him by abruptly calling off negotiations with bin Saud.

Holmes was now the frequent lunch companion of Sir Percy and Lady Cox. He said Cox had told him "many times" that he was "extremely glad my people have got the Al Hasa Concession as it is his desire that his friend bin Saud gets as fair a deal as possible and that his people secure in full measure the benefits of oil and mineral developments in their country."

Holmes added brightly, "I am only waiting for the reply of His Highness the Sultan of Nejd before closing up our new Company for Al Hasa." He also had another concession on the go in South Arabia as he reported to Rihani: "As regards the oil concession from Abdul Karim Fadhil, the Sultan of Lahaj, the Colonial Office are urging us to conclude with him at once as they have agreed our terms with the Sultan." Holmes was confident: "HMG is not in any way opposed to our Company, or Group. In fact we are, at the moment, negotiating with the British government to bring off a very big deal." He confided that "The British government are really fed up with the Anglo-Persian Oil Company and the company does not get all the support the people in the Persian Gulf might think."[13]

The message the Government of India's political officers, and Anglo-Persian personnel, were spreading throughout the Gulf, particularly since Holmes's departure, was something very different. There the Colonial Office statement had been distorted to become an allegation that Holmes himself was "not approved" by the British government and that E&GSynd was being condemned as a company of which "His Majesty's government do not approve."

E&GSYND FIGHT BACK

Word of this soon got back to the city where it caused quite a buzz in the business and financial circles of London. Hearing the rumors from the Persian Gulf, an affronted Edmund Davis and his directors in E&GSynd addressed a tart letter to the Colonial Office protesting the harm to their investments and business that could result from such slurs on their reputations. They wrote the duke of Devonshire they felt sure "Your Grace did not wish such an impression to be conveyed." Therefore, "as any such wrong impression may act in a most detrimental manner upon our syndicate," they "respectfully requested that you will kindly take such measures as you deem necessary" to correct the situation. As the duke of Devonshire well knew, Edmund Davis, chairman of the Eastern & General Syndicate and involved in some eighty-two companies and syndicates—he would later be knighted for service to British industry—had to be taken seriously.

There was a very real possibility that Davis and his colleagues might turn to the national press and accuse the current government of running a smear campaign aimed at protecting its own investment in Anglo-Persian and secretly assisting to block competition. Both Anglo-Persian and E&GSynd had advocated their cases to the various government departments. Anglo-Persian claimed they had followed the correct procedures, which E&GSynd, by going directly to the shaikhs, had not. In requesting approval for their transactions, E&GSynd pointed to the superior merit of the terms and conditions they offered.

The official letter of complaint from Davis and his directors brought things to a head. The duke of Devonshire, secretary of state at the Colonial Office, was called upon to make an adjudication.[14]

The duke wrote to Lieutenant Colonel Knox, still acting resident in the Gulf, and clarified the earlier statement. He sought to mollify the political officers and the Government of India by assuring them that the mandatory requirement to obtain permission before approaching any shaikh in the Persian Gulf still stood. He reaffirmed the Colonial Office's adherence to the protocol reached at Churchill's Cairo Conference that any communication with, or about, the shaikhs of the Arabian Peninsula must be solely through the political resident.

He then went on to say that Frank Holmes "avowedly and probably actually" had not known he needed government permission before he could speak to any shaikh in the Persian Gulf. The duke then took some pains to emphasize it was only "this irregular manner" in which E&GSynd's business offers were made that was not "approved." He stressed there was no objection to Holmes personally, nor to E&GSynd, which, the duke now declared, was "a substantial and reputable firm."

But the duke was enough of an imperialist to "feel very strongly" that the shaikhs should "not be allowed to disregard their solemn undertakings with impunity." Nor could a precedent be established by allowing "unapproved firms" to "secure concessions the terms of which had not previously been submitted to His Majesty's government."

Unlike Knox, the duke of Devonshire saw his duty as protecting the shaikhs, not the Anglo-Persian Oil Company. He thought Anglo-Persian's offer to the shaikh of Kuwait, as submitted by Knox and Wilson, left "much to be desired." In all fairness, he could not, he said, inform the shaikh that his rights and interests would be adequately protected under Knox and Wilson's proposals. The duke did not like E&GSynd's idea of introducing the shaikhs to *realbusiness* through becoming shareholders in the companies formed to exploit their resources. He preferred the old style of royalty based on export. But, he said, E&GSynd had "volunteered" to modify their contracts "in such a manner" to conform to his wishes. He told acting resident Knox that once he was "satisfied" the syndicate was "in a position and intend to carry out to the full" the obligations toward the shaikhs as they proposed, he could not "feel any strong objection to their candidature."

Nevertheless, the duke now rewarded Anglo-Persian by decreeing that, in view of their "prior application" and correct procedure, and if they were prepared to offer terms at least as favorable as those submitted by Holmes, he would withhold recognition of the E&GSynd agreements with the shaikhs of Bahrain and Kuwait.

He loaded the dice in favor of Anglo-Persian by directing that refusal of the shaikhs to sign with Anglo-Persian would only be considered if they attached a detailed list of their technical objections in order to allow Anglo-Persian the chance "to modify accordingly." If, after all this, Anglo-Persian could not agree with the shaikhs, then the duke would approve E&GSynd's contracts, suitably recast to his satisfaction, provided Holmes's terms were as favorable to the shaikhs as those rejected from Anglo-Persian.

The Colonial Office dispatched official letters of similar content to both Anglo-Persian and E&GSynd. Holmes added a postscript to his letter to Rihani: "Please write to the shaikhs and tell them not to give concessions until I see them. And inform them the British government will not in any way force them to give concessions to the Anglo-Persian or anyone else and that if the Anglo-Persian fail to obtain concessions from them within the next few months, the British government will support our application."

Having dealt with Bahrain and Kuwait, the duke of Devonshire next tackled the sultan of Nejd. He wrote Knox that he was now "not satisfied that it is either possible or desirable to maintain so strong a line of resistance to the claims of E&GSynd" as he was "first inclined to adopt." The duke had begun to suspect he may have been misled and was no longer convinced by the political officers' accusations. He told Knox he had learned that "Major Holmes has not, in fact, overstepped the bounds of legitimate commercial competition."

He adjudicated on the shaikh of Kuwait's request that he be sanctioned to act in concert with bin Saud, who had decided to award the concession in the Neutral Zone to Holmes. "In view of all the circumstances of the case," he wrote

Knox, "I do not feel disposed to stand in the way of the grant to E&GSynd by the Sultan of Nejd and the Shaikh of Kuwait jointly of a concession over the Neutral Zone." Nevertheless, the shaikh of Kuwait should be warned that he could not "draw the conclusion" that this implied anything positive in relation to E&GSynd and the Kuwait concession itself. He should be told that, in the duke of Devonshire's own opinion, the terms and conditions he personally had suggested to Anglo-Persian in replacement of those Knox and Wilson had tendered "are likely to be more favourable."

In "choosing the opportunity to reopen the matter" with bin Saud, the duke instructed Knox, he could say that "D'Arcy Exploration Company already possess the complete machinery necessary for refining and marketing the oil and that therefore it would be easier and lead to a quicker development of the territory if use could be made of this machinery to bring oil from Nejd on to the market."

But, obviously with Edmund Davis's letter in mind, the duke of Devonshire confided that the government's majority shareholding in Anglo-Persian "makes me anxious to avoid" any line of action that might be "represented" as a "championship" of that company "based on motives of direct financial interest." The duke said he had "reconsidered the whole situation" and had concluded "it is not desirable" for acting resident Knox to take any action that might "have the effect of causing the Sultan to cancel the Al Hasa concession" granted by him to Holmes. He had, he said, been informed that the syndicate were at present "engaged in negotiation with another company" for working the Nejd Concession," so cooperation with the D'Arcy Exploration Company may no longer be "a matter of immediate importance."[15]

BURMA OIL

The duke of Devonshire's cryptic reference to E&GSynd being "engaged in negotiation with another company" for working the Al Hasa concession did have a factual basis. Soon after Holmes's arrival back in England, the syndicate had held talks with the only company capable of facing down Anglo-Persian—Burma Oil.

Originally Burmah Oil, this was the first British oil company. Going into business after the 1885 annexation of Burma to India, the company took over the oil-gathering activities of Burmese village people on which it built a commercial industry and a refinery in Rangoon. Growth was greatly assisted by the monopoly granted in 1889 by the Government of India over the sale and extraction of oil products and protection from all foreign competitors throughout the Indian empire. Chairman Lord Strathcona financed Canadian Pacific Railways.

In Persia before locating a viable production, Englishman Knox D'Arcy needed financial assistance to keep him going and approached the British government.

The Admiralty, fearing the Persian concession might fall into the hands of American or Dutch oil trusts, persuaded Lord Strathcona and Burma Oil to back D'Arcy, and in May 1905 they formed Concessions Syndicate for operations in Persia. After D'Arcy's 1908 oil strike at Masjed Soleyman, Concessions Syndicate Ltd. was relegated to a subsidiary and the Anglo-Persian Oil Company formed. Burma Oil was the majority shareholder in Anglo-Persian until the Churchill-inspired British government purchase in 1914 of 51 percent of the stock. Burma Oil would continue to hold 23 percent of Anglo-Persian (later British Petroleum) until 1974 when the Bank of England would take over its shareholding.

By judiciously leaking a rumor that E&GSynd was in discussions with Burma Oil, Frank Holmes was undermining Anglo-Persian's power to monopolize the Persian Gulf in the eyes of many Whitehall officials. Holmes had dropped this little clanger during a friendly visit to one of his contacts at the Department of Overseas Trade, who had predictably forwarded the news to both the Colonial Office and the Foreign Office. As he had done when seeking supplies in the bazaars of Cairo and Alexandria during the war, Holmes had called in for a chat.

As he shrewdly figured, the ensuing "Personal & Confidential" note that quickly made its way to the Colonial Office and the Foreign Office reported on "the ruling fear of the Arabs, under Persian inspiration, is of economic penetration by HMG under the guise of Anglo-Persian in which, they well know, HMG has a controlling interest." It noted that "the personality of Sir Arnold Wilson is also cordially detested by the Gulf Arabs." Most importantly for Holmes, the confidential note revealed that he had "mentioned" the possibility that development of the Al Hasa concession might be financed by "20% of the shares held by bin Saud who cannot dispose of them until he has offered them to the syndicate, 41% to the Burma Oil Company and 39% to the Eastern & General Syndicate."[16]

THE NEJD OIL COMPANY

Regardless of whether or not it was ever seriously intended to bring Burma Oil into the Al Hasa concession, E&GSynd continued independently toward registering the new company and also finalizing arrangements for a team of geologists to travel to the Gulf. Rather than getting away in mid-September, Holmes was still in London in November waiting, as he wrote Rihani, for bin Saud to decide whether or not he would take up the additional shares.

Holmes explained to Rihani that if bin Saud chose to renounce his right to the additional 20 percent of shares, E&GSynd could sell them on the London market, "the proceeds of which will of course be utilised as working capital no matter whether proclaimed by the Sultan or the London people." Meanwhile, so that he could get started "at once" on the Al Hasa exploration work, the syndicate was now forming an exploration company with capital of "about sterling

100,000. But this Exploration Company will not have the right to remove ANY oil, either for sale or otherwise, from the Al Hasa territory."

Finally, Holmes and Dorothy left England on December 20. Rihani, who was waiting to meet them in Beirut, wrote Holmes that he had "been receiving news from Kuwait, Bahrain, Abdul Aziz bin Saud, from Aden and Assir for the past six months." Rihani hoped Holmes would be bringing "some good news from the Foreign Office about Yemen." The friendship between Holmes and Rihani was to be cemented by Rihani's younger brother, Albert, joining Holmes as the translator traveling with the geological party. Holmes had told Rihani: "Tell Albert he must learn how to use, fairly rapidly, a 'Remington' typewriter. I will arrange his salary with you when I leave London." If he proved to be a good worker, Holmes said, he would "hand Albert over to work with our new company in Al Hasa" after the geologists had completed their expedition.[17]

His Arab horses, gifts from bin Saud, had settled in well at Basra where Holmes had lodged them temporarily. "My mare had a fine colt foal on July 31st," Holmes told Rihani. "I have sent word to bin Saud. I enclose a photograph of the mare. I would like to see your mare. Is he not a great big-hearted man, His Highness the Sultan?" Holmes was so thrilled with his pure Arab horses that later in the year he bought a farm, MillHill, in Essex, to provide them a home.

As they were planning to visit Rihani at Frieke in the Lebanese mountains, on this trip to the Persian Gulf, Frank and Dorothy took a new route. They traveled from London to Paris and Marseilles and then by boat to Egypt. From Port Said they took a steamer overnight to the Palestinian port city of Haifa, then a train onto Beirut, the capital of Lebanon. A new service had opened providing road travel between Haifa, Beirut, Damascus, and Baghdad. The convoy departed Beirut on Thursdays for the Syrian capital of Damascus, left again early Friday morning, and arrived in Baghdad Saturday night. As Rihani remarked approvingly, the trip from London to Baghdad was now "mighty quick."

The duke of Devonshire, with his insistence on adhering to the old Colonial Model of concession with payment by royalty only, may have been surprised by the understanding of modern business displayed by the shaikhs of the Persian Gulf. As Holmes was departing London en route to the Persian Gulf, an advertisement appeared in the Arabic version of the *Times of Mesopotamia* regarding the Nejd Oil Company. It read:

> It is announced to the natives of Nejd resident in Iraq, India and the Persian Gulf that His Highness the Sultan of Nejd and its Dependencies has granted an oil concession, in the districts of Al Hasa and Katif an Ainain, to the Eastern & General Syndicate and, as the company is now issuing 300,000 shares of sterling pounds one each, it has been decided that one-fifth, that is 60,000 shares, shall be allotted to the Sultan personally and another one-fifth, that is 60,000, shall be allotted among his subjects resident in Nejd and its neighbourhood and Iraq, Syria, Hedjaz and all other countries.

And as the conditions of this concession are the most favourable conditions which up to date Governments have obtained, His Highness the Sultan is especially anxious that natives of Nejd should have full opportunity of joining in this important and useful enterprise. Notice is now given that, whoever wishes to take up the shares now offered, should forthwith register his name and pay cash to the Eastern Bank. We do not think that natives of Nejd, who are famous for their quickness to learn, and desiring to progress special advantages for their country, will abstain from subscribing. Nor do we think that they will be slow to take this opportunity. The remaining shares will be distributed shortly.

God is the giver of health and property.

Signed: Abdullah Sayed
Representative of His Highness the Sultan.

Arnold Wilson forwarded a translation to Lieutenant Colonel Trevor, now returned from six-month leave and back in his position, replacing Knox as political resident in the Persian Gulf. In high dudgeon Wilson commented scathingly:

I have ascertained that the Eastern Bank have only received one inquiry since the publication of this notice, and that only a casual one on the telephone. The manager has received no instructions with regard to receiving allotments. In fact, no communication whatsoever from the "New Company" and, until he does so, no application to the Bank will be entertained. I enclose three spare copies of this letter and enclosure, in case you may wish to send copies to Kuwait and Bahrain who may be interested. The publication of a notice of this sort in England, or India, would I believe render the company liable to prosecution![18]

NOTES

1. Archibald H. T. Chisholm, *The First Kuwait Oil Concession Agreement: A Record of the Negotiations, 1911–1934* (London: F. Cass, 1975), p. 5, contends Major More discussed Anglo-Persian's terms with the shaikh "without informing Anglo-Persian"; he does not take into account the resident's instructions to "talk up" the prospecting license proposal.

2. The wall around Kuwait Town stood until 1956 when it was dismantled to make way for new town planning. Kuwait's population was 120,000 including the nomadic tribes. Shaikh Salim died on February 22; Shaikh Ahmad's official date of accession is March 29, 1923. An unintended consequence of bin Saud's trade boycott of Kuwait was a healthy increase in trade for Bahrain. Most goods intended for Nejd, no longer able to move overland, passed through Bahrain and on to bin Saud's ports of Jubail, Qatif, and Ujair.

3. PRO/FO/371/8945, July 20, 1923, acting resident (Knox) to colonial secretary, the duke of Devonshire, "my obligation to assist Anglo-Persian"; Chisholm, *First Kuwait Oil Concession Agreement*, p. 210, the lawyer, Mirza Muhammad, was legal adviser in Basra to Anglo-Persian, his colleague, J. Gabriel, was attorney to Shaikh Ahmad of

Kuwait; IOL/R/15/2/96, vol. 1, June 14, 1923, political agent Bahrain (Daly) to acting resident (Knox), quoting Shaikh Hamad, "Anglo-Persian is too much mixed up in politics and Khazaal of Muhammerah is a tool in their hands."

4. IOL/L/PS/10, vol. 989, June 3, 1923, acting resident (Knox) to secretary of state India, "due to Holmes and his agent"; one or two officials did appear to appreciate the real situation, see for example, Philip Willard Ireland, *Iraq; A Study in Political Development* (New York: Russell & Russell, 1970), p. 198, quoting an India Office official on the treatment of bin Saud: "There has been recently a most unfortunate conjuncture of circumstances in our relations with bin Saud—the withdrawal of his subsidy, the expulsion of his Agent from Bahrain, and the change in the channel of communication must have had a deplorable cumulative effect." PRO/FO/371/8945, July 20, 1923, acting political resident (Knox) to duke of Devonshire, memorandum of June 2 meeting with shaikh of Kuwait, "in this we have succeeded!!"

5. IOL/L/PS/10, Vol. 989, June 3, 1923, acting resident (Knox) to secretary of state India, "no hurry to exploit (Kuwait)"; IOL/R/15/2/96, vol. 1, June 10, 1923, telegram, political agent Bahrain (Daly) to acting resident (Knox), "this seems to be a pity"; IOL/R/15/2/96, vol. 1, June 12, 1923, telegram, acting resident (Knox) to political agent Bahrain (Daly), "using your influence with Shaikh Hamad."

6. IOL/R/15/2/96, vol. 1, June 12, 1923, acting resident (Knox) to Shaikh Hamad, "inform that gentleman" and June 14, 1923, political agent Bahrain (Daly) to acting resident (Knox), "will not yield short of compulsion" and enclosing Hamad's reply, "by me or by our subjects"; June 26, 1923, acting resident (Knox) to Shaikh Hamad Bahrain, "to a man like Major Holmes."

7. Details of Holmes's 1923 Kuwait offer are in the original documentation in Holmes Personal Papers, also in Thomas E. Ward, *Negotiations for Oil Concessions in Bahrain, El Hasa (Saudi Arabia), the Neutral Zone, Qatar and Kuwait* (New York: privately published, 1965), p. 17, and Chisholm, *First Kuwait Oil Concession Agreement*, p. 23; see Chisholm, p. 97, citing August 17, 1923, Wilson to Anglo-Persian managing director London in which Knox concedes that Holmes's terms were superior; IOL/R/15/5/237, vol. 11, June 23, 1923, Shaikh Ahmad al Jabir as Sabah Ruler of Kuwait to political agent Kuwait (More), "from the justice of Government"; IOL/R/15/5/237, vol. 11, June 23, 1923, political agent Kuwait (More) to political resident (Knox), "a question which will be settled in London"; June 25, confidential, political agent Kuwait (More) to political resident (Knox), "Ahmad is a man of no strength of character."

8. IOL/R/15/5/237, vol. 11, June 28, 1923, ruler of Kuwait, Shaikh Ahmad, to political resident (Knox), "obedience is incumbent on me" and IOL/R/15/2/96, vol. 1, July 2, 1923, political resident (Knox) to undersecretary of state for India, London, repeat to viceroy India, "hostile" and "awaiting a visit from Wilson."

9. IOL/R/15/2/96, vol. 1, June 18, 1923, confidential, acting resident (Knox) to political agent Bahrain (Daly), "Wilson may, or may not, be correct" and "dangerous" and "why hurry?"; IOL/L/PS/10, vol. 989, July 9, 1923, acting resident (Knox) to duke of Devonshire, undersecretary of state Colonial Office, "force bin Saud into the open" and "a great deal of harm here"; IOL/L/PS/10, vol. 989, July 13, 1923, telegram from duke of Devonshire, undersecretary of state, Colonial Office to political resident (Knox), "unless specifically referred to by bin Saud" and "to secure cancellation of Nejd concession."

10. IOL/L/PS/10, vol. 989, July 17, 1923, acting resident (Knox) to duke of Devon-

shire, "in the event of bin Saud cancelling Holmes's concession"; IOL/R/15/5/237, vol. 11, July 14, 1923, confidential, political agent Kuwait (More) to acting resident (Knox) enclosing copy of Shaikh Ahmad's June 23, 1923, letter to Wilson, More told Knox, "I pointed out the advice was yours not mine" and added, "I fancy (Shaikh Ahmad) thinks that, when Major Holmes arrives in England his syndicate will bring pressure to bear on Government and get them to reverse their decision."

11. Rihani Papers, July 17, 1923, Rihani in Lebanon to Holmes in London: "Great opposition you have overcome and you will overcome yet in Kuwait." And August 1, 1923, Holmes to Rihani, "begin work in November at Al Hasa and the same time at Bahrain" and "did not think anyone could upset Anglo-Persian"; PRO/FO/371/8945, October 11, 1923, telegram, viceroy to secretary of state for India, "permission to accompany his father" and October 16, 1923, Wakely at India Office London to undersecretary of state, Colonial Office, requesting "advice of the Secretary of State for the Colonies as to the reply to be made" and October 23, 1923, Colonial Office to Government of India, "no objection is seen to Mr. Holmes visit."

12. Chisholm, *First Kuwait Oil Concession Agreement*, pp. 98–99, August 17, 1923, Wilson to Anglo-Persian managing director London, "sit tight and do nothing" and "has spoilt the market."

13. Rihani Personal Papers, August 30, 1923, Holmes in London to Rihani in Lebanon, "I am very friendly with Cox" and "the oil concession from the Sultan of Sahej" and August 23, 1923, Holmes to Rihani, "they will support us if we make an application through them."

14. IOL/L/PS/10, vol. 989, March 20, 1924, resident (Trevor) to secretary of state for the colonies (Thomas), "Holmes had been associated for 28 years in the service of his company—an engineering firm with connections in the Far East . . . one of the 82 companies with which Mr. Edmund Davis is connected"; IOL/L/PS/10, vol. 989, August 24, 1923, E&GSynd to duke of Devonshire, secretary of state, Colonial Office, "necessary to correct the situation"; Yossef Bilovich, "The Quest for Oil Bahrain, 1923–1930: A Study in British and American Policy," in *The Great Powers in the Middle East, 1919–1939*, ed. Uriel Dann (New York: Holmes and Meier, 1988), p. 254, draws on a Colonial Office Minute expressing fear of critical news coverage.

15. Rihani Personal Papers, September 20, 1923, Holmes in London to Rihani in Lebanon enclosing copy of a telegram September 19, 1923, from Ali Yateem at Bahrain to Holmes: "Sultan Ibn Saud has informed Shaikh of Kuwait that he has decided to give the Neutral Zone Concession to Eastern & General Syndicate Ltd. And strongly advising Shaikh of Kuwait to do likewise for the mutual interests of both. Shaikh of Kuwait forwarded Sultan Ibn Saud letter with his own letter to Political Agent of Kuwait with request to forward to His Majesty's government for immediate sanction"; Chisholm, *First Kuwait Oil Concession Agreement*, reprinted in full, pp. 99–103. September 6, 1923, duke of Devonshire, secretary of state, Colonial Office to the acting political resident (Knox) in the Persian Gulf, "not now convinced" and "engaged in negotiations with another company."

16. Anthony Sampson, *The Seven Sisters* (London: Hodder and Stoughton), 1975, pp. 70–71; Burma Oil still held 23 percent of Anglo-Persian; PRO/FO 371/8944, August 2, 1923, personal and confidential, Cecil C. Farrer, Department of Overseas Trade to Mallett at the Foreign Office reporting E&GSynd's discussions with Burma Oil. Mallett notes Farrer "has already written to the Colonial Office in the same sense."

17. Rihani Papers, October 17, 1923, Holmes in London to Rihani in Lebanon, "whether or not he would take up the additional shares"; November 22, 1923, Holmes to Rihani, "so we can get started at once on Hasa"; Rihani Papers, August 30, 1923, Holmes in London to Rihani in Lebanon, "a Remington typewriter" and December 15, 1923, Rihani in Lebanon to Holmes in Cairo, "good news from the Foreign Office about Yemen." Rihani was also proud of his Arab horse; he requested Holmes to bring him a special saddle and horse blanket from London.

18. Wayne Mineau, *The Go Devils* (London: Cassell, 1958), p. 184, Dorothy Holmes saying, "we had to buy MillHill in order to find a place for the two horses we got from bin Saud"; IOL/R/15/5/237, vol. 11, December 22, 1923, Arnold Wilson to resident (Trevor) attaching translation of *Times of Mesopotamia* advertisement.

CHAPTER 7

"A LITTLE AFFAIR LIKE THIS"— IRAQ OIL

As well as the concessions on the Arabian Peninsula, Frank Holmes was vying for the oil concessions of Iraq. He arrived in Baghdad before the Constitutional Assembly of Iraq opened on March 27, 1924, with its primary duty being consideration of the Anglo-Iraq Treaty through which future relations between Britain and the new state were to be governed. In addition, the boundaries of Iraq rather than Mesopotamia were yet to be ratified.

The Anglo-Iraq Treaty to be discussed by the Constitutional Assembly contained a clause requiring the new state to honor pledges made by Britain in all agreements and treaties she had previously signed as the Mandatory Power on Iraq's behalf. Once this treaty was signed, Iraq would be required to accept the validity of the Turkish Petroleum Company's claims—and also accept the "advise" of the high commissioner to award Iraq's oil rights to the Turkish Petroleum Company. Britain's reconstituting of Turkish Petroleum was a legal ploy used to lay a somewhat dubious claim to the concession held in that company's name before the war. The historical details were both complex and murky.

Armenian Calouste Gulbenkian founded the Turkish Petroleum Company. Born in 1887, Gulbenkian was a graduate in mining engineering from King's College London. He worked in the family's oil business in Baku and wrote scholarly articles and a highly respected book on Russian oil. Through a report he authored, he brought the oil possibilities of Mesopotamia to the attention of the Turkish sultan. In 1912 he formed the Turkish Petroleum Company with the Deutsche Bank and Royal Dutch Shell holding 25 percent each, and 50 percent held by the Turkish National Bank. As Gulbenkian personally held 30 percent of the Turkish National Bank, this gave him a 15 percent holding in Turkish Petroleum.

As they did to Frank Holmes, competitors branded Gulbenkian "a concession hunter." He would succeed in retaining 5 percent of the reconstituted company against pressure from every side, through the various incarnations of Turkish Petroleum and the subsequent Iraq Petroleum Company, until his death at age eighty-five, in Lisbon. In early 1914, as Turkey slipped deeper into economic distress, the Anglo-Persian Oil Company received 50 percent of Turkish Petroleum. The remainder, except for 5 percent held by Calouste Gulbenkian, stayed with Deutsche Bank and Royal Dutch Shell.

Britain's Public Trustee of Enemy Property confiscated Deutsche Bank's 22.5 percent during the war. Royal Dutch Shell held on to its allotment when the company submitted to British control in order to continue to operate on the high seas during the war. The managing director, Henry Deterding, became a British subject, and the company's headquarters moved from The Hague to London. Deterding was knighted in 1920 for his services to the British Empire.

In the April 1919 discussions between Britain and France that became known as the Long-Berenger Terms, the Deutsche Bank's shareholding in Turkish Petroleum, now held by the Public Trustee of Enemy Property, was given to France. In return France agreed to give up its claim to the Mesopotamian province of Mosul.

The Americans were not going to sit back and allow such a cozy carve-up of Mesopotamia's oil. Frank Holmes's old friend, Herbert Hoover, now US secretary of commerce, summoned the big American oil companies to Washington and launched a challenge to Britain and France for a piece of the pie. When Holmes arrived in Baghdad for the opening of the Constitutional Assembly of Iraq, the Turkish Petroleum Company was 22.5 percent French, 50 percent to Anglo-Persian, and 22.5 percent with the now British Royal Dutch Shell—giving Britain a substantial controlling interest—and 5 percent held by Calouste Gulbenkian.

A consortium of American companies, led by Standard Oil of New Jersey, responded with alacrity to Hoover's call. In 1922 they were offered entry, but on the proviso they ensured the State Department did not question the legal title of the Turkish Petroleum Company. For this guarantee Anglo-Persian offered 24 percent of its own shares with a trade-off that Anglo-Persian would receive free of charge 10 percent of oil produced by the concession. At this point, the State Department refused to recognize the legal validity of the claimed Turkish Petroleum concession (it would reverse this decision in April 1927). The British and French went ahead without the Americans.

That Iraq had a wealth of oil reserves was never in doubt. In 1920, while there was still strong argument about the form the future government of Mesopotamia would take, British officials including Prime Minister Lloyd George factored in anticipated income from oil when calculating the cost of administering the mandate. An editorial in the *Times* of London commented cynically that Lloyd George seemed to think Mesopotamian oil would pay the cost of administration: "We doubt it, for oil profits generally seem to find their way by some invisible pipeline into private pockets."

The Iraq Revolt later in that year resulted in an Arab form of government as now being developed, eventually to replace the British mandate. Therefore the cost of administration would fall on the new government of the State of Iraq, and not on Britain. Nevertheless, the British were not about to let the prospect of substantial oil profit slip from their grasp.

The Germans had already completed a considerable amount of work. Since the late 1800s German geologists had reported favorably on Mesopotamia and held several options, interrupted by the occasional military coup and the 1908 revolution. During the war the Germans ambitiously developed Baba Gurgur near Kirkuk, Tuz, Qaiyara, and Hit. They shallow-drilled, built walkways and galleries, conducted basic refining, and transported this fuel by road tanks for military use; they were planning a pipeline to the Mediterranean. Their work would never be acknowledged. When Anglo-Persian/Turkish Petroleum brought in a well at Baba Gurgur in 1927, they would have no hesitation in claiming it as their own "sensational" find and would be equally self-congratulatory at Qaiyara in 1928.[1]

Holmes had his contacts in Baghdad. As usual, his relations were stronger among the Arabs than with the British officials "advising" the fledgling Iraq government. Ameen Rihani was doing his best to support Holmes's endeavors and wrote him that he had received positive messages from the Najib. Rihani reported: "The Iraq Minister of Finance has written me saying you have been to see him and that he did what he could for you. The rest, he said, is for the High Commissioner and yourself to settle. I think he mentioned London also. I hope you scoop up something in Iraq."

In India, Harold Dickson was also hoping Holmes would do well. He wrote to Philby, who was already looking tenuous in his position as British adviser in Transjordan. Dickson put it to Philby:

> Holmes has been knocking about the Gulf for sometime . . . the Al Hasa concession is through and Holmes has, I believe, every intention of starting work before long. I am out of touch with him.
>
> How would you like to offer our services to Holmes's company as go between him and bin Saud, a sort of resident political officer to the Nejd Oil Company with headquarters at Hofuf or Qatif? I personally have been nibbling at this job for some time (one of my irons in the fire) I gather great things are expected of Holmes's concern.
>
> If you were to offer your services to Holmes, who is now in Baghdad, I have no doubt he would jump at the chance of getting you. The only thing is that it may not be a big enough job. Anyway both of us have got bin Saud's genuine interests at heart. With you as Chief Administrator Political and I as your second string we could, I think honestly, see that both bin Saud and the oil company were not let down.[2]

PRYING INTO THE POST

While the Iraq Constitutional Assembly opened its deliberations, Holmes continued to organize the survey that Dr. Arnold Heim had been hired to conduct. Heim was a lecturer in engineering at Zurich University and consultant geologist to many oil companies. With him would be a party including two Swiss assistants, also engineers. For Holmes, one of Heim's attractions may have been that he had no association with the Anglo-Persian Oil Company.

From Baghdad, Holmes traveled to Hofuf. There he found bin Saud recovering from a bad case of blood poisoning, "thinner, but in better health than I have ever seen him." (Curiously, rumors that bin Saud had tuberculosis, or cancer, had swept the Arab world followed by reports of his death.) Holmes and bin Saud spent ten days together: "I arranged all my business with him and bought sixty camels for the engineers to make their inspection trip," Holmes told Rihani.

Both Holmes's and Rihani's correspondence with the Arabs in the Gulf was of wider interest than that of the correspondents themselves. The political agents and the British officials in Iraq routinely intercepted cables and mail and advised of the content. Although it was almost six months since the duke of Devonshire had authorized the granting of a joint concession in the Neutral Zone to Holmes, Government of India officials in the Gulf had not passed on this important Colonial Office decision.

Through an intercepted Holmes-Rihani letter, Maj. James Carmichael More in Kuwait became aware that Holmes was expected soon. He wrote to the resident, Arthur Trevor, inquiring: "Should I now tell the shaikh there is no objection to his giving a joint concession with bin Saud in the common territory?"[3]

Maj. Clive Daly in Bahrain had, as usual, read the ruler's mail as it passed through the postal system located in his agency, among it a letter from Shaikh Hamad to Holmes. Hamad stated in no uncertain terms that he would not give "any concession" to Anglo-Persian because it was "common knowledge" that the government had "large share interests" in that company "which contains ex Government officials." He wondered whether government could be "unbiased and neutral arbitrators" in protecting his interests. Daly took personal offense at this criticism of the political officers and Anglo-Persian staff.

Not at all disturbed by revealing that he knew the contents of the shaikh's private correspondence, Daly tackled the ruler of Bahrain. Grandly, he rebuked Shaikh Hamad, telling him it was "unnecessary" to "give his views" to Holmes. It was "sufficient," he said, to have stated only that he "had received no further application from Anglo-Persian."

Unexpectedly for Daly his dressing down provoked a heated response from the usually mild-mannered Hamad. He told Daly the people of Bahrain believed that the Government of India deposed his father, Shaikh Isa, because he had given a concession to a company other than Anglo-Persian "which all Arabs regard as Government pure and simple." He said that if he now gave in to the pressure being exerted on him to agree to Anglo-Persian having the concession, he would be accused "by the whole family as well as other Arab Rulers" of having "sold his country in return for being made shaikh."

Major Daly soon discovered that Shaikh Hamad was correct in his reporting of public opinion in Bahrain. The people also believed that bin Saud's agent and friend, Abdulla Gosaibi, "would not have been deported from Bahrain if bin Saud had given his concession to Anglo-Persian" rather than to Holmes. Shaikh Hamad very clearly understood that the duke of Devonshire's adjudication meant that opposition to Holmes's application would be withdrawn in London if Anglo-Persian failed to clinch a deal in the near future. Daly reported to Trevor: "There is no doubt that Hamad has quite made up his mind to refuse to deal at all with Anglo-Persian."

Hamad's deposed father, Shaikh Isa, had not given up his fight to break the tyranny exercised by the Government of India's men in his country. Daly drew Trevor's attention to "the petition recently presented by Bahrain's Abdul Wahab Zayani on Shaikh Isa's behalf to the Viceroy of India."

Daly directed Trevor to note "the reference to oil, the pearl fisheries, etc." Daly described Zayani as "perfectly fanatical as regards foreign influence of any kind in Bahrain. He seems always to have been convinced that Great Britain intends by degrees to swallow up the resources of the islands. He will doubtless have made the most of his opportunity to spread tales that recent political events are not unconnected with oil concessions."[4]

TREVOR AND HOLMES

After visiting bin Saud, Holmes moved on to Bahrain where he and the permanent resident, Lieutenant Colonel Trevor, met for the first time. Trevor reported this meeting to the duke of Devonshire saying, in his view, Holmes "appeared to be quite frank and above board." Trevor said Holmes "impressed me favourably by his frank demeanour and his apparent desire to have all his cards on the table. Perhaps he is a bluffer but he did not strike me as one."

Trevor told the duke that Holmes objected to being called "an adventurer," a charge that was emanating from the political officers. "Holmes has been prospecting, mining, etc., all over the world, for twenty-eight years with the same company. He has never before been called an adventurer," Trevor said, adding that Holmes seemed to know who was spreading this allegation but "he would not say who was calumniating and slandering him everywhere." Trevor now proved to be more amenable than had Knox during his six-month assignment as acting resident.

Holmes raised the subject of Qatar, where he had planned to negotiate for a concession after Al Hasa in 1922. He told Trevor he had recently received "special messengers" and a personal letter from Shaikh Abdulla Al Thani, the ruler of Qatar, "imploring him" to come across and make an arrangement on the same terms as given to bin Saud. Holmes pointed out he was unable to act on Shaikh Abdulla's request because he had been "instructed by the Colonial Office" not to make an agreement with the shaikh of Qatar and he was "loyally obeying" this instruction. Trevor said he had confirmed, after "making some inquiries" in Bahrain, that "the shaikh of that place (Qatar) was very keen, on financial grounds to give a concession—but not to Anglo-Persian—for similar reasons to those activating Shaikh Hamad."

Trevor agreed with Holmes that he would, finally, now inform the shaikh of Kuwait he was free to grant to Holmes the Neutral Zone concession jointly with bin Saud. There was the matter of the boundaries of the Neutral Zone. In his October directive, the duke of Devonshire had remarked: "I observe that E&GSynd appear to consider that a large part of the actual territory of Kuwait is within the Neutral Zone."

Holmes puzzled over the duke's comment and compared his own map with that on file at the Baghdad Map Office. As far as he could see there was no difference. He had taken the two maps to Sir Henry Dobbs, the high commissioner in Iraq, who also could not see any difference between Holmes's map and that on file. Trevor could not see any difference either. The duke of Devonshire, it seems, had not caught up with the fact that, since Sir Percy Cox's virtuoso performance twelve months earlier at Ujair, a large part of the actual territory of Kuwait was indeed now within the Neutral Zone.

Trevor was happy for Holmes's survey party to begin from Kuwait, go down along the coastal belt to Jubail, and end up at Qatif. Trevor reported that Holmes had made all the arrangements for interpreters, camels, camping gear, and so on for the "men belonging to the Apex Company of Trinidad" who were experienced men. Major More was advised "an engineer and two geologists will be arriving" to make an examination of the Al Hasa concession and "if it be signed in time also of the Neutral Zone area." For the trip "sixty camels have been purchased in Hofuf and will be brought to Kuwait within a few days."[5]

TIME LIMIT

Trevor's meeting with Holmes, and his recognition of the "very determined opposition" to Anglo-Persian, provided an insight into the true situation prevailing in the shaikhdoms. Trevor remarked that a year ago, before he went on leave, he had not noticed such hostility toward Anglo-Persian "but it is very prominent now." Like Knox before him, Trevor did not make the connection between this opposition and the way in which the shaikhs had been treated during the preceding twelve months.

He did point out that "any ill-wishers of the British, or of Anglo-Persian, never lose an opportunity of impressing upon their hearers that it is a government concern and that to grant a concession to this company is tantamount to giving it to Government." But he did not explore why this possibility was so odious to the Arabs, even though he thought the hostility might be related to the "considerable number of ex-Mesopotamia military and ex-Mesopotamia political officers imported into Anglo-Persian."

He observed that "mischief makers" were no doubt "taking their cue" from the "ignorant stuff written in a portion of the English press about Britain, oil and Iraq . . . which has been copied and greatly enlarged on by the Egyptian and the Arabic newspapers." This, he thought, explained the local belief that "the desire of Government in the guise of Anglo-Persian to obtain the oil of Bahrain was at the bottom of the removal of Shaikh Isa." He also pointed out that the shaikh of Kuwait "whose mind has been thoroughly poisoned against Anglo-Persian has not the least idea of giving them the concession for his principality."

Nevertheless, Trevor reported objectively, this was the current position. He suggested a decision was required as to whether "we should keep the matter undecided for an indefinite period in the interests of Anglo-Persian" or whether "we should agree to the Shaikhs of Kuwait, Bahrain and Qatar closing with the offer of E&GSynd."

For the first time in official reports coming from the Persian Gulf, Trevor raised the injustice of the situation, saying, "it hardly seems fair to prevent them from granting the concessions and consequently from receiving the revenue there-from for an indefinite period." He said: "The principal object of Anglo-Persian . . . is to prevent other companies from coming in. Their object will be achieved if they simply wait and do not press for the concessions themselves."

Trevor was concerned, he said, for "the interests of the shaikhs." It would seem fairer "if Anglo-Persian, as the company first in the field, were requested to send an official to negotiate with the various shaikhs within a specified time." If their offer is then definitely refused, "we should authorise E&GSynd to conclude its concessions with the shaikhs concerned," he advised.[6]

NEUTRAL ZONE CONCESSION SIGNED

To witness the granting of the Neutral Zone concession on his behalf, bin Saud sent to Kuwait his representatives, all friends of Holmes. The visitors from Nejd included Holmes's companion between Hofuf and Basra, Hashem bin Ahmad. Also present was Dr. Abdullah Damlouji, who had "greatly assisted" at the signing of the Al Hasa document. Damlouji would go on to become bin Saud's first minister of foreign affairs.

The third witness was Shaikh Hafiz Al Wahba. The title *shaikh* was an honorific awarded to the originally Egyptian Wahba who had, some years before, been active in education and local politics in Kuwait. When he returned from Europe with his "wild political ideas," Shaikh Abdulla Al Khalifa asked Wahba to come to Bahrain and help him set up the first modern school there. Major Daly deported Wahba from Bahrain in 1921, allegedly for conspiracy against the Order-in-Council. He moved to Nejd where bin Saud had asked him to institute a modern education system for his territories. Shaikh Hafiz Al Wahba would go on to be Saudi Arabia's first ambassador to Britain. He and Holmes remained lifelong friends.

Though making little effort himself toward reaching agreement with the shaikh of Kuwait, or any of the other shaikhs, Arnold Wilson was still keeping track of Holmes. On the day bin Saud's representatives arrived, Wilson cabled the Kuwait political agent: "Let me know as soon as you can what the present position is between the Shaikh of Kuwait and Major Holmes."

Holmes's survey party was assembled. Rihani's brother, Albert, had arrived some weeks before, accompanied by Rihani's admonishment: "Take care of Albert. He is his mother's Darling. She wept her eyes out the day she learned he was going away." Holmes had reassured Rihani: "Albert is settling down but he is new to the rough life yet. He will soon become accustomed to a new viewpoint and find that the best man to help Albert is Albert himself. He will do well. I hope he is at the beginning of a very lucrative and pleasant career. He typed this and, as you can see, has improved very much."

There was a short delay caused by a raid on Kuwait's territory by Ikhwan tribesmen that caused the resident to comment, "The party may be allowed to start—provided His Majesty's government have not broken off relations with bin Saud." The prospecting party, on its sixty camels and with a twenty-five-man escort provided by bin Saud, finally departed from Kuwait on April 28 for what Trevor described as a "preliminary survey in Al Hasa, Nejd and the Kuwait Neutral Zone."

Holmes wrote to Rihani: "Albert went off like a proper Bedouin. I think he found the Arab food a little hard for him and also perhaps that I was not very sympathetic, but had to make him stand on his own feet. That boy has the makings of a very good man and will forge ahead in any work he follows. Tell his mother not to worry about him, he is looking better than when he came."[7]

Holmes was already certain he knew where the oil would be found; he had identified this on the maps he drew for Al Hasa, Bahrain, the Neutral Zone, Kuwait, and Qatar before he began negotiating for the concessions. It was obvious to Holmes that to carry through his plans for developing the oil fields, he needed a constant supply of water both for the drills and machinery and for the men working them. As he had told Dickson, Holmes thought an "enormous volume of sweet water" existed in the hinterland behind the Qatif Oasis in Al Hasa. Consequently, his instructions to the Swiss prospecting party were as much concerned with confirming his identification of the sources of water as they were with verifying his own positive opinion of the oil strength of his concessions. Like Holmes, Dr. Heim also clearly understood this basic need and would later remind the directors of E&GSynd that "drilling on oil in countries off from the rivers is impossible without the right for drilling on water."

In Whitehall, a "secret" report was circulating with the title "Persian Gulf Oil: Attitude of the Shaikhs to Anglo-Persian Oil Company." It read:

> The Shaikhs, one and all—with the exception of the Sultan of Muscat who has little knowledge of the matter—mistrust Anglo-Persian and advance a very plausible and indeed, from their point of view, sound argument for declining to give concessions to that company. They say HMG being so large a shareholder as to be identifiable with it cannot be relied upon as an impartial arbiter towards them and the company. Their view is that Anglo-Persian and Government are in effect one. This view is reinforced by the knowledge that many officers formerly known to them as officials of Government are now met in the shape of officials of Anglo-Persian—a transformation which curiously enough does not seem so suspicious to the Persians as it does to the Arabs.

The secret report continued:

> One wonders whether, in the light of the Arabs' opinion on this point, Anglo-Persian are not paying very heavily for the services of Sir Arnold Wilson and the many active and efficient army ex-officers who served with and under him in Iraq, and are now serving with him and under him in Anglo-Persian. Meanwhile, Anglo-Persian's British rival, E&GSynd, is pushing ahead while the former can do no more than wait and hope for a change of attitude on the part of the shaikhs concerned. Whether it is fair to the shaikhs to keep them waiting for the possible profits they might get from E&GSynd if they won't deal with Anglo-Persian is a question that needs decision. Major Holmes, it is to be noted, impressed Lieutenant Colonel Trevor face to face as eminently open and honest.

Soon after, the Colonial Office officially informed Anglo-Persian that its "claim to priority" would be recognized only until the end of March 1925. If agreement were not reached by that date, the shaikhs would be free to negotiate concessions with other "approved" parties, meaning E&GSynd.[8]

WATER

Heim agreed completely with Holmes's original assessment that water prospects in Al Hasa were good. In a preliminary "report on water and oil" written from Bahrain to the London office of E&GSynd, Heim said he thought "part of the Al Hasa Concession might become a great country for drilling on artesian water." Holmes had impressed upon him that "the question of drilling for water on Bahrain is of great importance." He and Holmes were "acting in perfect understanding," he wrote, adding that Holmes's directions were that "it will be to the syndicate's interest if I fill in the time until the next ship to Kuwait in examining and reporting on this question of water in Bahrain." This was certainly a matter of vital importance for Holmes, who had already raised with both Shaikh Hamad and the political agent the possibility of finding "sweet" water for the people of Bahrain.

The name Bahrain means "two seas" and describes the phenomenon of freshwater springs that bubble off the seafloor. Traditionally this water was gathered through hollow bamboo, by divers using goatskins, or by women wading out at low tide. A handful of natural springs existed, particularly on the Island of Bahrain from which the northern date palms were watered. The undersea springs could be both salt-contaminated and difficult to harvest in poor weather while the natural springs were uncertain. In 1922, the political agent had reported, "the town is in urgent need of water supply. At present all drinking water is brought in from a long distance at high cost." And again, in 1923, he commented, "Shaikh Hamad is desirous of introducing a water supply."

Since his first talks with Hamad, Holmes had assured him that although he would be drilling for oil, he could indeed produce potable water. Now they agreed that Holmes should produce water first, oil later. This would also allow the time to pass that the Colonial Office had allotted to Anglo-Persian as its exclusive negotiation period. Together, Shaikh Hamad and Holmes made an arrangement with Major Daly, agreeing that, for his work finding fresh water, Holmes would obtain "certain reports and maps which otherwise would be unobtainable."[9]

ANGLO-IRAQ TREATY SIGNED

After signing the Neutral Zone concession and seeing the prospecting party safely off from Kuwait, Holmes returned to Baghdad to organize the equipment for water drilling and for what he described as the Battle of Iraqi Oil.

He had been in touch with Iraqi officials while in Bahrain and Kuwait and was optimistic that with his Iraqi network, and Rihani's assistance, he might "accomplish much." In Baghdad, he said, he had "very strong friends and also very strong opponents." Wisely, he commented that "in a little affair of this sort [it is] as well not to overlook" the second group.

Furnishing financial references for the Iraq government was not a problem. But the Iraqis had said, more or less, that they needed something to use as argument in favor of E&GSynd against the high-powered Turkish Petroleum/Anglo-Persian Oil Company—and the high commissioner in Iraq, Sir Henry Dobbs. What they wanted, Holmes wrote Rihani, was for the British banks "to tell them they should give the concession to us." He added that of course the banks would never make such a statement "as it is beyond their province."[10]

He had applied for oil rights "dealing with the whole of Iraq." He was certain his terms were "very favourably received" by the Iraqis and had heard that "90% of the Government is with me." Despite the "full force" of high commissioner Dobbs being behind Turkish Petroleum, "which is the medium through which Anglo-Persian and the British government intend to receive control of Iraq's oil," Holmes told Rihani, the Iraq government had not awarded the concession to that company. For a marathon eight months the members of the Iraq Assembly resisted pressure from the high commissioner and Turkish Petroleum. Then they announced their intention to delineate several concession areas, rather than one monopoly.

British reaction was swift. They announced they would not support Iraq's own plan of several concessions over one monopoly. They launched a propaganda campaign warning that if the Anglo-Iraq Treaty was not signed, the British would withdraw from Iraq and, furthermore, would not support Iraq's claim for the province of Mosul.

Covering all possible outcomes, Arnold Wilson went to Constantinople. As Holmes put it, Wilson was armed with "a pre-war letter addressed to Turkish Petroleum from a pre-war prime minister of Turkey in which it is stated that Turkey was willing to grant a concession to the Turkish Petroleum Company provided suitable terms could be arranged covering Mosul and Baghdad provinces."

From the Iraqi viewpoint, Holmes wrote Rihani, the only conclusion to be drawn was that the Anglo-Persian Oil Company, through the person of Wilson, was "quite willing to sell out the Iraqis" and support the awarding of Mosul to Turkey, rather than to Iraq, "provided Anglo-Persian gets the oil rights of any territory handed over to the Turks." Anything Anglo-Persian now obtained from the Turks would "only be at the expense of Iraq," he said. In disgust, Holmes commented, "truly as unclean a deal as I have ever heard of and supported by the British government at that."

Although expressing disgust at the action of Wilson and Anglo-Persian in approaching the Turks, Holmes was not above suggesting that Rihani "write to one or two of your influential Turkish friends and lay my case before them so that it will find its way before the Turkish Minister and hold up Anglo-Persian until I can reach Constantinople."

Holmes would then let the Turkish government know E&GSynd "are prepared to offer better terms for any territory over which they wish to grant oil concessions." Once the Turks were aware that better terms were available, he told Rihani, "they will delay matters and give me a chance to apply for both the territory north of Mosul that Anglo-Persian are also after, and any territory they may secure in the Mosul area."[11]

Holmes had supplied bank references giving E&GSynd's financial status as "from immediate 5,000,000 sterling to drawing power of 15,000,000 sterling." It was also self-evident, Holmes wrote, that E&GSynd "offer liberal terms and are free from political bias." He added, "I am fairly certain that if this Anglo-Iraq Treaty is not signed in its present form, I will secure the concession." He firmly believed "the Iraqis are quite prepared to give me the Iraq oil concession, provided they have a free hand."

The very next week, on June 11, 1924, when only sixty-nine of the one hundred delegates were present, the Constitutional Assembly approved the Anglo-Iraq Treaty by a vote of thirty-seven for and twenty-four against with eight abstentions. On the boundary with Turkey the Assembly's resolution affirming the treaty declared: "This Treaty and its subsidiary agreements shall become null and void if the British government fail to safeguard the rights of Iraq in the Mosul vilayet in its entirety."[12]

THE SWISS EXPEDITION

Holmes had organized the geological expedition as a four-month survey. But Arnold Heim and his cold-clime assistants were unprepared for the oppressive summer heat of the Arabian desert. Swiss engineer Popham was to examine the territory inland, traveling by camel from Hofuf to Kuwait. He lasted forty-eight hours before flopping gratefully on a boat at Qatif and sailing to Bahrain.

In Bahrain, waiting to travel by sea to Kuwait, Heim wrote to E&GSynd explaining that Popham "was leaving Hofuf for going to Kuwait by caravan following another route further inland. After the second day of travelling, however, he cut off the whole voyage on account of unsupportable heat, some of his people having become ill. The temperature of the sand blowing winds I measured at 118 degrees Fahrenheit. It is true that the natives are used to travel also in summer time, but at night. Thus we could not do geological surveying."

Dr. Heim's enthusiasm for Holmes's water projects increased once he was safely out of the desert. From Kuwait Town he wrote to Holmes, "While waiting on the shaikh here I had all the time to think over the question of boring artesian water in Eastern Arabia." Although he had made "but a trip" near the coast, and was separated from his university resources, he was becoming "more and more inclined to believe that Kuwait and Hasa may become great countries for artesian bores." He thought the necessary geological and geographical features existed, besides which there were "geologically several factors resembling the great Artesian Basin of Eastern Australia." To really do the job, he would need an assistant "with the necessary instruments and for helping me in making physical and chemical tests of the waters in the field."

As Holmes had indicated that he wanted to undertake work the next season in the Red Sea, Heim said on that trip he would stop off and "have a look at the famous Artesian Bores of Northern Africa."

He was not nearly as optimistic as Holmes about oil in the Arab shaikhdoms.

Popham's turning back meant that no geological survey of the inland of Al Hasa had been undertaken and so remained, as Heim remarked, "as yet unknown." Despite this, he promptly wrote the E&GSynd directors in London that he believed "the result so far obtained is not encouraging for the coastal region in regard of drilling for oil."

Undeterred by the fact that he hadn't even seen let alone surveyed the Al Hasa concession area, Heim reported anyway. He confidently gave two reasons for his damning opinion. Any traces of oil, he said, had probably arrived on the wind. "The only seepages we have encountered are found along the shore and are formed of tar and asphalt blown in from an unknown source in the Persian Gulf. The rocks in situ did not show traces of impregnation."

And Heim, used to the alpine mountains of Switzerland, could not imagine what the deserts might mask. "Almost throughout the region of our journey—as far as it is not covered by sand—is formed of tertiary or older rocks, especially limestones, of perfectly horizontal stratification. No signs of anticlines were encountered." None of this had impressed Heim. Closing the letter, Heim told the directors he would now move on to "a careful examination of the northern part of the Kuwait Concession where a very important seepage near a long mountain range is known." After that, he expected "on July 1st to proceed to Iraq."

Arnold Heim's happy reverie about "great countries for artesian bores" was cut short by a cable from Holmes informing him that the directors, who had now received the pessimistic views expounded in his letter from Bahrain, did not want him to visit Iraq or Persia. They wanted him to return immediately to London. "The directors wish to consult with you in regard to the Al Hasa concession, large payments that are almost due, so they wish to discuss the position with you at the earliest moment," the cable said. Holmes arranged to meet Heim in Baghdad.

Ten days later, in London, Heim handed over his preliminary report together with a "plan for further investigations." He had now "studied the books of Pilgrim and of Pascoe" and concluded: "There is no other untested oil region of the world with such phenomenal promise—as Iraq!" While enthusiastic about the water possibilities in the "Persian Gulf Region," he was circumspect on the prospects for oil. He had little to add to his earlier assessment other than further hesitation about the Neutral Zone, Kuwait, and the crisp: "I would not advise drilling for oil on Bahrain Island."[13]

His long and detailed report, dated September 5, 1924, with results of chemical tests made at Zurich's Federal Institute for Research of Combustibles, was written in Switzerland over the next two months. Heim's was the fifth geological exploration of Kuwait. His opinion was that, while it may be "possible to develop an oil field of commercial importance," Kuwait is a "country of some possibility, but not of high promise." Heim believed there was "no reason to recommend the Neutral Zone Concession for oil." He also had "no reason to recommend the concession of Al Hasa for drilling on oil."

He added that "to drill on the dome of Jebel Durkhan, Bahrain Island," as Holmes advocated, "would not only be extremely expensive but also a pure gamble." In sum, Dr. Heim reported, "the countries of Eastern Arabia, rapidly traversed by this writer, do not present any decided promise for drilling on oil."[14]

A DESPONDENT HOLMES

From his discussions with Heim in Bahrain, and again in Baghdad as he passed through on the way to London, Holmes already knew that Heim's opinion on the oil possibilities of the concession areas was totally opposed to his own positive findings. He was worried that after the syndicate read Heim's deeply pessimistic report, combined with the doubt now on obtaining anything in Iraq, they might balk at committing the level of finance previously promised for development of these concessions.

After all, Heim had been hired to present his internationally acclaimed "expert" recommendation. All along Holmes had been alone in his conviction that Al Hasa, Bahrain, and Kuwait did indeed hide very rich oil deposits. Anglo-Persian did not think so. Neither did bin Saud. Now the expert that the syndicate had hired to give an independent opinion did not think so either.

Holmes went from cheerily telling Rihani at the beginning of June of a long and very optimistic letter about "Nejd Oil" from Dr. Abdulla Damlouji to an obvious depression at the beginning of July. Rihani did his best to encourage his friend. Referring to Arnold Heim's dependence on the theories of Guy Pilgrim and Edwin Pascoe in reaching his conclusions, Rihani told Holmes: "Geologists, like other men of science, are sometimes misled by a theory. Some of them, you will admit, often surrender themselves to a theory as if it were infallibility itself." He sympathized with Holmes, saying: "By all means I would get a second opinion."

Urging him to bolster the syndicate's resolve, he pointed out that if the Al Hasa concession was not actively developed, Holmes would not be able to "do much" anywhere else. "Your action in one part of the Peninsula today is sure to affect for good or for bad your plans in other parts," he wrote, adding: "You are soon going to Assir to negotiate a concession with the Idrissi who knows nothing about scientists and experts and expert opinions. If he doesn't see oil he doesn't see anything else but failure and downright shame on the part of the syndicate. And for that reason he's not going to give you a concession."

The Anglo-Persian Oil Company would have a field day, Rihani warned. "They will twist the circumstances and misinterpret them to make it understood in Iraq, for instance, or in Assir, that your syndicate's resources are inadequate or its staff or experts incompetent, or both." What might happen then, Rihani said, could be that "Iraq may get cold feet." Rihani completely agreed with Holmes that he should not give up "before a thorough search is made."[15]

As E&GSynd in London were digesting Arnold Heim's report, the shaikh of Kuwait was being treated to a VIP visit to Anglo-Persian's operations. He was personally escorted to the oil fields, and around the Abadan refinery by general manager T. L. Jacks, who now shared this title jointly with Arnold Wilson.

Jacks, born in 1884, had joined Anglo-Persian's managing agents in 1909 as an "oil assistant" at Abadan. He had risen to assistant manager by 1917 and would be joint general manager until Wilson's transfer to London in 1926. T. L. Jacks would then be appointed resident director in Tehran. Jacks and Wilson sincerely believed the shaikh's "first sight of such operations in actual progress" would so impress him that he would forget Holmes and immediately sign an agreement with them. Shaikh Ahmad, however, remained unmoved. Anglo-Persian was also trying to entice Shaikh Hamad of Bahrain to visit Persia, but he had not taken up the invitation.

Anglo-Persian's managers seemed unaware that the shaikhs were fully informed on the situation between the oil company and the Persian government. The first year the company made a profit was 1917. From that year's net profit of 344,109 sterling, Persia had been allocated a mere 3,829 as her share. The Persian government did not get even this token payment. The company withheld the 3,829 against what it claimed as the cost of damage resulting from the cutting of pipeline by Persian tribesmen. Persia had no return from the overall profits of the company, receiving only a royalty on the sale price of oil exported from the country, some of it passing to the British military at a heavily subsidized rate.

Revenue gained from Persia's oil was not reinvested in that country. For example, in 1923, Anglo-Persian used a substantial amount from its "General Reserve" to write off investments, outside Persia, that had already been abandoned. A second substantial amount went toward writing off other investments, also outside Persia, that were about to be abandoned. Together, these two write-offs totaled ten times more than the payment Persia received for the year. The situation, as it existed in reality—and as was further amplified by anti-Anglo-Persian sentiments "kept at fever pitch by the sedulous activities of Persian agents with whom the Trucial States were overrun"—certainly would not have swayed the shaikhs in favor of the company, no matter how impressed they might have been by the sight of operations.[16]

NOTES

1. Benjamin Shwadran, *The Middle East: Oil and the Great Powers* (London: Atlantic, 1956), p. 204, for detail of Royal Dutch Shell becoming a British company; Elizabeth Monroe, *Britain's Moment in the Middle East, 1914–1971* (London: Chatto and Windus, 1981), pp. 304–308, for the Long-Berenger Terms of April 1919. For in-depth background to the Iraq concessions, and American participation, see Shwadran, pp. 193–265; also p. 209, "Hoover took it upon himself to invite the big American oil com-

panies to Washington to interest them in Mesopotamian oil and work out a plan of action."
Also see Shwadran, p. 216, for 1922 Anglo-Persian offer to Americans in return for support of Turkish Petroleum exclusive concession, Teagle of Standard Oil New Jersey
approaches to the State Department, Secretary of State Charles Hughes refusal, also pp.
220–23, for the competing US claim of Admiral Chester; Shwadran, p. 239, March 25,
1920, the *Times* of London, "we doubt it. . . ." For early German involvement see Stephen
Helmsley Longrigg, *Oil in the Middle East: Its Discovery and Development* (London:
Oxford University Press, 1954), pp. 13–15, 45; p. 69, "sensational," p. 71, "Qaiyara."

2. Rihani Papers, March 10, 1924, Rihani in Lebanon to Holmes in Baghdad,
"scoop up something in Iraq"; Philby Papers, box XVII-6, February 13, 1924, Dickson in
India to Philby in Amman, "offer our services to Holmes."

3. IOL/R/15/5/237, vol. 1, February 22, 1924, "forwarded from Resident to Agent
Kuwait extract from Intercepted letter dated January 20, 1924, from Abdul Rahman al
Naqib to Ameen Rihani (received from Baghdad)." The letter informed Rihani that Shaikh
Ahmad of Kuwait had "refused unconditionally" Anglo-Persian's offer and that, in
Kuwait, "it is hoped the Shaikh will give the concession to the Eastern & General Com-
pany." The matter of intercepted mail would continue to bother Holmes. See Rihani
Papers, May 20, 1924, Rihani to Holmes in Baghdad, "I have already written you a long
letter on April 30th, and on the 7th, of this month. The first was enclosed with a letter to
Mullah Saleh, secretary to Shaikh Ahmad al Sabah, the second was sent care of Shaikh
Ahmad himself." June 3, 1924, Holmes in Basra to Rihani, "Your letters dated April 30th
and May 4th have not reached me. The Shaikh of Kuwait says that he did not receive (a
letter) and neither did Mullah Saleh."

4. IOL/R/15/5/237, vol. 11, February 28, 1924, political agent, Kuwait (More) to
resident (Trevor), "should I now tell the Shaikh?" Rihani Papers, April 4, 1924, Holmes
in Kuwait to Rihani, "enclosing translation of a letter from Shaikh Hamad Bahrain. You
will see that he says straight out he will not have anything to do Anglo-Persian";
IOL/L/PS/10, vol. 989, March 17, 1924, political agent Bahrain (Daly) to resident
(Trevor), "unnecessary to give his views" and "sold his country" and "would not have
been deported" and "quite made up his mind to refuse."

5. IOL/L/PS/10, vol. 989, March 20, 1924, enclosed with letter of March 28: political resident (Trevor) to secretary of state, Colonial Office (duke of Devonshire) note of
an interview with Major Holmes, representative of the Eastern & General Syndicate
"impressed me favourably"; IOL/L/PS/10, vol. 989, October 1923, duke of Devonshire,
secretary of state, Colonial Office to political resident (Knox), "E&GSynd have submitted
to me a draft from which I observe they appear to consider a large part of the actual territory of Kuwait is within the Neutral Zone"; and March 20, 1924, political resident
(Trevor) to Colonial Office (Thomas), "Shaikh of Qatar keen, but not to Anglo-Persian . . . ," IOL/R/15/5237, vol. 11, March 26, 1924, Frank Holmes in Bahrain to political
agent Kuwait (More), "60 camels will arrive shortly."

6. IOL/L/PS/10, vol. 989, March 28, 1924, resident (Trevor) to secretary of state for
the Colonies, re Anglo-Persian's desire to obtain concessions from the shaikhs on the Arabian
littoral of the Gulf, "hardly seems fair to prevent them from granting the concessions."

7. Rihani Papers, April 4, 1924, Holmes in Kuwait to Rihani in Lebanon, "bin Saud
sent . . . his representatives" and "Albert is settling down."

8. IOL/L/PS/10, vol. 989, May 16, 1924, telegram, resident (Trevor) to secretary of

state for the Colonies, "if failed to successfully reach agreement by that date"; IOL/R/15/5/237, vol. 11, April 4, 1924, telegram, Wilson to political agent Kuwait (More), "let me know as soon as you can"; IOL/R/15/5/237, vol. 11, April 7, 1924, telegram, resident (Trevor) to political agent Kuwait, "if HMG have not broken off relations with Ibn Sa'ud." And April 29, 1924, "The Prospecting Party left here on 28th April accompanied by a representative of Ibn Sa'ud. The Shaikh agrees that they will be quite safe"; IOL/L/PS/10, vol. 989, July 16, 1924, Political Department, secret, Persian Gulf Oil attitude of the shaikhs to Anglo-Persian, "whether it is fair to the Shaikhs to keep them waiting."

9. Heim Papers, June 10, 1924, Heim in Bahrain to E&GSynd, "for drilling on artesian water"; Administrative Report of Bahrain for the Year 1922 and for 1923 (Daly), "urgent need of water" and "desirous of introducing a water supply"; Rihani Papers, June 3, 1924, Holmes in Basra to Rihani, "certain reports and maps."

10. Rihani Papers, Holmes to Rihani, April 4, 1924, and May 15, 1924. On May 20, 1924, Rihani wrote to Holmes, "Let me know with whom you are dealing in Baghdad? Is it with the Prime Minister direct? Or with whom? I might drop a word for you if it is necessary."

11. Rihani Papers, June 3, 1924, Holmes in Basra to Rihani, "write to one or two of your influential Turkish friends . . . so that it would find its way to the Turkish Minister and hold up Anglo-Persian."

12. See Shwadran, *The Middle East*, p. 236, "when only 69 of the 100 delegates were present."

13. Arnold Heim Papers, June 25, 1924, Heim in Kuwait to Holmes in Basra, "resembling the great Artesian Basin of Eastern Australia" and "as far as it is not covered by sand"; July 2, 1924, cable from Holmes in Basra to Heim in Kuwait, "wish to consult with you"; and September 23, 1924, Heim in Zurich to Holmes, "not advise drilling for oil on Bahrain."

14. Zurich University, Exploration Department Technical Library, 24072 Geological Report No. 1, "The Question of Petroleum in Eastern Arabia, Koweit, Hasa, Bahrein" with eighteen plates, maps, sections, and photographs by Dr. Arnold Heim, docent at the University of Zurich.

15. Rihani Papers, June 3, 1924, Holmes in Basra to Rihani in Lebanon, "Dr. Abdulla is an optimistic animal" and July 2, 1924, Rihani in Lebanon to Holmes in Baghdad, "Geologists . . . are sometimes misled by theory."

16. Archibald H. T. Chisholm, *First Kuwait Oil Concession Agreement: A Record of the Negotiations 1911–1934* (London: F. Cass, 1975), p. 11, "first sight of such operations"; in 1921 Wilson was appointed Managing Director in Persia, Mesopotamia and the Gulf of Strick Scott & Co, managing agents for Anglo-Persian. From 1923 he was General Manager of Anglo-Persian at Abadan, jointly with T. L. Jacks, until his transfer in 1926 to head office in London; see Shwadran, *The Middle East*, pp. 153–64, this inequality remained unchanged until 1933, for example in 1931, from a net profit of almost 2,500,000 sterling the Persian government received a royalty of 134,750. In 1934, the percentage of income in the Iranian government's budget derived from its oil, and Anglo-Persian operations, was a minute 7 percent; PRO/FO 371/8944, August 2, 1923, Farrer at Overseas Trade to Mallett at Foreign Office, "sedulous activities of Persian agents"; Administrative Report Bahrain 1923 (Daly), "Persian residents in Bahrain who are responsible for anti-British propaganda in the Persian press, are persons of no standing, who engage in this pastime apparently out of a desire for notoriety."

CHAPTER 8

A COLONIAL
TAKES ON THE EMPIRE

The shaikhs of Kuwait and Bahrain were prevented from signing any oil concessions with Holmes for a further seven months, until the end of March 1925, by the Colonial Office ruling that Anglo-Persian had a "priority." Holmes's anxiety that after receiving Heim's report E&GSynd's directors might hesitate in advancing the Al Hasa concession was now overtaken by bin Saud's own dramatic diversion away from economic development to military matters.

Tensions had increased after the Stuart Knox–chaired Kuwait Conference had failed to settle the differences between bin Saud and his Hashemite neighbors in the Hijaz, Transjordan, and Iraq. Bin Saud and his Wahhabi Warriors began a series of campaigns around these borders that would culminate with the abdication of Sharif Husain, the capture of Mecca, and bin Saud's adoption of the title King of the Hijaz and Sultan of Nejd and its Dependencies.

BIN SAUD'S RAIDERS

The Administrative Report for 1924 detailed the offensives.

> On March 15th the Ikhwan raided along the border with Iraq. The Ikhwan numbered about 1000 men. They were joined by an indiscriminate mob of ill-armed Bedouin, attracted by the hope of plunder when word went out that the Ikhwan was going raiding. The morning after the initial raid, three British airplanes passed over them . . . they withdrew . . . considerable anxiety was felt in Kuwait . . . a panic ensued among the Iraqi tribesmen, and also among the villagers, who flocked into Kuwait for refuge. At dawn on April 15th, the Ikhwan raided ten miles south of Kuwait.

Bin Saud would soon unleash his tribesmen against Yemen, Transjordan, and the Hijaz. As the Administrative Report continued: "bin Saud despatched a force to

191

renew hostilities with Imam Yahya of Yemen. At the same time a large force was assembled to attack Transjordan. A raid ten miles south of Amman was repulsed with British aid. About the same time another body of Ikhwan from Riyadh was dispatched against Hijaz."

Recognizing the impossibility of trying to drill for oil in Al Hasa while these conflicts raged, Holmes arranged for the sixty camels to be cared for in Hofuf and stored the tents and camping equipment in Bahrain. He wrote to bin Saud, who was busy attacking the Hijaz town of Taif, which he would defeat by August, saying he would bring the prospecting party out again the next year to work on the Al Hasa concession.[1]

IDRISSI SQUABBLES

With his hands full in Al Hasa, Bahrain, Kuwait, and also Iraq, Holmes had left his bid for the Salif salt and Farasan oil concessions in the hands of E&GSynd's Red Sea representative, the eccentric Cdr. C. E. V. Craufurd. Since his last visit to Assir in January 1923 to deliver armaments at the behest of the Colonial Office, Holmes had directed communications with the Idrissi by letters to Craufurd. This had not been a satisfactory arrangement. As Holmes remarked to Rihani in May, "Craufurd, our representative in Aden, has done nothing. He has wired me now that owing to the disturbed state of Assir nothing can be done until November."

The disturbed state was a monumental Idrissi family feud that broke out in late 1923 following the death of the ruler of Assir, Muhammad bin Ahmad bin Idrissi.

The former ruler's brother, Mustafa Al Idrissi, was installed in Hodeida. He contacted Holmes in Baghdad, offering to grant him the Salif salt concession for terms that included military supplies. The former ruler's son, Ali Al Idrissi, was installed in Jizan. When Holmes arrived, Mustafa's revolt against Ali was in full swing. Holmes found Mustafa focused on securing upfront the armament factor in the proposed terms but little interested in finalizing the serious detail of the concessions.

On the assumption that, like Arnold Wilson with Iraq and Turkey, he should reach agreement with both contenders so that after the deadly family quarrel was settled he would be holding a valid agreement whoever proved victorious, Holmes moved on to meet with Ali.

Negotiations in Jizan were a mirror image of those he had just had in Hodeida. Ali offered Holmes both concessions, Salif salt and Farasan oil, to be operative after the civil war. Holmes concluded that both parties seemed more intent on getting hold of immediate supplies of ammunition, or cash with which to purchase them, than they were on concluding investment in their territories. Leaving the door open for further discussion, Holmes told both Mustafa and Ali he would come back later, and returned to Baghdad.[2]

KUWAIT WATER

In September, Holmes successfully sank an experimental water well in Kuwait, the first on the Arabian Peninsula to tap into the artesian water sources. This success was particularly noteworthy because Anglo-Persian's managing agents, Strick Scott & Co., had previously constructed with great fanfare a water desalination plant in Kuwait. Unfortunately, it never worked. While onboard ship back to London at the beginning of October 1924, Holmes wrote Heim of this success.

Heim was ecstatic—even though Holmes had not actually drilled where Heim had recommended. "Excellent. The new well at Kuwait! I suppose you would not mention it if the water would not be good. Thus a large new field will be opened," he congratulated Holmes. "As you know, I would have commenced at Jahra where I thought the prospect of striking good water would have been less risk. The better of course if you have it even at Kuwait!" he said. Still Heim could not resist offering his expert opinion: "Now you should make the next bore at Ujair where I think there is almost no risk! Then Jubail, etc. If Iraq will be successful, what I hardly doubt, an immense field will be opened and a large Artesian Company (I would propose naming it Artesian Orient Company) should be floated."

On the way to London Holmes stopped in Baghdad. There he discussed with the Arab officials his ideas for artesian water schemes. Even after signing the Anglo-Iraq Treaty, the Iraq government was continuing to employ delaying tactics in its talks with the representatives of the Turkish Petroleum Company. Holmes used the opportunity of talking about water proposals to follow up on his own bids for the Iraq oil concessions.

His meetings with Arab officials prompted the high commissioner of Iraq, Sir Henry Dobbs, to contact London urging intervention on the grounds that "Major Holmes is making advances to King Faisal." The Foreign Office received Dobbs's request, but curiously refused to act. They took the view that if the Iraq government did not want Turkish Petroleum, "it will not be possible to force them to grant a concession to the Turkish Petroleum Company."[3]

JUGGLING OIL, WATER, AND SALT

In London, there was little point in Holmes trying to advance any matters related to the Al Hasa concession or the Nejd Oil Company. Bin Saud's Wahhabi Warriors were in fierce battle against Sharif Husain's son, Ali, leading his father's Hijazi troops. Ali's forces would be defeated in October, his father would abdicate, the triumphant Wahhabis would enter Mecca, and a yearlong siege of Jedda and the remainder of the Hijaz would follow.

Holmes recruited Thomas George Madgwick to work on the Bahrain water project. An engineer, at the time in a consulting partnership in London, during

WWI Madgwick had been involved in the sinking of water wells in Gallipoli and Salonika for the Australian and New Zealand Army (the Anzacs). He also had experience as a petroleum engineer and had lectured in this field at Birmingham University. Without giving him any information about E&GSynd's various oil prospects across the Arabian Gulf, Holmes hired Madgwick to work on water in Bahrain. The documentation submitted for his visa application gave "water engineer" as his qualifications and experience.

By November, after less than a month at home, Holmes was once again aboard ship on his way back to Baghdad. He was acutely aware that his every move was watched and his mail intercepted. He created a secret code that he now used for all cables, and occasionally in letters. He warned the effusive Arnold Heim: "We must not mention one word to any single person about our business . . . we require to be most secretive in all ways. . . . Doctor, you must be most careful to keep everything to yourself (do not even tell your baby)."

He had not long arrived in Baghdad when he received a cable—in code—from his London office: "We consider of the utmost importance you obtain Salif Salt Concession as soon as possible as Italian activity liable to absolutely prejudice our own position. Suggest you proceed immediately to settle with Idrissi."

Still seeking arms and ammunition, Ali had turned to the Italians, and Holmes's colleagues feared the only item of value he could offer in trade would be the concessions. For Holmes, this was an unwelcome interruption. He was right in the middle of setting up water projects; salt was far from his mind. He replied, "Assistant Resident-Aden states present time is apposite [*sic*] for me to revisit Ali Idrissi. I have Salif salt concession position and I shall go to Assir immediately Ali Al Idrissi sends for me. Position in Iraq favourable I have opened negotiations for water schemes."[4]

The directors of E&GSynd were giving Holmes all the backup they could muster for his Iraq negotiations. Edmund Davis's son-in-law, Edmund Janson, continued to be a strong Holmes advocate and admirer. Janson wrote to Holmes,

> Enclosed is a copy of letter and cable I have sent to the Prime Minister of Iraq. We have got Lloyds Bank for myself, the National Provincial Bank for Mr. Davis and the Eastern Bank for Mr. Tarbutt to cable out to the manager at the Eastern Bank at Baghdad that Eastern & General Syndicate and its directors are capable of carrying through any undertaking which it may enter into. I think this should satisfy the Prime Minister.

Janson understood that the matter of secrecy was "high priority." For this reason he had "purposely not cabled to the Prime Minister direct" but "through Beirut." Janson said he thought as soon as Holmes had "got this business fixed up" it was "most essential" that he "get on with the salt business right away" because particular investors in E&GSynd were "pestering" him. In closing, Janson reminded Holmes, "don't forget the salt."[5]

Despite Holmes and Janson's best efforts a report appeared on November 22 in the *Baghdad Times* that said: "Major Frank Holmes who negotiated the Al Hasa Oil Concession with Sultan bin Saud is proceeding to Bahrain in about ten days. A Mr. Madgwick, who is to act as geologist for the E&GSynd, is due to arrive at Bahrain on December 18th and Dr. Arnold Heim, the prominent Swiss geologist, is expected in the first week of January. They will be equipped with an up-to-date drilling outfit."

Anglo-Persian had been feeling relatively secure. Jacks had proposed to "send Mackie to Kuwait at the end of this month *ostensibly* to inspect our oil agency there." Thanking Major More for his offer to put Mackie up in his own home, Jacks had had no sense of urgency that he should actively pursue the Kuwait and Bahrain concessions. It was "impossible to spare Mackie any earlier," he said.

When the article appeared in the *Baghdad Times*, Anglo-Persian swung into action immediately. They addressed a strongly worded complaint to the Colonial Office, saying: "We understood that, while our claims to priority cannot be recognised after the 31st March next, HMG would in the meantime prevent other applicants from instituting negotiations for the Bahrain oil rights. It appears however that E&GSynd is about to undertake actual drilling operations on the island." The company's London head office demanded that the political resident in the Persian Gulf "be instructed to prevent E&GSynd from beginning these operations and indeed from negotiating at all with the Shaikh of Bahrain for the present."

Madgwick arrived on January 1, 1925, to find the resident had issued orders that he could not enter Bahrain. Madgwick later described his welcome: "I found my landing at all on the Islands hotly contested by the Anglo-Persian Oil Company. The Resident had forbidden it, and it was only on the very active intervention by the Bahrain Political Agent that I was allowed ashore, more or less on sufferance."

The situation proved most unfortunate for Arnold Heim. On the eve of his departure, he received an urgent message from E&GSynd "that I should defer my voyage to Baghdad for two months owing to political reasons." As he wrote to Holmes, he understood that "some difficulties" might be caused by his presence, but he was anxious to carry through the plan "to continue investigations in Nejd, go to the Interior and study the question of boring artesian water." Heim suggested that he could reach Nejd via Bombay "so that my presence would not interfere and would probably not be difficult." Remembering his experience of Arabia's heat, Heim was extremely anxious "not to run into July." He wrote Holmes, "It would be a great pity to let the cool season pass again which would enable me to work three or four times more than before."

The Colonial Office sent a sharp directive to Anglo-Persian endorsing Holmes's boring of two artesian wells in Bahrain "on the opposite side of the island from that in which the presence of oil has been indicated."

At the beginning of February 1925, the Political Department circulated a further "secret" minute on oil concessions in Kuwait and Bahrain. It said:

Anglo-Persian have made no progress in their negotiations with the Shaikh of Kuwait. They have been equally unsuccessful in Bahrain. It is quite clear that innocent as may be the immediate plans of Major Holmes, the E&GSynd stand to gain much influence with the Shaikh of Bahrain if the water wells are successfully sunk. They should have little difficulty in securing an oil concession after the March 31st expiry of the Anglo-Persian claim to priority.[6]

ANGLO-PERSIAN EMPLOYS *AGENT PROVOCATEUR*

T. L. Jacks, Anglo-Persian's joint general manager with Arnold Wilson in Abadan, now launched an extraordinary escapade involving the Government of India's political agent in Kuwait, Major More, and the very special talents of Hajji Abdulla Williamson, former Arab horse trader, former Bedouin—and former British spy.

Jacks proposed that outright bribery of officials surrounding the rulers was the only sure way to get the shaikhs to sign with the Anglo-Persian Oil Company. In a letter to More, Jacks alleged that "Holmes appears to keep himself in sympathy with the Advisers of the Shaikh of Kuwait, as also the Advisers of the Shaikh of Bahrain, by means of presents suitably distributed." Displaying the Orientalist view taken by many British in the Persian Gulf of the people among whom they lived and worked, Jacks told More, "and I fully realise the influence this action has on the Eastern Mind."

Incongruously, he reasserted the general opinion of Anglo-Persian personnel about the negligible prospects on the Arabian littoral, saying that the Bahrain and Kuwait concessions "when and if obtained are of little, if any, real value until costly geological examination followed by actual testing of certain areas has been carried out." Nevertheless, with or without oil, Jacks had built his career on being tough with the natives. He had a reputation to uphold and he wanted to get the Kuwait concession—by fair means or foul.

He laid out a plot for getting it by foul. With his personal convictions about the workings of the "Eastern Mind," Jacks was unable to conceive that Holmes could have achieved his success with the shaikhs other than through corruption. "My own feeling in this matter is that the Advisers of the Shaikh of Kuwait will endeavour to hinder the shaikh from actually granting the concession as long as there is any prospects of their getting further bribes from Holmes," he explained to More.

Coming right to the point, Jacks told More: "It has occurred to me that, if we were definitely in a position to guarantee payment of suitable cash rewards immediately following the actual granting of a concession to us, we might successfully overcome the antagonism to Anglo-Persian which has been worked up by Major Holmes." Automatically assuming the Kuwait political agent would share his line of thinking, Jacks said, "I should very greatly appreciate your views in this connection and an indication of what you consider would be the total amount of suitable rewards."

While certainly not dissuading Jacks in his confidential reply, Major More did advise caution. He took a swipe at Holmes, saying, "I have no actual knowledge of Major Holmes having given any presents to anyone in Kuwait except to the shaikh himself. You no doubt know that he gave him a Kelvin motor launch last year but for some reason it has never gone well and the shaikh did not appreciate it a bit." The shaikh of Kuwait, said More, liked being "feted" as he had been on his visit to Anglo-Persian's refinery in Abadan. More suggested that as Jacks was one of the "pillars" of the Basra Turf Club, an invitation to a race meeting would be well received.

Artfully, he continued: "When I say that I have no knowledge of Holmes having given presents to anyone else in Kuwait I do not for a minute mean to say that he did not do so." He said that Holmes's "right hand man," Muhammad Yateem, had stayed in the home of the shaikh's personal secretary and secretary of Kuwait's Council of State, Mullah Saleh, "and probably he was not allowed to be out of pocket." He went on: "I should fancy that if Holmes did give such presents he would not do so himself but through some Arab agent of the Yateem type." This was, said More, exactly what he "presumed" Jacks intended to do if he proceeded with "trying the plan." Almost apologetically More added, "I am afraid I cannot advise you as to the amount to give in such presents."

Jacks was gratified that More shared his point of view and commented, "My own opinion is that if Holmes is to meet with any success he cannot help watering the ground which he would only do as you suggest through an Arab agent." Not to be outdone by More in the morality stakes, Jacks remarked piously: "I should only be inclined to adopt a similar action if assistance rendered led to our obtaining the concession whereafter presents promised would be distributed. I trust however that such action will not be necessary as it is distasteful to say the least of it and opposed to any procedure adopted by this company."

Jacks had just received a cable from Mackie informing him that the shaikh of Kuwait was "still procrastinating" and advising that "the only possible way of pushing matters is through the medium of Mullah Saleh." Jacks said that Mackie had told him More was "inclined to agree with this view . . . but considers it very dangerous for Mackie himself to approach Saleh."

Not to worry, Jacks had the solution to this predicament. Neither he, nor Mackie, nor More would risk getting their hands dirty, because Jacks already had the perfect candidate for the job. He would send Hajji Abdulla Williamson. "He is our Gulf Agency Inspector and he will go down ostensibly to look into agency matters in Kuwait. I have explained matters to him and I will let you know if he is able to obtain information by back-door methods calculated to assist us," Jacks promised.[7]

Writing to the resident, who was now Lt. Col. Francis Beville Prideaux, who had replaced Colonel Trevor, More described Hajji Abdulla Williamson. "He is an Englishman born in England who I believe ran away to sea as a boy. He led a roving life in America as an officer on an American whaling boat in the South Sea Islands, etc., and finally gravitated to Aden some twenty-five years ago."

The titles *Hajji* and *Abdulla*, More explained, resulted from Williamson having "'got religion,' as the British soldier calls it, and belonged to some queer sect or other, from which he became a Muslim, quite a genuine one I think. He has lived as an Arab ever since." Williamson had turned up "soon after we got to Basra and was used for intelligence, chiefly counter-espionage work, all through the war," More wrote. More did not fully detail for the resident the fact that Williamson was on a very similar mission right now, in Kuwait and Bahrain, while on the payroll of the Anglo-Persian Oil Company.

HAJJI ABDULLA WILLIAMSON

At the outbreak of WWI, the Bristol-born William Richard Williamson, alias Hajji Abdulla Fadhil, had lived "native" in Arabia and Mesopotomia for over twenty-five years. He had left England in 1885 as a thirteen-year-old cabin boy on a tea clipper bound for Australia.

If his adoring biographer is to be believed, Williamson jumped ship in San Diego, worked as a cowboy in California, mined for gold in Nevada, was shanghaied aboard a whaling ship in San Francisco, spent time in a Philippines prison, escaped to Hong Kong, and worked his way to Bombay onboard a P&O liner. Working his way out of Bombay as a sailor, he got his first glimpse of the Arab world at Aden. All this before he was nineteen years old.

He joined the Aden Constabulary, a branch of the Indian police. Under the personal sponsorship of the sultan of Lahaj, Fadhil bin Ali, William Williamson, now aged twenty, converted to Islam. He adopted his sponsor's name of Fadhil and began his Arabian adventures by joining a caravan going through the Yemen and Hijaz to Mecca.

When the war broke out, Hajji Williamson was in Basra with an Arab wife and children. In Mesopotamia he had married and divorced his first wife, an underage girl of Zobair. His second wife, also divorced, was a girl of the Shemba people. His third was from the Sadhun people of Basra. While he was in Baghdad, he took a fourth wife, an Indian Muslim from a family living in Mesopotamia.

He joined what his biographer describes as "the Secret Service of Advanced Intelligence . . . much of his work, especially counter-espionage, was done in secret." After the war, in 1919, Arnold Wilson's Mesopotamia Administration awarded Williamson the well-paid post of deputy collector of customs.

At the height of the Anglo-Persian Oil Company's efforts to rid itself of Frank Holmes, Arnold Wilson called in a favor. Wilson sent for Williamson in late 1924 and hired him under the intriguing title of "inspector of gulf agencies." Perhaps not surprisingly, his biographer remarks that while he was allotted a family bungalow at Abadan "most of his time was spent on the Arab side of the Gulf."

Arabian Adventurer, Williamson's biography published in 1951, does not mention this clandestine project aimed at undermining Frank Holmes on behalf of the Anglo-Persian Oil Company. The author does praise Holmes's "individual initiative in putting Arabia on the map as a great oil producing land." The author observes that Holmes and William Richard Williamson "detested each other" and remarks: "One imagines there would not have been room in the same desert for two such individualistic and forceful characters."[8]

WILLIAMSON'S SPECIAL SKILLS

Hajji Abdulla Williamson arrived in Kuwait during a most unusual cold snap in which the temperature plummeted below freezing and snow fell. This, however, did not cool his enthusiasm for the job at hand. In the manner of a French farce, J. B. Mackie left for Bahrain on the steamer that brought Williamson. Holmes was in Bahrain "but expected shortly in Kuwait" at which point Williamson would leave Kuwait and join Mackie in Bahrain. Within hours of his arrival Williamson was cabling Jacks, giving his address in Kuwait as "through the Political Agent" and reporting that "letters have been received from Holmes but I have been unable as yet to find out their substance but hope to get some information in a day or two."

He quickly followed with the first of his "back-door" information reports that began: "On my arrival here I found the whole of the leading class against the giving of the oil concession to Anglo-Persian." The rich pearl merchants, he said, worried that their pearl divers might find an alternative source of income. Currently unable to earn anything much between seasons, the divers were "forced to borrow from the merchants long before the season opens and therefore are always in debt and under the thumb of these merchants." Williamson explained that if Anglo-Persian, or another company, developed oil fields either at Kuwait or nearby, "the diving class would be enabled to earn money between seasons and so get out of debt and then command their own labour prices."

The majority of Kuwaitis were against Anglo-Persian and "for E&GSynd," Williamson reported. He said this was because "they have been given the impression by Holmes and his friends that Anglo-Persian is really the British government and as soon as they acquire the oil concession they will take over Kuwait as a British Possession and, as in Bahrain, interfere with the liberty of the people." Williamson's response may not have soothed Kuwaiti fears as he told them, "Bahrain was no criterion" because "it might happen anywhere when the ruler was weak and his people divided that a strong Political Agent might take over the direction and government of the place."

Williamson coerced a servant of the shaikh of Kuwait, who was being sent with letters from Shaikh Ahmad to bin Saud in Al Hasa, "to report carefully on all he can find out about Major Holmes and his movement in that part of the world." He also managed "several conversations and interviews" with Mullah Saleh, whom he described as "the principle man who matters and who if he comes over to our side will be able to do more to persuade the shaikh than all other advisers." He convinced Saleh to put in writing his thoughts on the objections to Anglo-Persian. This was not very enlightening, however, as Saleh's note amounted to the information that the shaikh of Kuwait intended "delaying any decisions."

Without giving any evidence, Williamson concluded that Saleh "has been promised for himself one per cent above all payments made to the shaikh and has already received a good present through Muhammad Yateem whom he put up when he was here with Major Holmes."

It does not seem to have occurred to Williamson that this "good present" may have amounted to the equivalent of room and board, traditionally given when leaving a host's home after an extended visit. Nor did he make the connection that Anglo-Persian's negotiator, Mackie, "put up" in the home of the Kuwait political agent, could just as easily be construed as conflict of interest.

While conceding that in his terms for the Kuwait oil concession, "Holmes made larger promises and better offers" than Anglo-Persian, Williamson recommended to Jacks "a 'grant' of at least Rs5000 or perhaps Rs8000 to be given privately to Mullah Saleh."[9]

THE BRIBE

Jacks was very excited by Williamson's report. He cabled More, asking him to tell Williamson to "indicate" to Mullah Saleh that "assistance leading to the signature of an agreement before March 31st will be substantially rewarded." More replied that Williamson was "conveying your message to the person you mention" but was of the opinion "and I agree with him, that it will not do much good unless a specific amount is named." Williamson was now on his way to Bahrain but "if necessary he can be back here in a few days."

Jacks was well pleased. As he told More, he believed that "If we could indicate to Saleh right away that we are prepared to recompense him for services rendered this indication might, pending definite sanction from London as to the amount, prevent him from flirting with Holmes should the latter again appear on the scene."

While they had no knowledge of the machinations in the Gulf involving the Kuwait political agent, Anglo-Persian's general manager, and the not-so-ex-spy, the officials in London were watching developments. The Colonial Office had already refused Anglo-Persian's request to prevent E&GSynd from conducting water-drilling operations on Bahrain.

They were now considering two other requests from the company. First, Anglo-Persian wanted the priority period extended beyond the March 31 deadline for the Kuwait negotiations. And, second, they wanted the Colonial Office to withdraw its insistence that exemption from payment of custom dues in any Kuwait concession be limited to ten years.

Meantime, Jacks, More, and Williamson were doing their utmost to outflank Holmes, by any means, before the deadline expired. Sure that he now had the trump card, Jacks cabled More on February 9: "We have received sanction from London to pay Mullah Saleh 100 sterling down and 500 more if the agreement is signed with us before March 31st. We have therefore wired to Williamson who is now in Bahrain to return to Kuwait at once and report to you. Would you kindly convey this information to him? We send you herewith Rs1500. Would you be good enough to hand it over to Williamson when he reports to you?"[10]

TARGETING MADGWICK

Whether or not it came via Williamson's "back-door" methods in Bahrain is not recorded, but Jacks now became aware that Madgwick was an engineer of wider experience than originally thought. Jacks responded to this information by launching a personal attack on Maj. Clive Daly, the Bahrain political agent, who had allowed Madgwick entry to Bahrain.

Sounding undiplomatic, as if it were the Anglo-Persian Oil Company running the Gulf and not the Government of India, Jacks stated grandly: "When the question of allowing or suspending the sinking of artesian wells in Bahrain was under discussion, one point to which I gave considerable weight was the question of the geological information which an experienced man might obtain for E&GSynd by this means." When he had "decided to waive my objections to the work," Jacks told Daly, he had been "greatly influenced" by Mackie's report "that you had seen Madgwick's papers" and thought he was "merely an artesian well expert and had been engaged on that work all his life and that as far as you knew he had no connection with oil."

Jacks had now heard "privately" that Madgwick was an "extremely well informed Petroleum Geologist possessed of wide experience in Russia, Rumania, Mexico and elsewhere and that he was actually at one time a Lecturer in Petroleum Technology at Birmingham University." Jacks contended that Holmes, "although for the moment debarred from active negotiations in Bahrain," was actually "leaving no stone unturned" to obtain valuable information. Jacks forwarded a copy of this letter to Major Daly's superior, the political resident Lieutenant Colonel Prideaux.

Daly strongly defended himself. "I gave Mackie all the information which we had received from the company about Madgwick's qualifications. I also said that I had watched him since his arrival in Bahrain and had learned nothing that led me to suppose he was associated with the oil activities of Holmes," he replied. "Mackie was of the opinion that it would be detrimental to Anglo-Persian's interests to oppose the sinking of the water wells as this would afford an opportunity to your rivals to antagonise the shaikh against you." Daly reminded Jacks that "you had yourself written to the Resident expressing the same opinion, i.e., that it might be a bad thing to oppose the work."

Daly was pretty annoyed by Jack's high-handedness and wrote: "I have also seen your letter to Lieutenant Colonel Prideaux in which you say I did not advise Mackie that I thought it inopportune to present Anglo-Persian's draft concession in view of the decision we had taken to do all we could to induce the shaikh to visit Abadan." With some irritation Daly pointed out: "I did not advise Mackie because he never consulted me. Nor did he, at any time, mention that he intended to present or had presented a draft concession. I only learned this after he had left Bahrain, from Shaikh Hamad, who told me that Mackie had given him some 'conditions' which I imagine to be a draft concession. As I knew nothing of the matter I was not in a position to express an opinion."

In his annoyance Daly scribbled angrily on Jack's letter the handwritten comment: "Anglo-Persian ought to be able to find out who geologists are. Jacks wrote me a long time before Madgwick arrived saying he was coming. We don't keep records of geologists at the Political Agency."

On March 24, one week before the expiry of Anglo-Persian's "priority," the Colonial Office refused Anglo-Persian's request for an extension for the Kuwait negotiations and also refused to amend the ten-year limitation on exemption from custom dues in any Kuwait concession.[11]

ASSIR-YEMEN CIVIL WAR

While Williamson was snooping into his affairs in Kuwait, Bahrain, and Al Hasa, Holmes was once again in Aden, the Assir, and Yemen. He had stayed in Bahrain with Madgwick long enough to locate the sites for two water wells and set up the drilling machines. Then, as he had promised Edmund Janson, who needed to mollify particular E&GSynd investors, Holmes returned to the Red Sea, this time with little enthusiasm.

Again he discussed the two concessions. Again he found the focus on the supply of ammunition. As Holmes had suspected, Ali Al Idrissi in Jizan was trying to strike a deal with the Italians for the Farasan oil concession in return for ammunition. Ali did want to talk to Holmes about the Salif salt concession, to be paid for in cash. But Holmes concluded that the ability to award this concession "is out of his control now . . . and the general belief is that he will be driven out of Arabia altogether within a few weeks."

Writing to Rihani from Jizan at the end of March, Holmes described the situation in the area: "I have been here more than a month . . . the awful chaos that Ali has allowed to come about. He has driven everybody from his country who could be of any help to him. There only remains one or two slaves to whom he will listen." Ali Al Idrissi's uncle, Mustafa Al Idrissi, had given up the internecine struggle and returned to Egypt. But Ali's troubles had compounded as civil war flared again between Assir and Yemen.

Holmes reported that the Zaidi Imam Yahya "and the tribes of Arabia have now driven Ali Al Idrissi to the point where he has lost the towns of Hodeida, Salif ibn Abbas and Lahaj. He may possibly have to give up Nudi." Ali had held his ground in Assir, but Holmes was not impressed. He confided to Rihani: "I have not seen a man handle his country in a more disastrous manner than in my experience with Ali. It is really painful." Prophetically, Holmes noted the "general opinion" that Abdul Aziz bin Saud and his Wahhabis would come and take "what remains of Ali's territory while the Imam Yahya will retain what he has recently conquered."

Rihani replied: "And what are your plans? You are too shrewd a businessman I think to hazard anything with any government that is rapidly going to pieces." Rihani, who had dreamed of a great Pan-Arab Federation, was truly despondent.

He remarked sadly, "The whole Peninsula is in the crucible at present. What will be the result? I can but guess. The Imam Yahya has conquered Hodeida. The Idrissi house is doomed. Old Sharif Husain's glory and prestige has gone. Bin Saud is looming up mightier than ever."

It was clear to Rihani that the ideology of Pan-Arabism was fading from the scene. Fellow supporter of a grand Arab Federation, Muhammad bin Ahmad bin Idrissi, the former ruler of Assir, was dead. Despite his unstinting support of Britain's war and his sponsorship of the Arab Revolt under T. E. Lawrence, Sharif Husain of Mecca had been double-crossed in the promised postwar Arab independence. In 1922, as reported to the British by bin Saud himself, Sherif Husain had even made an offer to *abdicate in my favour* [sic] if I would join him in helping to expel European powers from Arabia and cease to be a partisan of the British and a tool in their hands for intimidation of Iraq and weakening of the Arabs by setting one against another." The British had withdrawn support of Sharif Husain in favor of bin Saud.[12]

Rihani now became openly politically involved. He worked tirelessly toward genuine independence for Lebanon, Syria, and Iraq. He advocated deeper understanding and rapprochement between East and West and tried to counter the rising influence of the Zionist lobby seeking to establish an independent Jewish state in Palestine. He became a prolific writer, in both Arabic and English. He embarked on strenuous speaking tours in Europe and the United States and lectured in universities, both in the Arab world and abroad. When he died in 1940, at the age of sixty-four, his beloved *Muluk Al Arab* (Kings of Arabia, also the title of his popular book) paid their respects at his funeral in Beirut.

IRAQ SIGNS WITH TURKISH PETROLEUM

On March 24, 1925, the Iraq government signed an agreement with the Turkish Petroleum Company. Although this had actually been agreed upon in January, its passage through the Iraq Legislative Assembly was stormy. Two ministers resigned in protest at the terms of the agreement.

In the deal as originally reached, there had been a clear stipulation that the State of Iraq would hold 20 percent participation in the concession. This was not included in the agreement put to the Legislative Assembly. Nor could Iraq achieve any return through mechanisms such as taxation. The oil company designed itself as non-profit making through the device of selling the oil to its own members, at cost price.

This device left Iraq limited to a small royalty (four shillings per ton) calculated on the oil at extraction, for twenty years. Even this would only commence once a pipeline to the Mediterranean was completed. After this initial twenty-year period, payment would be based on the market value of oil, but averaged over ten-year periods.

Furthermore, the agreement the British reached with the Turks, on Iraq's behalf, decreed a 10 percent share to be paid to Turkey on oil extracted from the province of Mosul—not to be paid by the oil company but from Iraq's paltry royalty. Any real benefit to Iraq was further excluded through the provision that the company must remain a British company registered in Great Britain with its chairman to be at all times a British subject.

In order to ward off international criticism of British "oil grabbing," a vaguely worded sop was included allowing for the Iraq government, after four years and in consultation with the company, to put up for public auction by sealed bids twenty-four concession plots of eight square miles each. As some commentators of the time noted, the company had no intention of abiding by this agreement. Nor did they have any intention of meeting the provision, insisted on in the contract by the US State Department, barring Turkish Petroleum as a bidder at public auctions. By August that year, the strip of territory in northern Mosul transferred from Persia to Turkey in 1913, known as the "transferred territories" was also recognized as belonging to Anglo-Persian's concession area.[13]

NOTES

1. Administrative Report Bahrain for 1924, "Ikhwan raided along the border with Iraq . . . Ikhwan raided ten miles south of Kuwait"; correspondence about delaying work in Al Hasa is in Heim Papers, January 2, 1925, Heim in Zurich to Holmes in Baghdad and September 24, 1924, Holmes in Kuwait to Heim in Zurich, "one month ago I wrote Sultan Ibn Saud you were coming out again next year."

2. Rihani Papers, May 25, 1924, Holmes in Baghdad to Rihani in Lebanon, "disturbed state of Assir"; John Baldry, "The Powers and Mineral Concessions in the Idrissi Imamate of Assir, 1910–1929," *Arabian Studies* 11 (Cambridge: Middle East Center, University of Cambridge, 1975), p. 86, for these internecine struggles culminating in the Mecca Agreement with bin Saud.

3. Heim Papers, October 14, 1924, Heim in Zurich to Holmes in London, "I propose Artesian Orient Co"; R. W. Ferrier, *The History of the British Petroleum Company: The Developing Years, 1901–1932* (Cambridge: Cambridge University Press, 1982), p. 586, "making advances to King Faisal" and not possible to "force them" and p. 58, Sir Henry Dobbs as an Indian civil servant had passed through Persia in 1903 and had been quite taken with Reynolds and the work he was doing on behalf of the then D'Arcy Exploration Company. At the time he had commented, "Apart from the question of the existence of oil, the undertaking will certainly fail if the Concessionaires at home refuse to trust the man on the spot (Reynolds), and undermine his authority by listening to the tittle-tattle of interested persons." Apparently, as his attitude to Holmes indicates, he no longer believed in the value of the individual.

4. A copy of this code is in the Frank Holmes Papers in the possession of Peter Mort, UK; Heim Papers, November 3, 1924, Holmes to Heim in Zurich; Holmes Papers, November 5, 1924, coded telegram, E&GSynd to Holmes in Baghdad. Reply from Holmes at the Maude Hotel Baghdad.

5. Frank Holmes Papers, December 11, 1924, from Janson in London to Holmes in Baghdad. This letter also includes the following: "Dr. Mann came in a short while ago with regard to the Sicidair concession in Persia he said he had been acting for the Phoenix Oil and had been living at Tehran for over a year and had practically got the concession for anybody who would float a loan of two million pounds for the Persian government. I took Mann in to see Davis and he seems quite keen on getting this concession so we have formed a small group in which Davis, Tarbutt, myself and E&GSynd have put up seven hundred pounds to finance Mann to go and get the concession signed. . . . Adams cabled you some time ago regarding this.

6. IOL/R/15/5/238, vol. 11, December 8, 1924, confidential, Jacks at Anglo-Persian Abadan to political agent Kuwait (More), "*ostensibly* to inspect our oil agency"; IOL/L/PS/10, vol. 989, December 15, 1924, H. G. Nichols, Anglo-Persian London to undersecretary of state, Colonial Office, "prevent E&GSynd from beginning these operations"; Ward Papers, box 1, August 11, 1936, Madgwick in Canada to Ward in US: "my early experience in Bahrain . . ."; Heim Papers, January 2, 1925, Heim in Zurich to Holmes in Baghdad: "With 5-months engagement commencing in March, I would run into July again . . ."; IOL/L/PS/10/989, January 3, 1925, Colonial Office (Amery) to Anglo-Persian London, "on the opposite side of the island" and February 2, 1925, minute, "Secret," Political Department, Oil Concessions in the Persian Gulf—Kuwait and Bahrain "should have little difficulty in securing an oil concession."

7. IOL/R/15/5/238, vol. 111, December 19, 1924, secret, Jacks at Abadan to political agent Kuwait (More), "Eastern Mind" and "total amount of suitable rewards" and January 3, 1925, confidential, More to Jacks at Abadan and January 13, 1925, Jacks to More, "information by back door methods."

8. IOL/R/15/5/238, vol. 111, January 27, 1925, confidential political agent Kuwait (More) to resident (Prideaux), "chiefly counter espionage work." For a very admiring

Boy's Own–style biography, see the story of Hajji Williamson in Stanton Hope, *Arabian Adventurer* (London: R. Hale, 1951), p. 325, "Holmes . . . put Arabia on the map as a great oil producing land" and "detested each other" and "not room in the same desert for two such individualistic and forceful characters," p. 328 refers to Williamson as "interpreter, translator and general factotum" to Archibald Chisholm in Kuwait in 1934.

9. Dickson Papers, box 3, file 6, Kuwait Political Report 1925 (More), "during the last 10 days of January the cold was intense the thermometer falling below freezing for several nights in succession—and snow actually falling on one occasion"; IOL/R/15/5/238, vol. 111, January 22, 1925, Williamson in Kuwait to Jacks at Abadan, "a strong Political Agent might take over" and "be given privately to Mullah Saleh."

10. IOL/R/15/5/238, vol. 111, January 28, 1925, telegram, Jacks at Anglo-Persian Abadan to political agent Kuwait (More), "be substantially rewarded" and "prevent him flirting with Holmes"; and February 9, 1925, confidential, Jacks at Abadan to political agent Kuwait (More), "sanction from London to pay Mullah Saleh."

11. IOL/R/15/2/96, vol. 1, March 6, 1925, Jacks at Anglo-Persian Abadan to Bahrain political agent (Daly), "decided to waive my objection to the work"; and March 11, 1925, Daly to Jacks, "I have also seen your letter to the Resident" and "I was not in a position to express and opinion"; Archibald H. T. Chisholm, *The First Kuwait Oil Concession Agreement: A Record of the Negotiations, 1911–1934* (London: F. Cass, 1975), p. 12, "Both requests were refused on March 24, by the Colonial Office."

12. Holmes Papers, February 13, 1925, Dorothy Holmes in England to Heim in Zurich, "My husband is somewhere in the neighbourhood of Aden I believe now. . . . Things seem rather upset in that part of the world at the moment"; Rihani Papers, March 29, 1925, Holmes in Jizan to Rihani in Lebanon, "the awful chaos"; Rihani Papers, April 17, 1925, Rihani in Lebanon to Holmes in Baghdad, "the whole Peninsula is in the crucible"; Gary Troeller, *The Birth of Saudi Arabia: Britain and the Rise of the House of Sa'ud* (London: F. Cass, 1976), p. 178, bin Saud reported that Husain had declared, "ready to *abdicate in my favour*. . . ."

13. Heim Papers, February 17, 1925, Heim in Zurich to Holmes in Aden, "E&GSynd told me on Jan 21st that according to your news the Iraq Government has granted the oil concession to the Turkish Petroleum Syndicate and that you will cable from Aden about arrangements for me." Also Rihani Papers, March 29, 1925, Holmes in Jizan to Rihani in Lebanon, "I see that Anglo-Persian got the Mosul oil"; Benjamin Shwadran, *The Middle East: Oil and the Great Powers* (London: Atlantic, 1956), p. 248, "the company had never meant to abide by this agreement." Note that on April 20, 1932, the British Oil Development Company obtained a seventy-five-year concession covering all the lands in the *vilayets* of Mosul and Baghdad west of the Tigris and north of the thirty-third parallel. Ten years later this concession was transferred to the Mosul Petroleum Company, a subsidiary of Iraq Petroleum (renamed from Turkish Petroleum). On December 4, 1938, the Basrah Petroleum Company, also a subsidiary of Iraq Petroleum, obtained a concession covering all lands not included under previous concessions.

CHAPTER 9

WATER AND WAR

Watched by crowds of chattering locals, drilling for water in Bahrain went on day and night. At the site, Muhammad Yateem confided to Holmes: "They think you are mad. They call you the magician from England who expects to find sweet water in the bowels of the earth. They think it is all very funny." Holmes was quite happy to keep his colorful audience amused, and replied: "One day I'm hoping they'll wake up to find we *are* magicians." Holmes and Thomas Madgwick, assisted by two drillers, including one who had brought his wife along for an exotic experience, soldiered on.

Two days before expiry of Anglo-Persian's privileged negotiating period, a fountain of water gushed from the ground. The hundreds of villagers who quickly gathered wondered aloud at the tiny blue fish flapping and flopping on the ground where the water fell. It seemed to them that Holmes truly was a water wizard. As he had predicted, Holmes had tapped into an underground river. Shaikh Hamad threw a great feast with Frank Holmes as the guest of honor.

Immediately after his night of being feted by Hamad, Holmes headed back to the Red Sea. He was now under some pressure to obtain the Salif salt concession. The syndicate's chairman and chief investor, Edmund Davis, had accepted a British government offer to head up the prestigious British South Africa Company. The appointment required Davis to divest most of his directorships and attached investments. The loss of Davis came uncomfortably close to the failure to secure Iraq, or the Red Sea, and Arnold Heim's negative report on the Arab concessions. Although control of E&GSynd passed uncontested to Davis's son-in-law, Edmund W. Janson, and his longtime partner Percy Tarbutt, the transition period was causing a severe curtailment of the capital immediately available to the syndicate for ongoing work and development of existing investments.

Janson and Tarbutt had managed to bring in two new investors to the Eastern & General Syndicate, Alexander Lawrie & Company and Paterson Symons and

Company. The new shareholders were interested in the oil possibilities of the Farasan Islands. But, stretching Holmes's patience, they were very keen to obtain the still unawarded Salif salt concession. Finally, in April, Holmes would obtain a signed concession for Salif Salt. Unfortunately it would again slip from his hands, overtaken by political events.[1]

MADGWICK AND DALY

Since arriving in Bahrain, Madgwick had kept up a correspondence with his friend Thomas Ward, the New York–based proprietor of Oilfields Equipment Company, who was once described in *Petroleum Age* as an "Export Ace." They had worked together in the Trinidad oil fields in 1911 before the British-born Ward set up business as an equipment supplier. Madgwick relied on Ward for American consulting contracts.

He had written gaily to Ward: "Here I am, all amongst the pearls so to speak, trying to get my finger on the pulse—a very elusive one I am afraid—and just worrying myself to a skeleton how I can send a nice fat order for your equipment. Meantime," he asked Ward, "can I send you along a nice little line in pearls? Plenty of them around here and they start fishing soon. You'll have to get them strung, or whatever it is they do with them, but I expect Mrs. Ward knows all about that."

Thomas E. Ward was a true entrepreneur. He wasn't interested in a pretty necklace for his wife. He was interested in the business of dealing in Bahrain's pearls. This was more than Madgwick had expected. Suspecting he might have bitten off more than he could chew, he replied cautiously, "Pearls are a thing for experts to handle. Things are very dull in the business right now, although the fleet is out. I have made a few inquiries, based on your letter, as to what might be done but cannot find much encouragement."

He explained: "There are many merchants, Persians and Arabs, who buy direct from the boats and sell to the fancy Jews who come only for the season. They centre on Paris but whether New York could be independent of Paris I do not know. These pearls are second to none. Of course, the divers only find a 'pearl of great price' once in a blue moon."

Madgwick described for Ward the last days of Anglo-Persian's privileged negotiating period. "All this time the bazaar at Manama was haunted by sundry Anglo-Persian agents—including a picturesque renegade Englishman, one Hajji Abdulla Williamson." Madgwick told how Shaikh Hamad weathered the onslaught: "He stuck to his story that he wasn't going to go back on the word of his dear old father, etc., and so when the fateful March 31st 1925 arrived, they all packed up their traps and disappeared." He said that Anglo-Persian's permanent representative in Bahrain was a leading merchant, Yousuf Kanoo, "whose chief clerk, an Indian, I caught trying to burgle my copy of Holmes's secret code." Madgwick did not say whether or not he suspected the hand of Hajji Abdulla Williamson in the break-in.

He told Ward: "I see in recent numbers of the *Times of Mesopotamia* what purport to be the terms of the Turkish Petroleum Company . . . the British government diddled the Iraqis out of their property when others would have given much better terms. They have got it cheap with the only consideration being the four shillings per ton royalty. They have sold the Iraqis a pup."[2]

From Bahrain Madgwick reported to Eastern & General in London: "I gather the attitude of the Anglo-Persian people now is to urge the native rulers not to give any oil concessions as it would mean the over-running of the native rulers' territories with foreigners, etc. Also that Anglo-Persian is again causing trouble in London about us being here."

He acknowledged receipt of one thousand sterling and the request that he submit an "estimate of cash required to complete the Bahrain water contract." He told London that he had heard from Muhammad Yateem in Kuwait who was asking when Holmes would be back and for "other useful information to put up a good story why the Neutral Zone Concession has not been paid, on all of which I am unable to furnish data." He said he had received a cable from Holmes who "appears to be in some very out-of-the-way spot for the cable was much mutilated. I have only had one short letter from him in which he gives little account of his movements." Holmes's cable directed Madgwick to advise Yateem he "must return to Bahrain when Rs500 in addition to Rs1000 already paid to him is exhausted."

The scratching and scraping on money matters reflected the uncomfortable truth that, for the Eastern & General Syndicate, it was crunch time.

Passing on Holmes's message to Yateem, Madgwick remarked: "I wish Holmes were coming soon, I cannot say whether he has any very up to date news from this end. Practically, I am as wise in these matters as you are which is not saying much is it?" At this point Madgwick's job was concerned only with water. He was not in Holmes's confidence as to the plans for, or the location of, the Bahrain oil concession. In his letter to E&GSynd, pointing out that his four-month original contract was almost up, he said: "There was a suggestion from Major Holmes that you might like me to do a report in the Red Sea which I should feel honoured to be entrusted with."

Madgwick was taken aback to receive a letter from the Bahrain political agent, Maj. Clive Daly. Rather than joining in the congratulations on the discovery of an artesian water supply for Bahrain, Daly was proposing to cut the price agreed for the expert identification and actual sinking of the wells, suggesting that "two payment scales might be agreed." Daly's idea was that there should be "considerably less" payment if water was not found "in any given hole" and a different payment if water was "successfully tapped." Daly said he thought it would be unfair to expect to profit "from the boring of an empty hole" that would be "a total loss to the shaikh's government." He said, "from the experience gained from sinking the present well you will be in a position to locate water fairly accurately so that the risk of your syndicate being paid at the lower rate would be very small." After all, Daly said, the shaikh's government would not "desire to risk money boring holes unlikely to be productive."[3]

Madgwick angrily wrote to E&GSynd in London. "The Political Agent is just an autocrat here and means to get his pound of flesh. He has no idea of business and is firmly convinced that you are making a colossal fortune with this contract. He thinks he can get people from India to come and do the new wells for some ridiculously low figure and hence his rather naive suggestion that I should identify the locations, apparently without waiting for a contract to be signed." He warned: "If you fight him there will be nothing further doing here, the only way is to humour him. . . . Major Holmes is the only one who is able to talk to Daly."

Madgwick had soon realized that E&GSynd had interests in Bahrain other than water. As Hamad was about to depart on a visit to England, accompanied by Daly, Holmes had asked Madgwick to confirm with Daly whether he should "arrange to come to Bahrain to agree to the terms . . . or do so in London." Holmes also suggested to Daly he was ready to "undertake the installation of water supply and distribution of electrical plant with dynamo for Bahrain." Madgwick wrote to E&GSynd, "I am not in your council as to what you hope to do here . . . but Holmes and I could tie up things pretty well together, I fancy."

Madgwick's letter urging Holmes to come before Daly and Hamad left for England reached him at the Kamaran Islands, Albuquerque's sixteenth-century military base, from where he was trying to embark on detailed examinations of the Farasans. "My opinion of Daly is that, like many army people, his only idea of business is to be slim. He is a little tin god on wheels here all right and will not be overruled. . . . You are more used to the type and can handle him better," Madgwick wrote.

He was deeply offended by the devaluing of his engineering skills apparent in Daly's dismissal of "empty holes." He had thought Daly was impressed by "the pretty mushroom shaped water spout we had going," but told Holmes, "it was obviously not of use for me to quarrel with him." Now that he knew there was more at stake in Bahrain than just the water contract, he complained plaintively: "I am little in the confidence of the syndicate's affairs. I never attended a board meeting before leaving London. I do not know what the syndicate's real objective is."

He put a few ideas to Holmes for cutting costs on the new water wells, including utilizing "a percussive outfit of the American type," coincidentally, just the type as supplied by Thomas Ward. "Of course," he added enigmatically, "none of these ideas really take into consideration 'fishing' possibilities." On Holmes's electricity proposal, he said Daly had told him that "no new concessions can be given, other than the new water wells, until the oil concession is cleared." The political agent had declared that Holmes should not see Shaikh Hamad while he was in London because "it would cause trouble with Anglo-Persian."

Although Holmes was anxious to get back to London to clear reopening of the oil negotiations with the Colonial Office now that Anglo-Persian's "priority" period had ended, he hurried across to Bahrain to catch Hamad and Daly before they set off overland on May 2.

As Madgwick observed, Holmes could "handle" Daly and had no trouble reaching agreement for E&GSynd to bore fourteen water wells, seven at Manama and seven at Muharraq. As it turned out, Daly's two-tier pricing system never did become a problem as all fourteen of Holmes's water wells gave "a plentiful supply of drinkable water" and proved to be "an immense boon to the inhabitants." As he was back in the Gulf anyway, Holmes went on to make a personal visit with Shaikh Ahmad in Kuwait.

It had been a very long year, and Holmes's exhaustion showed in a letter to his wife:

Darling Dor,

Don't worry about grey hair. I have grown a bit fatter and also much greyer. I am beginning to be impatient at being out here. Days, weeks, months and years go by at a dazzling rate.

I wish I had a few more thousands of pounds then I would hand in my resignation and stick at home.[4]

HOLMES TELLS MADGWICK ABOUT OIL

As Holmes was about to set off for London from Kuwait, Madgwick reported a breakdown in the drilling engine in Bahrain. Holmes told him to get the spare parts in Basra and then meet him in Port Said.

With the embargo off at the end of March, the agreement on the water wells in Bahrain, and the prospect of an immediate water contract in Kuwait, Holmes could afford to keep Madgwick a few more weeks. There was also the opportunity presented by Major Daly being out of Bahrain escorting Hamad on an official visit to London. Holmes had no difficulty in persuading Madgwick to extend his contract by several weeks; as Madgwick had confided to Ward, he did not have a contract to follow. Holmes revealed to Madgwick his own opinion and geological observations of Bahrain and asked him to discreetly undertake an examination of the surface structures. It would seem Holmes was hoping that Madgwick's opinion would second his own. With this he could lessen the impact on the syndicate's investors of Arnold Heim's negative report.

Madgwick had a fondness for cloak-and-dagger and an eye for the main chance. Immediately on his return to Bahrain he wrote Ward. "There is (don't breathe a word) a possibility we might want to put down a geological hole in this part of the world." Madgwick was now viewing himself as integral to E&GSynd's plans. He told Ward: "When I left here for Egypt I didn't know whether I should return or not and now I am hoping to be called to London shortly. This is no place to spend the summer in and if any developments take place I should have to be in London."

In years to come Madgwick would inflate his claim to have identified the oil of Bahrain and would describe Guy Pilgrim as "no oil geologist," Hugo de Bockh as "that fancy Hungarian," Arnold Heim as "posing as an oil geologist," and Frank Holmes as "knowing nothing about oil." Yet, Madgwick had spent four months on Bahrain Island and had not noticed any indications of oil until Holmes sent him back from Port Said with instructions of where, and how, to look. Even then he would have nothing to add.

Madgwick's spurious claim to have "discovered" the oil in Bahrain would become more and more outrageous over the years, commencing after the 1932 Bahrain oil strike and continuing for almost a decade. He would barrage E&GSynd, Thomas E. Ward, the Gulf Oil Corporation, and Standard Oil of California with letters seeking "more generous treatment than I have so far received for my share in the work on the Bahrain Islands." Madgwick would claim he reached "the conclusion that Bahrain was probably a big oil field . . . before I left the Island" and that he "sold my enthusiasm to Holmes." He would further claim that it was he who "called Edmund Janson's attention to the really encouraging prospects" and that "it took some persuasion to get the syndicate to regularise the oil concession as the water seemed to be simpler and they knew nothing of the oil business."

In his outlandish claims, Madgwick would completely ignore the inconvenient fact that Holmes had originally mapped the Bahrain field in 1922 and concluded a concession in 1923, three years before Madgwick set foot on Bahrain. Nor would he ever mention that he did not arrive back in London until July 1925, six weeks after Holmes. By that time, rather than needing "some persuasion" as Madgwick would tell it, Janson and Holmes were already in advanced discussions with the Colonial Office for formalizing the Bahrain concession and negotiating that of Kuwait.

Madgwick had studied Pilgrim's work, lent by Major Daly for its assessment of water. What he actually said to Janson and E&GSynd in London in July 1925 was that, as far as Bahrain's oil was concerned, he had found it "impossible to go into more detail than Pilgrim had done." He said the approaching hot weather had deterred him, and getting in "better equipment" to work with "did not fall within his duties." As he would write Ward, "my work was done. I came merely to find water." The advice he did give to Janson and E&GSynd about Bahrain's oil was the fairly safe assumption that "only the drill could decide."[5]

DUKE OF DEVONSHIRE'S HANDIWORK

The Colonial Office assured Holmes and Janson in London that the British government now had no objection to E&GSynd negotiating for oil concessions in

Bahrain and Kuwait. There were, however, two conditions. First, negotiations could not be conducted directly with the shaikhs but only through the Government of India's political resident in the Persian Gulf and, second, the concession details must first be approved by the Colonial Office.

As neither seemed to present any difficulty, Holmes and Janson waited patiently to be advised if the Colonial Office wanted to amend any of the terms and conditions of the concessions as submitted eighteen months earlier to the shaikhs of Kuwait and Bahrain, and which the shaikh of Bahrain had accepted and signed.

Instead of reviewing Holmes's contracts the Colonial Office suggested replacing them entirely with the document originally penned personally by Victor Cavendish, Duke of Devonshire. This was the draft concession the duke had forwarded to Anglo-Persian in 1922 after Arnold Wilson and the acting political resident, Lieutenant Colonel Knox, had botched the job by presenting to the shaikh of Kuwait the one-sided combined contract of their own devising.

If this was what the Colonial Office really wanted, Holmes said, it was fine by him, but, he warned, the shaikhs might not like it as the terms were far less than Holmes had already offered. More than that, he told the Colonial Office, the shaikhs liked the idea of participation in the company through a shareholding much more than they liked the idea of royalties as applied to Persia and Iraq. The shaikhs of Bahrain and Kuwait were no different from bin Saud in this preference, Holmes said, and this was the system adopted by bin Saud for the Al Hasa concession.

The Colonial Office hesitated to authorize Holmes to offer shareholding, given the terms just imposed on Iraq, the original concession with Persia, and the agreement recently offered by the D'Arcy Exploration Company to the sultan of Muscat, none of which offered participation. It was finally agreed that Holmes would use the Colonial Office document for the Kuwait and Bahrain oil concessions and add the improved financial terms as previously offered.

The Colonial Office failed to notice one important difference in the two sets of documentation. E&GSynd's original concessions contained a specific clause in which the syndicate undertook not to transfer the concession to a non-British company. The only relevant clause in the Colonial Office document, now superseding E&GSynd's own, merely stated: "The right conveyed by this lease shall not be conveyed to a third party without the consent of the shaikh acting with the advice of the Resident in the Persian Gulf. Such consent shall not be reasonably withheld." This difference in clauses would return to haunt the Colonial Office.

The political resident was officially advised on September 3 that he should no longer obstruct Frank Holmes in concluding agreements with the shaikhs of Bahrain and Kuwait on behalf of E&GSynd. Once again Holmes headed back to the Gulf. Besides the oil concessions in Bahrain and Kuwait, he had the ongoing Bahrain water project to oversee; although Madgwick's contract had ended, the two drillers were still working there. There were new water projects to discuss with the shaikh of Kuwait and a check of the situation in Al Hasa, Iraq, Assir, and Yemen with which to attend.[6]

THE SHAIKH OF QATAR

Holmes made his calls to Kuwait and Bahrain. But as the formal Colonial Office instruction to open negotiations had not yet arrived, Holmes dropped across to Qatar for a nice little chat with the ruler, Shaikh Abdulla bin Qasim Al Thani. It was purely a courtesy call because, since his 1923 discussion with the resident about responding to the shaikh of Qatar's requests to conclude a concession, the Colonial Office had decreed that on "general political grounds" such negotiations were "unsuitable." The Colonial Office had advised both E&GSynd and Anglo-Persian of this decision.

Jutting out from the mainland, midway along the Gulf coast just a few kilometres south of the Bahrain Islands, Qatar was a limestone peninsula of about fifteen hundred square kilometers. It was arid and windswept, supporting mere clumps of date palms, some coarse grass, and the occasional stunted brushwood. At the most optimistic estimate its population would not top twenty-five thousand. Many among the settled population were the children of slaves brought to the Gulf from East Africa in the nineteenth century to work in the pearling industry. Qatar's biggest settlement was the fishing village of Doha with a population of maybe ten thousand.

When his father died in 1913, Abdulla had successfully governed Doha for seven years. At the time he was also a thriving pearl merchant and was not all that keen to accept his father's instruction that he take over the rulership, besides he had twelve brothers who were eager for the job. Nevertheless, he did obey his father's will and became ruler just in time for the outbreak of WWI. Wisely, he kept his head down and did not defer to either side. In 1915 the handful of resident Turks representing the Ottoman Empire, although never much in evidence, suddenly decamped from Doha, thus, by default, making Qatar independent. This was an intolerable situation for the British, who promptly stepped in and "persuaded" Abdulla to conclude a treaty in 1916 that placed Qatar on the same footing with Britain as were the surrounding shaikhdoms.

As with the other Arab rulers, Holmes had an easy personal friendship with Abdulla bin Qasim Al Thani. He visited in the early evening at Abdulla's hunting camp. Like the Arabs themselves, Holmes loved coffee and tobacco. He had bags of his favorite coffee beans shipped up on a regular basis from Yemen and was a dedicated chain-smoker, preferring Egyptian tobacco. As the men settled into comfortable positions among the cushions piled on the Persian carpet and prepared to relax with the tiny cups of fragrant Arabic coffee and the tall water pipes called *nargeela*, the ruler of Qatar's hunting dogs entered the tent.

Similar to a greyhound, salukis, or as the Arabs sometimes called them, "gazelle hounds," are depicted on seven-thousand-year-old pottery fragments, six-thousand-year-old Anatolian cave paintings, and in forty-five-thousand-year-old Egyptian tombs. Prized for their skill and speed—at full stretch they can hit fifty miles an hour—and considered clean animals by the Bedouins, these "royal" Arabian dogs were permitted to enter the women's quarters and at the hunting camps enjoyed the privilege of curling up loyally close to their master.

Holmes's eye fell on a handsome female with a coat of feathered golden hair. "I know this dog," he declared. "Her mother is Hoja." The gathering of Qatari notables around the tent sat up in amazement. Throughout the Middle East, the highly regarded saluki was never bought or sold. They were presented as a mark of great honor. The dog that Holmes was now regarding had arrived to the shaikh of Qatar in just such a fashion, as a personal gift from Shaikh Hamad of Bahrain. Forgetting the Arab protocol of delaying any talk of business, Abdulla Al Thani excitedly exclaimed to the assembly: "If this man can identify one dog among so many, surely he can identify where our oil is hidden," and openly proclaimed his intention of awarding Qatar's oil concession to Frank Holmes.

What Holmes neglected to mention was that he had presented this dog's mother, Hoja, to Shaikh Hamad. Over dinner one night Hamad had sadly remarked that Bahrain's once-pedigree Amiri dog pack seemed to be on the brink of extinction because of careless breeding. In London, Holmes had tracked down a breeder through the UK Kennel Club and purchased a fine young female that had subsequently been mated in Bahrain. Hamad called her Hoja, which means "lightheaded." She would revive the breed, and Shaikh Hamad would be famed across the Gulf for his royal salukis.[7]

Unfortunately for Holmes, the Colonial Office ruling prevented him immediately taking up Abdulla Al Thani's offer. Moreover, the shaikh of Qatar was himself restrained by his 1916 signature to the standard agreement not to grant any concession "without the consent of the High Government."

JOHN CADMAN

The Anglo-Persian Oil Company had a new man in charge. Currently deputy chairman, John Cadman would take over from Charles Greenway as chairman in 1927. As a longtime confidant of powerful people in government, he already wielded a lot of influence. Born in 1877, Cadman was a science graduate from Durham University whose first job was inspector of Scottish coal mines. He spent 1904–1907 setting up the Mines and Petroleum Department for the Government of Trinidad and 1908 with the British Royal Commission on Mines. In 1910 he was professor of mines at Birmingham University where he established the Department of Petroleum Technology.

In 1913 Winston Churchill sent John Cadman to accompany Adm. Edmond Slade on the mission to supply a report that Churchill could put to Parliament on the operations of the Anglo-Persian Oil Company.

It was Cadman and Slade's glowing report that silenced critics of Churchill's campaign for the 1914 British government purchase of 51 percent of Anglo-Persian. Soon after, Cadman was appointed director of the British government's Petroleum Executive and a member of the Inter-Allied Petroleum Committee. As Sir John Cadman, he joined the Anglo-Persian Oil Company in 1921, was made a director in 1923, and became deputy chairman in 1925.

Immediately upon joining Anglo-Persian, Cadman delivered a lecture in Chicago to the Second Annual Meeting of the American Petroleum Institute in which he rallied his audience against "fussy, ill-conceived legislation" in favor of workers that "placed artificial barriers between employers and employees" and the inclination of "certain countries" to impose taxation on crude petroleum and its products. He defended the Anglo-Persian Oil Company against what he called "misunderstanding and misrepresentation" surrounding British activities in Mesopotamia.

Referring to the "great controversy that has raged in the newspapers," Sir John Cadman assured his audience: "The ownership of the oil deposits in Mesopotamia will be secured to the newly constructed Irak State [*sic*] as part of the administrative arrangements under the Treaty with Britain, and the Mandate. The interests of the Arab State must be carefully safeguarded and are the first charge upon the government who is to exercise the Mandate." This was, of course, Britain. Until the day he died, Cadman would stick to the line that, in exploiting the oil of Persia and Iraq, at vast profit, Britain was actually doing these countries a favor.

As Holmes was visiting with his Gulf friends, Anglo-Persian's managing director in London reminded Cadman that "the question of concessions for Kuwait and Bahrain remain in suspense." He would be glad to know, he said, "the degree of importance you attach to these concessions." Cadman didn't think they were at all important. Like all Anglo-Persian's technical people, he didn't think there was any oil in Arabia. His personal opinion was "we should let the concession question drop for the time being." Almost as an afterthought, he said he would nevertheless bother to "ascertain the views of our technical people" regarding the possibilities of Kuwait and Bahrain.[8]

Maybe Cadman had not quite caught up with what was expected of him in the Persian Gulf, but the Government of India's political officers were as determined as ever to keep Holmes out of their area of influence.

CONTINUED OPPOSITION

In Bahrain, Capt. George Leslie Mallam was filling in as political agent while Daly was abroad. Continuing an established tradition among the political officers, he reported on Holmes's every move. He wrote the resident that he had attempted to find out from Holmes whether "any definite indications of the presence of oil on the islands" had been discovered.

Holmes didn't fall for it. He told Mallam that the Bahrain concession would be "pure wild catting." He put him off the scent, saying that because of the "present difficulties in connection with the Al Hasa concession," he wanted to bore in Bahrain in order to obtain "some indication of the geological conditions prevailing on the mainland." Holmes played it down even further by telling Mallam that E&GSynd were "prepared to lose as much as 40,000 sterling before giving up" on Bahrain.

Mallam told the resident he was sure Holmes wanted to get the Bahrain concession as "a feather in his cap" because of the "struggle he has had" in obtaining backing from the Colonial Office. He was positive, Mallam said, that Frank Holmes "has no definite prospect of finding oil in Bahrain."

Like his predecessors, Mallam was not about to grant Holmes an easy passage. He confided to the resident, "I presume that, although instructed by the Colonial Office to raise no objection to the conclusion of an oil concession on the lines of the draft agreement, there would be nothing to prevent my pointing out clearly to the shaikh the consequences which are likely to result from the grant of such a concession, and to advise him, purely from a Bahrain point of view?"

He would, he suggested, remind Shaikh Hamad that Bahrain's "real" industry was pearl diving. He would draw Hamad's attention to what could happen if the pearl divers found work "in a new industry such as oil, which if found in commercial quantities is likely to rival and even eclipse the pearling industry."

He would explain to Hamad "the danger of the divers giving up their old calling and engaging themselves in the more lucrative and less toilsome work of employees in a flourishing oil concern."

Mallam proposed that he would frighten Hamad by sketching a scene in which the oil could "peter out . . . leaving Bahrain a sucked orange." In an attached confidential note he added, "I cannot believe Shaikh Hamad did not accept the advice of His Majesty's government."

Meanwhile, Cadman had been brought into line. Anglo-Persian now claimed that as E&GSynd had sent a party of geologists under Arnold Heim in 1924, without having a concession, then they should be permitted to do the same.

They claimed they needed the information "to solve certain geological problems of a regional character" raised through the work of Professor de Bockh and his colleagues in Persia. In particular, they said, such information could, conveniently, be obtained in Kuwait. By this time, news had got out of the shaikh of Qatar's declaration at his hunting camp. The regional geological information they were seeking, Anglo-Persian now claimed, would be enhanced by a "reconnaissance" across Qatar.

The Colonial Office gave in and granted Anglo-Persian's request, provided that the individual shaikhs agreed. They reminded Anglo-Persian that "another concern" had been negotiating with the shaikh of Kuwait for "some time past" for an oil concession in his territory. Whether or not the shaikhs would require payment for Anglo-Persian's proposed geological venture was "one for settlement between your representative and the Shaikh," the Colonial Office stated, and added, "His Majesty's government are not in a position to give any undertaking on this subject."[9]

MADGWICK'S MISCHIEF

Unknown to Holmes, or any member of the Eastern & General Syndicate, Madgwick, in London for some four months and still with no prospects other than Holmes's Bahrain oil concession, was trying to create an opportunity for himself. He had written his friend Thomas E. Ward in New York of his pressing financial difficulties, including the cost of private schooling for his two children, and the lack of any offers of work. By October he was complaining: "I have been home since July and am not yet certain whether I return to the East but there is a chance. Things are very dull still." Madgwick seemed to imagine a possibility as he put it to the wealthy, entrepreneurial, and well-connected Ward:

> STRICTLY [*sic*] between ourselves, do you know anyone who would come in and test the area for oil where I have been lately? I think that a group over here will do it but I am by no means certain, and if you do, it won't do any harm to let me know.
>
> My friends have the concession all right and it might be wise to do it in their name, but it is really a testing spec. It is a case for test drilling under easy conditions as to accessibility and once you struck oil you have a big field proven. We ought to know soon if the people here will do anything and, if not, I should like to be able to suggest that my taking a trip over to see you would be worthwhile. I need hardly say that the big groups who operate in that part of the world are keen on it but I don't want to let them get in.
>
> This is a case where you and I could put by a little something for our old age because I think it is a good thing.

Perhaps conscious that he was betraying a trust, and with the only people with whom he currently had any chance of employment, Madgwick added: "Only just let me know generally what your ideas are. That is all I want at the moment. I will let you know any developments."[10]

HOLMES REGAINS BAHRAIN

It was eight weeks since the Colonial Office sent its formal instructions in the matter of the Bahrain and Kuwait oil concessions. The political officers could not delay longer. But still the resident could not resist giving Anglo-Persian another free kick. When he did finally go to Bahrain to tell Shaikh Hamad he could now conclude an agreement with Holmes, and to deliver the new E&GSynd concession document as approved by the Colonial Office, the resident took J. B. Mackie of Anglo-Persian with him.

He did not include this snippet in his memorandum to the Colonial Office. The resident did not understand that Anglo-Persian's belief that there was no oil made that company hesitate to commit to more than a geological survey. The resident also could not see that, in absolute contrast to Anglo-Persian's hesitation, Holmes's conviction that the Arabian littoral was rich with oil had always made him eager to get started immediately.

The resident took one last shot, proposing to the Colonial Office a four-point plan by which he could "oppose the signing of the concession" in Bahrain. He, would, he said:

(a) Press the Shaikh of Bahrain to refuse further consideration of the question for ten years. He would tell Hamad he should do this as a "sign" of his "real contrition" for breaching the treaty "engagement" signed by his father in 1913.

(b) He would refer to the current discussion of an Anglo-Saudi agreement and point out to Hamad that "while the result of negotiations between HMG and Abdul Aziz bin Saud is uncertain" Hamad could not be sure of "pleasing HMG by following bin Saud's lead."

(c) He would suggest to Hamad "that there is a possibility even of E&GSynd quarrelling with bin Saud."

And curiously, considering that it was Holmes who had found the water, he would tell Hamad that "now it is proved that good water can be tapped all over the northern half of his territory, the development of agriculture should be his chief interest, and the admission of an oil company will interfere both with this and with the pearling industry."

The resident wrote to Arnold Wilson warning him the draft concession had been sent to Baghdad for Arabic translation and was "expected to reach Bahrain in a few days time." He assured Wilson he had "telegraphed to the Colonial Office suggesting grounds on which Holmes's agreement may be held up indefinitely." But, he added, he and Wilson would have to face facts. If the Colonial Office did not agree with his four-point plan, "there is nothing more to be done."

While this final opposition to Holmes's revival of his Bahrain oil concession hung in the balance, Madgwick's mysterious hints of fortunes to be made were striking a chord with Ward. He was as curious as he had been about Madgwick's earlier suggestion that he take to a trade in pearls. He referred to "your suggestion in the matter of the Bahrain property" and declared he would "do my darnedest." He thought "it certainly sounds very promising" and was sure that

Madgwick "would not recommend anything but a good thing." He had a friend who "has been successful in organising a number of syndicates in Venezuela." Not unreasonably, Ward assumed that Madgwick was a principal in the prospect he was offering and assured him he would "do the needful with your property."

On December 1, the resident wrote to Arnold Wilson admitting defeat. The Colonial Office "after 25 days consideration of my four-point suggestion" had replied they could see no justification in "obstructing the grant of the Bahrain oil concession to the Eastern & General Syndicate."

On December 2, 1925, Shaikh Hamad bin Isa Al Khalifa and Frank Holmes signed an agreement granting the Bahrain oil concession to E&GSynd—a full two and a half years since the original agreement of May 1923.[11]

The political agent, Major Daly, recorded this event in the official 1925 Administrative Report for Bahrain. He noted "a concession for the exploitation of oil in Bahrain was granted by the local government to the Eastern & General Syndicate." Daly, like Anglo-Persian, and most of the British, did not believe there was any oil to be found in the Arab shaikhdoms. In the very next entry of this annual report, headed "Trade," Daly displayed little optimism for the future of the oil concession. He advised that in the absence of any stable industry other than pearling, "the Shaikh of Bahrain may be well advised to consider the possibility of the introduction of some industry." He suggested sailmaking might be just the thing.

BIN SAUD CONQUERS THE HIJAZ

The year 1925 closed with Abdul Aziz bin Saud's triumphant entry into Jedda on December 23 after a ten-month siege following the conquest of Mecca. On December 25 bin Saud officially announced that the Nejd-Hijaz War was ended, and on January 8, 1926, he adopted the title King of the Hijaz and Sultan of Nejd and its Dependencies.

Bin Saud and his Wahhabi Warriors had set the Hijaz ablaze. The British had encouraged the October 3, 1924, abdication of Sharif Husain of Mecca in favor of his son Ali. Sharif Husain moved first to Jedda and then to the Red Sea port city of Aqaba. This did not save Mecca. Bin Saud's fighters entered the traditional seat of the Hashemites on October 13. But still bin Saud was not satisfied. After the fall of Jedda, he set about rousing his followers to expel the elderly Sharif Husain from Aqaba by force and to mount an attack on the Hashemite territory of Ma'an, where Ali was residing.

Turning again on their one-time ally and partner in the Arab Revolt of WWI under T. E. Lawrence, the British ensured Husain's ignominious departure. Accompanied by HMS *Delhi*, the armored cruiser HMS *Cornflower* arrived in Aqaba harbor. From onboard, a message was sent to Sharif Husain. It read:

To King Husain

From

The Deputy Minister of the Foreign Office of Great Britain

The Government of Great Britain has learnt that the Sultan of Nejd has prepared a force to attack Aqaba where the Government of Your Majesty is at the moment. His Majesty's government therefore urges you to leave Aqaba so as not to become the cause of new problems between Britain and the Sultan of Nejd.

With no other choice Husain retired, an angry and bitter exile, to Cyprus. Ali handed Aqaba and Ma'an over to his brother, Amir Abdulla of Transjordan, and eventually decamped to join his other brother, Faisal, in Iraq. Five years later, after Sharif Husain suffered a severe stroke, the British "allowed" him to go to Transjordan to be with his sons. He died in Amman in 1931 at the age of seventy-five.

In return for British approval of his new title and conquests, bin Saud would sign the Treaty of Jedda in 1927 that recognized Husain's sons, Faisal as king of Iraq, and Abdulla as ruler of the Hashemite kingdom of Jordan.

In Assir, Hassan Al Idrissi, another member of the monumentally quarrelling family, had overthrown Ali Al Idrissi and embarked on a strategic move. Down the coast from the Idrissi town of Jizan, toward the Gulf of Aden, the Zaidi Imam Yahya was gathering military strength and threatening the Idrissi from Hodeida. In the other direction, up the Red Sea coast, Jedda and Yanbu were barely holding out under the siege by bin Saud's Wahhabi Warriors. Hassan Al Idrissi decided to secure at least one border and submitted to Abdul Aziz bin Saud.

In July 1926 the British Consul at Jedda would receive a letter from bin Saud stating that he had been "invited" by the Idrissi ruler of the Assir, the province between the Hijaz and Yemen, "to take over that government." The Mecca Agreement, signed shortly after between Hassan Al Idrissi and Abdul Aziz bin Saud, would transform the Assir into a virtual Saudi protectorate. This would leave the Idrissi free to concentrate on the threat posed from Yemen by the Zaidi Imam Yahya.

The Salif salt concession that Holmes had so laboriously worked for, and finally obtained in April 1925, fell through in the face of these developments.[12]

HARRY ST. JOHN PHILBY

One person who cheered Abdul Aziz bin Saud's conquests was Harry St. John Philby, whose Bedouin-escorted plunge across the desert in 1917 had increased the suspicion and tension between the two Arabian leaders, Sharif Husain of Mecca and bin Saud of Nejd. On the heels of the October 1924 capture of Mecca, Philby had moved to Jedda where, he had convinced a group of London financial backers, his personal acquaintance with bin Saud would serve to gain him a commercial foothold in that ruler's ever-expanding territory.

Since resigning his position as chief British representative to Amir Abdulla —the Hashemite ruler of Transjordan (just hours ahead of his intending sacking) Philby had been talking to a company called Midian, which was attempting to resume oil and mineral concessions in the Hijaz originally granted by the Ottoman government and discussed while Sharif Husain was ruler of Mecca.

After his retirement from the Indian civil service, Philby thought he had clinched a job with Midian for "embarking on negotiations with bin Saud" for ratification of these concessions. Midian, however, soon became embroiled in complications about succession to Turkish assets involving Anglo-Persian and its subsidiary the D'Arcy Exploration Company.

With only his civil service pension coming in, Philby was in need of immediate employment. He contacted Remy Fisher, a wealthy man with interests in the city of London. To induce Fisher to back him financially, Philby sent his wife Dora, whom Fisher admired, on a holiday to Fisher's home.

He "made it plain" to Dora, as Philby's biographer coyly put it, "that the exercise of her charms . . . would serve all their interests." Together with a few of his city friends, Fisher advanced Philby seven hundred fifty sterling to travel to Jedda to see what business could be initiated with bin Saud.

Philby managed a meeting with Abdul Aziz bin Saud in November 1925 while bin Saud was riding high on his takeover of the Hijaz. Philby launched into a long explanation about what had gone wrong in his job with Abdulla of Transjordan. He portrayed himself as a victim and vowed his current plight had come about only because of his unswerving personal loyalty to bin Saud.

He declared he had been "more or less forced" to retire from service with the British government because "I have always backed you, bin Saud, against Sharif Husain and his sons." Philby would repeat this story endlessly through the years and in this way exert his claim that bin Saud was obliged to him.

Although the new king of the Hijaz made no promises about concessions, Philby lingered in Jedda. After persistent requests, he finally received bin Saud's permission to found a branch of a British trading company in Jedda.

Philby would return in October 1926 to set up a business he named Sharqieh (Eastern) Limited—the Company of Explorers and Merchants in the Near & Middle East. On a salary of twelve hundred sterling per annum Philby would leave his wife in England, assuring her, "I have a job, certain for three years, and that ought to be enough to test the possibilities." Over the next decade, he would become an intractable and cunning foe in his efforts to obtain Holmes's Al Hasa concession.[13]

NOTES

1. Wayne Mineau, *The Go Devils* (London: Cassell, 1958), p. 183, "they call you the magician from England"; Percy Holmes Papers (Bolivia), undated handwritten letter from Percy Holmes's daughter, "blue fish came out with the water . . . from an under-

ground river." The Bahrain water well flowed on March 27, 1925; Chevron Archives, box 120791, December 28, 1928, E&GSynd corporate secretary (Adams), memorandum of Colonial Office meeting explaining that Davis was required to divest much of his directorships and shareholding in other ventures. See Thomas E. Ward, *Negotiations for Oil Concessions in Bahrain, El Hasa (Saudi Arabia), the Neutral Zone, Qatar and Kuwait* (New York: privately published, 1965), p. 24, the Salif salt concession on which Holmes was working prompted Alexander Lawrie & Company and Paterson Symons and Company to subscribe for shares in the syndicate.

2. Ward, *Negotiations for Oil Concessions*, p. 24, "worked together in the Trinidad oilfields in 1911"; Ward Papers, box 2, February 26, 1925, and June 3, 1925, Madgwick in Bahrain to Ward in New York, "pearls are a thing for experts to handle"; Ward Papers, box 1, August 11, 1936, Madgwick in Canada to Ward in New York, writing for Ward what Madgwick referred to as an account of "my early experience in Bahrain."

3. Holmes Papers, March 30, 1925, Madgwick in Bahrain to the secretary of Eastern and General Syndicate, London, March 30, 1925, "one short letter from Holmes"; Madgwick in Bahrain to Yateem in Kuwait, "I wish Holmes were coming soon"; Holmes Papers, April 3, 1925, political agent Bahrain (Daly) to E&GSynd via Madgwick in Bahrain, "profit from the boring of an empty hole."

4. Holmes Papers, April 6, 1925, Madgwick in Bahrain to E&GSynd London (Adams), "Holmes and I could tie up things" and April 9, 1925, Madgwick in Bahrain to Holmes in Kamaran, "I do not know what the syndicate's real objective is"; Administrative Report for Bahrain 1926 (Maj. C. C. J. Barrett), "immense boon to the inhabitants"; Holmes Papers, May 4, 1925, Holmes in Kuwait to Dorothy at MillHill, "Days, weeks, months and years go by at a dazzling rate."

5. Ward, *Negotiations for Oil Concessions*, p. 24, says Holmes met Madgwick at the Marine Palace Hotel, Port Said. Holmes "requested Madgwick to return to Bahrain and make a fairly close examination of the surface structure with possibility in mind of drilling a test well for oil"; see the Ward Papers for numerous letters of claims by Madgwick including his claim to have "discovered" Bahrain's oil. It is most unfortunate that Madgwick's invented version appears in Ward's own 1961 self-published record and is then imported unquestioningly into most subsequent works on Arabian oil discovery and development.

6. PRO/CO/ 727/11, June 9, 1925, minute by J. H. Hall, Eastern Division, Colonial Office. Holmes and Janson had two meetings with the Colonial Office on this matter, on June 9 and June 18, 1925; PRO/CO/727/11, July 24, 1925, minute by J. H. Hall, Eastern Division, Colonial Office records Holmes arguing in favor of offering shareholding; note that oil exploration did not begin in Oman until 1954, the first oil was produced in 1967, until then the country relied mainly on exports of dried limes, dates, frankincense, and tobacco, worth about one million sterling annually.

7. The story of *Hoja*, and the discussion in Shaikh Abdulla Al Thani's tent, were told to me by Husain Yateem, the nephew of Muhammad Yateem, in Bahrain in 1988; See also Holmes Papers, April 9, 1925, Madgwick in Bahrain to Holmes in Kamaran, "The puppy has evidently reached Bombay."

8. American Heritage Center, Cadman Papers, press release, Friday, December 9, 1921, "As John Bull Views It," an address by Sir John Cadman (formerly His Majesty's Petroleum Executive) before Second Annual Meeting American Petroleum Institute, Congress Hotel Chicago on December 8, 1921; Archibald H. T. Chisholm, *The First Kuwait*

Oil Concession Agreement: A Record of the Negotiations, 1911–1934 (London: F. Cass, 1975), p. 111, exchange of letters headed "Persian Gulf Concessions" between Anglo-Persian managing director (London) H. E. Nichols, October 1, 1925, and Sir John Cadman, deputy chairman, October 2, 1925.

9. IOL/R/15/2/96, vol. 1, October 11, 1925, acting political agent Bahrain (Mallam) to political resident (Prideaux), "more lucrative and less toilsome work of employees in a flourishing oil concern" and "like a sucked orange"; IOL/R/15/5/239, vol. 4, September 23, 1925, Colonial Office to Anglo-Persian London. See also Chisholm, *First Kuwait Oil Concession Agreement*, p. 12, and also IOL/R/15/2/875, vol. 39/4, June 11, 1932, resident (Biscoe) to secretary of state for the Colonies, memorandum, "Background to oil negotiations in Qatar."

10. Ward Papers, box 2, October 30, 1925, Madgwick in London to Ward in New York, "Strictly between ourselves."

11. IOL/R/15/1/649, C30, October 31, 1925, resident to secretary of state for the Colonies details plan for "opposing the signing of the Bahrain Concession" and November 2, 1925, resident to Wilson at Abadan, "Mackie with me"; IOL/R/15/1/649, December 1, 1925, resident to Wilson at Abadan, "I suggested grounds on which the agreement could be held up indefinitely . . . (but now) there is nothing more to be done"; Ward Papers, box 2, November 11, 1925, Ward in New York to Madgwick in London. Ward's friend was "John Alvin Young. He was formerly vice president of one of the companies with which I was associated."

12. Bin Saud captured Taif in August 1924, Mecca capitulated on October 13 (following the October 3 abdication of Sharif Husain), followed by Jedda and the remainder of the Hijaz in November 1925; Abdul Aziz and Moudi Mansour, *King Abdul-Aziz and the Kuwait Conference, 1923–1924* (London: Echoes: 1993), p. 109, for text of message sent to King Husain from HMS *Cornflower*; Gary Troeller, *The Birth of Saudi Arabia: Britain and the Rise of the House of Sa'ud* (London: F. Cass, 1976), p. 230, "invited to take over." Hafiz Wahba, *Arabian Days* (London: A. Barker, 1964), p. 165, "In 1919 the Turks evacuated the eastern part of Assir, with resultant chaos, and Abdul Aziz promptly annexed it. A year later the Idrissi tribe recognised his importance in Eastern Assir and ceded him the remainder of the province. (Later, in 1926, the Tihama of Assir came under Abdul Aziz' protection, and in 1930, by arrangement with the Idrissi, its titular sovereigns, was finally annexed to the Kingdom of Saudi Arabia.)" See Ward, *Negotiations for Oil Concessions*, p. 24, "unfortunately the Salif salt business fell through . . . and nothing more was done about the salt concession."

13. For details of Philby's relationship with Midian, Remy Fisher, and Sharqieh see H. St. J. B. Philby, *Arabian Oil Ventures* (Washington, DC: Middle East Institute, 1964), pp. 8–34, and Elizabeth Monroe, *Philby of Arabia* (London: Faber and Faber, 1973), pp. 140–59, "the exercise of her charms . . . would serve all their interests." Monroe, on p. 147, "he told bin Saud that his current plight was because of bin Saud" and E. C. Hodgkin, ed., *Two Kings in Arabia: Letters from Jedda, 1923–5 and 1936–9* (Reading, UK: Ithaca Press, 1993), pp. 48–49, Reader Bullard report from Jedda (1923–25), "Philby has quarrelled with three administrations, in India, in Iraq and in Transjordan. . . . Anybody who is inclined to regard Philby as a hero for resigning his job in Transjordan rather than carry out a policy he believed to be mistaken should know that he has been considering for years (resigning) from the Indian civil service. He'll get a pension of about 700 sterling a year and that, for a man not yet 40 years old, is not bad." Monroe, p. 148, "Philby returned to London in March 1926, pleased with this foot in the door, but owing Fisher and his group 750 sterling."

CHAPTER 10
BIG NEW YORK SHAIKHS

Following Edmund Davis's appointment as chairman of the British South Africa Company, Edmund Janson in London was so preoccupied with rearranging the Eastern & General Syndicate and its finances that Holmes's success in reacquiring the Bahrain oil concession brought little joy.

Janson was attempting to bring in new investors, or increase capital from the existing group. As the Bahrain concession entailed the immediate outlay of the syndicate's now precious available cash plus the expense of development, from Janson's point of view it only added to his present difficulties.

The strain showed in his response to confirmation that the concession had been signed. "Just received your cable re Bahrain," he noted in a bald postscript on a letter going to Holmes. As Janson noted, the water-drilling projects "seem to be the only proposition we have right now for getting immediate revenue."

News of Holmes's success with finding water in Bahrain spread wide. Lt. Col. Francis Prideaux, an old Persian Gulf hand, was now political resident. In a very friendly letter addressed to Holmes in Baghdad, he wrote that the governor of Bushire had asked him to inquire what Holmes would charge to bore water wells there. Prideaux said he couldn't be sure that "Anglo-Persian will not object that such an action infringes their rights" and somewhat wistfully suggested, "If so, perhaps you would simply advise the Governor how to proceed?"

Prideaux had been around the Gulf since 1876 and could no longer be bothered with intrigue. He liked and admired Holmes and inquired solicitously: "I hope you have quite got over your attack of illness now. When shall we next see you passing through? I am just off in the *Lawrence* to Sharjah, Dubai, etc., for about a fortnight."[1]

SEVENTH SURVEY OF KUWAIT

Anglo-Persian's resident agent provocateur, Hajji Abdulla Williamson, was with the Anglo-Persian geologists who arrived on January 7, 1926, for what would now be the seventh geological examination of Kuwait. Maj. James More, the Kuwait political agent, put the whole party up in his own home and made the arrangements for escorts, guides, and camel transport. He and Mrs. More treated it like a series of picnics and sallied forth in their car to join the geologists on the easier trips.

The men from Anglo-Persian were keen to get into bin Saud's territory, but Major More did not have the courage to allow this. Besides, Shaikh Ahmad of Kuwait blocked them very effectively by simply refusing to permit his escorting tribesmen to venture outside Kuwait's own boundaries.[2]

They stayed a week in Kuwait before moving across to Bahrain where Holmes and Muhammad Yateem had hired Ashraf Shamsuddin as office assistant. He watched the men from Anglo-Persian closely and reported to Holmes, now water drilling in Kuwait. "It is about a week they have been here. When they arrived they prepared Yousuf Kanoo's motor launch with a lifeboat, a big quantity of fuel and enough provisions for six days. Frequent inquiries cannot tell us why." Ashraf told Holmes: "This is certain. (1) They went everywhere you have been. And (2) soon after finishing work here they wanted to go to Al Hasa and Qatar and back to Kuwait along the coast. The *Shamal* (sandstorm) has delayed them three days. But I hear they will start at noon today."

The men from Anglo-Persian made it to Qatar where they did little more than lock Shaikh Abdulla Al Thani into an undertaking that he would not grant a concession to any other party for eighteen months. Although the Colonial Office had lifted the ban on Qatar, declaring in December 1925 "there was no longer any prima facie reason for discouraging oil development in the area."

Holmes was not in a position to follow up on Shaikh Abdulla's earlier enthusiasm. He was expecting to receive the Kuwait concession any day, and development of the Bahrain, Kuwait, and Al Hasa concessions, if he could ever get into that, had priority on E&GSynd's capital.

After a four-week survey, the Anglo-Persian report would again condemn the oil prospects of Bahrain. In Qatar, eighteen months would come and go without a sign of any Anglo-Persian activity. In Kuwait, the group would reach the same discouraging conclusion as had the six previous surveys. They would report "the total absence of direct geological evidence" and advise that "no convincing case" could ever be made for exploring the "unknown depths of Kuwait territory with the drill." In short, they would confirm the "unfavourable view" of oil prospects in Kuwait.[3]

JANSON SEEKS INVESTMENT

To finance development of the concessions held and those they hoped to soon finalize, Janson was talking to Cory Brothers, a Cardiff mining company that had been interested in the Farasan Islands in 1920 and had been a regular investor in many of Edmund Davis's syndicates. Janson suggested the possibility of a joint venture. On January 26, 1926, Clifford Cory put forward a counteroffer of a set of conditions for an "option on the concessions for Bahrain Island, Al Hasa, Neutral Zone, Kuwait and the Farasan Island." The detail made it more like a takeover than the "joint" venture Janson had in mind. He told Holmes: "Of course, it's no good to us."[4]

Janson took his joint venture idea to a number of London financiers, Burma Oil, Royal Dutch Shell, and, finally and reluctantly, to Anglo-Persian. By now the findings of the latest geological report were known, besides which neither Burma nor Shell were willing to compete on what they viewed as Anglo-Persian's home turf, particularly as the company was now in such a strong position with the addition of Iraq. Sir John Cadman refused to even think about it. He had the latest geological reports and these reinforced his own opinion that the Arab concessions were "not worth pursuing," as he had personally advised the company's managing director several months earlier.

Coincidental with Cadman's strengthening grip on the reins of Anglo-Persian, Arnold Wilson had been recalled to the London head office and given the title of managing director of the D'Arcy Exploration Company. With the results of the just-completed geological survey in hand, Wilson declared that Arabia appeared "to be devoid of all prospects" and pronounced "only Albania seems hopeful."[5]

With considerable exaggeration, Wilson wrote triumphantly to Major More in Kuwait: "Holmes's London principals have been bothering us for the last few months trying to sell us the Kuwait concession before they have got it! And all the other concessions too. They say that they have no intention of working themselves. All they wish to do is sell to someone else, but we are not having any." Wilson said it was "difficult for us to go in for it" because of the Colonial Office ten-year limit on waiver of customs duty. To agree to this, Wilson said, would "prejudice our position in Persia and that of our friends the Turkish Petroleum Company in Mesopotamia." Both operations were permanently exempted from customs duty.

Wilson claimed that "sooner or later" the Colonial Office and "others" would realize that "E&GSynd have, as they have admitted to me, no intention of working the thing themselves" and so would be left with no alternative but to allow Anglo-Persian to conclude an agreement "to our liking."

All this would have been news to Janson and Holmes, both of whom were running themselves ragged in efforts to find the means for "working the thing themselves."

Marked "Strictly Confidential," More immediately forwarded Wilson's letter to the pro-Holmes resident Francis Prideaux, with the comment, "I think the content of this private letter, which I received yesterday, will interest you." He was aware that Holmes was about to leave for Bushire "at your request about some schemes for sinking water wells there," More said, immediately putting a damper on that by adding, "He probably will not take it on because he does not really like water schemes anywhere."[6]

HOLMES—THE CIVIL ENGINEER

While Janson was dealing with the moneymen in London, Holmes was taking on a number of civil engineering projects in order to generate a cash flow.

He rented a house in Kuwait and managed the sinking of two water wells within the Kuwait town walls and the building of a new palace for Shaikh Ahmad. This was a shared experience for both men. Janson was urging Holmes to return to the Red Sea, but he felt it was important to stay where he was.

As he explained:

> The Arabs deal with persons and personalities. I am the only man who has the run of Shaikh Ahmad's house and private business affairs. He consults me on many things that have to do with every engineering job. I have to go with him to pass judgement and offer my opinion. I have just got out the plans for his new palace. He has had several costs of plans and has accepted the way I laid the building out. He likes me to see to inside arrangements and sends to my wife to purchase things in London for him and his wives (four).

Holmes was in and out of Baghdad purchasing building materials and discussing schemes for electric lighting and a telephone system to install in Bahrain. He was constructing a seawall, open to allow boat passage, and a six-foot-wide road in Bahrain. He was preparing plans for a bridge "that either lifts straight up and down or that swings. The simpler the better as it has to be operated by rather primitive Arabs."

Although Madgwick's contract had ended, Holmes and the two drillers were still working on the Bahrain water projects. Holmes reported in April that six of the fourteen Bahrain wells were almost completed. In the fifth well, he told Janson, he had "come into a very nice oil bearing shale. I was using three 2-foot drills at shallow depths. If I spot one or two more good sites for artesian wells, I could progress the oil testing." If he had a five-hundred-foot star drill, he said, he could, by himself, test for oil on Bahrain Island.

Holmes was working at a punishing rate. He did manage to get across to Assir where he again acquired the Farasan Islands oil concession, signed, this time, by Hassan Al Idrissi. He completed a water well for free public use and a water-storage tank in Bahrain and constructed a water reservoir with drainage and water supply fittings for the political agent's complex. The work on the Kuwait water wells continued as did his efforts to obtain contracts for artesian water in Iraq and an oil concession there.

Intent on involving Holmes in a project in French Sudan, Arnold Heim wrote repeatedly from Romania where he was on a consultancy for Shell. He drew up a contract saying: "The aim of this agreement is to obtain and to realise one or several concessions for hydrocarbons in French Sudan, especially South of Timbuktu, where Frank Holmes has discovered large deposits of asphaltum and sulphur." He urged Holmes, "regarding French Sudan, please don't forget your promise to send me as much information on the location as possible, so that I can become prepared." Holmes didn't have time to think about it.[7]

THOMAS E. WARD—ENTREPRENEUR

Neither Holmes nor Janson were aware of Madgwick's attempts, after Holmes had drawn his attention to the oil possibilities of Bahrain, to sell the concessions out from under them. In January Madgwick had again failed to inform Ward in New York that he was not a principal with E&GSynd and, more importantly, that the syndicate had no knowledge of the offer he was making. Instead, Madgwick assured Ward that "the thing to do, as you say, is to put up an attractive proposition to your friends and then we shall be all right."

Madgwick boasted about the success of "his" water drilling in Bahrain, commenting that "it is well to impress the Orientals." The shaikh "was hinting at rewards in the event of success so I hoped at least for a nice coffee pot," Madgwick said. "I don't want a saluki," he added emphatically.

He had written again in March, apologizing for not replying sooner, saying, "My friends have been away." Whom he was referring to is not clear, as Janson was certainly in London. Madgwick told Ward: "Roughly, the situation is this. E&GSynd have a concession over the whole of the Province of Al Hasa, on the south side of the Persian Gulf, together with one over the Islands of Bahrain." His sweeping statement that "the syndicate are not themselves interested in oil and are willing to hand the whole thing over to anybody who will work it," would surely have angered Holmes and Janson had they known.

Still implying that he was a principal in E&GSynd, Madgwick said that in his opinion, "Bahrain is the most interesting. . . . I advised them that a deep test is justified on present evidence." He said that the mainland should be "looked over" although the only part he had seen was Kuwait, "a separate country, not included in the concession but which could be acquired." Telling him to get hold of Pilgrim's work, Madgwick assured Ward, "You will have no difficulty about terms with the syndicate." The thing to do, he said, was to "let me know and I will at once get what information you ask for and come over to New York with it."

In anticipation of receiving all the technical information from Madgwick, the energetic Ward began talking to American geologists, including those of the Gulf Oil Corporation, about oil possibilities in the Persian Gulf. Neither Janson nor Holmes had the faintest idea their interests in Arabia were being discussed across the Atlantic. In April, Madgwick told Ward that possible interest from the Gulf Oil Corporation was "a happy omen" and repeated his assurance that "my friends here are willing to hand it over on easy terms as they do not want to do anything themselves."

Ward was now arranging a consultancy job for Madgwick with the Canadian government in Calgary. Meanwhile, he had a four-week job in Egypt but didn't think this would be a problem because "there is no hurry about going out to the Arabian Gulf now as the hot weather is coming on and the winter could be more convenient. But we could have everything cut and dried before that."

Now that he had ensured employment in Canada, Madgwick attempted to dampen Ward's enthusiasm for the ventures in Arabia that he had previously worked hard to foster. "Now listen," he wrote. "I hope to have a few valuable ideas about Russia to put before you." He had "access to a source of intelligence in Russian matters that cannot be equalled," he told Ward, adding, "I have at least two Russian ideas worth considering."

When he reached New York in August 1926, thirteen months after leaving the Persian Gulf, Madgwick found Ward still very interested indeed in Bahrain and keen to follow through on the preliminary efforts he had already made with the American oil industry. Saving face, Madgwick wrote to Janson. Implying that discussion of E&GSynd's business had come about by accident, Madgwick told Janson he had discussed Bahrain "in general" with friends in New York. "Mr. Ward is associated with influential people who have been extremely successful in organising companies in foreign fields," he informed Janson enticingly. "I think it could result in business and therefore urge you not to lose time in forwarding particulars of terms that you would consider, if they came into it, also full details of the concessions—and any geological information that you may have."

Janson's interest was piqued. He wrote directly to Ward, saying Frank Holmes was scheduled to visit Canada and the United States "on other business" and could "call on" Ward in New York. Belying Madgwick's assertion that E&GSynd "did not want to do anything themselves," Janson stressed to Ward that "we have decided to begin boring ourselves" in the Arabian concessions. It was because the syndicate had "such a huge tract of land," he said, that they might be interested in being "associated with another group." He told Ward: "Major Holmes will be empowered to conclude a deal, if the prospect attracts you."

For his part, Madgwick dismissed E&GSynd, the syndicate he had until recently so much wanted to be a part of, saying: "Their idea of drilling themselves is a farce. They have no geological advice and they will not put up the heavy plant which alone will do the job satisfactorily." Grandly, he advised

Ward: "I think you will find that if you relieve them of all further expenditures you will get good terms from them." Pompously, he intoned, "Holmes is a decent sort. Of course knows nothing of oil," adding, "I can quite believe that Holmes is a useful person to keep in the fore when dealing with the natives."

Madgwick's opinion was that "most, if not all, the Al Hasa concession is likely to prove worthless. Bahrain Island is the main thing. If they have Kuwait that will require very careful surveying as there are possibilities there, and also the Farasan Islands in the Red Sea."

Finally Madgwick informed Ward he was not, in fact, a principal in E&GSynd. "I have *NO* [*sic*] arrangement with Janson about where I come in. I think you had better arrange for me to be on your side. No immediate cash consideration but preferably an interest, with a possibility of realising it should I need the money. Do the best you can for me." Still enthusiastic, Ward replied, "If the information we secure from Major Holmes is favourable we shall go ahead with our negotiation with the New York syndicate and, of course, you will be taken care of, through me." He asked Madgwick to send him "your written opinion on Bahrain" together with "any other data you believe would assist discussion."[8]

Via Canada, where he did not contact Madgwick, Frank Holmes, accompanied by his wife, arrived in New York on September 21, 1926.

HOLMES IN NEW YORK

Ward and Holmes hit it off immediately. "I like Holmes very much and he has created a favourable impression being plain spoken with facts at his fingertips. He leaves nothing to conjecture," Ward reported to Madgwick.

Holmes laid out the sort of arrangement he would discuss with the Americans. He stressed to Ward: "You should make it absolutely clear to your friends that the first deal must be made with the Bahrain oil concession." Once a group had committed to Bahrain, Holmes said, the possibility of an option over all the other concessions "obtained, or to be obtained by Eastern & General Syndicate, on the two areas, Persian Gulf and Red Sea," could be discussed with the same group. Holmes told Ward he had with him, to show what he called "the really big New York Shaikhs," some rock samples and "substance" samples, and he would detail his own technical findings and speak of the oil traces he had found, particularly in the Bahrain wells.

He summed up six years of his work into an outline of the current position "of the oil concessions that have been, and are in progress of being, negotiated in Arabia and the Red Sea." It read:

1. Bahrain Concession, area 320 square miles, (under British protection) covering the Bahrain Island in the Persian Gulf. Concession completed.

2. Al Hasa Concession, area 40,000 square miles. Granted by Sultan bin Saud, King of Hedjaz and Sultan of Nedj Arabia. Concession complete but owing to Sultan bin Saud being at war, annual payments of 3,000 sterling for past two years are in arrears.

3. Neutral Zone Concession, area 2,700 miles. Granted jointly by bin Saud and Shaikh of Kuwait. In same position as Al Hasa Concession.

4. Kuwait Concession, area 3,600 miles. Being negotiated with Shaikh of Kuwait at the present time.

5. Red Sea Area: Farasan Oil Concession. Approx. 300 square miles. This concession has been granted but the concession grant contains two points not agreeable to the Colonial Office, so our representative (Craufurd) is at present in the Red Sea adjusting this with Hassan Al Idrissi, the Ruler of the Territories in which the Farasan Island which the concession covers is situated.

Note: The concession having a clear title at the moment, is the Bahrain Concession. This concession is of considerable prospective value. The Farasan concession is, in my opinion, of equal, if not greater prospective value than the Bahrain Concession.

(a) The payments due on the Farasan Concession would amount to about 12,000 sterling to the Eastern & General Syndicate, which would satisfy certain sums due on this concession as time progresses. At present all payments due have been made. This concession agreement is the same as the Bahrain concession except that two points mentioned above are being altered. These points refer to sale of oil won from the concession and no doubt will be altered at an early date.

(b) There is a certain position established and being confirmed in regard to the Mosul area (Iraq). This could come in also but is not sufficiently advanced to speak on definitely.

Madgwick's "written opinion on Bahrain" did not, as Ward had hoped, provide any material that he could use to interest his American contacts. Madgwick began by admitting that he had not actually done any survey work and so couldn't draw a map, couldn't even pinpoint the center of Bahrain Island. Nor had he collected any sample specimens, fossils, or rocks, except for a single coral.

His two-page "Opinion on the Oil Prospects of Bahrain" was basically a rewrite of Pilgrim's 1908 report. He said he could not offer anything new on the asphalt occurrence examined by Pilgrim because, since then, the pits had "largely fallen in." He repeated his advice that "a deep test is essential" and although he personally could not indicate the place, "it ought to be possible to make a location which should definitely determine whether or not oil does occur in payable quantities."[9]

CRAUFURD—ARMS AND THE FARASAN CONCESSION

While Holmes was seriously negotiating in New York and including the possibility of the Farasan concession that he had secured after so much effort, the eccentric Cdr. C. E. V. Craufurd was at the center of a fiasco playing itself out on the Red Sea that would result in the loss of this concession for E&GSynd.

The renewal of Holmes's contract for the Farasan Islands oil concession had proved useful to the British in their diplomatic struggle with the Italians. The resident in Aden cabled London that because the Zaidi Imam Yahya was getting arms from the Italians—the Italians had just supplied Sana'a with two heavy guns—the supply of arms to the Idrissi was now "necessary."

Holmes's concession would make it possible to disguise an influx of weaponry to the Idrissi "as a means of keeping the Farasan Islands concession in British hands" and for "the immediate protection of the concession area." The resident pointed out that a substantial fringe benefit would be that British armament manufacturers would get the business. The British made it known to the French, Belgian, Italian, and Japanese governments that they were lifting their ban on the export of arms to Yemen and Assir, but, of course, "such operations would be made within the terms of the Arms Traffic Convention."

An extraordinary episode of comic opera dimension now unfolded involving E&GSynd's representative in Aden, Commander Craufurd, the Shell Oil company, the Colonial Office, the Foreign Office, British officials in Aden—and one hundred boxes of ammunition.

Temporarily breaking away from his obsession with the search for King Solomon's Mines, Craufurd threw a monkey wrench in everybody's works. In his opinion, Holmes's offer was unnecessarily generous for a "native ruler."

So, after Holmes's departure, Craufurd unilaterally altered some of the terms and conditions and held back the second payment while he waited for the Idrissi to agree to the altered document. News of the resulting furor in Assir reached Egypt where it provided an opportunity for the Alexandria office of Anglo-Saxon Petroleum, a subsidiary of Shell, to declare that they held a valid concession over the Farasans, signed in 1923 by Mustafa Al Idrissi who now lived in Egypt. To induce agreement from the current ruler, Hassan Al Idrissi, Anglo-Saxon announced they were happy to supply arms and ammunition to Assir as part payment.

Further complicating matters, the Italians sent their governor of Eritrea to hold talks with the Idrissi rival, the Zaidi Imam Yahya, that resulted in the Italo-Yemen Treaty. Following this diplomatic coup, talks with the British were held in Rome during which the Italians pressed a claim to trade with the Farasan Islands on the grounds that this would help the economy of Eritrea. The Foreign Office declared it could not "allow" any European power to establish itself on either the Farasan or Kamaran islands.

In view of the Italian threat to the Farasans that Britain believed to be imminent, the Colonial Office cabled the resident in Aden, authorizing E&GSynd to immediately supply one hundred boxes of bullets to the Idrissi "from any source." Rather than wait for a shipment to arrive from Europe, the resident raided the Residency's own stores and sent Craufurd off from Aden with one hundred boxes containing one hundred thousand rounds of Le Gras ammunition to deliver to Hassan Al Idrissi.

While Craufurd was winding his way to Jizan, Mustafa Al Idrissi lodged an official request, also signed by Hassan Al Idrissi, that the resident supply the representative of Anglo-Saxon with one hundred thousand rounds of Le Gras ammunition for delivery to Jizan in part payment for the same oil concession.

The Aden resident threw up his hands at the dilemma this presented. He cabled the Colonial Office for a decision on which of the two British companies, E&GSynd or Anglo-Saxon, actually held the concession. The reply was an example of true British diplomacy. The Colonial Office sanctioned the resident to issue ammunition to Anglo-Saxon but stressed that this did not mean the British government recognized the claims of that company.

The resident was now obliged to lend Anglo-Saxon one hundred cases containing one hundred thousand rounds of ammunition from the Residency's stock. There was just one problem. The Residency was bare and Craufurd could not return his one hundred cases because Hassan Al Idrissi, not surprisingly, had sent his tribesmen out to meet Craufurd en route and had impounded the lot.

With an almost Solomon-like wisdom, the Aden resident declared that the one hundred thousand rounds of ammunition now in Hassan Al Idrissi's hands should be considered as sent on behalf of Anglo-Saxon. Pushing their luck, Anglo-Saxon requested that the resident now send a British gunship to accompany a geologist to Farasan. Aden replied that it was "unable" to assist.

Anglo-Saxon did begin a geological examination of Farasan. Demanding additional payments and a further three thousand cases of ammunition and three thousand rifles, Hassan Al Idrissi would order all work to stop in mid-1927. At the close of that year he would denounce the agreement held by Anglo-Saxon as "a forgery." In late 1928, under the protection of a British sloop, HMS *Dahlia*, the company would evacuate the Farasans entirely.[10]

HOLMES—CONNECTION TO GULF OIL

After the fiasco of the bouncing boxes of bullets, Craufurd returned in high dudgeon to Aden where he lodged an unsuccessful complaint about the methods of Anglo-Saxon. Janson sent word that the Farasan concession was lost to Frank Holmes on the other side of the world and, according to Ward, "working like a Trojan." He and Holmes had made a trip together to Toronto and had "understood one another very well," Ward wrote Madgwick, adding admiringly, "Holmes knows how to hold his own and protect the best interests of his associates."

Holmes had had great hopes that the Gulf Oil Corporation would be keen on developing the Bahrain field as a joint venture with E&GSynd. He felt he had a connection to Gulf Oil. The founder and majority shareholder in this company was the Mellon family. In 1926, when Holmes arrived in New York, Andrew Mellon was US Secretary for the Treasury. In that position Mellon was working with Secretary of Commerce Herbert Hoover, Holmes's friend from their Western Australia and China days.

It was Herbert Hoover who provided Holmes with a glowing personal reference to present to the American oil companies attesting to his character and professional abilities.

In meetings with the head of Gulf Oil's New York office, William T. Wallace, and the company's geologist, Holmes had shed his usual caution—secretiveness almost—about giving any detail of his Arabian concessions. Although the two Gulf Oil men appeared most excited about oil prospects in Venezuela, Holmes had thought they could be convinced of the validity of his opinion on his Arabian concessions. He had cabled Janson telling him to establish his credentials so that his New York bank would be in "a position to say Holmes has full power to deal and to close a deal. Edmund Davis' help would be of much benefit." He warned Janson that he suspected the Gulf Oil Corporation would "want Farasan included in any deal."[11]

Now Craufurd's ineptness had lost Holmes the Farasan concession, and consequently, the Gulf Oil men, who, as Ward reported to Madgwick, "seemed to lose interest now that the Farasan Concession is no longer available. They would prefer propositions in the Red Sea to anything in the Persian Gulf."

Not about to give up, Ward arranged a meeting with geologist Norval Baker of the Foreign Producing Department of Standard Oil New Jersey. Also present was that company's adviser of Near Eastern work, C. Stuart Morgan, who was very interested indeed in what Frank Holmes had to say about Arabia.

STUART MORGAN—STANDARD OIL NEW JERSEY

An Englishman, Morgan had come to Standard Oil New Jersey from a job with the Anglo-Persian Oil Company. He had landed that job via the usual route of being one of Arnold Wilson's enthusiastic administrators in Mesopotamia and following his boss into Anglo-Persian. Under Wilson in Mesopotamia, Morgan and Harry Philby had worked together. They had a mutual friend who was extremely well connected through his former successful business in Baghdad, in which he had held the subagency for Standard Oil.

In fact, Morgan owed his current position to this mutual friend, T. D. Cree, who would shortly explain to Philby that "Stuart Morgan left Anglo-Persian after some trouble with the chairman. I personally recommended him to Walter Teagle, head of the Standard Oil Company of New Jersey." (After the US federal court antitrust decision of May 1911 decreeing that Rockefeller's Standard Oil Company must be broken up, several separate entities emerged. Although these were ostensibly individual companies, they continued in close cooperation. Among the companies spawned, the former holding company Standard Oil of New Jersey was the largest, with almost half of the total net value; it eventually became Exxon. Standard Oil of New York, with 9 percent of net value, eventually became Mobil. Standard Oil of California would eventually become Chevron.)

Cree was now involved in Sharqieh—the company of explorers and merchants in the Near and Middle East—which, he told Morgan, was "formed by a small group, of which I was not a member, with the idea of employing Philby to get concessions for oil and minerals in bin Saud's territory." Cree told Morgan that "Philby has gone to Jedda and is making remarkably good progress. He is a great personal friend of the King, Abdul Aziz bin Saud, whom he recognised long ago as the coming power in Arabia."

At the meeting with Stuart Morgan and Standard Oil New Jersey's geologist, Holmes showed his rock and substance specimens and his maps. Urged on by an enthusiastic Ward, he spoke openly of his findings and didn't hold back on his prediction that the entire area was certainly rich with oil. Ward was disappointed that the response was not more enthusiastic. He wrote Madgwick that Standard Oil New Jersey had "heard of Arnold Heim's report on Bahrain and in fact it is rather interesting to observe how closely some of our friends follow the investigations which one would normally believe to be private." He again asked Madgwick to send him "a more detailed report" on Bahrain that he could use to further interest Standard Oil of New Jersey.

Ever the opportunist, Madgwick replied that he didn't have any more information. He said, "Without some time and expenditure of money being made much more definite ideas on the detailed geology of Bahrain will not be possible."

However, it just so happened that as his Canadian contract might soon expire, he would personally be "delighted" to undertake a report on Bahrain if the Standard Oil New Jersey people would "reach some satisfactory arrangement" with him good enough to "relieve me from further financial worries."

But Morgan already had what he wanted. He would soon secretly forward details of Holmes's information, and a copy of his map, to Cree and Philby in an attempt to cut out Holmes and initiate Standard Oil's own move into Saudi Arabia.[12]

Apparently having exhausted the immediate possibility of a joint venture with either British or American backing, the members of Holmes's Eastern & General Syndicate in London confirmed their financial commitment to going ahead independently with developing the Bahrain field. On October 27, 1926, after more than a month in New York, Holmes and Dorothy sailed for London.

CHARLES BELGRAVE—THE BAHRAIN ADVISER

Changes had taken place in Bahrain. There was a new political agent, Major C. C. J. Barrett. Major Daly had left the island after dramatic events that occurred in August 1926 just as Holmes was departing for New York. The head of the Manama Town police was shot and wounded by his own men. This was the force that Daly had personally recruited and imposed in 1924. Two days later Daly himself was wounded by gunshot and stabbed several times with a bayonet. He recovered, but two Indian "officers" of the Bahrain Levy Corps were shot dead.

Criticism had been growing, particularly from the Colonial Office, of the autocratic methods of the political agent and the political resident and their tendency to treat Bahrain too much like a native state in India. While he was escorting Shaikh Hamad on the ruler's visit to London, Whitehall officials had warned Daly and insisted a civilian adviser must be appointed for Bahrain.

The man the India Office hired for this job was Charles Belgrave; he took up his appointment on March 31, 1926. Oxford-educated, he had been with the British Camel Corps in the Sudan and the Colonial Service in East Africa. He had spent several years in the Siwa Oasis of Egypt where, he reported, "I was the only white man."

His self-stated preference was to "work among Arabs" rather than in Africa. Belgrave was very well connected. His mentor was Sir Reginald Wingate, Lord Kitchener's successor as head of the Egyptian army and governor-general of the Sudan. After Kitchener's death in 1917, Wingate became high commissioner of Egypt. Just before leaving for Bahrain, Belgrave married the daughter of an English baron. His family and hers, he said, "and their various branches had known each other for generations."

Daly and Prideaux interviewed Belgrave for the job. According to Belgrave, they told him that the ruler of Bahrain "could not depend permanently on the sole advice of the Political Agent who had guided him during the difficult days after his father's abdication." They said the ruler of Bahrain now wanted "someone belonging to him, whom he could trust and rely upon."

HOLMES BEGINS WORK ON BAHRAIN

Via Iraq, Holmes arrived back in Bahrain to discover that even though he had the Bahrain concession signed, sealed, and delivered, Anglo-Persian was still out to play a spoiler role, this time through their local agent, the Kanoo family. Barrett, the new political agent, was proposing that in competition with Holmes, Kanoo now bore water wells across Bahrain ostensibly to "provide water for agricultural and domestic purposes." Holmes objected that this contravened his exclusive right under the exploration license of his oil concession. Ingenuously, Barrett wrote the resident that there was no problem because Khalil Kanoo had assured him that he "has no connection with Anglo-Persian, even though his uncle is their agent." "Furthermore," said Barrett, "Kanoo has given me a written undertaking that he will at once inform the Government of Bahrain if any traces of oil are found in any of the wells he sinks." To counter this new scheme, Holmes offered to drill more water wells at "practically cost price," an offer that Barrett depicted as not very magnanimous because, he claimed, Holmes had made "a handsome profit on each of the water wells."[13]

Ward very much wanted to get some sort of deal going. After all, he had been working on it for almost two years, since Madgwick first wrote him in mid-1925, although E&GSynd did not know this. He announced that he would come to London.

While Ward's letter was passing from the green of his Essex farm to the parched interior of Bahrain, Holmes was reporting to his London colleagues: "I am extremely glad that the Board of Directors have decided to drill for oil. I hope for success. I have the huts erected at the oil seepage site, a fresh water well sunk and everything, including a motor road to the seepage, ready for the drillers to take up their residence as soon as the plant arrives."

What Holmes was doing out there in the middle of Bahrain's desert sands was of interest to people other than his London colleagues. He wrote: "Since I have made the road to the seepage and built the huts there, four Anglo-Persian geologists have been three times on 'visits' to Bahrain port. On two of these 'visits' the geologists have gone to the oil site and taken away samples of the bitumen and oil sands. The work at the seepage has exposed interesting things among them a larger showing of oily bitumen. I hear they are now reporting that Bahrain is well worth boring."[14]

HOLMES ACCEPTS GULF OIL INTEREST IN BAHRAIN

When Ward's letter reached him, Holmes was not in the least interested in Standard Oil New Jersey. His antagonism was aroused because the State Department had just reversed its refusal to recognize the legal validity of the Turkish Petroleum Company's concession. Stuart Morgan had personally taken a leading role, lobbying on behalf of the American interests trying to get a share in Iraq's oil. Standard Oil New Jersey was now part of the American consortium in Turkish Petroleum, a company that Holmes viewed as dealing dishonestly with the Iraqi people.

But he was keen to do business with the Gulf Oil Corporation, which had not joined Turkish Petroleum and was associated with the Mellon family linked to Herbert Hoover. He saw this as presenting the possibility of immediately developing his Bahrain concession without further financial difficulty. More importantly, because of Andrew Mellon's high standing with the US administration, Holmes believed the political difficulties could be overcome.

His reply was passed to Ward from E&GSynd in London. It read:

> We have received a cable from Major Holmes at Bahrain in which he advises that he received a letter from you by which it appears that Mr. W. Wallace, Chief Vice President of Mellon's oil interests in New York is still much interested in our oil business. We authorise you to offer Mellon our entire interest in the Bahrain concession for one hundred thousand dollars cash, payable half on signing agreement, remainder in two years.
>
> If this deal is completed we will give Mellon the fullest consideration regarding any of our other oil interests the position of which Holmes disclosed to you—but Bahrain must be dealt with first—and owing to the necessity for immediate drilling there Mellon decision must be telegraphed quickly. Under the terms of the Bahrain concession drilling is to be commenced during this year and our arrangements to that end are well advanced.

Even so, Holmes's natural caution prevailed. He was not about to let a group of American geologists run free over his Arabian concessions—which was what Stuart Morgan was pushing for—without first having a firm agreement on development. His instructions continued:

> Major Holmes points out in his cable that the offer now made to Mr. Wallace can only be subject to Mellon coming to a decision on the information he has given you, and in Mr. T. G. Madgwick's report which was passed to you. There would be no time to make an examination themselves of the territory in view of the necessity for drilling to commence immediately.
>
> With reference to our other oil interests we believe that Major Holmes went into the position with you and explained the present status of the concessions in which there has been no change since his visit. If however you require further information we shall be glad to supply it, but for the reason already given, it is essential that Bahrain should be dealt with first.

Ward was buoyed and replied that Gulf Oil was still "much interested in the situation discussed while Holmes was here." He attempted to create a job for his old friend, saying: "Professor Madgwick insists that a deep test is necessary and that he is prepared to cooperate in every way with Gulf Oil's interests." Both E&GSynd and the Gulf Oil Corporation simply ignored Ward on this point. To Madgwick, Ward wrote, "I assume you will be prepared to submit more detailed advise to the Gulf Corporation in the event you are called upon to do so? The Gulf Corporation is very much interested and I hope we are now approaching some real business." And, "By the way, I never asked you how it was that you were unable to make a survey. . . . I should have asked you the location for the test well."[15]

Unable to proceed until the oil drill ordered from Ward arrived in Bahrain, Holmes went to Kuwait where he was drilling four water wells. The grinding pace of work was getting to him. He confided to Dorothy: "Son writes, he seems restless and thoroughly hard up. I hope he doesn't expect us to fund too much for him. I want every shilling I am likely to get for the next year or two, otherwise I will get an overdose of these less healthy places."

WARD PRESSURES THE DEAL

Ward wasn't satisfied with the prospect of Bahrain only. In a succession of cables and letters he began agitating for an arrangement that would include all Holmes's Arabian concessions. In June, E&GSynd responded with a terse statement.

> Our cable of April 20th and letter of same date set out our terms upon which we were prepared to sell our interest in the Bahrain concession. The drilling at Bahrain will be proceeded with immediately the plant arrives and naturally our offer was not intended to be, nor can it be, kept open until the drilling has demonstrated the value of the concession. We must therefore reserve to ourselves the right to withdraw our offer at any time after the end of this month if the negotiations have not been completed by then.

E&GSynd reiterated that "Farasan was acquired by the Shell group and Kamaran, for political reasons, cannot be granted to us by the Government at the present time."

E&GSynd again emphasized to Ward the current situation.

> Nothing has been done so far to re-establishing the status of the Al Hasa concession the yearly protection fee or rental upon which is, as you are aware, in arrears. But in the event of the Bahrain concession business being put through with your friends, we would be prepared to approach Abdul Aziz bin Saud with a view to validating the concession upon the best terms possible. But it would be useless doing this unless your friends were prepared to say definitely to what cost they would go to secure such validity. The same remarks apply to the Neutral Zone concession that, although signed, has never become operative. And applies also to the Kuwait concession for which acquisition negotiations have been suspended for some time, in view of the position of the other concessions.

They made it clear that "in the event of the Bahrain concession being sold through your agency, we would reserve 15% of the $100,000 cash purchase price for you and your friends, including Mr. Madgwick, the commission to be deducted from payment of one-half at signature of an agreement and the remainder within two years."

Ward wrote to Madgwick confirming that "if we are successful in bringing off the deal with the Eastern & General Syndicate Ltd, the commission of 15% will be divided equally between you, me and John Alvin Young. I hope this will be satisfactory to you."[16]

Unable to get his own geological team into Holmes's Arabian concession, Stuart Morgan and Standard Oil New Jersey dropped out of Ward's New York group of "friends" interested in discussions with E&GSynd. Morgan, however, already had a Plan B to target Holmes's Al Hasa concession.

Madgwick finally replied to six urgent letters from Ward. He finally admitted: "I am afraid you have all my information about Bahrain. I know you always think I can give you much more about Bahrain, but I can't. It is very difficult to make things clear, but all I went out for was to do water work. I did oil on the side. The strongest line I can give you is that I have checked over Pilgrim's published work and can endorse it from the oil point of view. I do not say he is an oil man but he is a geologist."

He had not done any surveying in Bahrain, Madgwick said, because "Anglo-Persian had it as a closed preserve until the end of March in order to make terms with the shaikh, which they couldn't. After that I was alone and too busy. I did not make a location. I decided that a proper topographical survey was necessary to work out the details of the geology and until this was done no location could be made."

Madgwick was exceedingly put out at being constantly frustrated in his attempts to "relieve" himself "from further financial worries." He turned nasty when Ward sent him copies of Holmes's letters detailing the positive developments at the location he was working in Bahrain, without benefit of a "topographical survey" as Madgwick insisted was first required. Perhaps forgetting he had once described Holmes as "a decent sort," he now fumed: "Somehow I can scarcely see much resulting from anything with Holmes in it, can you?"

Janson personally wrote to Ward again stressing that E&GSynd wanted to deal with Bahrain only, as a single entity. "It is no good our entering into an agreement to sell the other concessions as we do not hold them, at least until we have paid up the back rents," he pointed out. He told Ward, "You will be able to discuss the matter in person with Major Holmes when you are in London in July." Meantime, Janson said, no further negotiation could be conducted on behalf of the Gulf Oil Corporation until Frank Holmes returned to London.[17]

NOTES

1. Holmes Papers, December 3, 1925, Janson to Holmes in Bahrain, "just received your cable"; and January 5, 1926, resident (Prideaux) handwritten letter to Holmes, "Perhaps after you have done for Bahrain all that they want you'll be able to sell your machines at a cheap figure to avoid the expense of carrying it to Iraq or further. . . . Bushire has been considering a project to bring water a distance of about 40 miles in pipes into Bushire from a lake near the mountains which is fed by a perennial flow."

2. See preface; the surveys were (1) Guy E. Pilgrim in 1904–1905 for the Geological Survey of India; (2) E. H. Pascoe, March 1913, for the Geological Survey of India; (3) Pascoe and Lister James in October 1913 with John Cadman and Admiral Slade on the mission sent by Churchill; (4) Lister James and G. W. Halse for APOC in January 1917; (5) Dr. Arnold Heim and his Swiss Expedition for Frank Holmes, 1924; (6) Prof. Hugo de Bockh, two extensive surveys for APOC, 1923–24 and 1924–25, the second with chief geologist Lister James, G. M. Lees, and F. D. S. Richardson; (7) B. K. N. Wyllie and A. G. H. Mayhew supervised by G. M. Lees for APOC, January–February 1926; (8) P. T. Cox for APOC, five months in Kuwait, including drilling, 1931–32.

3. Holmes Papers, February 1, 1926, Ashraf in Bahrain to Holmes in Kuwait, "After your departure on January 18th . . . the Anglo-Persian personnel arrived . . . including Hajji Abdulla Williamson, English Geologists and one Govanis a botanist." IOL/R/15/2/875, vol. 39/4, June 11, 1932, resident (Biscoe) to secretary of state for the Colonies, memorandum, "Background to oil negotiations in Qatar"; see Archibald H. T. Chisholm, *The First Kuwait Oil Concession Agreement: A Record of the Negotiations, 1911–1934* (London: F. Cass, 1975), pp. 13, 111–18, and R. W. Ferrier, *The History of the British Petroleum Company: The Developing Years, 1901–1932* (Cambridge: Cambridge University Press, 1982), p. 567, for the result of this January–February 1926 Anglo-Persian expedition.

4. See John Baldry, "The Powers and Mineral Concessions in the Idrissi Imamate of Assir, 1910–1929," *Arabian Studies* 11 (Cambridge: Middle East Center, University of Cambridge, 1975), p. 82, for Cory Brothers and their 1920 interest in the Farasans. Note that Thomas E. Ward, *Negotiations for Oil Concessions in Bahrain, El Hasa (Saudi Arabia), the Neutral Zone, Qatar and Kuwait* (New York: privately published, 1965), p. 27, is substantially in error when he states it was Madgwick who approached Cory Bros., "one of his London clients." Obviously using Ward as the source, this same error is repeated on p. 62 by Angela Clarke; Holmes Papers, January 26, 1926, from Cory Brothers, Cory's Building, London to Edmund Janson, Eastern & General Syndicate, London, "Bahrain Island concession and the Neutral Zone are complete. Regards El Hasa, owing to the paramount chief having been at war, this concession may have technically lapsed, but you have no doubt that on the rents up to date being paid, the concession will be renewed without difficulty. . . . Kuwait, the Chief there, you say, is undergoing a period of mourning during which time he will be unable to deal with business matters for about three months, but that when this period has ended you do not contemplate any difficulty in getting this concession confirmed . . . Farasan Islands, this you say will have to be held up until the British Government have made a settlement with the different Chiefs"; and January 29, 1926, Janson to Holmes in Bahrain, "no good to us."

5. Chevron Archives, box 120791, December 28, 1928, E&GSynd (Adams), "memorandum" of Colonial Office meetings and for Janson's various approaches, see also Ward, *Negotiations for Oil Concessions*, p. 24; Ferrier, *History of the British Petroleum*, p. 555, "devoid of all prospects" and "only Albania is hopeful." Note that Holmes is in the Gulf during 1925–26, which makes impossible the claim by Daniel Yergin, *The Prize: Epic Quest for Oil, Money, and Power* (New York: Simon & Schuster, 1991), p. 282, and Angela Clarke, *Bahrain Oil and Development, 1929–1989* (London: Immel Publishing, 1991), p. 63, that Holmes "toured the London clubs and offices searching for financial investors and become 'an interminable bore.'" This error originates from an article in the

January 1939 edition of *American* magazine, "Is John Bull's Face Red!" by Jerome Beatty, who clearly states he did not speak to Holmes for the article. Beatty writes, "for six years Holmes went from office to office in London trying to get the government and the oil barons to take a chance on Bahrain." Beatty also quotes an unidentified Englishman saying, "Holmes was the worst nuisance in London. People ran when they saw him coming!" At no point could Holmes have spent six years around the offices and clubs of London. Beatty's original artistic error was repeated in the May 1947 *Fortune* magazine in "The Great Oil Deals" and again in the unpublished thesis for Princeton University, "Origin of American Oil Concessions in Bahrain, Kuwait and Saudi Arabia," by Frederick Lee Moore Jr. 1948. The error is continued in most subsequent publications.

    6. IOL/R/15/5/238, vol. 111, April 12, 1926, Wilson in London to political agent Kuwait (More), "no intention of working the thing themselves" and IOL/R/15/1/638, D73, May 4, 1926, Kuwait political agent (More) to resident (Prideaux), "this . . . will interest you"; Yossef Bilovich, "Great Power and Corporate Rivalry in Kuwait, 1912–1934: A Study in International Politics" (PhD diss., University of London, 1982), p. 123, details Wilson stringing out the discussions with Janson even though he was already in receipt of APOC's latest damning report on the oil prospects of Bahrain and Kuwait and therefore had no intention of reaching an agreement with E&GSynd. Perhaps Wilson was fishing for information on Holmes's successful negotiating postures, or his geological findings, or perhaps it was pure self-aggrandizement by Wilson.

    7. Holmes Papers, see correspondence including February 1, 1926, from Cotterell & Greig Ltd. Baghdad, Basra, and London to Frank Holmes at Eastern Bank Bahrain and contract with the Bahrain government (Shaikh Hamad) dated April 6 to build the seawall and road, also February 1926 to Richardson & Cruddas Engineers, Baghdad, in which Holmes asks for quotes on swing bridges; Holmes Papers, April 12, 1926, Holmes in Bahrain to Janson in London, "a very nice oil bearing shale." Note that Chisholm, *First Kuwait Oil Concession Agreement*, p. 14, is in error when he locates this well "showing slight traces of oil" in Kuwait and then states, "this closely guarded secret greatly encouraged" the Gulf Oil Corporation; Heim Papers, November 13, 1926, and December 2, 1926, "Agreement" between Frank Holmes and Dr. Arnold Heim; this document is written on Heim's typewriter, neither party has signed it.

    8. Holmes Papers, January 4, 1926, Madgwick in London to Ward in New York, "put up an attractive proposition to your friends . . . I don't want a Saluki"; Ward Papers, box 2, March 16, 1926, Madgwick in London to Ward in New York, "what information you ask for and come over to New York with it"; Ward, *Negotiations for Oil Concessions*, p. 26, "in the early part of 1926, after receiving Madgwick's first letters on the subject . . . discussed Persian Gulf oil possibilities with a number of American geologists"; Ward Papers, box 2, April 25, 1926, Madgwick in London to Ward in New York, "we could have everything cut and dried before that"; Ward Papers, box 2, July 21, 1926, Madgwick in London to Ward in New York, " at least two Russian ideas" and box 2, August 10, 1926, Madgwick in New York to Janson in London, "could result in business" and "any geological information you may have"; Ward Papers, box 2, August 19, 1926, Janson in London to Ward in New York, "we have decided to go on boring ourselves" and box 2, September 8, 1926, Madgwick in Calgary to Ward in New York, "Holmes is a useful person to keep in the fore when dealing with the natives" and "I have *NO* [*sic*] arrangement with Janson."

    9. Yergin, *The Prize*, p. 282, "greasy substance"; Ward Papers, box 2, October 1,

1926, Ward in New York to Madgwick in Calgary, "facts at his fingertips . . . leaves nothing to conjecture"; Ward Papers, box 1, September 23, 1926, Holmes in New York to Ward. Curiously, in his book Ward states that Holmes's detailed position paper was given to him just prior to Holmes's departure for London, not as the date clearly shows one month earlier on Holmes's arrival to New York. Moreover, the text as it appears in Ward's book, pp. 34–35, differs substantially from that of Holmes's original document. Ward has rewritten this document and added text of his own; Ward Papers, box 2, September 23, 1926, Madgwick in Calgary to Ward in New York, "did not actually do any survey work."

10. See detail of this episode in Baldry, "Powers and Mineral Concessions," pp. 76–107; see also Holmes Papers, correspondence from Hassan Saleh Jaffer c/o K. B. Ali Jaffer, Aden Camp, to Frank Holmes including one dated June 13, 1926, warning, "Said Al Arabi has reported to the Aden authorities that Craufurd has not paid the instalment to him in accordance to the instructions given by Hassan Al Idrissi."

11. Chevron Archives, box 120791, memorandum for historical research project, notes of telephone conversation with Mrs. Mary Stearns October 20, 1955, "Holmes had worked with Herbert Hoover in Australia at some time pre WWI and gave Hoover's name as reference . . . Loomis checked on Holmes with Hoover"; Ward Papers, box 2, September 28, 1926, cable, Holmes in New York to Janson, "Davis help will be of much benefit . . . want Farasan included in the deal."

12. Philby Papers, box XXX-7, contains letters between Cree, Morgan, and Philby including September 9, 1927, Cree to Philby, "Stuart Morgan left Anglo-Persian after trouble with Greenway, he is now Adviser of Near Eastern work for Standard Oil. I personally (Cree) recommended him to Teagle, head of SOC of New Jersey." And August 26, 1927, C. F. Mayer Company letterhead (C. F. Mayer was vice president of Standard Oil Company of New York). "T. D. Cree was formerly connected with Messrs Blockley Cree & Co of Baghdad who represented us, C. F. Mayer, as sub agents in Mesopotamia for a number of years"; Ward Papers, box 2, October 22, 1926, Ward in New York to Madgwick in Calgary, "they have heard of Arnold Heim's report" and October 25, 1926, requesting "a more detailed report"; and November 11, 1926, Madgwick in Calgary to Ward in New York, "if the Standard Oil people would reach some satisfactory arrangement."

13. IOL/R/15/2/96, vol. 1, January 7, 1927, political agent Bahrain (Barrett) to resident, "at once inform the Government of Bahrain if any traces of oil are found in any of the wells he sinks."

14. Ward Papers, box 2, February 23, 1927, Janson to Ward, "no support if we introduced American capital" and March 9, 1927, Ward to Janson, "our people are very disappointed with the negotiations" and March 11, 1927, Ward to Holmes, "so that something can be done with our New York friends" and April 4, 1927, Ward to Madgwick, "they have decided to go ahead and drill on Bahrain." Ward Papers, box 2, March 28, 1927, Holmes in Bahrain to E&GSynd London, included in letter from E&GSynd to Ward dated June 9, 1927, "everything ready for the drillers . . . as the plant arrives" and April 9, 1927, "Anglo-Persian geologists . . . now reporting Bahrain well worth boring."

15. In April 1927 the state department reversed its decision not to recognize the validity of Turkish Petroleum's claim to the old Mesopotamian concession thus paving the way for a consortium of American oil companies to join the Turkish Petroleum Company; Ward Papers, box 2, April 20, 1927, secretary, E&GSynd (Adams) to Ward in New York, "cable from Major Holmes" in reply to your letter re Mellon's interests in our concessions"

and May 20, 1927, Ward to secretary, E&GSynd (Adams), "still very much interested in the situation discussed with Holmes" and Ward to Madgwick, May 24, "in the event you are called upon" and June 2, 1927, "how it was you were unable to make a survey. . . ."

16. Holmes Papers, June 7, 1927, Holmes in Kuwait to Dorothy in Essex, "every shilling I am likely to get for the next year or two otherwise I will get an overdose of these less healthy places." Ward Papers, box 2, June 9, 1927, E&GSynd to Ward, "our offer (cannot) be kept open until the drilling has demonstrated the value of the concession"; Ward Papers, box 2, July 1, 1927, Ward to Madgwick, "the 15% commission will be divided equally between Mr. John Alvin Young, Mr. T George Madgwick and Mr. T E Ward. I hope this will be satisfactory to you."

17. Ward Papers, box 2, June 21, 1927, Madgwick to Ward, "you have all my information about Bahrain" and "can scarcely see much resulting from anything with Holmes in it"; and June 15, 1927, E&GSynd to Ward, "no good our entering into an agreement to sell the other concessions as we do not hold them" and June 16, 1927, "no further negotiations . . . until Holmes returns to London."

CHAPTER 11

NO HONOR
AMONG OILMEN

"**O**ld Crauford is lecturing before some society in London claiming to have found the long lost Ophir, the reputed spot from which the Queen of Sheba drew her 30 tonnes of gold which she took to Solomon on her visit," Holmes wrote from Kuwait to Dorothy about an article he came across in the *Times of Baghdad*.

Unfazed by his fiasco with the Farasan Islands oil concession, Craufurd had quickly returned to his true passion—the search for King Solomon's Mines. "He says Ophir is about 40 miles east of Aden and that it took him twenty years to find," Holmes told Dorothy. As an experienced gold miner Holmes was skeptical of both Craufurd's claim and the original biblical story. "If Sheba took 30 tonnes from it, that would make it richer than the Transvaal gold mines. The Rand only turned out about 50,000 tonnes during the whole past 30 years," he remarked.

Holmes might not have been so amused had he known that Craufurd's obsession was about to set in motion a chain of events that would eventually cost him his Al Hasa concession and deliver to the Americans the richest of all the Arabian oil fields.

What seemed to Holmes a harmless hobby for the eccentric C. E. V. Craufurd would involve American engineer Karl Twitchell with Harry Philby and the executives of Standard Oil of California. All would plot, scheme, intrigue, and double-cross in their efforts to secure Holmes's Al Hasa concession.

The American consul at Aden, James Loder Park, forged the first link in the chain. Anxious to promote US trade and influence in the Yemen in order to compete with the British, and with the Italians arriving in the wake of the Italo-Yemen Treaty, Park successfully urged the Imam Yahya to invite American philanthropist Charles Crane to visit the Yemen.

CHARLES CRANE IN YEMEN

A former US minister to China, Charles Crane, now approaching seventy years of age, had a long involvement with the Arab world originally inspired by time spent in Egypt as a young man. With a fortune inherited through the Crane bathroom fittings business, Crane's philanthropy was well known, particularly his substantial involvement with the charity work of the American Missions and especially their hospitals in Bahrain and Basra. He was the "Crane" of the King-Crane Commission appointed by President Woodrow Wilson in 1919 to report on the form of postwar government wanted by the people of Palestine, Syria, and Iraq. King-Crane's findings that the scheme devised by Britain and France for the future of the Middle East could not be approved, most particularly so in the case of the Balfour Declaration setting up a Jewish homeland in Palestine, never saw the light of day at the Versailles Peace Conference.[1]

CHARLES CRANE IN JEDDA

Earlier, when the Arab world had begun to buzz about Crane's forthcoming visit to the Yemen, two men with personal memories of Crane heard the news while they were in Riyadh. Dr. Shahbendar of Jerusalem had just been released from a French prison on a charge of helping to direct the Syrian revolution. The other was the Grand Mufti of Jerusalem whom Crane had come across in 1919 surrounded by Syrian refugees on the Mount of Olives. Crane had donated generously to the refugee cause. Both men urged Abdul Aziz bin Saud to also invite Crane. The task fell to Philby, who wrote in glowing terms of what bin Saud was setting out to achieve in his expanding territories. Philby, being Philby, did not miss the opportunity to suggest that he and his company, Sharqieh, might be the ideal conduit for any good deeds Crane might have in mind for bin Saud's territories.

Crane made it clear that he had no interest in commercial matters anywhere in the Arab world as he replied: "I have not been in business in a good many years and have always felt it wiser to have my relations with Asiatic people on another basis." Nevertheless, he said: "I am much interested in what bin Saud is trying to do for his people and believe that he promises to help them more than any other Mosleman [*sic*] of my time. If he could feel that I could be helpful also, it would have much to do with determining my point of view." Crane said he would like an opportunity "to go into the new and promising conditions under bin Saud."[2]

On his way to Yemen, Charles Crane did stop in Jedda. This was not his first visit. Four years earlier he had been Sharif Husain's houseguest. As a thank-you gift to Sharif Husain, he had shipped everything required for a handsomely tiled bathroom and sent an American engineer from Paris to install it. On this second visit, the new king of the Hijaz, Abdul Aziz bin Saud, did not come to Jedda to meet him. He sent his son instead. As Crane recorded, "Prince Faisal bin Abdul Aziz came from Mecca to welcome me and I was the King's guest." Faisal brought bin Saud's gift with him. It was a fine Arab mare, which Crane named Saudia and shipped back to New York before continuing his journey to the Yemen.

In a gracious letter of thanks, Crane told Abdul Aziz bin Saud: "Dr. Grayson, the celebrated companion of President Wilson and himself a noted horseman, will take care of the mare until I return." He went on to explain that, in his opinion, three areas needed to be looked into as the basis on which bin Saud could set out to develop his people and his territories. According to Crane these were "(1) Hygiene (2) Education and (3) Agriculture."

Crane would return in early 1930 at which time bin Saud would personally fete him in Jedda as a celebrated guest. Trying to steal the credit as usual, Philby would complain to Crane's biographer: "Mr. Crane never so much as sent me a postcard of thanks for the part I played in enabling him to meet the great bin Saud, apparently the only crowned head which he had not had the honour of meeting!"[3]

PHILBY, CREE, AND MORGAN

Stuart Morgan of Standard Oil New Jersey certainly had an appreciation for the part that Philby might play with "the great bin Saud." As Ward was leaving for London to discuss with Holmes and Janson on behalf of the Gulf Oil Corporation the same possibilities that Standard Oil New Jersey had just turned down, Morgan was writing to his friend T. D. Cree. He had arranged the appointment of Sharqieh as Standard Oil's agent in bin Saud's territories, saying, "I am awfully glad, Cree, to have been of service and trust it will develop to the advantage of Sharqieh."

Morgan was astute enough to recognize from Frank Holmes's presentations in New York that he did indeed know what he was talking about when he showed his maps and said there was oil in Arabia. But Morgan had been stymied in his attempts to send a team of Standard Oil's own geologists to examine Holmes's concessions.

Now he wrote to Cree, warning emphatically that his letter was strictly confidential and the content must not "go beyond the Sharqieh Board and H St John Philby."

Referring to the "broader implications" of the connection between Standard Oil and Sharqieh and saying the "channel opened" by the agency would be only the "jumping off" point, Morgan confided to Cree: "Major Frank Holmes has been over here to talk with me." What he learned from Holmes, Morgan said, had prompted him to "take a deeper interest in Al Hasa and the Trucial Coast."

He asked Cree to get Philby's reactions as to whether "bin Saud would like to have a geological reconnaissance of parts of Arabia." Again he warned that his letter was confidential, saying the Anglo-Persian Oil Company and his former boss, Arnold Wilson, were "friendly to me and I to them."

Morgan said he understood that while bin Saud was "quite friendly on broad lines of policy to Great Britain," he was "prejudiced" against Anglo-Persian coming into his country because of "that company's government connection and a feeling that out of it might arise a wedge for interference of an administrative character in his country." He said he also understood that "the Foreign Office will certainly not want the USA to get any political foothold in Arabia. And I heartily agree with them."

Now he put his proposal to Cree. If bin Saud's interest could be raised, he said, everything "could be done by an Arabia Company—or through you—in such a manner that it would have a two fold attraction (a) being an advantage to Standard Oil as a source of production and (b) be non-controversial politically because it would be a British or Arabian concern." Morgan requested that "your people canvass with Philby" and promised to write again shortly "on this particular point."[4]

Promptly writing to Philby, Cree explained that taking on the Agency for Standard Oil "is a very important question because we must realise that if we handle the Standard Oil Company's products they will ultimately expect us to do political and exploitation work for them too. The lines on which Stuart Morgan suggests this might be done are very interesting and might suit us well. The arrangement might also be agreeable to bin Saud." Cree could not have made it more clear when he advised Philby: "The objections are that we will not be working for British interests. This does not weigh very heavily with me—the interests of the company as a moneymaking concern come first. But this is not every one's opinion."

Cree could not have known Philby very well. Philby was the last person likely to have any ethical or ideological concerns about British interests. Cree was soon reporting to Morgan: "Philby's opinion is that bin Saud might be prepared to welcome the entry of Standard Oil as an American concern with no political axe to grind in Arabia and that a British facade as suggested by you would not be necessary. This would simplify matters as Sharqieh would then act as your agents on terms to be arranged and if is American interest for which we are to work, we would just as soon do it openly as undercover."[5]

It was Morgan who needed to keep his dealing with Cree, Philby, and Sharqieh under cover. It was only a matter of weeks since the American oil companies had succeeded in gaining participation in the oil wealth of Iraq through the resurrected Turkish Petroleum Company. The original Turkish Petroleum Agreement, which the British and French claimed to be a still valid contract, included self-denial provisions dating from 1912. No activity could be initiated by members, except collectively through the Turkish Petroleum Company, within the borders of what was the Ottoman Empire of the time. Kuwait was set aside from this provision because in the Foreign Office Oil Agreement of March 19, 1914, relating to the original Turkish Petroleum Company, paragraph 10 specifically excluded Kuwait on the basis of the 1913 Anglo-Turkish Convention. The "self-denial" area was eventually marked out as a red line on a map and became known as the Red Line Agreement.

With Standard Oil New Jersey, Stuart Morgan had been heavily involved in pressuring for American entry into Turkish Petroleum. This had been delayed by refusal of the State Department initially to recognize the legal validity of the concession and then to allow American acceptance of the monopolistic Red Line Agreement. The British and French had gone ahead, obtaining the concession from the Iraq government in March 1925. Led by Standard Oil New Jersey, the Americans continued to agitate until, in April 1927, the US secretary of state withdrew official objection. As Stuart Morgan well knew, the area "forbidden" for individual activity by the members of Turkish Petroleum included all of bin Saud's territories. Moreover, Morgan was lobbying for the appointment of secretary of the Near East Development Corporation, the entity put together to represent the combined American interests in Turkish Petroleum. He knew the scandal that would erupt if his secret instructions to Philby and Sharqieh became known.[6]

NEGOTIATIONS WITH WARD

Over four weeks in London, Ward's stubborn insistence that he be given more than Bahrain to talk about with the Gulf Oil Corporation paid off. Ward, Holmes, and Janson hammered out an understanding that Ward would first put forward the Bahrain concession as an outright purchase. After that, he could also discuss an option on Kuwait that would kick in if Holmes could finalize that negotiation and options on Al Hasa and the Neutral Zone, provided Gulf Oil understood that these needed to be revalidated by Holmes.

Full of energy, Ward arrived back in New York on September 15, 1927, hoping to have his proposals on the agenda of Gulf Oil's board meeting scheduled for the first week of October.

Ward seems to have been more enthusiastic than was E&GSynd. He was still waiting three weeks later for the promised documentation and power of attorney. After receiving an almost frantic letter from Ward, Frank Holmes replied: "I regret the delay in sending you the letter of particulars which we decided upon when you were here in regard to the Bahrain concession. I was in the office yesterday and the lawyers were preparing your Power of Attorney. If it doesn't go forward by this mail it certainly will at a very early date." He calmed Ward down, saying: "If you think it absolutely necessary I could easily run over to America to assist you at the final stages of the negotiations by supplying you with first-hand information about Bahrain."

He told Ward he could tell Gulf Oil that arrangements would be made "so that the geologists will be quite as comfortable in Bahrain as they would be in New York. The food may not be quite as varied or good; but the Gulf Oil people surely could see from me that a person living out there does not altogether fade away." The equipment purchased so far through Ward's company was not up to scratch, he warned, "I hear great complaints from my drillers."

E&GSynd stated categorically to Ward "there is no doubt that your friends will strike oil at Bahrain." They stressed emphatically: "The position in regard to the Bahrain Concession is perfectly clear and it is this area which we want to deal with in the first instance. Major Holmes could then make every effort to secure the Kuwait area and to arrange satisfactorily about Al Hasa and the Neutral Zone."

E&GSynd repeated their warning to Ward: "It cannot be made too clear to your American friends that there is a possibility that, owing to non payment of back rents, or at least until such back rents are paid, bin Saud's government may refuse permission to geologists to examine the Al Hasa concession and Neutral Zone areas. Major Holmes does not anticipate any such development."

The major back rent referred to by E&GSynd was nine thousand sterling on the Al Hasa concession. It was debatable, however, as this figure was the total of the annual three thousand sterling "protection fee" covering operations at the concession. Because of bin Saud's military activities in the area, Holmes had not been able to begin operations there. The same had happened for the Neutral Zone, signed in May 1924, where neither the twenty-five-hundred-sterling initial payment nor the three-thousand-sterling annual protection fee had been paid because of the inadvisability of working in the area while bin Saud's Wahhabi Warriors raged through.[7]

Ward forwarded the package from E&GSynd, including a sketch map "specially prepared and marked up by Major Frank Holmes" to William Wallace at Gulf Oil. If Wallace wanted an option to buy the Bahrain concession, he had to agree before the end of November on how Gulf Oil's geologists would work with Holmes; they would be given nine months to complete a survey. If they failed to meet this condition before the end of November, Holmes would go ahead as planned and begin drilling for the oil of Bahrain exclusively on behalf of E&GSynd.

If Gulf Oil accepted the option and sent geologists, Holmes would remain with them in Bahrain "to facilitate the geological party in every way and make arrangements with the local authorities." For Holmes's services, Gulf Oil would pay E&GSynd two hundred sterling per month plus expenses, and his London colleagues would "render assistance in negotiations with the Colonial Office." The price for the concession was set at "$50,000 as and when your geological party reports favourably on Bahrain and $50,000 one year from the date of that report."[8]

MORGAN SENDS HOLMES'S MAP TO PHILBY

As the Gulf Oil Corporation was considering the Bahrain concession, the October 14 spectacular rush of oil in Turkish Petroleum's Iraq concession hit the world's headlines. Close to Kirkuk, at Baba Gurgur where the Germans had worked during WWI, a column of oil burst with a deafening roar through the stillness of the night.

It was heard, and seen, five miles away. A river of black drenched the surrounding countryside bringing with it clouds of poisonous gas. The site was near what the Mesopotamians had called the "Eternal Fires," the field of associated gas that had been permanently alight since time immemorial and was thought to be the "Fiery Furnace" mentioned in the biblical story of Daniel. Initial output of this well was estimated at sixty thousand barrels per day.

Convinced of the merit of what he had heard from Holmes in New York, Stuart Morgan stepped up his confidential correspondence with T. D. Cree who assured him: "Philby knows all the facts of the situation, as explained by you . . . Let me have any information affecting the possibility of your friends taking a direct interest in bin Saud's territories."

Following the Iraqi strike, in which Standard Oil of New Jersey and Standard Oil of New York now had an interest through Turkish Petroleum, Morgan offered Cree some strong advice. "Why not make your position watertight by coming to an arrangement with bin Saud that the business end of any petroleum project or development should be conducted by your people on his behalf? You will have your own arguments which will occur to you, and which you can put forward to the King, as to why it is in his own interests to do so." He pointed out that "such an arrangement would safeguard your firm in any event, i.e., whether Turkish Petroleum, Royal Dutch Shell or some American concern make a move at looking towards a concession in bin Saud's territory."

Morgan secretly forwarded to Philby a copy of Holmes's map together with a copy of the concession document signed between bin Saud and E&GSynd in which Frank Holmes laid out the dimensions of the Al Hasa field. Sending this material obtained during Holmes's visit to New York, Morgan impressed upon Cree the need to ensure that his name, and that of Standard Oil, remained well concealed. In turn, Cree warned Philby: "Stuart Morgan and Standard Oil fear to make a move that might embroil them with their associates in the Turkish Petroleum Company. For this reason, we cannot at this stage use their name."

Philby had been unable to act on the urging of Stuart Morgan to use his self-proclaimed personal friendship with King Abdul Aziz to "aim at so controlling matters that anyone desiring concessions would have to come to you." Bin Saud had been pinned to Mecca and Medina where he was striving to placate his critics.

For more than twelve months, Philby had cooled his heels in Jedda, trying, without much success, to sell on commission such items as Sunlight soap. At the end of 1927, Philby secured his first government contract from bin Saud's finance minister, Abdulla Suleiman, for two million nickel coins to be minted in Britain. Philby grandly assured his London backers that Sharqieh would profit by two thousand sterling. But as the contract involved Sharqieh advancing the payment for the coinage, Cree was worried.

By December he was sharing his anxiety with Stuart Morgan: "The capital of the company is not sufficient. . . . I am meeting with a good deal of difficulty in finding the fresh capital necessary in London where everyone is hesitant about a business of this nature. The possibilities of which are only evident to a few people with actual experience in Arabia." Morgan replied reassuringly that everything would come good very soon. According to "a conversation I had with one of the directors of our Standard Oil Company of New York . . . matters are going to crystallise very shortly," he wrote.[9]

GULF OIL INSISTS ON LOWER TERMS

It seemed to Holmes that the Americans at the Gulf Oil Corporation were as mean, if not more so, than the Anglo-Persian Oil Company had been toward the Arab shaikhs. In rapid succession Ward informed E&GSynd that Gulf Oil required that in the new contracts that Holmes would draw up for Kuwait and for revalidation of Al Hasa and the Neutral Zone he must "secure a reduction" in the agreed royalty from five shillings and threepence per ton to four shillings. Moreover, this reduced royalty would be paid "in lieu of any shareholding interest whatsoever for the Sultan and the Shaikhs." Furthermore, Holmes should ensure the "entire elimination" of export tax "both now and for all time."

The Gulf Oil Corporation was adamant, Ward emphasized, that "shareholding participation for the Sultan and Shaikh as per the original concession for the Al Hasa and the Neutral Zone, as well as the export tax, is in no way satisfactory to the Gulf Corporation. They want both the shareholding participation and the export tax entirely eliminated from the picture."

Meaner still, the Americans now wanted Holmes to go back to the shaikh of Bahrain and try to "secure a reduction" in royalty agreed in the already signed concession. Ward claimed, "I am doing my best to carry on the negotiations . . . it was expected that we would have to trade."

He overlooked the fact that it was not Ward, or Gulf Oil, or E&GSynd, but the Arabian shaikhs who would lose through what he described as "having to trade."[10]

The response to Gulf Oil's requirements was formal. E&GSynd suggested that "discretion be permitted Holmes in negotiating with regards to reducing the royalty payable to the Shaikh of Bahrain as our Board of Directors fear friction may be caused, both with the shaikh and with the Colonial Office if too great insistence is put upon this condition."

From the very beginning, E&GSynd had wanted to deal with the Bahrain concession only. Now in response to Gulf Oil's demanding stand, the syndicate's board insisted all three "Mainland" areas, Kuwait, Al Hasa, and the Neutral Zone, should be discussed separately. They stated that "Holmes concurs in this and adds that negotiations would be expedited by treating these as separate concessions, each being the subject of a separate contract."

For his part, Holmes wrote personally to Ward, saying: "As regards the confidence that Mr. Wallace says he has in my ability to perform, I would like you to make it clear to him that I do hope he also realises that I may not be able to procure all that he hopes for in connection with the whole of the business in the Persian Gulf."

Holmes said that driving too hard a bargain could be counterproductive. "We must not lose sight of the fact that failure can also be the result, after the most painstaking efforts," he warned. William Wallace of Gulf Oil must be made to understand, he told Ward, "that the Arabs are more or less untutored children and it is often extremely difficult to convey to them our point of view and some time is often required to effect this." Nevertheless, he told Ward, "to convey to Mr. Wallace that I am going out to Bahrain and the Gulf with every confidence of success."[11]

Ward stuck to his guns. He could not, he said, "emphasise too strongly the importance that the Gulf Oil Corporation attaches to the securing of all the concessions. Again and again it was stated that they can only look upon Bahrain as part of the picture." But E&GSynd would not agree to either Gulf Oil's terms or that all E&GSynd's Arabian interests be included in a single contract.

Saying that the expectation was that "the cost of revalidating the Mainland contracts and securing the Kuwait concessions will be kept down as low as possible" and "we are all reposing our confidence in Major Holmes with a feeling that he will be successful in making the requisite adjustments at a reasonable figure," Ward finally put forward Gulf Oil's compromise of two separate contracts. The first was a straightforward agreement covering the Bahrain concession. The second was an option on Mainland concessions covering Kuwait, for which Holmes was yet to conclude the concession, and Al Hasa and the Neutral Zone subject to revalidation.

But the Americans had not quite finished yet. They now wanted E&GSynd to obtain, in writing, "the consent of the Sultan and Shaikhs—and of the British Colonial Office—to the transfer of E&GSynd's rights in the Bahrain and Mainland concessions to either a British or Canadian corporation" to be set up as the nominee of the Gulf Oil Corporation.

This time, Edmund Janson sent a sharply worded letter to Ward saying he was busy negotiating a big electrical contract for E&GSynd, and that the last reply sent by the board of E&GSynd was the "final to you regarding the contract with the Gulf Oil people." Irritably, he told Ward: "Surely the Gulf Oil people must understand that it is quite impossible for us to guarantee that any government department will sign a statement to be operative in the future. That they will agree when the time comes, I have no doubt whatever." He said that E&GSynd had now done everything they were prepared to do in the matter.

Janson knew the Gulf Oil Corporation "would get oil," he said, but if the deal fell through, "we will start to bore for oil ourselves in Bahrain, immediately." He reminded Ward that Holmes had left some time earlier for Bahrain "so as to get everything ready for the geologists" and so could start his drilling at any time. Janson said he hoped the deal came off because it would be "in all our interests" and Ward had put a lot of hard work into it, but Ward must also bear in mind that "we cannot handle our Government Offices like you can yours in America."

Gulf Oil capitulated. Two separate agreements were signed on November 30, 1927, just in time to remit the payments due to the shaikh of Bahrain on December 2, 1927.

Ward wrote the news to Madgwick. He was sending the geologists to speak to him about local conditions on Bahrain, he said, adding: "I am sure you realise it is the desire of our friends to keep the arrangements they have made confidential." Telling Madgwick "to render an account for the consultation. Of course I mean in dollars and not in sterling," he appealed to his friend to curb his envy of Holmes.

> I hope you will speak kindly of Major Holmes in every way for we have got to look up to the Major as our responsible representative and support him. A great deal depends upon the support given to Major Holmes in carrying out the instructions given to him. There is every possibility of Bahrain and the Mainland concession developing into quite a big thing. I might mention that during our visit to London, Major Holmes spoke very highly of you and very strongly recommended you to Edmund Janson and the other E&GSynd Directors.

E&GSynd were arranging passage for the geologists by "the quickest way." This would involve London to Marseilles, then boat to Beirut and Syria, then by car across the desert to Baghdad, train to Basra, and boat to Bahrain. "The next best way, should there be any delay in catching the steamer at Marseilles," they advised, "is boat to Port Said, railroad to Haifa, car to Beirut then across the desert to Baghdad, etc." Either way, the trip from London to Basra would take a minimum of fourteen days.

On December 24, 1927, Gulf Oil geologist Ralph Rhoades with two assistants, all handpicked by Wallace, left New York on their way to join Frank Holmes in Bahrain.[12]

AMERICAN GEOLOGISTS IN BAHRAIN

Holmes cautioned his New York colleagues: "We must all proceed with the utmost secrecy lest we arouse the activities of our powerful competitors. The geologists must be instructed not to reveal their business in any manner whatsoever." As Holmes intended to keep quiet about the Americans being from the Gulf Oil Corporation, he advised Wallace to ensure they were "in possession of a letter from the head of one of the large banking concerns in your country stating the financial standing of the people they represent. I do not anticipate that this letter will be required but if the Shaikh of Bahrain or bin Saud inquire as to the standing of my associates, this would be a ready means of reassuring them." The geologists should also bring "their tennis racquets and golf clubs as there will be plenty of opportunities for using both, especially in Bahrain," he said.

Holmes went to a great deal of trouble to ensure the Americans were "properly" introduced in Bahrain. He took them first to the political agent, Colonel Barrett, who he described as the "real" power, before visiting the ruler of Bahrain, Shaikh Hamad, and his brother Shaikh Abdulla. He ensured they made a formal call on Charles Belgrave, the adviser. Holmes was pleased. He reported that the Americans had been "well launched" and had made a good impression on "both the Arab ruling family and the local European people." He commented: "Many of these little things may seem superfluous, but we have to adapt our methods to suit the Arab minds and habits."[13]

William Wallace of the Gulf Oil Corporation expected his geologists to need the full nine months for their survey. The company's chief geologist feared that was not nearly enough and wanted the Americans to have eighteen months. When they began in February 1928, the Americans found they needed to do little more than verify the substantial work Holmes had already completed.

Taking time out for such enjoyments as Arabian dinners and falcon hunting with Shaikh Hamad and Shaikh Abdulla and tennis and golf with the Europeans, the Americans completed their fieldwork in six weeks, between February 6 and March 19. The Gulf Oil executives in New York declared "very gratifying" the "speed with which the survey is being accomplished." They noted that "Ralph Rhoades speaks very highly of the work done by Major Holmes" and they were, said Ward, "extremely satisfied with the work."

While the Gulf Oil geologists were socializing and surveying in Bahrain, T. D. Cree was beginning to grasp the full import on Standard Oil of the Red Line Agreement accepted when the New Jersey and New York companies joined Turkish Petroleum.

He explained to Philby: "It means that Standard Oil cannot enter bin Saud's territory on their own but only as a member of the Turkish Petroleum." Because Shell had obtained the Farasans, Cree worried that company might make a move and stressed Morgan's advice. He told Philby: "I can only leave it to you to watch events closely and if you are able to fix matters so that the necessary approaches and negotiations can only be done through you, it will give us a big pull."

In a letter to Morgan, Cree said he had

one or two indications from Philby that the question of concessions will come to the fore shortly. I feel convinced that if any are given Philby will get them. You know the King's objections to Anglo-Persian. His objection to Turkish Petroleum may be equally as strong. Do you think Turkish Petroleum would come in under another name and under your people's auspices? I can quite see that you may not care to answer this question. If so please ignore it and I will quite understand.

Morgan did not ignore it. He replied promptly, saying, "I think it will be possible. But I again emphasise that the best thing is for Philby to arrange with King bin Saud that he, Philby, will handle all concessionary matters. Then, whoever it may be, would have to go to him. And as I said before I will, in the meantime, find a channel which your suggestion could crystallise." After receiving this letter, Cree assured Philby that "a way will be found by which Standard Oil will handle the business either under their own or another name. I hope you will be in a position to make a move in the question before much longer."[14]

On April 10, 1928, Rhoades delivered to Wallace a favorable nine-page report on Bahrain. Praising Holmes's work at the site, he reported, "The presence of ashphalt deposits and seepage along the western flank of the Arabian Gulf geosyncline strongly suggest that oil bearing formation underlie, more or less continuously, the western part of the Gulf and the adjacent mainland as far south as Bahrain. . . . It is concluded that a test well favourably located on the Bahrain anticline may reasonably be expected to encounter oil." Rhoades remarked that they could easily proceed to this because of the available water supply, which came from the wells that Holmes had sunk in the area.

Rhoades advised Gulf Oil to exercise their option on E&GSynd's Bahrain concession. He also advised as a precaution against "unpleasant competition" negotiating an option covering the remaining area of the Bahrain Islands. The "unpleasant competition" was a reference to Anglo-Persian. Holmes had heard from Belgrave that, following the three visits by the four geologists and their collection of samples from Holmes's site, Anglo-Persian was now showing an interest in the area outside the one hundred thousand acres covered in Holmes's concession.

POLITICAL OFFICERS' SUSPICION

The political resident, Lt. Col. Lionel Haworth, arrived in Bahrain for the purpose of ascertaining what, exactly, might be the dimension of Holmes's involvement with the Americans. Holmes explained to Wallace: "The fact of our geologists being American makes the Political Agents very anxious to know all about them and my relation to the people in America. They have guessed wrongly and are quite convinced that your men come from the Standard Oil Company." He was confident he could "hold my own with the Resident, but we shall see," and while he was sure "all this questioning and agitation" would die down in time he wanted to "keep it under control until our 'campaign' has been brought to a successful issue."

He was surprised to find the latest political resident reasonably pleasant because "from what I heard from the Bahrain Political Agent, I expected to find him a little aggressive." Remarking to Holmes that Anglo-Persian was "dreadfully" annoyed by the mere suspicion that Holmes may have secured American financial backing, the resident, Haworth, inquired whether it was too late to make the Bahrain Oil Concession "a real strong British concern." Holmes replied diplomatically that he would welcome any suggestions to forward to his London colleagues. He warned Wallace that when the competition "really find out that we have strong financial strength behind us" they will redouble their efforts to "outrun us." Still, Holmes commented cheerfully, "this is all part of the game."

Haworth was a man who paid attention to detail. Although he did not get any information on Holmes's actual arrangements with the Gulf Oil Corporation, he considered the possibility that there might be more to it than extending a line of finance, as Holmes implied.

Haworth went to the files and examined the contract between E&GSynd and the shaikh of Bahrain; now based on the document personally composed by the duke of Devonshire as insisted upon by the Colonial Office. He picked up the inconsistency in this document and wrote the Colonial Office that "the agreement . . . contains nothing to prevent the rights under it from being transferred to an American, or other, foreign concessionary." In future, Haworth suggested, it might be wise to ensure all such agreements contained a specific British nationality clause.

Flushed with the success of imposing just such a limitation on Iraq, the Colonial Office believed they knew how to create a "real strong British concern." In June 1928 (the month Herbert Hoover was nominated as the Republican Party's candidate for the forthcoming US presidential elections) the colonial secretary himself would reply, in confidence, to the resident. His Majesty's government would require any company involved with the development of the Bahrain concession to be a British company registered in Great Britain or in a British colony, he said. Furthermore, as decreed for Iraq, it would be required that as large a percentage of the local staff as possible should be British subjects, and that the chairman and managing director—and a majority of the other directors— and the local general manager must also be British.

The colonial secretary included for the resident's benefit a draft of an appropriate clause dictating that "neither the company, nor the premises, liberties, powers, and privileges . . . nor any land occupied for any of the purchases of the lease, at any time to be, or become, directly or indirectly controlled or managed by a foreigner or foreigners or any foreign corporation or corporations." The definition of *foreigner* was "any person who is neither a British subject nor a subject of the shaikh." A foreign corporation was any corporation "other than a corporation established under, and subject to, the laws of some part of His Majesty's Dominions and having its principal place of business in those dominions."

After warming up Wallace with encouraging words about the geological survey, Holmes broke the news that there would be no downwards revision of the terms as he had originally agreed upon with the shaikh of Bahrain.

He wrote: "In my previous letter I omitted to mention that when I approached Shaikh Hamad with a view to securing a reduction on the royalty payable per ton of oil, he was much perturbed. He said that it would give those of his people who opposed E&GSynd getting the concession an opportunity of saying that he had made a mistake in giving us the concession. Besides, he did not decide the amount of royalty to be paid. My company had fixed that with the Colonial Office." Holmes closed this subject firmly, saying: "I saw it was not wise to press the Shaikh of Bahrain any further."[15]

NOTES

1. Holmes Papers, June 7, 1927, Holmes in Kuwait to Dorothy in Essex, "Old Craufurd . . . claims to have found the long lost Ophir"; Briton Cooper Busch, *Britain, India, and the Arabs, 1914–1921* (Berkeley: University of California Press, 1971), pp. 320–41, for the King-Crane Commission.

2. Crane Papers, box XVI-2, document headed "Unpublished Biography of Charles Crane" by Edgar Snow: "Dr. Shahbendar of Jerusalem . . . on a visit to Riyadh to see King bin Saud arranged for Crane's coming trip to Mecca, a recommendation seconded by the Grand Mufti, whom Crane had found surrounded by Syrian refugees, on the Mount of Olives." And Crane Papers; August 27, 1926, Charles Crane to Philby, "Thank you for your kind letter. Perhaps I can answer it more fully after I have again visited the Red Sea. I have not been in business in a good many years and have always felt it wiser to have my relations with Asiatic people on another basis. I plan to arrive in Cairo around December 1st and my address will be the American Legation."

3. Crane Papers, "Unpublished Biography of Charles Crane" by Edgar Snow for material on Crane and James Loder Park and Crane's visit to Yemen. And Crane Papers, letters dated December 24, 1926, January 9, 1927, and April 20, 1927. E. C. Hodgkin, ed., *Two Kings in Arabia: Letters from Jedda, 1923–5 and 1936–9* (Reading, UK: Ithaca Press, 1993), pp. 24–25, for Crane's gift of a bathroom to Sharif Husain. Crane had a personal friendship with the revered Dutch Oriental scholar Snouck Hurgronje, through this introduction, on this second visit to Jedda he stayed in the house of Muhammad Nassif, "the wealthiest man in Jedda." Crane arrived in Sana'a on January 23, 1927. H. St. J. B. Philby, *Arabian Oil Ventures* (Washington, DC: Middle East Institute, 1964), p. 75, "As I told Antonius some years later Mr. Crane never so much as sent me a postcard of thanks for the part I played in enabling him to meet the great ibn Saud: apparently the only crowned head which he had not had the honour of meeting!"

4. Philby Papers, box XXX-7, Morgan had arranged the appointment of Sharqieh

as Standard Oil agents following a letter of July 2, 1927, from Cree saying that kerosene and petrol consumption was rapidly increasing because, under bin Saud, "the country is opening up in a remarkable way." Cree said there were two hundred motor cars in Jedda and there should be "over 500 cars" within a year. "The only brands on the market," Cree had said, "are Shell and a little Fiume oil." And August 26, 1927, Morgan to Cree, "Frank Holmes here to talk to me," "Agency . . . only the jumping off point," "whether bin Saud would like to have a geological reconnaissance," "your people canvass Philby."

5. Philby Papers, box XXX-7, September 9, 1927, Cree to Philby, "the lines on which Stuart Morgan suggests this might be done are very interesting . . . we will not be working for British interests" and September 21, 1927, Cree to Morgan, "if is American interest for which we are to work, we would just as soon do it openly as undercover."

6. See Benjamin Shwadran, *The Middle East: Oil and the Great Powers* (London: Atlantic, 1956), p. 244, "State Department initially refused to approve the American's acceptance of the self denial clause." See also Shwadran, p. 384, for background to exclusion of Kuwait in the Red Line Agreement. For surrendering 22.5 percent of its holdings to give the Americans a share, Anglo-Persian received an overriding 10percent royalty on TPC output. The Red Line Agreement covered the entire Arabian Peninsula, Palestine, Jordan, Iraq, Syria, and Turkey. Persia and Kuwait were not included. The companies of the American group were organized in February 1928 as the Near East Development Corporation. In 1929 the name of the Turkish Petroleum Company was changed to Iraq Petroleum Company. Note: A romantic myth exists, fueled by Gulbenkian's own memoirs, that he personally took a "red pencil" and delineated the boundaries of the self-denial agreement, saying, "this is the Ottoman Empire as I remember it." This is very doubtful. See, for instance, Daniel Yergin, *The Prize: Epic Quest for Oil, Money, and Power* (New York: Simon & Schuster, 1991), p. 204, "months earlier, the British using Foreign Office maps and the French with maps from the Quai d'Orsay had already fixed the same boundaries." At various times the American group would include Standard Oil New Jersey, Standard Oil New York, Gulf Oil Corporation, Atlantic Oil, Mexican Oil, Sinclair Oil, and Texas Oil Company. Gulbenkian would still have his 5 percent.

7. Ward Papers, box 2, September 21, 1927, Ward to E&GSynd urgent request for documentation and power of attorney; Ward Papers, box 1, October 5, 1927, Frank Holmes at his farm MillHill to Ward in New York, "tell the geologists" and "great complaints (about the equipment) from my drillers"; Ward Papers, box 2, October 7, 1927, E&GSynd (Adams) to Ward, "no doubt your friends will strike oil at Bahrain" and, "it cannot be made too clear . . . the non payment of back rents."

8. Ward Papers, box 2, September 14, 1927, E&GSynd to Ward, "With reference to the interviews we had with you here, will you as arranged, kindly pass on to Mr. Wallace of the Gulf Oil Corporation the enclosed memorandum in regard to the terms suggested for the sale to the Corporation of our oil interests in Arabia. You have in your office copies of the concessions and other documents which Major Holmes left with you last year, but we enclose for convenience further copies of the Bahrain Concession and those for Al Hasa and the Neutral Zone. For your convenience also we set out the principal conditions of the concessions but for the full obligations etc you must of course refer to the concessions themselves." And October 7, 1927, Ward to Wallace at Gulf Oil Corporation, "including a sketch map 'specially marked by Major Frank Holmes.'" Note: After Gulf Oil bought into Bahrain, Ward insisted to E&GSynd that he should also receive an

ongoing 15 percent commission on "a royalty of one shilling per ton to be paid to the Eastern & General Syndicate Ltd on production in excess of 250 tons per day."

9. Philby Papers, box XXX-7, October 9, 1927, Cree to Morgan, "Philby knows all the facts of the situation as explained by you" and October 27, 1927, Morgan to Cree, "make your position watertight by coming to an arrangement with bin Saud that the business end of any petroleum project or development should be conducted by your people on his behalf"; Philby Papers, box XXX-7, November 10, 1927, Cree to Philby, "fear to make a move that might embroil them with their associates in the Turkish Petroleum Company" and December 7, 1927, Cree to Stuart Morgan, "difficulty in finding fresh capital' and December 19, 1927, Stuart Morgan to Cree, "matters are going to crystallise very shortly." See Elizabeth Monroe, *Philby of Arabia* (London: Faber and Faber, 1973), p. 148; far from making a profit, Sharqieh was forced to carry its outlay on this order on its own account. Monroe details how Philby spent years attempting to reclaim this, and a number of ensuing debts, from bin Saud's administration. Sharqieh paid its first dividend to shareholders, 10 percent, in 1940, almost fifteen years after its inception.

10. Ward Papers, box 2, October 25, 1927, E&GSynd (Adams) to Ward and October 26, and November 27, 1927, Ward to E&GSynd (Adams). If Holmes did not reduce the shaikh of Bahrain's agreed royalty of five shillings and threepence down to four shillings a ton, then E&GSynd's own one shilling royalty would kick in only over production of 750 tons per day, not as originally stated on a flat 250 tons per day. In addition, Ward advised, "After due consideration I agreed to recommend to you that the cost of revalidation of Al Hasa, Neutral Zone and Kuwait is shared equally between Gulf Oil and E&GSynd."

11. Ward Papers, box 2, E&GSynd to Ward November 9, 1927, and November 26, 1927, "discretion be permitted Holmes in negotiating with regards to reducing the royalty payable" and "treating as separate contracts"; Ward Papers, box 1, November 5, 1927, Holmes at his farm MillHill to Ward, "I do hope Wallace also realises that I may not be able to procure all that he hopes for in connection with the whole of the business in the Persian Gulf."

12. Ward Papers, box 2, Ward to E&GSynd (Adams) November 22, 1927, Ward to E&GSynd (Adams) to Ward, "not accept terms or one contract" and same date Ward to E&GSynd, "Again and again it was stated that they can only look upon Bahrain as part of the picture" and Thomas E. Ward, *Negotiations for Oil Concessions in Bahrain, El Hasa (Saudi Arabia), the Neutral Zone, Qatar and Kuwait* (New York: privately published, 1965), p. 42, re two separate contracts and "to obtain, in writing, "the consent of the Sultan and Shaikhs, and of the British Colonial Office . . ."" and "we cannot handle our Government Offices like you can yours in America"; Ward Papers, box 2, November 30, 1927, Ward to Madgwick, "I hope you will speak kindly of Major Holmes in every way" and December 20, 1927, "I mean in dollars not sterling," and December 1, 1927, E&GSynd (Adams) to Ward, "the quickest way"; the three geologists sent by Wallace were Rhoades, Shalibo, and Eastman.

13. Ward Papers, box 1, November 5, 1927, Holmes at MillHill to Ward, "proceed with the utmost secrecy" and October 5, 1927, "bring their tennis racquets and golf clubs"; box 2, February 19, 1928, Holmes in Bahrain to Wallace, "made a good impression" and "suit the Arab mind and habit."

14. Philby Papers, box XXX-7, March 6, 1928, Cree to Philby, "fix matters so that

the necessary approaches and negotiations can only be done through you" and April 8, 1928, personal, Cree to Morgan, "if any concessions are given Philby will get them" and April 19, 1928, Morgan to Cree, "best thing is for Philby to arrange with King bin Saud that he, Philby, will handle all concessionary matters" and May 1, 1928, Cree to Philby, "a way will be found by which Standard Oil will handle the business."

15. Ward Papers, box 3, March 29 and May 17, 1928, Ward to E&GSynd, "Ralph Rhoades speaks very highly of the work done by Holmes"; Ralph Rhoades report on Bahrain to William Wallace at Gulf Oil Corporation is reprinted in full in Angela Clarke, *Bahrain Oil and Development, 1929–1989* (London: Immel Publishing, 1991), pp. 71–76; p. 72, states Gulf Oil's expectation of nine months, preferably eighteen months, for the Bahrain survey; Ward Papers, box 2, March 12, 1928, Holmes in Bahrain to Gulf Oil, "dreadfully annoyed . . . real strong British" and "all part of the game"; Yossef Bilovich, "The Quest for Oil in Bahrain, 1923–1930: A Study in British and American Policy," in *The Great Powers in the Middle East, 1919–1939*, ed. Uriel Dann, for Tel Aviv University (New York: Holmes and Meier, 1988), p. 256, citing IOL/LPS/10/993/3299, April 2, 1928, resident to Colonial Office, "nothing to prevent" and p. 257, citing IOL/LPS/10/993/3299, June 19, 1928, secretary of state for the Colonies (Amery) to resident (Haworth), also reprinted in full in Clarke, p. 82, "require any company involved with the development of the Bahrain concession to be a British company registered in Great Britain or in a British Colony" and "neither the company, nor the premises . . ."; Ward Papers, box 2, March 12, 1928, Holmes in Bahrain to Gulf Oil "no revision of the terms agreed with the Shaikh of Bahrain."

CHAPTER 12

DOUBLE-DEALING

As he sifted through the glowing newspaper reports of rich oil finds in Iraq by Turkish Petroleum, of which his company was not a member, William Wallace of the Gulf Oil Corporation was not a happy man.

The results of his recent business decisions weren't looking so good. Both he and Standard Oil New Jersey had been offered Holmes's concessions; Standard had passed, electing instead to stay with Turkish Petroleum, and now that concession was gushing. Shell had beaten him to the Farasan Islands in the Red Sea and was also with Turkish Petroleum in Iraq. Anglo-Persian was rolling in oil, in both Persia and Iraq. His rivals seemed to be beating him hollow.

What he wanted, and fast, was for Frank Holmes to get him into the game by revalidating the Al Hasa concession and securing the Kuwait concession. And, while he was about it, Holmes should overcome his reluctance to lower the terms of the concession as originally agreed with the shaikh of Bahrain.

Communication was by tortuous means. Holmes wrote to E&GSynd in London, who passed his letters, or extracts of letters, to Ward, who then paraphrased the content and passed that on to William Wallace. Replies to Holmes followed the same circuitous route. Cables from Holmes were in E&GSynd's own code and had to be translated in London before being forwarded to New York. When the Gulf Oil geologists arrived in Bahrain, they had their own code and that could only be translated in New York.

Viewed from the plush comfort of his New York office, William Wallace may not have thought so, but Holmes was making an effort. He had arrived back in Bahrain in late 1927 to find that what looked suspiciously like civil war was breaking out in bin Saud's territory. Abdul Aziz Al Gosaibi, bin Saud's representative in Bahrain, airily dismissed the fierce Bedouin raiding along the borders of Iraq and Kuwait. These were merely "skirmishes" that would be "only a temporary affair," he told Holmes.

Via London and through Ward, Holmes wrote to Wallace at Gulf Oil reporting that he had contacted bin Saud and received a favorable reply. Holmes had explained that he had "interested an exceedingly strong American oil group to cooperate with my company to develop Bahrain and to obtain a renewal of the Al Hasa concession, provided that an arrangement can be made with Your Majesty to permit this." He detailed the cooperation: "The Americans are prepared to take over the technical control and to find, jointly with my company, the funds necessary to thoroughly examine, prospect and develop the most favourable areas both in Al Hasa and the Neutral Zone. A British company would be formed to operate the Al Hasa area with both American and English money. The management both technically and commercially would be American."

Holmes had asked to meet bin Saud in order to "arrange fresh terms and liberal payment." He suggested this might be (a) cash payment to be agreed upon, or (b) drill a certain number of wells for water at this company's own expense, or (c) payment partly in cash and partly in drilling water wells. He advised water wells at Dammam, Ras Tanura, and Ojair. He said: "Ras Tanura would require careful geological work in selecting the drilling site owing to the presence of much sand. Of course it would be left to Your Majesty to decide . . . but the exact site of the well in any locality would need to be left to the geologists." He had assured bin Saud that once agreement for revalidation was reached there would be "no delay in beginning the work of making a geological examination of the whole area. A geological party is ready and could be quickly sent. I am ready to come at once to any place agreeable to Your Majesty."

Holmes told Wallace that Gosaibi had commented when he brought bin Saud's reply that "any influence with bin Saud is personally from you, Holmes. Bin Saud believes Frank Holmes will do his best for him and for his people." Gosaibi thought bin Saud was pleased Holmes had brought his country to the "notice of rich Americans who are always clever for oil."

A liberal dose of flattery was not beyond Holmes. He consistently buttered up Wallace while trying to educate him in the intricacies of the Arabian shaikhdoms. That he could state "we are associated with Americans" and the fact that "Americans are here in Bahrain" helped him immensely with the Arab rulers, Holmes told Wallace. "More especially with His Majesty bin Saud. He rules over an immense area. There is no Arab ruler that approaches him in prestige or power."

Despite bin Saud's prestige and power, Holmes continued explaining encouragingly to Wallace, "bin Saud has a remote fear that sooner or later he may become embroiled with the British and consequently lessen the independence of his country."

Holmes laid it on, saying bin Saud's particular fear was "the peaceful penetration of his country by trade and the appointment of foreign Consuls, etc.; actually, he has the British in view. So you can readily understand that if the oil industry of his country is developed by Americans, who have no political axe to grind, he risks less of his country's independence." Even so, Holmes warned, it wouldn't pay to get too cocky. "On the other hand," he said, "bin Saud does not wish to be without the British influence entirely. As Gosaibi said to me, he is certain the mixture of American and English capital will intrigue bin Saud immensely."

Bin Saud's very friendly reply was: "We would have been pleased to fix up the time for having an interview with you by the present opportunity. But with great regret we have to inform you that the present business that is before us does not allow us to do so at the present moment. But we hope that after our being free from our present business we would be able to fix a time in future for your interview. Accept our most sincere greetings."

"The present business," Holmes explained soothingly to Wallace, "refers to the raiding along his borders."[1]

THE IKHWAN REBELLION

After the early years of excitement and the triumphs of the Nejd-Hijaz War that culminated in the capture of Mecca and ouster of Sharif Husain, bin Saud's Ikhwan tribesmen did not share his vision of their future as one of quietly settling down as farmers in agricultural colonies (*hijar*).

As John Bagot Glubb ("Glubb Pasha"), of Arab Legion fame, put it in his memoirs: "For more than twenty years the Ikhwan had gone forward from victory to victory, fired not only by love of war and plunder but also by genuine, not to say fanatical, religious enthusiasm. Their simple creed had made them for twenty years well nigh invincible and had sufficed to overthrow both the House of Rasheed in Central Arabia and Sharif Husain in the Hijaz."

The Ikhwan had taken on the glorious mission of rooting out decadent popular Islam, anywhere they could find it, and converting other Muslims to the puritan Wahhabi creed. Now, to their disgust, they saw their leader, Abdul Aziz bin Saud, dealing with the ungodly British and openly consorting with other foreigners. They were angered in May 1927 when bin Saud signed the Treaty of Jedda with the British. They were incensed by bin Saud's subsequent accommodation of neighboring Transjordan and Iraq, which they viewed as pure British satellites under the Hashemites. Ikhwan discontent erupted into outright rebellion against bin Saud in October 1927 and flamed into conflict with bordering Iraq and Kuwait.

Since Sir Percy Cox's removal of two-thirds of Kuwait's territory at the 1922 Ujair Conference, Kuwait was left without the tribes on whom it had traditionally relied to provide an outer perimeter of defense. Kuwait Town was left entirely unprotected. Taking advantage of this vulnerability, the Ikhwan demanded that the inhabitants of Kuwait Town adopt the Wahhabi creed or face the consequences.

The Administrative Report for 1927 recorded:

A Nejd caravan of 19 men and 35 camel loads of merchandise, said to be worth Rs6000 and to include 15000 Maria Theresa dollars in silver on its way from Kuwait to Qasim, was intercepted before it was clear of Kuwait territory by a detachment of 45 men sent by Abdulla bin Jiluwi the Governor of Hasa. The detachment confiscated everything and took the men in chains to Hasa. On three or four other occasions detachments visited the frontier and the southern portion of Kuwait territory ostensibly to enforce bin Saud's trade boycott.

The ostensible enforcement would soon turn lethal. The political agent, Maj. James More, would record: "two Nejd merchants were spotted bringing three camels across the border into Kuwait. At the very gates of Kuwait Town, their throats were cut by a Negro slave of the Governor of Hasa who shouted the warning, 'Thus shall my master teach you, oh people of Kuwait, that his boundary extends up to your very walls.'"

Kuwait proper came under attack in the first week of December. "As soon as news of the raid reached Kuwait town," the political agent recorded, "a force of over 100 men was collected and rushed out in cars but arrived too late to intercept the raiders, who got away with a considerable number of camels."

There was "much anxiety," he said. "The town wall was manned nightly until the close of the year. Mercenaries were enlisted to supplement the garrison at Jahra which was brought up to the strength of over 300 men not counting the Bedouin camped in the vicinity."

Kuwait was now totally dependent on Britain for its very survival. Finally, Britain revealed the extent to which it was prepared to protect Kuwait by declaring that any threat to the town itself would be repelled by them.

Tribes in Iraq were attacked. A group of fifty Ikhwan rushed the camp of Iraqi workmen building a police post, killing all the workers and police guard. A series of bigger raids on Iraqi tribesmen followed. Unlike the limited support for Kuwait, the British defended Iraq vigorously. Disregarding any border, Royal Air Force planes took to the air, flying into bin Saud's territory to conduct a campaign of fierce strafing and bombing of the camel-mounted Ikhwan rebels.

The results were similar to the Arabian tribes' first encounter with British bombs, modern aircraft, and armored cars described memorably by John Bagot Glubb. "They did the worst thing they could have done, they rallied around their war banners," he reported of the 1924 encounter south of Amman. "The resulting compact mass of men, camels and horses made an easy target for the spitting machine guns. Soon men and animals were falling on one another in inextricable confusion. The plain was strewn with dead and dying men and animals. The Royal Air Force were sick with the killing."

At the beginning of 1928 as Gulf Oil's geologists were traveling to Bahrain, Holmes was buoyed by a request to supply bin Saud's father with an "electric plant for light and an elevator capable of carrying two men." He took this as proof that bin Saud was willing "to do business" and did not, as he had feared, view either him or E&GSynd as "defaulters, more or less."

In the following weeks he kept bin Saud updated about progress in Bahrain. "The first group of American geologists have arrived and are now busy making a detailed Geological and Trigonometrical Survey of the Bahrain Islands," he wrote. "As soon as you are free to discuss business and when such arrangements can be made, another geological party will be ready to examine the Al Hasa Territory," Holmes pledged. "I can again assure you that ample funds are now available to thoroughly examine and test the concession area. And my principals are prepared to act liberally towards Your Majesty with regard to rearranging the concession agreement," he continued. "Meantime, if there is anything I can do to help you, such as artesian well boring or any kind of engineering work, please call me," he offered, adding that he hoped to be contacted "at an early date" for discussion of the concession.[2]

In March, as the geologists were completing their work in Bahrain, Holmes reported that the resident was trying to fix a meeting with bin Saud in order to discuss the "border problems." If it happened, he would accompany the group and "take the opportunity to discuss our concession directly with bin Saud." If the meeting didn't happen, Holmes said, "I fear that little can be done anywhere on the mainland during the next few months." The "current political situation" and approaching hot weather would make fieldwork difficult and "preclude direct efforts for the time being," he said. By April, Holmes reported that bin Saud was on the way to Jedda to meet a British representative "to discuss the raids by some of his tribes on Iraq and Kuwait."

Bin Saud sent a message regretting he was unable to invite Holmes before his meeting with the British representative. He said that as soon as he had "disposed of the British representative, Sir Gilbert Clayton," and after the approaching pilgrimage season (the Haj) he would return to Riyadh and call together the tribal leaders involved in the raids and border troubles in order to "adjust their difficulties." He said he would call Holmes as soon as possible to discuss "our business of concessions." He asked for a personal and confidential letter about whether or not Holmes thought "the prospects of obtaining oil in Bahrain were good" and any other information he might have. He also wanted "particulars of Caterpillar Tractors, building material and many little things."

WALLACE PUSHES FOR KUWAIT

As the talks would take a month at least and then bin Saud "will have to arrange his internal affairs in keeping with the arrangements he makes with the British," Holmes told Wallace he could not see any good in "my remaining here, seeing that I cannot do anything with bin Saud until perhaps October." He intended seeing the shaikh of Kuwait before leaving, but it was "better to settle with bin Saud before approaching the Shaikh of Kuwait about his concession," he said.

Gulf Oil didn't buy it. Wallace was a bird-in-the-hand man. Besides he was chafing at the bit to get *something* going in the Middle East with which he could enter the conversations now so frequently focusing on Iraq at oilmen's social gatherings. "The uncertainty with respect to bin Saud might continue indefinitely and in the meantime we might lose Kuwait also," he wrote Holmes. As Kuwait and Bahrain were politically identical, he claimed, he couldn't see why Holmes should have any hesitation in approaching the shaikh of Kuwait.

Holmes's reply to Wallace was a lecture in realpolitik. The difference between Bahrain and Kuwait was in the degree of British influence, he said. A succession of treaties, and the conditions they contained, "have placed such power in British hands that, to all intents and purposes, they practically rule the Bahrain Islands," he explained.

Holmes said: "There is also a British Agent in Kuwait whose duty is to advise the shaikh as regards foreign intercourse and, under their Treaty, no other nation can send a representative to Kuwait." But the British position was not the same as in Bahrain because "the Political Agent has no power to interfere in the internal affairs of Kuwait. The shaikh has not alienated the right of trying in his own court any persons residing inside his dominions, foreigner or otherwise. And the Political Agent does not run the legal, commercial and religious Courts."

But this might change, he warned, because the British had "expended much money and energy" in defending the town of Kuwait from the rebellious Ikhwan. He had heard the British were "a little tired of the present state of affairs." It was expected they would ask the ruler of Kuwait to decide "to whom, in future, he intends looking for protection, whether bin Saud, the Iraq government or the British government." Holmes thought the shaikh of Kuwait would choose the British, who would then "lay down their terms. They will demand the same powers on which they rule Bahrain."

Wallace should understand, Holmes said, that "the principle ruler in Arabia is bin Saud and he, more or less, dominates all other Arab rulers. The shaikh of Kuwait, being in fear of bin Saud, will not do anything that may arouse bin Saud's ire or touch his dignity." For this reason the shaikh of Kuwait would wait for an indication of bin Saud's wishes in regard to the Neutral Zone. In Holmes's opinion, he would also look for a lead "by waiting to see what bin Saud does with his own territory" in the matter of concessions.

"If I can secure bin Saud's signature to a concession, on the terms now laid out by Gulf Oil in the agreement with E&GSynd, which I feel confident I can, I am almost sure the Shaikh of Kuwait will be prepared to sign an agreement on the same terms." Holmes said that, all things considered, he judged it wiser to wait before taking up negotiations with the shaikh of Kuwait.

Holmes was still drilling water wells in Kuwait, although these were suspended at about five hundred feet so that the American equipment, supplied by Thomas Ward, could be replaced with more reliable machinery from Europe. He would be in Kuwait anyway, but once more he repeated his advice to the Americans that on the matter of concessions "it will be better to negotiate with bin Saud in the first place, and when we have agreed with him, then to deal with the Shaikh of Kuwait." From faraway New York, Wallace redoubled his insistence that Holmes immediately open negotiations in Kuwait.

Considering the sweep of the projects Holmes was attempting to bring to fruition, much of the American behavior appeared trivial. Ward was pressing for a share of E&GSynd's royalty. He warned London that "the Courts in this country have maintained that the commission in such a transaction as that which I have concluded on your behalf applies upon the full consideration, viz both cash consideration and royalties. For an introduction only, resulting in the successful termination of negotiations, our Courts maintain that a minimum of 10% should be allowed. On your behalf, I not only brought the negotiations to a successful termination but also accomplished the introduction to the Gulf Oil people." There was a tussle over the latest geological map that Holmes wanted to use in discussing an expansion of the Bahrain concession area but that Ralph Rhoades was clinging to as Gulf Oil's property. With such irritations nipping at his heels, Holmes left for Kuwait via Basra.[3]

KUWAIT NEGOTIATIONS, ON GULF OIL'S TERMS

At the shaikh of Kuwait's palace, the one that he had helped to build, Holmes settled into his room and then made his way to Shaikh Ahmad's meeting room. When he entered the *majlis*, he was a little taken aback to see "eleven persons in all" waiting to hear what he had to say. The group gathered there on April 24, 1928, included Ahmad's *wazir* (principal adviser), Mullah Saleh, "the Foreign Secretary, four chief advisers and one or two others." The number of people caused Holmes to hesitate.

As he well knew, it was the tradition of the Arabian Peninsula that each person invited by the shaikh to attend an important meeting was expected to share responsibility for decisions taken at that meeting. In return, they could expect to receive a share of any benefit resulting from that collective decision making.

Describing this meeting to Wallace, Holmes said Shaikh Ahmad began with a speech of welcome and then announced to the gathering: "I had come to discuss an oil concession covering his territory. I had proved a friend of the Arabs and would be more likely to help the Arabs than any other party and that I had no political aspirations, being purely commercial as had been proved in Bahrain." Scattered through Ahmad's opening address were repeated references to Holmes's company being "prepared to pay well, both in cash and in kind, for what was required, that is the concession."

The gathering was given the definite impression, Holmes said, that there would be "a large sum of money the moment the concession was signed." Struggling with a 70 percent loss of revenue in taxes previously collected from tribes in the area excised by Sir Percy Cox at the 1922 Ujair Conference and compounded by the fierce trade boycott, a large sum of money was exactly what Kuwait needed.

Holmes was sure that, as the shaikh of Kuwait had been in "constant communication" with bin Saud during the previous months, "bin Saud has told him to ask for plenty." This would have been a smart move on bin Saud's part because it would serve to up the ante for his own Al Hasa revalidation. And Holmes had repeatedly advised Wallace that they should deal first with bin Saud.

But even without such influence, the reality was that in his earlier negotiations with Bahrain, with bin Saud, and with the shaikh of Kuwait, Holmes had been generous—far more generous than Anglo-Persian had ever been. What the shaikh of Kuwait had stated was fact. Because of his absolute conviction that there was oil in these areas, when he was negotiating on his own account, for his own E&GSynd, Holmes had been prepared to pay well to get his concessions. Now, however, Holmes's generosity was curbed by the instructions of the Americans in the Gulf Oil Corporation on whose behalf he was negotiating.

In front of the eagerly expectant group, Holmes had to shift tactics. He said that Kuwait "showed indications" that it might be developed to "an actual value." But, the company with which he was now in cooperation wanted to negotiate a concession on the basis that "if the territory *actually* became valuable" then the shaikh and his people would be "thoroughly protected and liberally rewarded." Holmes told the assembly not to expect "excessive initial payments for testing and proving whether or not the area can be made valuable." Holmes told Wallace he had carefully explained that now any concession agreed upon "would have to be on a royalty basis." He said he had "eventually got the shaikh to agree to this." He added: "I find with the Arabs, as with ourselves, it is better if anything disagreeable has to be told, the sooner it is told the better."

Later, back in the private rooms of the palace, Holmes tried to warn his friend Ahmad. He said, "Filling peoples' minds with the prospect of receiving large initial payments was a dangerous thing for Ahmad, because it would not eventuate." Holmes believed he could have "there and then made an agreement, provided, I was prepared to pay a very large amount for the concession."

Holmes said he noticed that when he "hinted" there was an "American tang" to the company with which he was now cooperating, the Kuwaitis exhibited a "greedy attitude." He said that, while the Arab rulers certainly suspected "every move of the British," such suspicion was "largely removed when American money is suggested." He added, "Unfortunately, I can't make too much of this fact openly." He had parted "very amicably" from the shaikh of Kuwait, promising to draw up a draft concession and return later in the month. He said he had not pressed the shaikh to "name a sum."

Amazingly, for the three years since 1925, the officials of the Government of India in the Gulf had neglected to inform the shaikh of Kuwait that Frank Holmes and E&GSynd had been approved by the Colonial Office as "candidates for oil concessions" and were authorized to conduct negotiations at any time. Seemingly having no knowledge of the meeting at the shaikh's palace, the Kuwait political agent, Major More, cabled the resident, Lionel Haworth, on May 24. He inquired: "Should I comply with Holmes's request and inform the shaikh that there is no objection to him negotiating an oil concession with E&GSynd if he is willing to do so?" Helpfully, More said he believed the shaikh of Kuwait "would not be willing." Haworth promptly replied that the shaikh should be immediately informed of the Colonial Office decision of three years earlier, "although we take no responsibility we give the matter our blessing."[4]

PHILBY'S CLAIM TO "INFLUENCE"

Meantime, the talks between the British and bin Saud were not going well. Far from quickly "disposing of" Sir Gilbert Clayton, bin Saud found the British representative insisting on Iraq's right to build fortifications and police posts along the border. Bin Saud was equally insistent that this was provocative violation of previous agreements. The talks were suspended until after the approaching pilgrimage and after Clayton returned to London to consult with his colleagues.

Two people who were not unhappy at the prolonging of bin Saud's discussions with the British were Harry Philby and T. D. Cree. Waiting until after the pilgrimage for the talks to resume meant that bin Saud remained in Mecca and Jedda for some weeks. In jubilation, Cree passed to Stuart Morgan the news, as he heard it from Philby: "Philby's influence with the King and his government is stronger than ever. The King was in Jedda for sometime and Philby was in daily consultation with him on political questions. Bin Saud even went to Philby's house and had tea with the visiting Mrs. Philby, the only European house he has ever entered in Jedda."

This was pure bravado from Philby. In fact he had had almost no contact with bin Saud who was rarely in Jedda while he was attempting to politically consolidate his territories. Earlier that year bin Saud had expelled the Indian Hindu traders living in Qatif and prohibited foreigners holding land in Al Hasa. Because this brought the entire date trade to a halt, the Indians were later given "limited permission" to return. He had also been extremely preoccupied with arranging practical assistance from the British for his fight with the Ikhwan rebels. Midyear he had received three thousand boxes of ammunition and twenty-seven hundred rifles, delivered through Bahrain, from the Government of India.

Not having ready access to bin Saud was foiling Philby's attainment of the two things he most wanted, which were (a) concessions and (b) assistance to transform himself into an "explorer" by crossing the Rub al Khali, the desert known to Europeans as the Empty Quarter. In reality, because of the Ikhwan criticism of his indulgence toward foreigners, bin Saud was not in a position to grant either, even had he wanted to. Yet, Philby continued to claim that he had strong influence over bin Saud, particularly in the king's dealings with the British. The British themselves rarely fell for Philby's self-proclaimed importance.

One British official termed him "an arch humbug." Another commented: "He's got a mad streak in him." The Foreign Office went along with the verdict of its consul in Jedda who said that Philby "is a nuisance rather than a power of evil. The king, I am convinced, though he likes and admires him, rarely takes him seriously."

Although Philby kept up a facade of success, Sharqieh was not generating business. He kept busy writing exotic adventure books, mostly starring himself. He managed to get three books out of the one twelve-month period, 1917–18, of his (unsuccessful) mission to persuade bin Saud to cooperate with Sharif Husain and his associated Bedouin-escorted dash across the desert. He had just completed the third of these under the title *Arabia of the Wahhabis*. He produced a steady stream of stridently controversial newspaper articles criticizing British Middle East policy in general and the mandates in particular.

Philby's belief in his own uniqueness never wavered. He was certain his association with bin Saud would bring him fame and fortune. He was unsympathetic to his wife's struggle at home in England to feed, clothe, and educate the children. Philby wrote her: "I am far too immersed in the pursuit of my ambitions. My chief aim being to secure the immortality to be gained by the accomplishment of some great work. . . . Everything seems to indicate that the climax is not far off."

Philby would be furious in February 1931 when Bertram Thomas stole his limelight by crossing the Rub al-Khali from the southern coast on the Arabian Sea to the Persian Gulf. Thomas had been Philby's junior in the Mesopotamia Administration and translator/assistant in Transjordan before becoming *wazir* to the sultan of Muscat. Philby would not set out until January 1932 and his crossing would be luxurious compared to that of Thomas. Fourteen Arab guides, charged with protecting his life with theirs, and thirty-two camels carrying three months of provisions would accompany Philby.

Harry St. John Philby's talent for self-promotion may have met its match in bin Saud himself. The Bahrain political agent observed dryly: "bin Saud thoroughly understands the use of the press and a constant stream of Egyptian, Syrian and Iraqi journalists pours through Bahrain to visit bin Saud and be suitably feted by him. He has even paid Bahrain school masters to write him up."[5]

BIN SAUD'S RIYADH SPEECH

Sir Gilbert Clayton returned to Jedda "for a short and fruitless meeting with bin Saud." Without an agreement, he returned to London where he was appointed high commissioner of Iraq, replacing Sir Henry Dobbs. Men whose sole experience had been in India and who viewed Iraq as an Indian province rather than an Arab state had previously administered Iraq. Clayton came from the British Administration of Egypt. Unfortunately, he did not have time to set a new style. A year later, on September 11, 1929, he died in Baghdad of a heart attack during a game of polo.[6]

From the abortive conference with Clayton, bin Saud returned to Riyadh. There he summoned some eight hundred shaikhs and chiefs, townspeople and Bedouin. In October, in a wing of his palace put aside for the occasion, he would masterfully address the gathering. He detailed his capture of Riyadh as a young man and moved through each of his achievements to dwell on his present goal of unifying the entire Arabian Peninsula. He closed a long and emotionally rallying speech by inviting those present to choose another member of the Al Saud family to take his place as ruler.

As Shaikh Hafiz Wahba observed dryly, the gathering "found it difficult to believe that he really meant what he said because he had reached his position after waging a series of wars—on his own uncle's family, on the Rasheeds, on the Turks, on the Hashemites and on the Yemenis. It seemed hardly likely that after all this he would so easily relinquish his throne, when he had not even been challenged to do so." But he achieved the desired effect. The assembly broke into loud acclaim and, to a man, pledged to obey him implicitly.

In Riyadh, bin Saud was preaching to the converted. The rebels simply ignored his summons. In the weeks that followed, the Ikhwan claimed they were the "Defenders of the True Faith" and "Supporters of the Law," which Abdul Aziz bin Saud was attempting to destroy. They charged that he wanted only personal power and conquest. They alleged he was a friend of infidels and a willing accomplice to the activities of foreigners in his territories. They soon began to swarm over the Kuwait and Iraq frontiers, and around Qasim and Ha'il, with sporadic but deadly attacks.[7]

KUWAIT TERMS "DISAPPOINT" GULF OIL

In New York, William Wallace of Gulf Oil followed events and concluded it didn't look good for business. When bin Saud's talks with the British were suspended, Wallace cabled Holmes: "The press carries news that Clayton's negotiations with bin Saud were in nowise conclusive. In consequence Clayton has returned to London for conference with his government. This means that if you cannot take up discussions with bin Saud until after conclusion of Clayton's negotiations, then you are apt to be long delayed. This is good reason why you should further push the Kuwait negotiations." He continued pressuring Holmes, implying that Gulf Oil had never been all that interested in Bahrain and telling him that he needed to "justify a venture so far from home," which he wanted to do by having more territory than just Bahrain.

Holmes was dealing with three sets of immediate negotiations. He was seeking to extend the period of the exploration license in Bahrain, as required by Gulf Oil, and negotiate an additional area, plus the concession for Kuwait. An agreement for the additional area of Bahrain had been forwarded to the Colonial Office, but not before Charles Belgrave, the adviser, had tried to flex his muscles by inserting a 100 percent increase in the payment figure. Holmes had spotted this and mailed off the untouched original document rather than the copy returned from Belgrave.

The Americans were not helping either in their minute questioning of every expenditure. In a confidential letter to Ward, E&GSynd attempted to make the Americans understand that "in the nature" of certain payments "there *is* no account, or receipts and disbursements." They gave the example of "the wife of Major Holmes, who purchased, at the request of the Shaikh of Kuwait, sundry dresses and other goods for the members of his household and we met the bill to the extent of 200 sterling. This is all the information we can give you. If you want more information you will have to wait until Holmes returns."

Following the April meeting with Shaikh Ahmad and his Council of State, details had been ironed out in a series of meetings with Mullah Saleh, and by July Holmes was ready to submit his concession to the shaikh of Kuwait. "The opposition have been most virulent and are still very active. Their attitude has been more to prevent the signing rather than endeavouring to secure a concession for themselves," Holmes reported.

Major More was still convinced the shaikh of Kuwait would not deal with Holmes. When Shaikh Ahmad showed him a draft of the agreement, More wrote the resident, Haworth, saying: "I gather the shaikh has only been persuaded to consider it at all by the promise of a Lincoln motor car. I do not think there is much likelihood of his granting the concession as he told me that he considered the E&GSynd a very 'weak' concern." More claimed the "failure" of the water-drilling operations "although they still say they hope to get water any day" had "undoubtedly detracted" from Holmes's prestige. He confided to the resident Haworth that "a good many people in Kuwait think they were never trying for water but were secretly trying for oil instead. I, of course, do not know what the shaikh thinks about this but he certainly *must* [*sic*] have heard the suggestion."

The draft for Kuwait was based on the Bahrain oil concession, More pointed out, but was "very *much more* generous to the shaikh [*sic*]." Unhappily, he was forced to conclude, "if the shaikh should wish to grant the concession I can see no grounds on which we should object." There was one exception, he said. Article IX could not be accepted. This read "all imported native labour will be subject to and obey the laws of the State of Kuwait." More believed this clause was the result of previous acrimonious discussion with the shaikh of Kuwait during which the British had tried, unsuccessfully, to bring Persian, and possibly other "foreign" residents in Kuwait under British jurisdiction in the same way as applied in Bahrain. This clause could not be accepted, More commented, "as imported native labour would probably be largely Indian."[8]

Not being in any way familiar with the traditions of the Arab shaikhdoms, William Wallace was shocked when he received the details, although the expression he actually used was "quite disappointed."

Holmes explained the costs involved in securing the concession. These included "Rupees 4,000 to the State Adviser (Mullah Saleh) when signed concession is delivered to us. Rupees 4,000 is to be divided among four members of the Council of State. Rupees 2,000 to lesser members of the State Government, that is roughly Rupees 500 each to the Government Office people." He had already paid out some six thousand rupees to the shaikh. Still to come, after sixty days, was the first year's rent of thirty thousand rupees plus payment of "Rupees 45,000 among the five highest State officials of which amount the Shaikh of Kuwait takes Rupees 20,000 for his private purse, distinct from the Kuwait State purse. After the first year, the annual rent will drop to Rupees 20,000."

For the benefit of the folk in New York, Holmes explained carefully: "It must be noted that while all above payments are charged to this Kuwait Oil Concession, part of the payments to the state officials (excluding the Shaikh of Kuwait) include payment for their future help in securing additional Kuwait territory and revalidation of the Kuwait half of the Neutral Zone Oil Concession." He did not even bother telling the Americans that, at one point in the negotiations, he had had to dissuade the shaikh of Kuwait and his Council of State from demanding a provision that the concession was granted to Holmes "personally" and would be invalidated if transferred to any company or other interest.

Holmes had met Gulf Oil's fundamental requirement that the agreement be based on royalty only. There was no profit participation for Kuwait. And he did succeed in raising the original area of the concession from three hundred sixty square miles to six hundred forty square miles with an option of acquiring a further six hundred forty square miles.

E&GSynd did not react well to Gulf Oil's "disappointment" with the terms of the Kuwait concession. They had no sympathy with what they viewed as their American colleagues' timidity. Wallace had refused to authorize a two-hundred-sterling reimbursement related to the extension of the exploration license for Bahrain. Contending that this expenditure was incurred "solely on your behalf," E&GSynd contemptuously pointed out that they would not have asked for an extension of the exploration license. They would have been sufficiently confident, they said, to move directly to the next step of the prospecting license as allowed for in the Bahrain Agreement. E&GSynd said they believed the terms now reached by Major Holmes for Kuwait were "more than acceptable" and advised Gulf Oil to quickly agree in order that the Kuwait oil concession could be submitted to the Colonial Office for authorization.

Somewhat haughtily, they suggested that if Gulf Oil wished to "discuss personally any questions arising out of the Kuwait negotiations with Major Holmes," he would travel to New York from London where he was due to arrive shortly for the purpose of following up with the Colonial Office.

Gulf Oil Corporation said that was exactly what they wanted to do. On September 28, 1928, as bin Saud was congratulating himself on the success of his Riyadh speech, Frank Holmes was returning to New York.[9]

GULF OIL BACKS AWAY FROM BAHRAIN

When Holmes arrived in New York he found that the only subject the Gulf Oil Corporation wanted to discuss with him was how soon—and how cheaply—he could lock in the Kuwait concession. Wallace and Gulf Oil had been steadily losing interest in Bahrain ever since Holmes had stated, and E&GSynd had confirmed, there was no possibility whatsoever of reducing the terms already agreed upon with the shaikh of Bahrain. But when he left London, Holmes had no idea that Wallace had already informed Ward that the Gulf Oil Corporation would not be going ahead with development of the Bahrain concession.

While Gulf Oil had been engaging in penny-pinching correspondence with E&GSynd, they had neglected to inform their British colleagues that they were negotiating to join the Americans of the Near East Development Corporation in the Turkish Petroleum Company. In July, while Holmes was submitting the offer for the Kuwait concession to Shaikh Ahmad, the Gulf Oil Corporation signed the Red Line Agreement, with its very clear map. Although it seems unlikely, Ward claimed that not until mid-August did they become aware Bahrain was within the exclusion zone. Even so, he and Gulf Oil kept their British "partners" completely in the dark until Frank Holmes actually arrived in New York at the end of September.

Informing a shocked Holmes of Gulf Oil's decision to pull out of the agreement to develop Bahrain, Ward contended the company had decided "with reluctance" that they were "duty bound" to "honour" the Red Line exclusion clause. Gulf Oil's dominant interest now would be Kuwait, on the grounds that this was outside the Red Line Agreement. The claim to be adopting a high moral stance was tarnished by the also-stated intention to hold on to the options for the Al Hasa and Neutral Zone concessions—both of which were clearly inside the Red Line Agreement—but both Ward and Gulf Oil chose to ignore this obvious contradiction.[10]

Wallace brought the company attorney, James M. Greer, to the meetings with Holmes and Ward, and the two executives spelled out in no uncertain terms what was now required of Holmes. They stressed that Holmes should "urgently secure the Kuwait concession," on terms that Gulf Oil would now dictate. They approved the gift of "an automobile," although not a Lincoln but one "purchased in England, of such type and make as will, in Major Holme's judgement, fairly comply with his undertakings towards the Shaikh of Kuwait."

They told Holmes he must exert every effort to revalidate the Al Hasa and Neutral Zone concessions. He should secure an extension of the exploration license, due to expire on December 2, and ensure the additional area to the Bahrain agreement. E&GSynd should grant to Gulf Oil a six-month extension of their option on the Bahrain contract, for nominal consideration of one dollar. With these items taken care of, Gulf Oil would proceed to on-sell its interest in Bahrain. To Holmes's apprehension, the Gulf Oil men said they intended offering it to either Turkish Petroleum or the Anglo-Persian Oil Company.[11]

Holmes didn't hang around. He was out of New York in six days. But before leaving he made it quite clear that securing revalidation of Al Hasa and the Neutral Zone was no easy task, as the Americans seemed to think. As to Gulf Oil's intention to on-sell its interest in Bahrain, Holmes convinced them to agree that he should first "see what could be done" in London "by him and his syndicate," if necessary, including with Turkish Petroleum and Anglo-Persian.

Once again Holmes's supposed "partners," Ward and Gulf Oil, were keeping him in the dark. They didn't tell him they had already offered the Bahrain concession to Turkish Petroleum, before signing the Red Line Agreement way back in July, two months before Holmes arrived in New York.

In a bitter quirk of fate, Gulf Oil had taken the Bahrain concession to none other than Stuart Morgan, now secretary of the Near East Development Corporation of Turkish Petroleum.[12]

NOTES

1. Ward Papers, box 2, December 23, 1927, Holmes in Bahrain to Abdul Aziz bin Saud and January 16, 1928, bin Saud (the Government of Hijaz and Nejd and Dependencies) to Holmes in Bahrain and February 19, 1928, Holmes in Bahrain to Wallace in New York. Knowing the Arabs would not respect him if he said physical danger had prevented the starting of work at Al Hasa, even if a deadly conflict involving the feared Wahhabi Warriors was raging all around the oil site, Holmes said he had emphasized the effect of Arnold Heim's negative report. He felt bad about this and explained to Wallace: "This is not quite right as Heim requested that he be allowed to revisit the area. I had to stress that Heim's report was unfavourable in order to make this the reason for our non-payment of the annual fee. I had no other reason I could give." Having used Heim as an excuse, Holmes was faced with the problem of explaining why he and his company had now changed their minds. He told Wallace he had used Madgwick for this in his letter to bin

Saud by saying: "I told my company that at least one other opinion should be obtained. They agreed and Madgwick was engaged to make an independent report on Bahrain . . . he reported favourably."

2. John Bagot Glubb, *The Story of the Arab Legion* (London: Hodder and Stoughton, 1948), p. 70, "Ikhwan . . . for twenty years . . . well nigh invincible"; Hafiz Wahba, *Arabian Days* (London: A. Barker, 1964), see chap. 8 for the Ikhwan rebellion; Administrative Report for Kuwait 1927 (Major More), "took the men in chains to Hasa" and "The town wall was manned nightly" and for 1928, "thus shall my master teach you"; Glubb, pp. 61–64, "strewn with dead and dying . . . RAF were sick with killing . . . (we) followed the retreat and took prisoners"; Ward Papers, box 2, February 19, 1928, Holmes in Bahrain to Wallace in New York, "request to order for bin Saud's father"; Ward Papers, box 3, February 22, 1928, Holmes in Bahrain to Abdul Aziz bin Saud, "anything I can do to help, such as artesian well boring or any engineering work."

3. Thomas E. Ward, *Negotiations for Oil Concessions in Bahrain, El Hasa (Saudi Arabia), the Neutral Zone, Qatar and Kuwait* (New York: privately published, 1965), p. 53. March 25, 1928, Holmes in Bahrain to Wallace, "opportunity to discuss our concession directly with bin Saud" and April 16, 1928, "I cannot do anything with bin Saud until perhaps October"; Ward, p. 61, April 21, 1928, Holmes to Wallace, "building materials and many little things" and "when we have agreed with bin Saud, then deal with the Shaikh of Kuwait"; Angela Clarke, *Bahrain Oil and Development, 1929–1989* (London: Immel Publishing, 1991), p. 79, for the map incident; Ward Papers, box 3, March 23, 1928, Ward to E&Gsynd, "the Courts in this country."

4. Ward, *Negotiations for Oil Concessions*, p. 69, May 6, 1928, Holmes in Bahrain to Wallace describing the meeting with the shaikh of Kuwait and "eleven persons in all" and "did not press him to name a sum"; IOL/R/15/5/238, vol. 111. May 24, 1928, political agent Kuwait (More) to political resident (Haworth), "should I comply with Holmes's request and inform the shaikh?" and May 26, 1928 (Haworth) to (More), "we give it our blessing."

5. Philby Papers, box XXX-7, June 9, 1928, Cree to Morgan, "Philby's influence with the King . . . is stronger than ever"; Administrative Report Bahrain 1928, "3000 boxes of ammunition and 2700 rifles"; Clive Leatherdale, *Britain and Saudi Arabia, 1925–1939: The Imperial Oasis* (London: Frank Cass 1983), p. 194, "arch humbug"; E. C. Hodgkin, ed., *Two Kings in Arabia: Letters from Jeddah, 1923–5 and 1936–9* (Reading, UK: Ithaca Press, 1993), p. 48, Reader Bullard saying, "He's got a mad streak in him." Elizabeth Monroe, *Britain's Moment in the Middle East, 1914–1971* (London: Chatto and Windus, 1981), pp. 149–57, "secure the immortality . . ." and p. 176 for Philby's fury at Bertram Thomas; Administrative Report Bahrain 1928, "Bin Saud thoroughly understands the use of the press. . . ."

6. See Robert O. Collins, ed., *An Arabian Diary: Sir Gilbert Falkingham Clayton*, (Berkeley: University of California Press, 1969), p. 266, "Clayton came from a different Imperial tradition, the Egyptian, where the role of British officials had for long been that of Advisers rather than Masters."

7. Wahba, *Arabian Days*, pp. 138–39, the Riyadh gathering was held on October 19, 1928. Wahba comments, "The offer of abdication was of course not accepted. Abdul Aziz was a great man and everybody recognised it. Besides they found it difficult to believe . . ."

8. Ward, *Negotiations for Oil Concessions*, p. 85, June 4, 1928, Wallace to Holmes in Bahrain, "justify a venture so far from home"; Ward Papers, box 1, July 6, 1928, E&GSynd to Ward, "the nature of certain payments"; Ward, p. 87, July 3, 1928, Holmes in Kuwait to London and New York, "prevent the signing"; IOL/R/15/1/638, D73, July 20, 1928, political agent Kuwait (More) to political resident (Haworth), "no grounds on which we should object."

9. Ward Papers, box 3, August 3, 1928, Ward to E&GSynd, "Wallace's first inclination was to suggest our cabling Holmes that the Kuwait concession discussions with the shaikh had now reached the point where the terms were not acceptable to us. However, after calm and considerate deliberation, Wallace came to the conclusion that it might be wiser to accept now the very best possible bargain we can arrive at with the shaikh and not run the risk of having Holmes come out with the attendant possibility that our competitors might obtain the territory during his absence." See exchange of letters Holmes, E&GSynd, Ward, Wallace in Ward, *Negotiations for Oil Concessions*, pp. 86–99, "includes payment for future help in securing . . ."; Archibald H. T. Chisholm, *The First Kuwait Oil Concession Agreement: A Record of the Negotiations, 1911–1934* (London: F. Cass, 1975), p. 15, "concession granted to Holmes 'personally'"; Ward Papers, box 3, July 26, 1928, E&GSynd to Ward in New York, "would not have asked for an extension" and "discuss personally any questions . . . with Major Holmes."

10. Ward is coy on this point. On p. 104 of *Negotiations for Oil Concessions* he says, "at the end of August 1928, arrangements were made for Holmes to come to New York." In fact this was arranged at the end of July; see above, as Ward correctly continues, "It had already been decided to have Holmes come to New York to discuss a draft of a Kuwait concession for submission to the Shaikh of Kuwait." Note that Gulf Oil signed the Red Line Agreement on July 31, 1928. On p. 106 Ward notes that Holmes arrived New York September 28, 1928, and claims, "Owing to Holmes's illness and the need for discussion of the impact of the Red Line Agreement *the problem had not been conveyed to him in correspondence*" (my italics); Ward, p. 103. Ward also makes the unsupported statement "the Gulf Company was determined to go ahead (in Kuwait) having become convinced of the possibility of finding oil in Kuwait." Jerome Beatty, "Is John Bull's Face Red!" *American*, January 1939, offers another opinion. "Gulf Oil Corporation was Andrew Mellon's Company. Mr. Mellon was Secretary of the Treasury . . . and was about to be made Ambassador to England. If Gulf hit oil in (British controlled) Bahrain there would be political complications and he was getting rid of all business that might touch politics." Beatty is in error, though this opinion has entered the literature. Mellon did not became Ambassador to Britain until four years later, in 1932.

11. Ward Papers, box 2, October 3, 1928, Wallace to E&GSynd, "gift of an automobile"; Ward, *Negotiations for Oil Concessions*, pp. 106, 109, for Greer and Wallace dictates to Holmes during these meetings.

12. Clarke, *Bahrain Oil and Development*, p. 98, states Gulf Oil offered its Bahrain concession to Turkish Petroleum on July 27, 1928, just prior to the Red Line Agreement being effected. Ward's version is so contradictory that it cannot be relied upon. In *Negotiations for Oil Concessions*, p. 106, Ward says Holmes and E&GSynd were to see what could be done *in London* with Turkish Petroleum and Anglo-Persian. On p. 107 he says that Stuart Morgan of Turkish Petroleum was "*contacted in New York* but only *after Holmes had departed and* no response was received from Holmes in London." On p. 110

he says, "the proposition to take over Bahrain was laid before Turkish Petroleum *prior to Major Holmes visit to New York.*" See also Ward Papers, box 3, December 11, 1928, Ward to Holmes in Bahrain, "the Bahrain contract, the offer on which I understand was *submitted by Mr. Wallace to Turkish Petroleum immediately following the conference which you and I had with him in New York*" and January 7, 1929, Ward to E&GSynd, "negotiations with Turkish Petroleum were opened on October 16, 1928. . . ."

CHAPTER 13

THE NATIONALITY CLAUSE

easoning that Stuart Morgan was well connected with Turkish Petroleum and the Standard Oil group of companies—and had previously shown such lively interest in Holmes's presentations of his Arabian concessions—Gulf Oil had gone directly to Morgan for "advise." Morgan was willing.

He was still in discreet correspondence with Cree and Philby. In a recent letter Cree had explained to Philby:

> Stuart Morgan again pushed the point that Standard Oil could not act without the assent of the other members of Turkish Petroleum. I pointed out that because of the dominance of Anglo-Persian, Turkish Petroleum would have no chance of getting into the country. But Standard Oil might succeed in its own name and they could make their own arrangements. Morgan agreed that this could be done, but said that opinion in Turkish Petroleum and Anglo-Persian was that prospects of getting large production in bin Saud's territory was looked on so unfavourably that those two companies would prefer to leave it alone.

At least for the moment, this remained a moot point because the raging Ikhwan rebellion ensured that Al Hasa was closed to everybody, including bin Saud himself.

In letters to E&GSynd, Ward waxed lyrical in praise of Frank Holmes's recent visit to New York. Holmes had been presented with a fait accompli draft agreement "embodying the ideas of Mr. Wallace" that he was ordered to present to the shaikh of Kuwait. Although there had been "considerable discussion of various terms and conditions," Ward reported "splendid cooperation" had gone into this draft. The Gulf Oil Corporation was "confident" that Holmes would now proceed directly to Bahrain from London and "conclude the extension of the original concession and also secure the additional Bahrain territory" and then move on to Kuwait. "We are all anxiously looking forward to the conclusion of the Kuwait negotiations and to the early agreement between His Majesty bin Saud and Major Holmes on the Al Hasa and Neutral Zone properties," Ward rhapsodized.

Some of this enthusiasm may be explained in two letters of the same date in which Ward laid out his expectation that Madgwick would make his own arrangements but would receive "10% commission on the proceeds and arrangements with Gulf Oil to apply on the entire consideration, both cash and royalties." His expectations now included "a payment of $5,000 to cover expenses up to September 30th 1928 followed by annual payments of $2,500 to cover further expenses." E&GSynd dashed his hopes with their reply. They said:

> Had you asked in the first place for a salary this would have debarred commission. The Board feels they must point out that they cannot agree that Madgwick was at any time in the position of a principal, and Major Holmes states that, from the first, you arranged to look after Madgwick.
>
> It is well that you should realise that any information Madgwick may have obtained in regard to Bahrain was secured during the period in which he was receiving a salary and expenses from this syndicate, in other words Madgwick's "report" was paid for by us.[1]

BRITISH ONLY

William Wallace and the lawyer for the Gulf Oil Corporation had spelled out to Holmes in no uncertain terms what was expected of him and E&GSynd. He was obliged, they told him, to secure an extension of the Bahrain exploration license, due to expire at the end of the year, and ensure the additional area. He must also get official sanction for the Kuwait concession and exert every effort to revalidate Al Hasa and the Neutral Zone. Unaware that Wallace and Ward were already hawking it around, Holmes was proceeding on the understanding that he would first "see what could be done" in London "by him and his syndicate" about buying out the Bahrain concession before Gulf Oil offered it elsewhere, possibly to Turkish Petroleum and Anglo-Persian.

Edmund Janson and Holmes considered the possibility that they might be able to revert to developing Bahrain on their own, just as they were doing before the enthusiastic Ward stepped in to broker a deal.

On his return from New York, Holmes and Janson called on the Colonial Office and explained that the first geologist they had sent to Bahrain, Dr. Arnold Heim, had reported unfavorably, but since then "other geologists have been sent, and although conflicting, their reports are for the most part very favourable." They said they now wanted to send again "the original geologist"—Heim—in order to give him an opportunity of reconsidering his unfavorable report. They were satisfied, they told the Colonial Office, "in view of the geological information now available," that Heim would report favorably this time around.

Had there been some thought of involving the Americans in the Bahrain concession? the Colonial Office asked. Janson and Holmes replied that they were at present negotiating "for finance" with American and Canadian interests in order to capitalize prospecting operations and ultimate commercial development in Bahrain. The men at the Colonial Office were still smarting from the American success at forcing their way into a share of Iraq's oil through joining the Turkish Petroleum Company. Speculating that if the capital were American then it was conceivable that "real control" would pass into American hands, the Colonial Office wanted to know "if the discussions were successful" would "the preponderance of capital be American?"

Janson went into a long explanation of how E&GSynd had "expended some 60,000 sterling on exploratory work in Bahrain" and had sincerely attempted to obtain development capital from British investors. Gambling that apprehension at the prospect of American involvement might jolt the Colonial Office into influencing British investment in developing the Bahrain concession (as had happened in 1905 with Burma Oil and the D'Arcy Exploration Company in Persia), Janson tried to turn the situation to his advantage. He said he supposed it was possible the British company could eventually be "the merest facade" behind which the Americans "would pull the strings."

Unfortunately, the gamble backfired. Janson and Holmes were sharply reminded that "under the terms of the Agreement" the concession could not be conveyed to another party without the consent of the shaikh "acting on the advice of the Resident." The Colonial Office claimed that when grant of the concession was first discussed they had been assured that "control would be, and would remain, purely British." Gamely, Janson insisted, "it has proved wholly impossible to raise the necessary capital" in Britain. He was peremptorily instructed that, before negotiations with the Americans "went further," he must formally present any proposal for financing to His Majesty's government and "invite their concurrence."

In a memorandum forwarded to the secretary of state for India, the Colonial Office referred to the British nationality clause as used in Iraq and explained that they would now be insisting on this as a primary condition in any proposed oil concession for Kuwait. The Colonial Office assured the Government of India that Janson and Holmes would be told "if they are prepared" to give a similar undertaking for Bahrain, that is, "that the Company shall at all times be and remain a British Company," then "His Majesty's government will be prepared to recommend to the Shaikh of Bahrain that he grant the extension of the Exploration License for the further year as the syndicate desires." E&GSynd, the memorandum noted, "are still hopeful of obtaining British capital."[2]

STANDARD OIL OF CALIFORNIA

Perhaps Ward was fearful his commission would slip away. Perhaps the Gulf Oil Corporation thought January 1, 1929, the date on which they must exercise or lose their option, was too close for comfort. Across the Atlantic, without informing his London colleagues, Ward was trying to sell Gulf Oil's option on Bahrain to Standard Oil of California, which was not a member of Turkish Petroleum and therefore not affected by the Red Line Agreement. At the same time, William Wallace at Gulf Oil was authorizing Stuart Morgan to offer the option on the Bahrain concession "across the counter" to the upcoming board meeting of Turkish Petroleum in London. Wallace said it had cost him $50,000 to send his geologists to confirm the work Holmes had already put into Bahrain.

In a chatty little letter to Holmes dated November 2, Ward made no mention of either Standard Oil of California or the forthcoming Turkish Petroleum board meeting. Holmes was on his way back to the Persian Gulf and confident he would "hear word that the Colonial Office have recommended the extension of the Bahrain exploration license." He had arranged to see the shaikh of Kuwait.

While Holmes was onboard ship somewhere between Marseilles and Suez, Ward cabled London requesting "confirmation" that extension of the Bahrain exploration license had been taken care of and expansion of the concession area signed. Most importantly, he said, he urgently needed to know that E&GSynd would approve Gulf Oil's demand for a six-month extension of their option, for one dollar.

For the first time, Ward revealed to his London colleagues that he had taken all the material and information along to Standard Oil of California in pursuit of a sale. Now he expected E&GSynd to send him "urgent telegraphic" approval for transfer of Gulf Oil's option to Standard Oil. He assured London that "Standard Oil of California is quite separate from other Standard companies and is of the highest standing, equal to Gulf Oil in ability to perform all obligations under contracts, and is in nowise identified with Turkish Petroleum Company."

When E&GSynd learned Ward was discussing their business with Standard Oil of California, they sent an "urgent telegraphic" message, all right. To Ward's consternation, it stated that for the impact this could have on the discussions with the Colonial Office, Ward's approach to Standard Oil of California "will be fatal."

It is difficult not to see the influence of Stuart Morgan, "well connected with the Standard Oil group of companies," behind the enthusiasm of Standard Oil of California for purchasing Holmes's Bahrain concession. This was a deeply conservative company, headquartered in San Francisco. They had not exhibited any notable skill at discovery. From 1920 to 1928 they lost more than $50 million in unsuccessful drilling in Latin America, Alaska, and the Philippines. They had not been able to develop a single barrel of foreign commercial production, an experience that left many of their directors skeptical about exploration abroad. They had not had the slightest inclination to join their sister oil companies in the postwar battle with the British to let the Americans in on Iraq oil.

Yet, they were suddenly keen to buy into tiny, insignificant Bahrain. Something, or someone, made them so sure that they would not waiver even in the face of the renewed "expert" opinion that there was no oil in Bahrain, as repeated by the members of Turkish Petroleum, Shell, and Anglo-Persian. Standard Oil of California remained so keen that, in pursuit of the faraway Bahrain concession, they would hold on with absolute tenacity through all the obstacles that the Colonial Office would throw in their path.

As to Stuart Morgan himself, as agreed upon with William Wallace, on October 30 he laid out the full detail of arrangements for all the Arabian concessions between E&GSynd and Gulf Oil to the board of Turkish Petroleum . . . in which the Anglo-Persian Oil Company was the dominant shareholder.

A colorful story later circulated about the deliberations of this meeting. One of the Dutch directors was said to have leaned back in his chair, leisurely filled his pipe from an oilskin pouch, and announced that the general Arabian area was "not unknown" to his company, Shell, which had sent geologists there "long before." They had encountered considerable difficulties and unpromising political conditions, the story went, but no oil. One of the British directors agreed wholeheartedly. Stuart Morgan was said to have grasped this opportunity to suggest that, if Turkish Petroleum as a whole was not interested, then would they consider allowing one, or all, of the American companies to take it on as a separate venture? According to Ward, who recounted the story in his memoirs, the Dutch and British directors would not agree to this. The Dutch director closed his final argument "with a slow shake of his head," Ward wrote, "and the words: 'No, I'm afraid there is no oil in Arabia.'"[3]

THE COLONIAL OFFICE LEARNS ALL

The details revealed at the board meeting of Turkish Petroleum were quickly passed on. Now the Colonial Office knew everything. The American involvement was not merely to supply financing for a British company to develop the Bahrain concession, as Janson and Holmes had intimated, but it was a joint venture; an American takeover, even. The Colonial Office made its move. They decreed that the extension of the Bahrain exploration license could only be obtained on the basis that a British nationality clause be inserted—retrospectively—in the already signed concession.

In high dudgeon E&GSynd lashed out at their American colleagues, accusingly cabling New York, "Colonial Office apparently have full knowledge our arrangements with you." Incensed, they dashed off an extremely irritated letter to Ward about "outside influences brought to bear on the Colonial Office." Specifically, they charged:

> We cannot but assume that when particulars of our interests were given to the Turkish Petroleum Company, *which company is managed by the Anglo-Persian Oil Company* [sic], Anglo-Persian placed all the facts before the Colonial Office and drew attention to a probability that the properties would cease to be under

direct British or Canadian control. It is obvious that such representations have been made to the Colonial Office for upon our approaching them in reference to the extension of the Bahrain Exploration Licence, it was apparent that they already knew the full detail of all our arrangements with the Gulf Oil Corporation.

Just in case the Americans hadn't quite got the message, they followed with a cable forty-eight hours later to Ward, who was still anxious to sell to Standard Oil of California. E&GSynd were angry and told Ward in no uncertain terms: "It is now unavoidable that any extension should contain new Colonial Office conditions. This has come about because of disclosures made to the Colonial Office by others. Probably as a result of your offer to Turkish Petroleum Company— unknown to us—of your interests in our properties."

E&GSynd proposed that there was an alternative to extending the exploration license and this was to immediately take up the prospecting license "over the area selected by Holmes." E&GSynd had earlier alluded to American timidity in wanting to extend the exploration license while pointing out that, under the terms of the Bahrain concession, they were automatically entitled to move up directly to the prospecting license, without any further endorsement by the Colonial Office. Now they again urged this step as a way of circumventing the need for Colonial Office sanction.

They urgently warned Ward against making any moves toward Standard Oil of California until a letter from Holmes reached London. "Holmes fears the Colonial Office will now insist on a similar clause in the prospecting license . . . he thinks that as the application for the exploration license was made here, the application for the prospecting license should also be made from here, otherwise the Colonial Office will think we are avoiding them." While they were trying for a satisfactory solution, they said, they could not impress strongly enough on Ward that "hasty action on your part will only result in loss to all concerned."

The warning was too late. Ward was already charging full steam ahead "losing no time at this end." He had already met with Standard Oil of California's president and vice president and had a meeting scheduled with the company's geologists. He had invited Madgwick, who would be visiting from Canada, to come along "to discuss the reports and maps prepared by Rhoades and his party"—the material that built on the work Holmes had conducted before Gulf Oil's geologists arrived in Bahrain.

Seemingly totally insensitive to his London colleagues, Ward informed them that he also had another iron in the fire. "As you are no doubt aware," he told E&GSynd, "Sir John Cadman, Chairman of the Anglo-Persian Oil Company, is now on his way here and we shall discuss with him the matter of Turkish Petroleum and the attitude of the Colonial Office while he is in the USA."

Ward went right ahead and approached Cadman at the annual convention of the American Petroleum Institute in Chicago. Ward reported that when "Sir John

Shaikh Hamad Al Khalifa, ruler of Bahrain (1874–1942).
Photo taken circa 1915. (Directorate of Heritage and Museums, Bahrain)

Shaikh Mubarak Al Sabah, ruler of Kuwait (1896–1915).
Photo taken circa 1905.

Yahya bin Hamad Al Din, ruler of
Yemen and Sufi Imam of the Zaidis.
(From Ameen Rihani, *Arabian Peak and
Desert: Travels in al-Yaman* [1930]. Courtesy
Dr. Ameen A. Rihani)

The staff of Strick Scott and Co. Ltd., Muhammara, 1917.
(Courtesy British Petroleum [BP] Archives, University of Warwick)

The following generation in 1985: (L to R) the rulers of Bahrain, Qatar, Oman, Kuwait, Saudi Arabia, and the United Arab Emirates.

(Media pack, Gulf Cooperation Council)

The Cairo Conference: Churchill's Forty Thieves, 1921.

M. E. Lombardi, vice president and director of
Standard Oil of California, retired in 1941.
(Chevron Texaco Archives, San Francisco. Images copyrighted
by Chevron Corporation and used with permission)

Lord George Nathaniel Curzon (1859–1925) with his wife.
(Courtesy British Petroleum [BP] Archives, University of Warwick)

Lloyd Hamilton (Standard Oil of California) and Abdulla Suleiman signing the Al Hasa concession on May 29, 1933.

(Chevron Texaco Archives, San Francisco. Images copyrighted by Chevron Corporation and used with permission)

William Knox D'Arcy (1849–1917) signed one of the earliest oil concessions in Persia in 1901.

(Courtesy British Petroleum [BP] Archives, University of Warwick)

Harry St. John Philby (1885–1960) in 1938.
(Chevron Texaco Archives, San Francisco. Images copyrighted by
Chevron Corporation and used with permission)

Sir John Cadman, chairman of both the Anglo-Iranian Oil
Company and the Iraq Petroleum Company from 1927 to 1941.
(Courtesy British Petroleum [BP] Archives, University of Warwick)

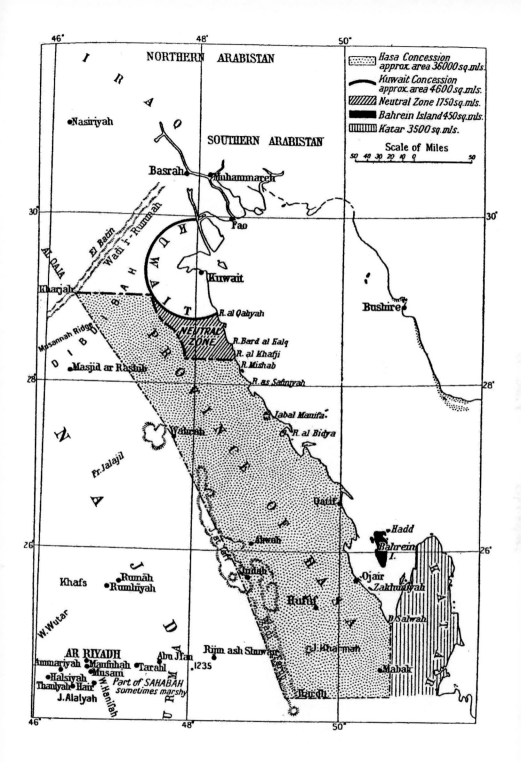

Frank Holmes's map of the Al Hasa concession prepared in 1922.

(From Ameen Rihani, *Ibn Sa'oud of Arabia: His People and His Land* [1928].
Courtesy Dr. Ameen A. Rihani)

Group at Bahrain circa 1930. Fifth and sixth from the left are Frank Holmes and Shaikh Hamad; eighth from the left is Holmes's assistant, Muhammed Yateem.

Fred Davis and Bill Taylor in Bahrain, 1931.

Frank Holmes (1874–1947), called by the
Arabs *Abu Al Naft*—Father of Oil.
(Author's collection)

Charles Greenway, chairman then president of the Anglo-Persian Oil Company (1914–1934).
(Courtesy British Petroleum [BP] Archives, University of Warwick)

Dorothy and Frank Holmes in Kuwait, 1924.
(Author's collection)

Thomas Boverton Redwood, Britain's oil expert.
(Courtesy British Petroleum [BP] Archives, University of Warwick)

Archibald H. T. Chisholm (1902–1988).
(Courtesy British Petroleum [BP] Archives, University of Warwick)

Admiral Sir Edmond Slade (1859–1928).
(Courtesy British Petroleum [BP] Archives, University of Warwick)

Arnold Heim in 1924.

(Swiss Federal Institute of Technology, Zurich.
Courtesy Image Archive ETH-Bibliothek, Zurich)

Ameen Rihani in 1911.

(Courtesy Dr. Ameen A. Rihani)

King Faisal, Gertrude Bell, and others on a picnic outside Baghdad.
(Middle East Centre Archive, St. Antony's College, Oxford.
Gertrude Bell Collection Ref2/37)

Arnold Heim's 1924 expedition approaching Qatif.
(Swiss Federal Institute of Technology, Zurich. Courtesy Image Archive ETH-Bibliothek, Zurich)

Muhammed Yateem, the personal
assistant to Frank Holmes.
(Courtesy Husain Ali Yateem)

Shaikh Ahmad bin Jabir Al Sabah, ruler of Kuwait (1885–1950).
(From Ameen Rihani, *Around the Coasts of Arabia* [1930].
Courtesy Dr. Ameen A. Rihani)

Ali Yateem standing in front of the first artesian well discovered in Bahrain
by Frank Holmes. (Courtesy Husain Ali Yateem)

King Abdul Aziz bin Saud (1880–1953).
(Chevron Texaco Archives, San Francisco. Images copyrighted by
Chevron Corporation and used with permission)

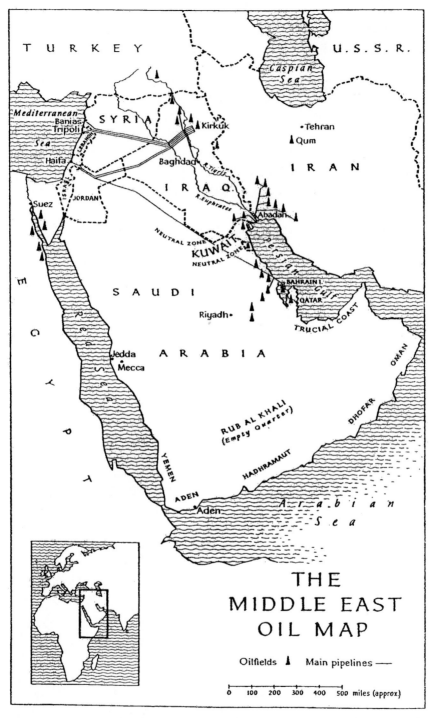

Development of the Arabian oilfields in 1958, and the surrounding areas of
Iraq, Iran, Yemen, and Aden.

(From Wayne Mineau, *The Go Devils* [London: Cassell & Co., 1958].
Courtesy Cassell & Co., an imprint of The Orion Publishing Group)

was offered the Bahrain option he refused on the grounds that the Anglo-Persian geologists were not impressed by the oil prospects on the Persian Gulf littoral."

Curiously, Stuart Morgan's old boss, Walter Teagle of Standard Oil New Jersey, was with John Cadman that day. In what later became known as "the billion dollar error," he, too, knocked back the Bahrain offer. Ward didn't seem too disturbed. He was more than happy to continue on his own "forwarding negotiations with Standard Oil of California."[4]

BAHRAIN TRANSFER

Ward and Wallace would not agree that they had let the cat out of the bag and so provoked the Colonial Office ruling. Wallace said he was "wholly unable to understand the attitude displayed by E&GSynd. I cannot help but feel they are failing to respond in good faith and are merely attempting to drag out the discussion with the Colonial Office until their option with us expires in eight weeks."

For his part, Ward went on the attack. Any "qualifying conditions" now demanded by the Colonial Office, he told E&GSynd, "must not affect your contractual obligation with Gulf Oil and Gulf's right to exercise options through a British company even though Gulf thereby is indirectly in control through its nominee vis a Canadian or British company." He demanded an immediate cable granting "consent to assignment of contract to Standard Oil of California." E&GSynd did not reply.

Ward also wrote to Holmes in Bahrain, complaining, like Wallace, that "E&GSynd does not appear to be willing to either extend our option or to favourably respond to the idea of substituting Standard Oil of California for the Gulf Oil Company in respect of the Bahrain contract. For my part I shall be very glad when the matter is settled and I am compensated." He added: "We are all very anxious to hear the result of your further discussions with the Shaikh of Kuwait. I am sure you will be successful."

To E&GSynd he cabled peremptorily: "Do you, or do you not, authorise me to give your consent in writing to assignment by Gulf Oil to Standard Oil of California."

On December 12 E&GSynd forwarded the consent. The cable read: "We have no objection to transfer Gulf Oil Corporation option agreement to California Standard—but that company must understand that, if they exercise the option, the Colonial Office conditions are compulsory for constitution of the ultimate company. We are powerless to refuse such conditions."

Perhaps not grasping the true dimension of E&GSynd's warning—or perhaps Ward and Wallace did not tell them of the warning—the very next day, Standard Oil of California set about registering a company in Canada. By December 21, an agreement of assignment from the Gulf Oil Corporation to Standard Oil of California would be finalized, with a payment of $157,000.[5]

BLAME GAME

Ward and Wallace had no concept of the labyrinth that was British officialdom. They were pretty pleased with themselves, imagining they had neatly passed over the Bahrain concession to Standard Oil of California. Having freed themselves of that, as they thought, they switched their attention elsewhere.

Blithely, Ward cabled London. "Kindly telegraph immediately the date on which Holmes expects to complete the Kuwait negotiations. Also inform us if he has secured the agreement with the Shaikh of Kuwait along the lines drawn up in New York. If not, what changes are incorporated?"

E&GSynd's reply was terse. "We have received no further news from Holmes regarding Kuwait. He is obviously at a standstill unless the Colonial Office conditions regarding foreign control are acceptable to Gulf?" They repeated their grievance.

> It must have been obvious to you (we think the matter was discussed fully with you by Major Holmes) that the situation as regards the Colonial Office was one of some delicacy necessitating the avoidance of anything calculated to make them inclined to advise the Rulers concerned to prevent the properties coming under any other than British or Canadian control.
>
> We had hoped to so pave the way with the officials in Arabia so that, at the proper time, the Colonial Office objections, if they had any, could have been overcome. You were also aware of the great opposition we have always experienced from Anglo-Persian.
>
> We cannot but assume that when particulars of our interests were given to the Turkish Petroleum Company, which company is managed by Anglo-Persian, all the facts were laid before the Colonial Office and attention drawn to a probability that the properties would cease to be under direct British or Canadian control. It is obvious that some such representations have been made. . . .
>
> Hence has arisen the new condition to prevent foreign control, direct or indirect, which it is reasonable to suppose is the direct result of your negotiations with the Turkish Petroleum Company—to which we were not privy.

Ward and Wallace would have none of it. Ward confided to Wallace, "I cannot understand the delay with the Colonial Office . . . ," and wrote London that he and Wallace were "at a loss to understand why any delay has occurred."

As if the powerful British Colonial Office did not even exist, Ward and Wallace claimed "conditions under which the assignment of the Kuwait concession is to be made were clearly set forth in our agreement with you of November 1927 and were fully discussed with Major Holmes during his visit to New York this October." Ward asked for Holmes's current address so that "Wallace could communicate with him directly."[6]

COLONIAL OFFICE DILEMMA

E&GSynd tried again in London. They rallied an argument they hoped would breach the walls. They reminded the Colonial Office that the Bahrain Concession Agreement signed in December 1925 was the one given them by the duke of Devonshire, replacing E&GSynd's own 1923 documentation. The Colonial Office failure to notice one important difference in the two sets of documentation was now returning to haunt them.

Instead of E&GSynd's clause in which the syndicate undertook specifically not to transfer the concession to a non-British company, the duke of Devonshire's document read: "The rights conveyed by the Lease shall not be conveyed to a third party without the consent of the shaikh acting under the advice of the Resident in the Persian Gulf."

But more important, E&GSynd argued, was the next line, which stated: "Such consent shall not be unreasonably withheld." Moreover, they pointed out, in the duke of Devonshire's document that the Colonial Office had required E&GSynd to use, "no stipulation as to the British character of the Assignee is imposed."

E&GSynd now ran through the factors. They had been unable to find British investors to join in the development of the Bahrain concession or any other of their Arabian interests. Consequently, they had signed two agreements with the Gulf Oil Corporation, the first applicable to the Bahrain concession and the second, known collectively as the Mainland concession, related to the other oil interests. Both these agreements stipulated that any subsequent companies formed for the purpose of development must be British or Canadian.

The Colonial Office felt they had a trump card in that they believed they could refuse to recommend an extension of the exploration license to the shaikh of Bahrain. Reminding the Colonial Office that they had already spent some sixty thousand sterling on the Bahrain concession, E&GSynd pointed out that the "new conditions you now seek to impose upon us" before you will recommend an extension or renewal will not be acceptable to Gulf Oil, "with whom we entered into agreements in good faith."

E&GSynd told the Colonial Office that Gulf Oil would "inevitably" sue E&GSynd, an action that would force the syndicate into liquidation "with the loss of the whole of the capital invested in the Arabian oil propositions."

They left the Colonial Office to contemplate the scandal such an eventuality would create among the business houses in the city.

The Bahrain political agent, Col. Cyril Barrett, now claimed to have guessed all along about American involvement. "I know Holmes had paid visits to the USA to purchase machinery for his venture," he wrote the political resident. "It was strongly suspected therefore, before I received your letter, that his syndicate had a strong American tendency."

Barrett had firm ideas on the dangers of American tendencies. "American influence is already strong in Bahrain owing to the activities of the American Mission, the members of which, in addition to their excellent medical work, miss no opportunity of pushing American trade. It is the opinion of the British colony in Bahrain that they are very efficient unofficial Consuls," he warned.

"I am of the opinion that any increase of American influence is strongly to be deprecated and that no further concession should be granted to E&GSynd unless they give unreservedly an undertaking that the company shall at all times be, and remain, a British company and not degenerate into a mere facade to cover the working of American interests." Barrett predicted darkly: "This undertaking is of utmost importance as the agreement gives the syndicate such large powers that in time they would practically control the government of these small islands."

Barrett thought the solution very simple as he said, "Shaikh Hamad will be guided solely by the Resident's advise," and he "did not anticipate any opposition from him (Hamad)." Three weeks earlier Holmes had paid ten thousand rupees to the shaikh for the extension but this was no problem, Barrett said, because Charles Belgrave, the adviser, "has been careful" while acknowledging receipt to state that "he awaits further communication on the subject."[7]

COLONIAL OFFICE SOLUTION

The Colonial Office took a week to seriously consider E&GSynd's charge that continued official intransigence would result in their being sued by the Americans and forced into bankruptcy. The Colonial Office knew well from past experience that if this did occur a sympathetic press could create a most unwelcome backlash in the city's financial and political circles.

Four Colonial Office heavyweights called a meeting with E&GSynd, telling them to bring their lawyer. They opened by saying the syndicate had erred in granting to the Gulf Oil Corporation "rights which they were not free to dispose of except with the consent of the Shaikh of Bahrain." They said E&GSynd could not "assume" that His Majesty's government would instruct the resident to advise the shaikh of Bahrain to agree to the transfer. They then sketched a highly improbable scenario in which they said that even if the shaikh of Bahrain was "instructed" to agree to the transfer, it could not be assumed he would "be prepared to accept" that instruction from His Majesty's government.

Vowing the Colonial Office was "not desirous of forcing the syndicate into liquidation," they presented their ingenious solution. The best thing to do now, the Colonial Office said, was nothing at all. Instead of E&GSynd insisting on its right to an extension of the exploration license, everything related to the Bahrain concession should simply be allowed to expire.

As a consequence the syndicate would possess no rights on which Gulf Oil could exercise an option and, therefore, Gulf Oil could have no claim on

E&GSynd and, happily, no grounds on which to sue. After this solution was effected, the syndicate could then reapply to the Colonial Office and the shaikh of Bahrain for a completely new concession—one that would, of course, contain the British nationality clause and thereby "preclude Gulf Oil from reaping any advantage from the new concession."

They simply brushed aside E&GSynd's response that this would still leave them out of pocket and bankrupt because they had no more chance now of raising development finance from British investors than they had had before. The Colonial Office was delighted with the solution they had devised. They advised that E&GSynd should now give "further consideration" as to whether or not they really did want to press their application for an extension of the exploration license. The Colonial Office said they felt sure that after consideration the syndicate "might very well prefer to abandon their rights" under the current concession and, "possibly, in due course," apply for a new concession as outlined.[8]

BAHRAIN PETROLEUM COMPANY REGISTERED

On December 27—the day before the Colonial Office was meeting with E&GSynd and advising them that the best thing they could do to avoid being sued by the Gulf Oil Corporation was to do nothing—Standard Oil of California was authorizing Thomas Ward to exercise the option with E&GSynd, due to expire on January 1. Standard Oil of California instructed Ward to exercise the option covering the "so called Bahrain Concession" in the name of the Bahrain Petroleum Company (Bapco), registered in Canada and wholly owned by Standard Oil of California.

Ward was over the moon. With supreme optimism he inquired of E&GSynd how soon he could expect "the formal and effective written consent of the Shaikh of Bahrain" to the transfer of the concession to Standard Oil of California.

With considerable naïveté for an Englishman, Ward said he also expected to quickly receive "the formal, written and effective assurances of the British Colonial Office itself, and/or of any other proper and necessary government officials, that it and/or they will not raise any objection to the assignment of the Bahrain concession."

At the time the Colonial Office was advising Holmes's colleagues that the best thing to do with the Bahrain concession that he had fought so hard for—twice—was to let it expire, Holmes was in the Maude Hotel Baghdad writing to a nephew who wanted help finding a technical mining job. Commenting that "to me mining is a very fine profession and is a man's job," Holmes said he would see what could be done through "one of our mine managers" at the London office. A little wistfully, he explained: "I am not in the way at the moment of being able to place employees in mining. Right now, my work is dealing in the main with the financial side of mining work."[9]

MADGWICK INTERFERES

In the midst of all this, Madgwick arrived in London. His status had finally been changed to that of a permanent employee of the Canadian government and he had come back to collect his wife, daughter, and maid. Before leaving he had written to Ward saying he intended to visit E&GSynd to see what compensation he could get out of Janson. Madgwick's sense of being hard done by was steadily increasing. He genuinely believed that, somewhere between E&GSynd, the Gulf Oil Corporation, and Standard Oil of California, he should have made a lot of money, much more than the one-third commission that Ward was promising him. Janson referred him back to Ward.

On his return Madgwick reported just how angry Janson and the syndicate were about the leak to Anglo-Persian through Turkish Petroleum. Ward promptly wrote to London claiming that his original letter saying that the contact with Turkish Petroleum had occurred before Holmes's arrival in New York "was incorrect." What had happened before Holmes arrived, he said, was that Wallace had spoken to the "American Associates" in Turkish Petroleum, not to Turkish Petroleum per se. As Wallace was confident ensuing discussions "did not go any further than the Board Room," the leak to the Colonial Office could not possibly be their fault, he claimed.

Ward gratefully leapt on the explanation Madgwick soothingly put forward. Madgwick said neither Ward nor Wallace had done anything wrong. Although it did not accord with Holmes's advice to Wallace at the time, that everybody in the Persian Gulf had "guessed wrongly and are quite convinced that your men come from the Standard Oil Company," Ward and Wallace chose to believe Madgwick. He said it was most likely that "Turkish Petroleum, through their association with Anglo-Persian already had the information in regard to the association with E&GSynd very soon after the arrival in Bahrain of Gulf Oil's geologists."

Ward and Wallace were also easily convinced by Madgwick that the Colonial Office's uncooperative attitude was directed personally at the members of Holmes's syndicate and, furthermore, that E&GSynd were simply incapable of dealing with British or Government of India officials.[10]

HOW TO HANDLE THE BRITISH

Ward took Madgwick along to see Standard Oil of California's legal adviser, introducing him as "the first person to report evidence of petroleum on the island of Bahrain." He told the legal adviser that Madgwick had "discovered the oil evidence while he was drilling wells for water there."

The legal adviser was the powerful Judge Frank Feuille. Born in Texas in 1860, his specialty was Latin American law, particularly land leasing, subsurface rights, mining, and oil. He had been consultant to Standard Oil of California since 1920 and would hold the position until 1940.

When he was told, by Madgwick, of obstacles being put in the way of Standard Oil of California's takeover of the Bahrain concession, Feuille wrote to the company's Land and Lease Department that handled all contracts and concessions. "We have the remarkable case in which the Colonial Office has forced a British company, under British control, to lose a valid concession it had obtained from the Shaikh of Bahrain Island." That he was not given clear information is obvious from the legal opinion he added. He wrote: "The fact that the syndicate is prevented from transferring its concession to a foreign company would not in itself affect the syndicate's right to hold the concession for itself." There had never been any suggestion from the Colonial Office that E&GSynd did not have the right to retain the concession for itself.

Judge Frank Feuille had been legal adviser to the Panama Canal from 1910 to 1920. At Standard Oil of California was Washington representative Francis B. Loomis. He was a former State Department undersecretary. When the secretary of state fell ill, it was Loomis who negotiated the Panama Canal concession. Francis Loomis and Frank Feuille, both now with Standard Oil, were thus long-time colleagues.

Alerted by Feuille, Loomis sent out some personal letters. He started with a friend at the State Department. "I think you will be interested in seeing this," he wrote. "The Colonial Office is evidently not willing that an American company should engage in any activities in the matter of oil prospecting or production in the Island of Bahrain, which is under their political control. The Anglo-Persian Oil Company is doubtless behind the action of the Colonial Office in this matter." To another friend at the Department of the Interior, he wrote: "It would seem that Great Britain is not acting in a manner that would entitle her to reciprocal advantages and treatment as provided for in our Leasing Law."[11]

The American oil companies had recent experience of British officials, and of the Anglo-Persian Oil Company, from their efforts to break into Iraq. If, as they were told, the people at E&GSynd were incapable of dealing with their own government departments, this could be fixed. They would send their own American representative to London "to take part in the discussions with the Colonial Office."

Both companies had strong links with the US administration. The Gulf Oil Corporation had Andrew W. Mellon, current Secretary of the US Treasury and Standard Oil of California had Feuille and Loomis. The Gulf Oil Corporation, wanting the Kuwait concession, and Standard Oil of California wanting the Bahrain concession, agreed to combine in "the presentation of the matter to our State Department."

Ward wrote to instruct Holmes on these developments. He said: "Following a conference with Professor Madgwick on his return here, it has been decided to send a representative to London to assist in the Colonial Office discussions."

Holmes had cabled repeatedly that he needed to know whether or not Gulf Oil accepted the nationality clause that must now be inserted in the Kuwait concession. Without such acceptance, he kept telling the Americans, there was no point in his continuing the discussions in Kuwait.

About to dispatch his own man to sort out the Colonial Office and confident of strong ties to the State Department, Ward was encouraged to take a tough line. He informed London that what was expected was nothing less than "compliance" from E&GSynd to its contract with Gulf Oil. Moreover, Ward asserted grandly, the Colonial Office condition that Holmes kept so tiresomely referring to was "at variance with international agreements between our two governments regarding oil matters. . . . Until our USA government states such conditions are justified, Gulf Oil does not propose to seriously consider them."

On receipt of Ward's cable, Holmes departed immediately for London. He left so fast that, behind him, the Kuwait political agent and the political resident were left wondering. The resident reported to the Colonial Office that Holmes had arrived in Kuwait on February 6, left on the sixteenth for Baghdad, and departed for London on the twenty-third.

According to a puzzled Major More, Holmes had not renewed negotiations for an oil concession with the shaikh of Kuwait. In fact, he said, "apparently, Major Holmes did not mention the word 'oil' to the shaikh during this visit." He had agreed to waive all amounts outstanding on the two water wells drilled to date, even though payment under the contract was not conditional on success. More reported that Holmes had even offered to drill another well "entirely at his expense and to rely on the shaikh's generosity in the event of finding good water."

Suspicious as ever, Major More concluded the answer to the riddle lay in "the nature of the rock reached at a depth of 700 feet." He reported that Holmes was being most mysterious about "wanting to get a geological opinion about some rock."

He told the resident, Frederick Johnson: "I think there can be no manner of doubt that, he at any rate, thinks he has found some formation which suggests the presence of oil and that it is worth the syndicate's while to drill at their own expense in another place in order to carry out further investigations." Displaying the usual skill at information gathering at which the officers of the Government of India so excelled, the resident passed this news on to the Colonial Office, adding, "Major Holmes has instructed his assistant driller at Kuwait to send him in England, by air mail, a sample of the rock."[12]

NOTES

1. Philby Papers, box XXX-7, October 15, 1928, Cree to Philby, "Standard Oil might succeed in their own name"; Ward Papers, box 3, October 4, and three letters all

dated October 5, 1928, Ward to E&GSynd, "10% commission on the entire consideration" and "$2,500 per year from that date." The reply is box 1, November 2, 1928, E&GSynd to Ward, "The Board are agreeable to your party (Ward, Madgwick, Young) having 10% interest on both cash and royalty. As regards payment 1000 sterling to you personally and further at 500 sterling per annum, they cannot agree to this . . . they will add 250 sterling to your commission following the exercise of the option on any or all of the four concessions . . . had you asked in the first place for a salary . . . (we) cannot agree that Madgwick was at any time in the position of a principal."

2. IOL/R/15/1/649 C30, November 8, 1928, Colonial Office to the secretary of state for India and copies to the India Office, the Foreign Office, and the Petroleum Department of the Board of Trade; memorandum of meeting at the Colonial Office with Janson and Holmes, "if (they) agree to a similar clause for Bahrain" and "still hopeful of obtaining British capital."

3. Ward Papers, box 1, November 12, 1928, Holmes at sea onboard the P&O "SS *Malwa*" handwritten to Ward, "arranged to see the Shaikh of Kuwait"; Thomas E. Ward, *Negotiations for Oil Concessions in Bahrain, El Hasa (Saudi Arabia), the Neutral Zone, Qatar and Kuwait* (New York: privately published, 1965), p. 109, November 23, 1928, cable, Ward to E&GSynd; Ward Papers, box 1, November 28, 1928, cable E&GSynd to Ward, "will be fatal"; Standard Oil of California's poor discovery record is in Frederick Lee Moore Jr., *Origin of American Oil Concessions in Bahrain, Kuwait and Saudi Arabia* (BA thesis, Department of History, Princeton University, 1948), p. 62, and Anthony Sampson, *The Seven Sisters, the Great Oil Companies and the World They Made* (Hodder and Stoughton, 1975), p. 104, "and there was anyway a glut of oil." This story and subsequent remarks attributed to Stuart Morgan appear in Ward, p. 107, who attributes it to Moore's 1948 Princeton thesis. Angela Clarke, *Bahrain Oil and Development, 1929–1989* (London: Immel Publishing, 1991), p. 81, repeats Ward's version while giving Moore, erroneously cited as p. 32, as the source. In fact, the Turkish Petroleum 1928 board meeting is not mentioned anywhere in Moore's thesis.

4. Ward Papers, box 1, December 20, 1928, E&GSynd to Ward, "outside influences brought to bear" and "particulars of our interests were given to the Turkish Petroleum Company, which company is managed by Anglo-Persian." Ward Papers, box 1, cables, E&GSynd to Ward November 28, November 30, December 3, 1928, "as a result of your offer to Turkish Petroleum Company—unknown to us—of your interests in our properties" and "hasty action"; this was Standard Oil of California president K. R. Kingsbury and vice president M. E. Lombardi. See also Ward, *Negotiations for Oil Concessions*, p. 117, December 3, 1928, cable from Wallace to Ward, "E&GSynd are failing to respond in good faith . . . until the option period expires"; Ward, p. 110, November 24, 1928, Ward to E&GSynd, "Sir John Cadman coming here . . ."; Sampson, *Seven Sisters*, p. 104, "billion dollar error"; Ward, pp. 117–19, "Sir John refused . . ." and "forward negotiations with Standard Oil of California."

5. Ward, *Negotiations for Oil Concessions*, p. 112, cables, November 28 and November 30, 1928, Ward to E&GSynd, "your contractual obligation with Gulf Oil and Gulf's right to exercise options"; Ward Papers, box 3, Ward to Holmes in Bahrain, December 11, 1928, "E&GSynd does not appear to be willing . . ."; Ward, pp. 119–21, Ward to E&GSynd December 11 and December 12, 1928, "Do you, or do you not authorize . . ."; Ward Papers, box 2, December 12, 1928, Ward to Gulf Oil enclosing cable

same date from E&GSynd, "that company must understand Colonial Office conditions compulsory . . ."; Clarke, *Bahrain Oil and Development*, p. 86, for Canadian registration of Bahrain Petroleum Company; The figure of $157,000 is given by Clarke.

6. Ward, *Negotiations for Oil Concessions*, p. 125, December 20, 1928, cable, Ward to E&GSynd, "telegraph immediately the date on which Holmes expects to complete the Kuwait negotiations"; Ward Papers, box 1, December 20, 1928, E&GSynd to Ward, "Hence has arisen the new condition to prevent foreign control" and enclosing copy of their statement made to the Colonial Office; Ward, p. 126, December 21, 1928, Ward to Wallace, "cannot understand the delay with the Colonial Office"; Ward, p. 135, December 28, 1928, Ward to E&GSynd, "conditions clearly set forth."

7. IOL/ R/15/297.ii, December 19, 1928, E&GSynd to secretary of state, Colonial Office, "reference to our application for permission to negotiate a renewal of the Exploration Licence granted by the Shaikh of Bahrain"; IOL/R/15/1/649 C30, December 17, 1928, political agent Bahrain (Barrett) to political resident (Haworth), "do not anticipate any opposition from him (Shaikh of Bahrain)."

8. IOL/ R/15/297-ii; memorandum of a meeting at the Colonial Office, December 28, 1928, with E&GSynd (including G. S. Pott, solicitor to the Syndicate), "might very well prefer to abandon their rights."

9. Ward, *Negotiations for Oil Concessions*, pp. 127–35, how soon receive "formal and written consent . . ."; letter in the possession of Carole Davidson, Western Australia, dated December 31, 1928, Frank Holmes at the Maude Hotel Baghdad to R. Murray Davidson, the Mexican Corporation, Fresmillo, Lacatecas, Mexico, "I have only today received your letter dated November 10th. It has been following me around . . . a fine profession and a man's job."

10. Ward Papers, box 3, January 4 and January 7, 1929, Ward to E&GSynd; box 2, January 10, 1929, Ward to Madgwick in Sussex, England; box 3, January 23, 1929; and three letters dated January 31, 1929, Ward to E&GSynd. Madgwick claimed to Ward that E&GSynd had agreed to pay him 5 percent of the option sale price (over and above the commission to be shared with Ward) and also an ongoing 5 percent on royalties. Madgwick told Ward to go ahead and deduct on his behalf from the payment to be made by Standard Oil of California. Wisely, Ward asked E&GSynd to confirm this in writing. No confirmation was ever received. (But emboldened by what he thought was Madgwick's success, Ward renewed his own claim to the $5,000 plus $2,500 per year.) Madgwick eventually fixated on Holmes as the cause of his not getting the compensation he believed he deserved.

11. Ward Papers, box 1. In a letter to Dr. Heald, chief geologist, Gulf Oil (May 14, 1935), Madgwick claimed this visit to E&GSynd as "in the winter of 1928/29, I again stepped into the breach when a deadlock had arisen. I twice visited NY and also introduced some semblance of order into London . . ."; Chevron Oil Archives, box 120791, January 31, 1929 (W. F. Vane, manager, Land and Lease Division), reporting content of Feuille's letter also details Ward's introduction of Madgwick as "discovered (the oil) while drilling for water in Bahrain"; Clarke, *Bahrain Oil and Development*, p. 94; February 6, 1929, Francis Loomis to Paul T. Culbertson US Department of State, "Colonial Office evidently not willing" and February 13, 1929, Francis Loomis to Judge Edward C. Finney, first assistant secretary, Department of the Interior, "Great Britain is not acting in a manner that would entitle her to reciprocal advantages and treatment."

12. Ward Papers, box 3, February 4, 1929, Ward to Holmes in Bahrain, "decided to send a representative to London to assist with Colonial Office"; Ward Papers, box 3, February 15, 1929, Ward to E&GSynd, "until our USA government states"; IOL/R/15/5/238, vol. 111, February 17, 1929, confidential, political agent Kuwait (More) to political resident (Sir Frederick William Johnson, temporary), "Holmes was being most mysterious about wanting to get a geological opinion about some rock"; IOL/R/15/1/638-D73, resident (Johnson) to colonial secretary, "send him in England, by air mail, a sample of the rock."

CHAPTER 14

DIPLOMATIC DANCE

The grievances against the British were outlined at a March 20 meeting with US Secretary of State Frank B. Kellogg, arranged by Francis Loomis of Standard Oil of California. Accompanying Loomis was William Wallace of the Gulf Oil Corporation.

Although the Government of India allowed almost no "foreigners" to even visit Bahrain, the two men ignored the possibility that British reaction might have been the same no matter what the nationality of a non-British company trying to set up there and charged that it was Americans who were being discriminated against.

Kellogg had been ambassador to London during the American's saga over Iraq oil; it still took him eight days to initiate any follow-up at all from this meeting. In a mildly written instruction to the US charge d'affaires in London, dated March 28, Kellogg said the State Department "desired" him to discuss this case "informally" with the appropriate British authorities. The charge d'affaires was told to refer "in conversation" to US legislation as being extremely liberal in regard to foreign-controlled petroleum operations in its territories. In a follow-up, he was instructed to also tell the British that the US Department of State would be "glad" to obtain a statement of British government policy regarding the holding and operation by foreigners of petroleum concessions in territories "such as Bahrain." Not until April 3 did the US charge d'affaires call on the Foreign Office to raise the matter "informally," as instructed.

Although the charge d'affaires and the US secretary of state did not know it, there was now no difficulty over Bahrain. On March 20, the very same day that the two oil company executives were briefing Kellogg in Washington, the Colonial Office in London was advising various government departments that it was withdrawing opposition to American investment in the development of Bahrain's oil concession. On that date the Colonial Office officially declared their decision "to abandon the idea of opposing the introduction of American capital to operate the concession and to concentrate upon obtaining such a degree of British control as may be practicable." The Mines Department replied that, anyway, in their opinion, it was doubtful "whether any considerable oil production could be anticipated in Bahrain."

The Foreign Office judged that "Britain would not be on strong ground in insisting on the exclusion of US capital from this particular concession." And even the Admiralty conceded "in view of the absence of a British control clause it is impossible to stop the transfer." Only the Government of India remained obdurate.[1]

PERSIA CLAIMS SOVEREIGNTY OF BAHRAIN

The seeming sudden turnaround on opposition to the involvement of Americans in Bahrain appears to have sprung from a root far different than a desire not to antagonize American friends, as the State Department came to believe.

Since November 1927, Persia had been actively pursuing a claim to sovereignty over Bahrain that they said had existed since 1602. Among other evidence, two events from the nineteenth century were cited. The first was the so-called Treaty of Shiraz agreed upon between Persia and Britain, although never ratified by either party. The second was letters written in 1860 by the amir of Bahrain of the time, Muhammad Al Khalifa.

In his letters to the shah of Persia and to the governor of the Persian province of Fars, Muhammad stated that Bahrain and its inhabitants were "always under the protection of Persia" and arranged to pay tribute. This might not be irrevocable proof, however, as fearing incursion by the Wahhabi Warriors of bin Saud's grandfather, Muhammad had prudently addressed exactly the same protection-seeking letter to the Turks. The Persians had replied first, and Muhammad had urgently endorsed submission to them. But the Turks replied shortly after with a better offer. Muhammad responded by immediately declaring himself a subject of Turkey and agreeing to pay tribute to the pasha of Baghdad.

The Persians had taken offense to a clause in the Treaty of Jedda, signed in May 1927 between the British and Abdul Aziz bin Saud following the capture of the Hijaz. The offending clause read: "bin Saud undertakes to maintain friendly and peaceful relations with . . . Bahrain, who is in special treaty relations with His Britannic Majesty's government." A Persian protest addressed to the British minister at Tehran stated "the article is contrary to the territorial integrity of Persia." The Persian government requested the British government to "promptly take necessary measures to relieve its implications."

The next day the Persian government invoked Article 10 of the Covenant of the League of Nations guaranteeing the territorial integrity of all member states. In a note to the secretary general, Persia asked the League to "guarantee her undisputed rights over Bahrain." In January 1928, Britain defended its presence on Bahrain to the League by stating it was there "to ensure that the pacific development of the island and the prosperity of its Arab inhabitants are not troubled by unjustified advances coming from her neighbours and having in view the subordination of the inhabitants to foreign domination."

With a completely straight face Britain was not referring to *itself* as the "foreign dominator" but to Persia. In a superb display of diplomatic doublespeak, Britain defined Bahrain's political status thus: "The principality is an independent Arab State under the protection of His Majesty's government, but not a British Protectorate." An Orwellian legal ruse was then employed. As the political agent recorded: "Owing to the claims to sovereignty advanced by the Persian government, Shaikh Isa Al Khalifa on behalf of himself and his islands gave His Majesty's government authority to represent the Bahrain State in rebutting Persian pretensions before the League of Nations." This was the same Shaikh Isa whom the British had so acrimoniously forced to abdicate in 1923; they had left him with the title of ruler of Bahrain while transferring the total exercise of authority to his son, Shaikh Hamad, as deputy ruler.[2]

The Persian claim to Bahrain was still simmering in 1929 when the Colonial Office became aware of the extent of American interest in E&GSynd's concession. It is possible the Colonial Office calculated that an American presence on Bahrain—suitably hedged and strongly controlled by Britain—might help to bolster its position before the League of Nations against Persia's claim to sovereignty, despite the United States not being a League member.

Certainly, driving a respectable British company like E&GSynd into liquidation by refusing to allow American investment could hardly be construed as "ensuring the development and prosperity" of the inhabitants of Bahrain, as Britain claimed to be doing in her submission to the League. Besides, even the Foreign Office judged that, in international eyes, "Britain would not be on strong ground in insisting on the exclusion of US capital." From Britain's point of view, there probably was nothing to lose anyway, as they considered it doubtful "whether any considerable oil production could be anticipated in Bahrain."

THE BATTLE OF SIBILA

Meanwhile, in bin Saud's territory the Ikhwan revolt had ratcheted up another notch to a dangerous new phase. Attacks by the rebels had become more bold as American philanthropist Charles Crane had discovered. Crane was visiting the American Mission Hospital in Basra, which he had assisted Dr. John Van Ess in setting up. With him was American missionary Henry Bilkert. They were motoring down the Basra Road near the Kuwait border when they were fired on by a group of white-turbaned Ikhwan on camels. Bilkert and the local driver were killed but Crane survived.

The incident hit the headlines in New York, prompting Ward to write

Holmes: "We are all anxiously awaiting news from you, especially since the newspaper reports gave us the news of the tragedy in the party of Mr. Charles R. Crane." Traveling the Basra Road was not necessarily a foolhardy act by Crane's party. The British, under John Glubb (later "Glubb Pasha" of Transjordanian Desert Legion fame) had for weeks been pursuing the camel-mounted rebels, mercilessly bombing and strafing from the air. After the incident involving Crane, Glubb added to his air force offensive with a flotilla of machine-gun-fitted trucks and a detachment of armored cars.

For some months, bin Saud had been gathering the loyal tribes to his flag. On March 24 the warriors of Abdul Aziz and the warriors of the Ikhwan camped, four miles apart, at the wells of Sibila. Abdul Aziz's men outnumbered the rebels three to one. For several days the leaders of both sides attempted conciliation, each presenting their grievances and arguments to a panel of the most learned and faithful jurists of the *Ulema* (leaders in the *Sharia* or Islamic body of law). With this process still ongoing, on the morning of March 29 the armies of Abdul Aziz mounted a traditional *Sabah* (Dawn); silently, using only their swords, they rushed the enemy's camp in a surprise attack that scattered their opponents. By noon, the Ikhwan were in full flight.

After what became known as the Battle of Sibila, Abdul Aziz set the policy by which Saudi Arabia would be governed to the present day. He announced that all regulations and prohibitions concerning personal conduct and social interaction, including fatwas (rulings based on the Islamic law), would pertain to the entire population, that is, both Nejd and Hijaz, and would in future "be referred to the Holy Books and decided by the Ulema." He decreed that the ruler is entitled to absolute obedience. Satisfied that he had dealt with the rebels militarily and with all his subjects politically through the announcement of this national policy, bin Saud returned to the Hijaz.

Any complacency was soon shaken, however, as he realized that the rebellion was breaking out again, more fierce than ever. And this time it would involve all the areas of Holmes's mainland concessions toward which so many covetous eyes were being cast, Kuwait, Al Hasa, and the Neutral Zone.[3]

LOOMIS MEETS HOLMES

On March 4, 1929, as Loomis was arriving in London for the purpose of tackling the British officials, accompanied by a Maj. Harry G. Davis representing Gulf Oil, Herbert Hoover replaced Calvin Coolidge as president of the United States. This must have given heart to the two prospective "fixers" of the Colonial Office. Hoover's activities on behalf of the American oil companies in Iraq and his advocacy of an Open Door policy for Americans were well known. Besides, as Loomis knew, Frank Holmes and Herbert Hoover were old friends. When

Loomis arrived in London, he also "made some inquiries" about E&GSynd. He wrote to M. E. Lombardi, vice president of Standard Oil of California, saying that the syndicate members "make quite a good impression and from what I learn, their standing seems to be good . . . they seem to have standing as mining men and mining engineers."

Loomis reported that he was impressed when he met Holmes for the first time. "I had a long conversation with Major Holmes who seems to be a very straightforward, truthful person, though rather optimistic. He is somewhat of the promoter type but I think an honest one." He spent three hours with Holmes "discussing the whole Bahrain matter." He had, he said, obtained from Holmes "a considerable number of facts that will be of use to us in a practical way in case we should operate on Bahrain Island."

He went on to say:

> I gather from what Major Holmes says that the important property the syndicate has in which the Gulf Oil Corporation is interested is the Kuwait Concession." Holmes seems to be very familiar with the country on the mainland and in fact with all this region. He says there is a line of seepages about three hundred miles long in the Kuwait territory and thinks there are plenty of good areas to be had. He is also trying to get E&GSynd to back him in an effort to secure all the Mosul territory that is left over after Turkish Petroleum chooses its selected areas. Holmes seems to have close affiliations and influence with the Native Rulers in the Persian Gulf and Mosul regions.
>
> It occurs to me, if we take over the Bahrain concession, ultimately we may want a good deal more in that general region. It might be worth while to keep in touch with Major Holmes.

Loomis's understanding was that the action of the Colonial Office "was brought about wholly by the intervention of the Anglo-Persian Oil Company. I am informed this company is very unpopular and not on good terms with several of the native shaikhs and does not find it at all easy to extend its holdings." He said this analysis "quite coincides with statements made to me by a former Persian Minister of Finance who called on me in London some few years ago and offered to obtain for us the oil rights of the three Northern Provinces of Persia. At the time he said there would be a revolution in the country if these were given to Anglo-Persian. Major Holmes by the way thinks they are extremely valuable and would be well worth acquiring."

Two ships each week and a wireless station gave reasonable communications with Bahrain, Loomis said. "I didn't know this before, but there is an official 'adviser' to the Shaikh of Bahrain, a Mr. C. D. Belgrave, an Englishman who is very loyal to the shaikh. Holmes says Belgrave will work very wholeheartedly and harmoniously with us if we take the concession. This adviser holds a position quite apart from that of the Political Resident and his main business is to advance the commercial and economic fortunes of Bahrain. Major Holmes lays stress upon the fact that there are no difficulties with either the shaikh or his officials."[4]

ASSUMPTIONS SHAKEN

Lt. Col. Cyril Charles Johnson Barrett, promoted to political resident for his last six months prior to retirement, stepped up his opposition. He advised his superiors in the Government of India that "we should resist control of this concession by American interests." If this was "quite unavoidable" then "British subjects should certainly be local representatives," he said. Barrett repeated his confident belief that "the shaikh, acting on our advice, would veto a transfer." He seemed to be suggesting one of the perennial favorites of officers of the Government of India in the Persian Gulf—scare tactics—when he suggested raising "misbehaviour of American drillers who were formerly employed by Anglo-Persian at Abadan" as a means of "prejudicing the shaikh against persons of their stamp." Although, he warned, "the Shaikh of Bahrain is not averse to drawing his royalties from any source." Barrett piously claimed that his opposition to the transfer of the Bahrain concession was purely "for the sake of British interests."

The Foreign Office drew up a proposed reply to the US State Department request for a statement of British government policy regarding the holding and operation by foreigners of petroleum concessions in territories "such as Bahrain." Mindful that when they had solved Bahrain where "existing arrangement gives no sure ground on which to oppose transfer" they would need to deal with the Americans again over Kuwait and possibly bin Saud's territory, the proposed reply was cagey: "His Majesty's government feel bound to reserve to themselves the right to consider, on its merits and in light of circumstances obtaining at the time, each proposal for holding or operation of petroleum concession by foreigners in territories such as Bahrain islands, and find themselves unable to make any statement of their policy on questions such as American Embassy desire to have." The proposed text was forwarded to the Government of India with the direction that the "matter is urgent and cannot wait for settlement of the question of the future status of Bahrain," a reference to the ongoing Persian claim to sovereignty.

The Colonial Office decision followed by the proposed Foreign Office response in a matter related to an area of influence the Government of India considered its own generated a flurry of responses. The viceroy himself replied: "We have throughout assumed that the Eastern Company's [*sic*, assume E&GSynd] agreement had to be read with our basic Treaty rights. And that Article 3 of our Treaty of 1892 gave monopoly rights to Great Britain against foreigners. And our special oil understanding of 1914 gave us control over selection of even British Companies."

The Government of India had believed that "existing Treaties seemed to suffice against external encroachments on our position in Bahrain." If this was not correct, the viceroy thought, then the future could be decidedly shaky.

"Our general position is seriously compromised if His Majesty's government feel that this contention is untenable because practical control of a petty government like Bahrain might easily pass into the hands of a powerful foreign financial institution," he wrote. "It will be difficult to prevent ground being cut from under our feet all along the coast if we give way over Bahrain because our exclusive agreements are all on the same model."

Nevertheless, the viceroy conceded that there might be matters of "wider consideration the importance of which we do not appreciate" that could preclude "a rigid stand being taken on our Treaty rights in the case of Bahrain." For the protection that it may give "elsewhere," the viceroy endorsed the wording of the Foreign Office reply while stressing the vital need to "secure maximum British control and insist in particular on British personnel locally."[5]

Knowing what precarious ground they were on legally, and internationally, with their assumption of total control of all matters related to Bahrain, the British officials themselves seemed surprised at just how much they were actually getting away with. The secretary of state for India confided to the viceroy that "the right of His Majesty's government to impose certain restrictions has not been questioned, either by the Americans or the syndicate." And the India Office marveled at their luck that neither the Americans nor the syndicate had "discovered the lacunae in our legal position of which we ourselves are conscious after a careful diagnosis of our case."

There were those who believed they had a moral right, if not exactly a legal one, because, as Barrett put it, "by sacrifice of men and money" the British had "made the Persian Gulf safe for enterprises of this character." They believed they "had every right" to impose conditions "on the conduct of commercial enterprises in those regions."

The secretary of state for India advised the viceroy that the solution was insistence on American compliance with five conditions, which, "if we can secure them," would probably be "the best we can hope for." He continued, "If they are accepted, and fully implemented . . . they should give us a degree of control." He added hopefully: "It is also, of course always possible that on closer investigation by the American company the oil prospects may prove insufficient to justify their continued interest."

Charles Belgrave, the adviser in Bahrain, offered a timely reminder that he wanted included in the conditions. "Suggest no imported Persian labour until Persian question settled," he cabled. Barrett was direct: "I should have preferred to exclude the Americans altogether but as Secretary of State has decided that this is impossible the arrangements outlined seems to keep in our hands as much control as we can expect."

Barrett laid "great stress" on the condition that dictated that "all dealings of the company with the Ruler of Bahrain" must pass first through the (British) chief local representative who, in turn, "should address the Political Agent."

Presumably, the political agent would then decide whether or not the shaikh of Bahrain should actually receive the message intended for him from the oil company.

This condition was too much even for the viceroy of the Government of India, who commented, "actually, this would be going beyond even the practice in Indian States." The viceroy suggested a well-practiced Government of India tactic for getting this condition. "It would seem very desirable," he advised, "to arrange that the Shaikh of Bahrain himself should insist on this stipulation."

No doubt the shaikh of Bahrain would have been very interested to be approached for this purpose for while the barrage of letters and cables were crossing London to India, and London to the Gulf, he was the one person kept completely out of the loop. Although it was his country—and his oil concession— that had been the subject of such highly charged discussions, meetings, cables, and letters for at least eight months, nobody had bothered talking to him about it.

Belgrave, whose sole job it supposedly was to take care of the shaikh's personal and financial interest, reassured Barrett, "the shaikh has heard nothing about the question of company's nationality." Belgrave airily dismissed this disloyal, but deliberate, oversight by offering his opinion that the shaikh "would have no feelings on it as long as money resulted." In a confidential note to the Government of India, not circulated to the British officials in London, the resident, Barrett, affirmed that "the shaikh has no inkling of our suggestions" and guaranteed "so long as he gets his royalties he is not concerned with the nationality of the company."[6]

AMERICANS IN LONDON

Although their diplomatic initiative had achieved little—the Colonial Office decision to allow the Americans into Bahrain was taken before the approach from the charge d'affaires, and the Foreign Office's vaguely worded statement on British oil policy in the Persian Gulf promised exactly nil—the Americans were convinced they had slain Goliath, as subsequent State Department claims attest.

From the day he arrived in London on March 4 until the day the British conditions were agreed upon in New York on October 28, the man the Gulf Oil Corporation sent to "assist" E&GSynd with the Colonial Office, Maj. Harry Davis, never got near the British officials.

The Colonial Office firmly refused to even discuss Davis's primary interest, Kuwait, on the grounds that "the Department of State had made no reference to Kuwait in its informal note presented to the British government through the

American Embassy in London." The Colonial Office conducted negotiations with Janson and Holmes who reported the content to Major Davis at his London hotel. Davis then cabled to Gulf Oil in New York, and they relayed the message to Standard Oil in San Francisco.

At one stage, the Colonial Office put forward another ingenious solution. This one involved a convoluted exercise under which John Cadman of Anglo-Persian would induce Turkish Petroleum to take over the Bahrain concession (and possibly Kuwait) and then designate the operating companies formed by Gulf Oil and Standard Oil of California as its nominees. In other words, the operating companies set up by the two American oil companies would become subsidiaries of Turkish Petroleum and, by default, answerable to the Anglo-Persian Oil Company. Nobody wanted to play ball with this one and after it had caused several weeks' delay, it was abandoned.

Gulf Oil's William Wallace and Maj. Harry Davis had almost religious faith in the efficacy of their country's diplomacy. Without any evidence or claims to this effect from the embassy, Davis sincerely believed that individuals at the American Embassy in London were "talking" to their counterparts at the Foreign Office and being truly effective. Both he and Wallace were certain they had initiated a diplomatic offensive that was besting the British and now had them running scared. They were so sure of the effectiveness of their US diplomats that they frequently threatened to invoke "diplomatic pressure" during the months of negotiations over the British conditions for Bahrain.

Several times Davis earnestly remarked to E&GSynd that he could always "resort to diplomatic channels." Wallace went so far as directing Davis to "impress" upon Holmes and Janson that they should "warn" the Colonial Office that "the effort now being made is for the purpose of avoiding further diplomatic representations, which may have far reaching effect."

Francis B. Loomis of Standard Oil of California had given up early, declaring the entire process to be Gulf Oil's fight. Neither he nor his former companion in the Panama Canal negotiations, Judge Frank Feuille, had any illusions about the success of the American diplomatic initiative.

On the Foreign Office "policy" statement, Feuille remarked to Loomis: "The reply is limited to a concrete case (Bahrain) and it is quite indefinite as to policy. You will note that the Foreign Office's reply to our State Department leaves the former free to impose such conditions as it may deem fit . . . and leaves the door open for a retreat by the Foreign Office. Our State Department should request a more definite reply."

Throughout this period the object of the whole exercise—oil—seems almost to have been overlooked. Holmes could wait no longer and in July had contacted Ward. "I am most anxious to hear what the Gulf Oil people thought of the samples from the Kuwait No. 1 Water Well that I handed over to Major Davis in London," he wrote. He told Ward the "indication of oil in this well caused me much worry for fear something may leak out that these indications had been encountered."

Holmes had underestimated the spying capacity of Major More, who had reported on exactly this in April to the resident, to the Government of India, and to London. For his part, Ward seemed more interested in the bookkeeping than in evidence of oil in Kuwait. He replied almost dismissively: "With regard to the sample of water from No. 1 Well, you will realise there is nothing Mr. Wallace can do until the matter now in the hands of Major Davis is satisfactorily disposed of." More importantly, he added: "It will be a great relief for us all when the transaction is completed. I hope there will not be any further delay in sending a Statement of Account."[7]

AMERICANS AGREE UPON BAHRAIN CONDITIONS

After months of communication between all parties in London, in the Persian Gulf, and in the Government of India, the Colonial Office finally arrived at five conditions for authorizing transfer of the Bahrain concession from E&GSynd to Standard Oil of California, via the Gulf Oil Corporation. They were:

1. The "Britishness" of the company defined as registered in British territory.
2. The right of the British government to nominate one director to the board.
3. As many employees as possible to be either British or Bahrain subjects.
4. The company to guarantee not to obstruct British plans for proposed sites for aircraft landing and a seaplane station in Bahrain.
5. The appointment of a British chief local representative, approved by His Majesty's government, as the "sole person empowered to deal direct with the local authorities and the population of Bahrain."

For the important position of chief local representative, on which they "laid great stress," the Colonial Office nominated Maj. Frank Holmes, with whom they had a long association, beginning with Churchill and including missions such as his task of supplying arms and ammunition to the Yemen. The Colonial Office was strongly advocating Holmes but knew better than to go ahead without consulting the officers of the Government of India. Writing to say whom the Colonial Office wanted to appoint, Barrett asked the political agent in Bahrain and the adviser: "Will Holmes's health permit him to stay five years continuously in Bahrain?" Belgrave replied: "I don't think Holmes could stand five summers."

Barrett was inspired to write the viceroy of India: "Holmes is not an ideal choice for the position. He is an elderly man whose health broke down a year and a half ago with bad effects on his temper. I doubt whether he could stand the summer climate of Bahrain for five years." (Holmes was to confound them by living, in very good health, another eighteen years.) Barrett said that although Holmes spoke only unschooled colloquial Arabic, he did have "a wonderful way with the Arabs." He added that "in a small community" Holmes was "a social acquisition" and was "probably the best choice that can be made, as we have taken his measure."

Still uncertain regarding their legal grounds, the Colonial Office insisted that Holmes and Janson travel without delay to New York to "personally discuss the whole position" and obtain the American signatures to the "five conditions required to approve the transfer." Advised of this, Holmes said he had "been given to understand" if the Americans wanted any "alterations or readjustment" of the five conditions now laid out in writing by the Colonial Office, these would only be considered after he and Janson had visited New York. They were both certain, he said, that it would be unwise to "approach or even attempt to discuss" the document "until we have carried out their wish and seen you personally."[8]

ARRANGEMENTS WITH HOLMES

The meeting in New York on October 19, 1929, included Gulf Oil's legal adviser, Jems Greer, William Wallace, and Major Davis returned from London. Standard Oil of California was represented by Francis Loomis and legal adviser Judge Frank Feuille; Holmes and Janson represented E&GSynd. Ward was also present. The matter of Frank Holmes as appointment as chief local representative loomed large.

Before leaving London Major Davis had informed the Americans that "Holmes has stated that he will not agree unless satisfactory to Wallace and only if it can be so arranged that he will not be prevented from continuing negotiating for Kuwait and other concessions." While the Americans were quite happy at the prospect of Holmes in this position, a guaranteed period of employment was a benefit that no American company ever offered. They viewed as poor business practice a "hard and fast" five-year term of appointment as set down by the Colonial Office. Loomis and Wallace discussed their reservations with Janson and Holmes. Taking a British point of view, both of them were in favor of fixed appointments and couldn't see any reason why that clause should be removed.

There was no doubt, Loomis and Wallace said, that "both our American companies have the highest and most sincere regard for Major Holmes and are convinced that he is the best man available for the post on account of his proved abilities." Yet, the Americans were incapable of contemplating a secure period of employment for anyone. They kept at it. Holmes did not seem too bothered by the fuss. His attitude was understandable, as Loomis explained to his Standard Oil colleagues: "Janson finally explained to us, privately, that the Colonial Office had offered to make Holmes the Petroleum Adviser to the Iraq government at a salary of 5,000 sterling per year." Although the Iraq offer was extremely attractive, Loomis said, "Because of his long connection with E&GSynd and all the Arab shaikhdoms . . . Holmes felt he ought to stay with the syndicate."

The Americans came to the conclusion that the Colonial Office wanted Holmes because of the "undoubted confidence Abdul Aziz bin Saud, the Shaikh of Bahrain and the Shaikh of Kuwait have in him. They view Holmes as a strong factor in assisting the British government to quiet existing unrest and to maintain peaceful conditions in those regions and particularly to uphold British prestige in view of the change in British policy in permitting American capital, and consequently American influence, to secure a footing in the Persian Gulf."

The final arrangement was the appointment of Holmes, for five years or as long as the arrangement was "mutually satisfactory," at $25,000 a year plus expenses, paid one-third each by Gulf Oil, Standard Oil of California, and E&GSynd. Both American companies declared themselves "on the whole" satisfied with this arrangement and Standard Oil of California believed "Major Holmes should be able to take care of the interests of Gulf Oil and the E&GSynd in relation to Kuwait, as well as doing all he is able to do for us in Bahrain."[9]

Ward had remained adamant that he had a legal right to both a share in any cash payments to be received and also in the overriding royalties that might come out of both the Bahrain and Mainland agreements. Although he never did get the $5,000 in back expenses he claimed, Janson did sign for the commission and royalties. Ward's request for a $2,500 annual fee didn't come up again because all future E&GSynd communications and business matters were conducted directly with both the Gulf Oil Corporation and Standard Oil of California. From this date, apart from his regular check, Thomas E. Ward (and by extension Madgwick) had no further dealings with E&GSynd. In 1931, just eight months before the first commercial oil flowed in Bahrain, Ward would write ruefully to Madgwick: "As far as I am concerned Bahrain has been a calamity. I would have been much better off today if I had never heard of the place."

Having fulfilled the requirement of obtaining, in person, the two American oil company's agreement to all conditions and so facilitating the transfer of the Bahrain concession to Standard Oil of California, Janson and Holmes headed back to London. As they crossed the Atlantic, the US stock market recorded its worst trading day ever, wiping out $14 billion by 3 PM on October 29, 1929, a day that went down in history as the beginning of the Great Depression.

The Colonial Office still held all the cards. Throughout this entire period of negotiation, and formal and informal government-to-government contact, the Colonial Office had succeeded in keeping any discussion of Kuwait off the agenda. When news of the completion of the transfer of the Bahrain oil concession was published in mid-1930, the Persian government was provoked to submit another note to the League of Nations. Persia declared the Bahrain concession invalid and "reserved the right to reclaim, or to demand restitution of all profits resulting eventually from that concession, without prejudice of all damages relating to it."[10]

NOTES

1. That all non-British nationalities were being discriminated against, see, for example, Administrative Report for Bahrain for the Year 1927, "three Germans wished to disembark at Bahrain on SS 'Bandra' on June 18 but the master of the vessel was informed that no Germans are allowed to land at this port . . . another German named Karl Lindner wished to land in Bahrain in connection with pearl shell business but he was not allowed to do so." And for the Year 1929, "the Bahrain Government decided not to allow Russian steamers to call at Bahrain." For a systematic revision of the American claim that "State Department intervention was the decisive factor in facilitating the American entrance into Bahrain" see Yossef Bilovich, "Great Power and Corporate Rivalry in Kuwait, 1912–1934: A Study in International Politics," pt. 111, "British Nationality Clause versus the American Open Door" (PhD thesis, University of London, 1982). Bilovich says this claim was "first initiated by the State Department in 1945 and since repeated"; see also Benjamin Shwadran, *The Middle East: Oil and the Great Powers* (London: Atlantic, 1956), pp. 347–74, "In the memorandum to the 1947, Senate Special Committee Investigating Petroleum Resources (US Senate American Petroleum Interests in Foreign Countries, pp. 23–24) the State Department declared: 'Here again the prompt and positive action of the State Department has secured results favourable to an American-owned company. By securing the entry of American oil interests into Bahrain the way was paved for some American interests to obtain concessions in nearby Arabia'; also Shwadran, p. 388, concerning Kuwait "as in the case of Bahrain, the State Department claimed credit . . ."; Bilovich, pp. 151–54, "abandon the idea of opposing" and "impossible to stop."

2. For details of this episode see Muhammad G. Rumaihi, *Bahrain: Social and Political Change Since the First World War* (London: Bowker, 1976), pp. 6–9, and Abbas Faroughy, *The Bahrein Islands (750–1951); A Contribution to the Study of Power Politics in the Persian Gulf, an Historical, Economic, and Geographical Survey* (New York: Verry, Fisher, 1951), pp. 102–103; Mahdi Abdalla Al Tajir, *Bahrain 1920–1945: Britain, the Shaikh, and the Administration* (London: Croom Helm, 1987), p. 6, "under the protection of . . . but not a British Protectorate"; Administrative Report for Bahrain for the year 1928 (Barratt), "Shaikh Isa . . . gave HMG authority . . . to rebutt Persian pretensions."

3. St. Antony's College, Middle East Center, Private Papers Collection, Charles R. Crane Papers; among other generous endeavors, Crane assisted Dr. Harrison to build a hospital in Muscat and Dr. John Van Ess to build one in Basra. Van Ess of the American Mission is author of *The Spoken Arabic of Iraq*, 2nd ed., with rev. and additional vocabulary (London: Oxford University Press, 1917), first produced for the use of the British Military and Administration in Mesopotamia. Reprinted thirteen times, it remained the standard handbook until well into the 1960s; Ward Papers, box 3, February 4, 1929, Ward to Holmes in Bahrain, "anxiously awaiting news from you"; Hafiz Wahba, *Arabian Days* (London: A. Barker, 1964), pp. 138–40, for the battle of Sibilla and political policy decrees of Abdul Aziz bin Saud, see also David Howarth, *The Desert King* (London: William Collins, 1965), and H. C. Armstrong, *Lord of Arabia* (London: Penguin Books, 1924).

4. Chevron Oil Archives, box 120797, April 15, 1929, Loomis at the Carleton Hotel London to Lombardi in San Francisco, "Holmes seems to be a very straightforward, truthful person" and "ultimately we may want a good deal more in that general region . . . keep in touch with Holmes."

5. IOL/R/15/1/649-C30, May 1, 1929, telegram, political resident (Barrett) to Foreign and Political Department, Simla, India, cc political agent Bahrain, "prejudicing the shaikh against persons of their stamp"; IOL/R/15/297-ii, May 10, 1929, telegram, secretary of state for India, London, to viceroy, Foreign and Political Department, Government of India, Simla, "unable to make any statement of their policy on questions such as American Embassy desire to have"; IOL/R/15/1/649-C30, May 12, 1929, telegram, "Important," viceroy, Government of India, Simla, to secretary of state for India, London, "we have throughout assumed . . . control over selection of even British companies" and "difficult to prevent ground being cut from under our feet all along the coast."

6. IOL/R/15/297, vol. ii: August 15, 1929, telegram, secretary ofsState for India, London to viceroy, Simla, cc political resident, "the right of HMG to impose certain restrictions has not been questioned." Bilovich, "Great Power and Corporate Rivalry," p. 150, "men and money," citing Colonial Office minute, July 17, 1929, and "lacunae in our case," citing an India Office minute, August 15, 1929; IOL/R/15/297, vol. ii: August 16, 1929, political resident to political agent Bahrain (now Capt. Charles Geoffrey Prior), " by sacrifice of men and money"; and August 17, 1929, urgent, the adviser Bahrain Government (Belgrave) to political resident, "no imported Persian labour until Persian question settled"; and August 19, 1929, telegram, political resident to Government of India, Simla, cc secretary of state for India, London, "have preferred to exclude the Americans altogether" and "all dealings of the company with the Ruler of Bahrain . . . address the political Agent"; and August 20, 1929, telegram, viceroy, Government of India, Simla, to secretary of state for India, London, cc political resident, "going beyond even the practice in Indian States"; IOL/R/15/297, vol. ii, August 17, 1929, urgent, the adviser, Bahrain government (Belgrave) to political resident, "shaikh has heard nothing"; IOL/PS/10-10R 13188.9, vol. 993, political resident to Government of India ONLY [sic], "shaikh has no inkling."

7. Chevron Archives, box 120797, May 27, 1929, Davis in London reporting to Wallace, Colonial Office comment, "the Department of State had made no reference (to Kuwait)" and "Memorandum of Conference at Colonial Office on May 10, 1929, Holmes, Janson, Hall"; and May 24 and May 27, 1929, Major Davis in London to Wallace concerning Colonial Office suggestion that the two American oil companies become "nominees" of Turkish Petroleum; and July 17, 1929, Davis in London to legal adviser, Gulf Oil (Greer), "prompt action yesterday afternoon by American Embassy has resulted in Colonial Office arranging date for conference with E&GSynd"; and July 20, 1929, Wallace in New York to Davis, London, "impress" on Janson and Holmes that they should "warn" the Colonial Office; and August 2, 1929, Davis in London to Wallace in New York, "resort to diplomatic channels"; and June 5, 1929, Feuille to Loomis, "The Foreign Office reply is limited to a concrete case (Bahrain) and it is quite indefinite as to policy"; Thomas E. Ward, *Negotiations for Oil Concessions in Bahrain, El Hasa (Saudi Arabia), the Neutral Zone, Qatar and Kuwait* (New York: privately published, 1965), p. 163; July 8, 1929, Holmes in London to Ward, "most anxious to hear what the Gulf Oil people thought of the samples from the Kuwait No. 1 Water Well" and July 19, 1929, Ward to Holmes, "hope there will not be any further delay in sending a Statement of Account."

8. IOL/R/15/297, vol. ii, August 15, 1929, secretary of state for India, London, to viceroy, Government of India, copies to political resident, political agent Bahrain, detailing the five conditions; IOL/R/15/297, vol. ii, August 16, 1929, telegram, resident (Barrett) to political agent Bahrain, cc Belgrave, "will Holmes's health permit five years continuously in Bahrain?" August 17, 1929, urgent, Belgrave to resident (Barrett), "Don't think could stand five summers"; August 19, 1929, telegram, resident (Barrett) to Government of India, Simla, cc secretary of state for India, London, "wonderful way with the Arabs" and "we have taken his measure"; Ward, *Negotiations for Oil Concessions*, p. 167, September 20, 1929, Holmes in London to Ward, "unwise to 'approach or even attempt to discuss' the document 'until we have carried out their wish and seen you personally.'"

9. Chevron Archives, box 120797, August 2, 1929, Davis in London to Gulf Oil cc Standard Oil California and October 24, 1929, Gulf Oil to E&GSynd and October 24, 1929, Standard Oil of California (Loomis to Berg) discussing the Colonial Office insistence on Holmes as chief local representative and Loomis saying, "although the Iraq Oil Adviser position was extremely attractive . . . Holmes felt he ought to stay with the syndicate"; and January 17, 1930, Gulf Oil's Wallace to Judge Feuille, "they view Holmes as a strong factor"; and January 18 and January 22, 1930, Feuille to Loomis and January 23, 1930, telegram, Loomis to Wallace and January 24, 1930, Wallace to Feuille all detailing the financial arrangements pertaining to Holmes's appointment and duties. Note that the Americans had selected a Mr. Montague Grant Powell whom they wanted to place as either British director or chief local representative.

10. Ward, *Negotiations for Oil Concessions*, p. 173, "all future E&GSynd communications and business matters were conducted directly with both the Gulf Oil Corporation and Standard Oil of California"; Ward Papers, box 2, October 29, 1931, Ward to Madgwick, "Bahrain has been a calamity. I would have been much better off today if I had never heard of the place. . . . [I]t was a coincidence of course that the last discussions so far as E&GSynd and myself was concerned took place the week the grand crash started on Wall St exactly two years ago. I feel the experience again while I write you. I literally closed my eyes to the Wall St conditions when I should have had them wide open, so that I could concentrate on the E&GSynd matter. The net result was that it ultimately cost me more than $300,000. Under other circumstances it would have been different but it just had to happen when the crash was on the way"; Faroughy, *Bahrein Islands*, p. 103, for this second Persian protest. The Persian response to the transfer of the Bahrain concession was dated July 23, 1930.

CHAPTER 15

PIPELINE POLITICS

W hen Holmes returned from New York and London, oil concessions were not top priority on Abdul Aziz bin Saud's mind as he fought to contain the rebellion that threatened to destroy his work of thirty years. The British concluded that "bin Saud cannot control his own followers. . . . He has lost control of his own tribes." British intelligence sources reported that over the previous twelve months bin Saud has "become increasingly unpopular and has temporarily ceased to exist as a political factor."

The British were now openly afraid that the Ikhwan rebellion would bring about a change of leadership. This would leave Britain's mandates in neighboring Palestine, Iraq, and Transjordan exposed to attack by fanatical tribesmen if the Ikhwan succeeded. More importantly, loss of bin Saud's authority, particularly in Nejd, would interfere with plans to build the Turkish Petroleum Company's pipeline from the oil field in Iraq to the port of Haifa in Palestine from where the oil could be transported to the markets of Europe.

The high commissioner of Iraq warned the colonial secretary: "If some means cannot be devised of controlling Ikhwan raiding, the whole desert, at least as far north as a line from Rutbah to Damascus, will become untenable—and the pipeline and railway will have to take a northern route through Syrian territory." This would take the pipeline out of the areas under British control and place it predominantly within the French mandate, something distinctly unpalatable to Britain.

From June 1929 construction of the Iraq oil pipeline had taken on immediate urgency. Under the terms of its 1925 concession Turkish Petroleum was given thirty-two months in which to select twenty-four plots, each of eight square miles. The Iraq government would then be free to make its own selection of twenty-four plots to be offered on the open market by sealed bids. The oil company's selection should have been completed by November 1927 but they had obtained a twelve-month extension. In May 1928 they sought a further extension.

While the company was delaying, an independent British-Italian syndicate, the British Oil Development Company, approached King Faisal to purchase oil leases that should soon become available. British Oil Development's offer included a promise to build a railway from Iraq to the Mediterranean. Turkish Petroleum moved quickly to keep the newcomers out of the game. If they were granted a two-year extension on final selections, they countered, they would immediately commence the survey for the pipeline—and build a railway to the Mediterranean. Simultaneously, the company initiated discussions with the Iraq government aimed at modifying the terms of the original concession and, prudently, declared they had changed their name from Turkish Petroleum Company. From now on, it would be known by the patriotic "Iraq Petroleum Company."[1]

During the months Holmes had been completely absorbed mediating between the Colonial Office and the Americans over the Bahrain concession, fighting in Central Arabia almost separated bin Saud's two territories of Nejd and Hijaz. Riyadh was cut off from the Persian Gulf and the Red Sea and for two months there was no communication between Riyadh and Mecca.

One response from bin Saud was to revive the idea of setting up a series of wireless stations across his territories. Holmes had been involved in approaching Guglielmo Marconi for an estimate back in 1925. Bin Saud had shelved the project then because of Ikhwan objections to such modern inventions. Now he saw it as a matter of some urgency. He invested five hundred sterling in sending Harry Philby to England to tie it up. As Philby did so often, he claimed this to be entirely his own idea and "an important contribution to the consolidation of the Saudi Kingdom."

Philby's wife, Dora, plagued by debt, had resorted to taking in lodgers. Yet Philby spent his time in England in a "pleasurable round" of calling on members of the newly installed Labor government and promoting his self-declared "special relationship" with bin Saud. Philby's self-importance knew no bounds. At a meeting in London with Standard Oil New York, Philby declared he was "authorised by King Abdul Aziz to offer the Farasan Islands concession, abandoned by Shell, to your company." Standard Oil of New York was not interested. But Philby's partner in Sharqieh, T. D. Cree, in possession of Holmes's maps and technical findings passed on by Stuart Morgan, hadn't lost sight of the goal. "Is there any chance of the Al Hasa concession being available?" he asked Philby.[2]

BRITISH END IKHWAN REBELLION

With the stakes now being the monopoly of Iraq's oil, the British were determined to bring the Ikhwan troubles to an end in order to ensure the construction and security of the oil pipeline. Every assistance was extended to bin Saud. The Government of India supplied arms and ammunition; two-thirds of the cost was met by the British taxpayer. The British positioned an entire unit of aircraft and armored cars to patrol the frontiers of Iraq and Kuwait, cutting off access to grazing by the tribes and preventing their obtaining even basic supplies.

Some of the tribes originally resident in Kuwait's "outer green circle" sent a delegation to Shaikh Ahmad of Kuwait. Reminding Ahmad that they were formerly "his" tribes, they asked for the right to water and supplies along with "your protection." They said they had been "enticed away under the name of religion by bin Saud" but now "as erring children, we wish to return to our old home and be under our old Ruler." The shaikh of Kuwait was receptive. In his view, this was the natural return of people who had counted themselves as Kuwaitis until Sir Percy Cox's awarding of their territory to bin Saud at the 1922 Ujair Conference. King Faisal of Iraq felt much the same about the tribes seeking shelter in his territory. Both Faisal of Iraq and Ahmad of Kuwait sent the rebels money, food, arms and ammunition, camels, and horses. Ahmad personally donated one thousand riyals, and the tribesmen were soon seen moving freely in Kuwait Town itself.

The British would have none of this. They set up a meeting between themselves, a "representative" of Kuwait—meaning Major More, the political agent—and bin Saud. Once again, the shaikh of Kuwait was not invited to a discussion that concerned his own territory. "It was agreed," the British recorded laconically, "that the rebel tribes would be evicted from Kuwait and returned to Nejd."

In no uncertain terms the shaikh of Kuwait was instructed that he must forbid any tribesman to cross the Kuwait frontier, even for water. He was to refuse to guarantee safety to a camp where the tribes proposed to leave their women and children. If the tribesmen did cross the frontier for any reason whatsoever, the shaikh was to tell them that the British Royal Air Force would fly out and bomb them all. They were, said the British, in "treaty relations" with bin Saud and so would not discuss any requests from the tribes. Britain claimed to be driven by the gentlemanly obligation of keeping their word to bin Saud, something they had never obviously done for the shaikh of Kuwait.

As though herding sheep, the British relentlessly pushed the tribes out of Iraq and Kuwait into an area where bin Saud, fortified with British-supplied modern armaments, could finish them off. As the India Office reported to the Colonial Office: "The rebellious forces are to some extent hemmed in . . . they have their backs to the Iraq-Kuwait frontier . . . in deference to a request for cooperation from bin Saud a zone has been cleared some 15 miles deep along the frontier of Kuwait . . . bin Saud has further been assured that, should the rebels cross the frontier line into Iraq or Kuwait, steps will be taken to eject them, and that instructions have been given for the concerting of military measures to this end."

In the last quarter of the year, Abdul Aziz bin Saud led his army, vastly outnumbering the rebels, to deliver the knockout blow. The Ikhwan were caught in the deadly crossfire of a dozen British-supplied machine guns, the existence of which Abdul Aziz had carefully concealed. They were pursued across the desert by automobiles and trucks carrying mounted guns, also supplied by the British. British ground forces situated inside the Saudi frontier prevented any escapes.

The British were not, as they later claimed, merely observers in the defeat of the Ikhwan. On December 30 the rebellious Ajman and Mutair tribes surrendered to the Royal Air Force, on the Iraqi side of the border. The British told them to go back, and after much discussion they did so, but into Kuwait territory. Several days later, British planes and armored cars caught up with them and, through the simple expedient of dropping bombs in a circle around the village where they were sheltering, pinned them until the political agent in Kuwait could arrive to "persuade" them to surrender to bin Saud.

On January 9, 1930, the three rebel leaders surrendered to the Royal Air Force and were sent by air to Basra. The British then rounded up stragglers and used armored cars to ensure they also surrendered to bin Saud. "Escorted" by the commander of a British warship and the political agent in Kuwait, the three leaders were flown from Basra to bin Saud's war camp and ceremoniously handed over to bin Saud by the Government of India's political resident. In an elegant understatement, bin Saud's minister recorded, "the King thanked the British for their friendship and kindness of which they were continually offering fresh proof."[3]

IRAQ-SAUDI TREATY OF FRIENDSHIP

In order to clear the way for the oil pipeline, the British insisted that Abdul Aziz bin Saud and King Faisal of Iraq sign what the British termed a "Treaty of Friendship." There was, however, one big problem. Neither ruler would agree to visit the other. The British set up the meeting onboard the sloop HMS *Lupin*, moored in the Persian Gulf. An advance party went onboard to arrange the details, beginning with how the two steamers separately carrying the Iraq and Saudi parties to the *Lupin* would be welcomed. Ingeniously the British suggested that two gangways be lowered, one on the port and one on the starboard, so the two rulers could board simultaneously and meet in the middle of the ship. The fact that the *Lupin* possessed only one gangway put paid to that plan.

A very British ruse was employed. King Faisal was told that in the British navy, the senior man always comes onboard last. Abdul Aziz bin Saud, who, as a desert ruler, was "unlikely to know naval customs," was told navy protocol dictated that the senior man come onboard first. The British noted with satisfaction that "each monarch believed he had gained precedence."

The Treaty of Friendship was duly signed under the watchful eye of the British party led by the high commissioner of Iraq, Sir Francis Humphreys. Although the stated purpose of the conference was to draft a new frontier agreement, which so directly concerned Kuwait, once again the shaikh of Kuwait was not invited. After the 1915 Anglo-Saudi Agreement and the 1922 Treaty of Ujair, this was the third agreement that involved Kuwait's borders and, predictably, the result was the same. Kuwait gained nothing at all. The tight limits to Kuwait's territory as imposed at Ujair were confirmed. Adding insult to injury, bin Saud was still not required to lift the ruinous blockade on trade with Kuwait that would continue for another seven years, in the face of very little objection from the British.

On his way back from the *Lupin*, bin Saud ignored Kuwait but appeared in Bahrain as triumphant as any conqueror. The Bahrain political agent, Captain Prior, recorded this as "the outstanding event of the year." Bin Saud, he wrote, "received an amazing ovation on his arrival, the shaikhs having gone out to meet him as soon as the steamer was sighted." The day was not so joyful for the political agent. The Residency had sent him a message saying bin Saud was not coming and he had gone to tell the palace. "As bin Saud landed almost immediately afterwards a most unfortunate impression was created," he noted.

The political agent concluded that bin Saud's visit was a disaster for British prestige and warned that "nothing could have demonstrated more exactly the tenuity of our hold on Bahrain."

On March 24, a new agreement was signed with the Iraq Petroleum Company giving sole rights to all land east of the Tigris, covering an area of thirty-two thousand square miles; the company guaranteeing prompt construction of the pipeline. In June 1930 a new treaty with Iraq was signed by which Britain renounced its mandatory rights and recognized full sovereignty; a two-year timetable was set up for Iraq to join the League of Nations. Iraq's "Rule by Advisers" would finally come to an end. The British had termed it *dyarky* or "double rule." It was the system they set up when they ostensibly recognized Iraq as independent, but used the Anglo-Iraq Treaty to lock in a twenty-five-year "alliance." Under dyarky, all Iraqi institutions had double staffs, with an Iraqi national as the nominal head but the British adviser in reality holding all executive power.

Recognition of bin Saud's independence and agreement to raise British representation at Jedda from a consulate to a legation were further "fresh proof of British kindness."[4]

PHILBY'S FIRST CONTACT WITH TWITCHELL

Despite practically keeping himself for almost five years through selling virulently anti-British press articles, Philby immediately sent his wife in England to Hugh Dalton, the undersecretary of state for foreign affairs, to request that he, Philby, now be appointed the first British minister to Jedda. "If I had the job," Philby wrote his wife, "I should of course run the show. And expect the Foreign Office to leave me to do so without questioning my discretion as they must surely realise now that in my judgement of the Arab situation I have never made a mistake." The job went to Sir Andrew Ryan.

Philby had traveled back from England via Constantinople, Damascus, and Jerusalem, where he could not resist interfering. Among other activities, he urged the Palestinian Arabs "to take the British position into account and not to pitch Arab demands too high," and proposed a constitution in which Arabs and Jews would have proportional representation "under a continuing British mandate." The colonial secretary had immediately disowned Philby, explaining that he had "no authority to enter into negotiations with anybody or to act on my behalf in any respect whatever."

Nevertheless, Philby remained convinced that he had all the answers, and wrote his wife, "I believe a commission consisting of Israel Cohen (a prominent Zionist and historian of Jewry) and myself could solve the Palestinian problem."

He misunderstood the gist of a letter written by Karl Twitchell, the engineer sent to Yemen by American philanthropist Charles Crane. Philby thought Twitchell wanted to buy Ford vehicles and immediately sent a sales pitch. "When I was recently in Constantinople the director of the Ford Company there showed me your correspondence in connection with the economic development of the Yemen in which you are actively engaged on behalf of Mr. Crane," Philby wrote Twitchell. "I have had the pleasure of acquaintance with Mr. Crane for a number of years. I saw much of him here in 1927 just before he left for his very interesting visit to the Imam Yahya and his country. I am therefore much interested to know that he is taking such active steps to help that country in its development."

Philby said he would supply the vehicles "to the Yemen government at actual cost price to me by way of starting the market going for such vehicles." In return for this generosity, he said, he would "ask that the Imam should grant us permission to open a Ford Agency on the understanding that 10% discount should be allowed to the government in respect of all orders placed by it. I think you will agree that this is a fair offer and on hearing from you that you accept it, and on receipt of a cheque on account, I shall be happy to arrange for the immediate shipment of the vehicles mentioned." He and Twitchell should meet to discuss this business venture, Philby suggested.

A bemused Karl Twitchell replied, "I have heard of you from various sources and it would be a pleasure to meet you." But, Twitchell continued, "I am afraid there is a misunderstanding for I have absolutely nothing to do regarding agreeing to your proposal to send two tractors, one Phaeton and one truck, if I will send you a cheque on account." Although they already had Ford vehicles purchased from Aden, Twitchell said, Philby should send his proposals directly to the Yemen government.

Instead of the Yemen government, Philby wrote to Charles Crane. "My company has always been anxious to open up work in Yemen if and when conditions become favourable," he said. "It seems to me that in view of what Twitchell seems to be doing in the matter of roads, the time may not be very far distant when we can start selling cars and tractors there. My company holds the Agency of Ford, Firestone, Standard Oil, Singer Sewing Machines, Franklin, etc. etc."

This was not exactly true. Philby had an arrangement to collect commission on anything he sold. There was no Ford agency because cars were forbidden to all but the royal house until 1934. Philby urged Crane to use his influence to push the US government to diplomatically recognize Abdul Aziz bin Saud. He advised: "Forget that little business of your being shot at here on the Basra-Kuwait road last year."

Although the Ikhwan rebellion against bin Saud was at its height, Philby wrote Crane again:

> There are many other directions in which American cooperation may prove of lasting benefit to Arabia and the first step must obviously be the establishment of diplomatic relations. For instance, the government is now calling for applications for three important posts (1) a mining engineer to study and report on the mineral resources of the country (2) an engineer to examine Jedda harbour and draw up a scheme for the construction of a suitable wharf and (3) an engineer to draw up a comprehensive scheme for the electric lighting of Jedda and Mecca.

He was still hopeful of personal benefit from Crane, and continued. "I have had some correspondence with Twitchell on the subject of the possibility of pushing Ford cars and tractors in Yemen. In my *History of Arabia*, I refer to the steps you have taken to promote the development of Yemen. I only wish that Wahhabi Arabia had secured your cooperation when you were here offering it with both hands. But you came slightly before the country was ripe to appreciate you." Three months later he wrote again: "It is now practically certain that the King will be in the Hijaz all next cold weather. I hope this will encourage you to pay us another visit." Knowing that Charles Crane believed agriculture to be the savior of poor countries, Philby added the sweetener. "You will find that much progress has been made in many directions and that the King is particularly keen on encouraging the development of agriculture."[5]

SECRET ORDERS

As Holmes was preparing to receive Standard Oil of California's geologists in Bahrain, the company's executives in San Francisco were already calculating how far they could go without noticeably double-crossing their new colleagues at the Gulf Oil Corporation.

M. E. Lombardi was Standard Oil of California's vice president concerned with actual oil production. He had taken part in the early discussions about buying Gulf Oil's option on Bahrain. In the Gulf Oil office, he had seen Holmes's map of the Persian Gulf and he had spotted the areas of petroleum interest beyond Bahrain that Holmes had marked. He knew by Francis Loomis's letters from London after talking to Holmes that he had also noticed. Lombardi wrote to Loomis: "On two or three occasions when we were discussing the Bahrain matter you mentioned the idea that there might be other desirable localities in the general area of the Persian Gulf in which we might become interested."

Lombardi said he didn't think Standard Oil was "morally bound" to give Gulf Oil "a free hand" except perhaps in a very tight area of Kuwait. He said he wanted to advise the men he was sending to Bahrain to "cautiously sound out Major Holmes on the situation with regard to the rest of Arabia and nearby territory." He wanted Loomis's opinion on what he should instruct his geologist about "how far he can go when talking to Major Holmes in suggesting that we might be interested in other territory besides that in the Bahrain concession." He selected two of his own men for this mission, geologist Fred A. Davis and the superintendent of the Producing Department, Bill Taylor.

Before departure, their task was outlined as checking the conclusions reached by Gulf Oil's Ralph Rhoades in his examination of the work undertaken by Holmes in Bahrain. It was expected that this would take no more than ten days. After that, according to their instructions, they were to get into Al Hasa and Kuwait for the purpose of discreetly reporting on the petroleum sites that Lombardi had seen on Holmes's map in the office of Gulf Oil. They were also instructed to bring back as much information as they could about regional oil possibilities in general.

Obediently, Davis and Taylor hung out with oil field workers in the London bars and reported back the gossip. On arrival in Baghdad they quizzed the American consul. "The government here feel they are not getting a square deal so the Iraq Petroleum Company are not going over so good. The American Consul is very frank and states that the Iraq Petroleum Company is very high handed in their methods," the two men wrote their San Francisco office. They managed to get a tour of Iraq Petroleum's camps where they were "treated royally and told nothing," but concluded: "There is plenty of oil here." Leaving Iraq they assured their colleagues: "Major Holmes will remain in Bahrain until our arrival and will stay with us for a time at least."

Francis Loomis was doing his part. He contacted Dr. Arthur Wade, the British geologist whose favorable 1912 report had originally stimulated interest in the Farasan Islands. Wade replied that the most up-to-date information on the Persian Gulf was the Anglo-Persian Oil Company's surveys. But he had "run to earth an old miner who was for a long time in Kuwait and only returned about two years ago. He is coming to see me one day this week and may have some useful and fairly modern information to disclose. I do not know what you have in mind, nor how well informed you may be with regard to the Kuwait area, but it occurred to me that this man would be a useful person to meet. I have known him for over 20 years." He would be willing to undertake any work Standard Oil might require in the region, Wade said, adding, "The harder and more difficult it may be the better I shall like it." If "for any reason whatever" Loomis wanted to employ another man, he would help to find someone. "Although," he warned, "it may be hard to find one with any experience of that part of the world today. The old hands I know are either dead or scattered abroad."[6]

STANDARD OIL OF CALIFORNIA GEOLOGISTS

As he had done with the Americans from Gulf Oil, Holmes ensured that the Standard Oil men were properly introduced. "Major Holmes came out to the boat here to meet us, brought us in, and has established us in a comfortable building here where we get all the breeze there is and are as comfortable as it is possible to be. We aren't attempting to do any field work these first few days but are spending our time getting acquainted with general conditions through conversations with Major Holmes and meeting through him the various influential men around, both native and British," Davis reported back.

He was impressed when he met bin Saud's Bahrain representatives, the Gosaibi brothers. He told the men back in San Francisco: "bin Saud is, as you probably know, the King of practically all of Arabia and the man to whom even the so-called independent sheiks look for approval. Abdul Aziz and Hassan are two of the five Gosaibi brothers who are the biggest pearl merchants here, who have considerable influence with bin Saud personally and who should be of great assistance in obtaining concessions on the Arabian mainland."

For his part, Holmes seems to have been feeling out the situation. His response was to play dumb to the American's barely disguised eagerness for information on possibilities beyond their rights solely to the Bahrain concession.

"At the present time he appears to be suffering from shellshock causing his conversation to be rather disjointed at times. The first evening we were with him he cautioned us to check up on him when he lost the thread of conversation," a seemingly naive Davis reported of the first meeting with Holmes.

From their initial imagining of Holmes as a colonial yokel, perhaps even a shell-shocked one at that, the two Americans slowly began to realize that he was running rings around them.

"When we left Baghdad we had no idea that Major Holmes was familiar with conditions in Iraq or was paying any attention to them. But on talking with him we discovered that he is very much interested, has been following it closely and is well acquainted and has considerable influence with the government officials who will have the outlining of the Iraq government future oil policy," they wrote their head office.

From what they had seen and heard, Davis said, Frank Holmes "appears to be on extremely friendly terms with all the influential Arabs and we believe he could go farther toward obtaining concessions from them, both in Arabia and Iraq, than any other European or American here."

Holmes had tried to interest William Wallace of Gulf Oil in "pushing for Iraq ahead of Kuwait," they reported, "but Wallace has refused, probably because of Gulf's participation in Iraq Petroleum that precludes their having any other interests in the zone." They said that Holmes had tried to bring Iraq possibilities to the attention of Francis Loomis when they were together in London, "but was prevented from doing so by the attitude of Major Davis, Gulf Oil's representative in London."

They found out that Holmes had studied British Admiralty maps closely during WWI and from these "as well as personal observations in the field, he knows of many seepages along both the Persian Gulf and Red Sea coasts of Arabia and believes there are many areas worthy of exploration." Right now, they said, Holmes was restricted by his contractual obligation with Gulf Oil and that company was dictating "the order and manner in which his work out here should be conducted." Wallace was insisting on the immediate securing of the Kuwait concession, regardless of local conditions, Davis and Taylor reported.

NOT SOLELY BAHRAIN

No matter how much Standard Oil of California wanted Frank Holmes to use his expertise to compete on their behalf for concessions in Iraq, Kuwait, and bin Saud's territory, he was not free to do so. Holmes's position was officially that of chief local representative of the Bahrain Petroleum Company, with the cost of his salary split three ways between his own E&GSynd, the Gulf Oil Company, and Standard Oil of California. Under this arrangement Standard Oil of California had a call on Holmes solely for facilitating the development of the Bahrain concession, which was their only current business in the area. The remaining two-thirds of his work was to be fulfilling the terms of the second arrangement, the Mainland Agreement, signed in 1927 between E&GSynd and Gulf Oil. That agreement obliged E&GSynd to expend every effort in concluding the Kuwait concession, and the Al Hasa and Neutral Zone concessions also, for transfer to the Gulf Oil Company.

Just how much a disadvantage they were at because of their complete ignorance of the politics, culture, and conditions of the Persian Gulf was just beginning to seep through to the executives at Standard Oil of California. And they were having doubts about their relationship with the Gulf Oil Corporation.

Lombardi wrote Loomis: "The visit of Davis and Taylor to the Persian Gulf has not turned out in an entirely satisfactory way. There seems to be a lot of restrictions thrown around them that we did not anticipate. Of course the primary object of their visit is to determine the advisability of drilling on Bahrain Island and to gather as much practical information about the country, labour conditions, material and supply conditions, etc., as possible. Nevertheless we also desired them to get a general view of the situation which would enable us to determine whether or not we desired to acquire more territory than simply the islands of Bahrain."

Lombardi thought Gulf Oil was not being helpful. Their attitude, he complained to Loomis, "seems to show they are not particularly anxious for us to see more than just the islands of Bahrain. It is my recollection that you brought up with me some time ago the question of further activity by our company in the Near East. I think we agreed that we have never indicated to Gulf Oil that we expected to confine ourselves solely to Bahrain."

It had to be established, Lombardi said, whether Gulf Oil expected Standard Oil "to be satisfied only with the Bahrain concession. Or whether so far as they are concerned it is agreeable for us to look over other possible oil territory in the Near East, including Arabia, Iraq, Persia and Turkey. Of course we will not interfere with the present negotiations of Gulf Oil regarding the Kuwait concession." Lombardi saw clearly that the three-way condition under which Holmes was working was "fundamentally unsound" because of the "possibility of conflicting interests among the principals." He asked Loomis to ascertain from Wallace as to whether or not Standard Oil of California "have any call on Major Holmes's services (and if so to what extent) for any ventures in the Near East outside the Bahrain concession."

In Bahrain, Fred Davis and Bill Taylor were getting edgy. They had quickly concluded that Holmes's original marking out of the Bahrain oil field was correct, so much so that they did not believe any additional area was necessary. Taylor wrote: "Personally, I think the 100,000 acres we've already got will cover all the oil in the islands, but the additional acreage may keep someone else from causing trouble if we get lucky and find oil." The two Americans were anxious to move on. "We thought when we arrived on May 14th that we would get away in two or three weeks at the most, but here we are yet and expect that it will be necessary to stay until about 21st June at least . . . we still hope to go to Kuwait." They described how they were fed up with the attitude of the British on Bahrain. "We find a great deal of jealousy here from the British residents. But when everything is signed up we can tell 'em all to go to hell. These damned British Agents are sure 'upstage.' We go to tea and all that sort of thing but they remain very aloof."

LOMBARDI'S PLAN FAILS

Lombardi became so frustrated that he telephoned Wallace. In echo of the justifications that Anglo-Persian had employed, Lombardi said that in order to understand the geology of Bahrain, they needed to examine the Arabian mainland. He claimed: "To study the geology of Bahrain Island, it is necessary for our men to visit other portions of Arabia, particularly the mainland, near Bahrain and as far north as Kuwait. My men have finished looking at the island itself and have nothing else to do until they get permission to go on the mainland." Stating that he had no desire to "embarrass your negotiations for the Kuwait concession," Lombardi told Wallace "what we want now, while our men are there, is to visit those localities that are geologically important in connection with Bahrain."

Wallace replied that he did not want any Standard Oil men in Kuwait because it might "put the idea of competition into the mind of the Shaikh of Kuwait." He said he would cable Holmes suggesting the men "go to Baghdad and also visit the Persian oil fields." Wallace seems to have been as ignorant of local conditions as Lombardi in assuming that the Americans could simply drop in to Iraq and Persia without permission from the British. His total lack of comprehension was evident when he assured Lombardi that he would have Holmes "finish at Bahrain and Kuwait in about thirty days." He would then instruct Holmes, he said, to "immediately start across Arabia to interview bin Saud and get permission for Standard Oil of California to visit the mainland."[7]

Davis and Taylor thought they would be in Baghdad to celebrate the Fourth of July with the American consul. They didn't get there in time. Nor did they get anywhere else at all, not to Kuwait, not to the mainland, not to Persia. Before sailing out, on July 21, geologist Fred Davis expressed his disappointment. He wrote to San Francisco: "I am very sorry about this, for when I left the States I had very ambitious hopes of coming back from over here with a great fund of general information on oil, both geology and operations." In the company magazine, Taylor and Davis described their time on Bahrain as a "tour of inspection to the Bahrain Islands." They wrote: "For six weeks we were busily engaged in the pleasant occupation of familiarising ourselves with the character and customs of the islands' inhabitants, as well as its physical state."

Fred Davis would eventually become chairman of the Arabian American Oil Company (Aramco) in Saudi Arabia. In his foreword to Philby's 1964 *Arabian Oil Ventures* Davis would unkindly dismiss Holmes as a person "with whom the company dealt in its introduction to Bahrain." He would not mention Lombardi's secret instructions but would instead claim that his own recognition of the oil possibility of Saudi Arabia began "from the day we first set foot on Bahrain." (As to Davis's time in Bahrain, a letter to Wesley Owen preparing the 1975 *Trek of the Oil Finders: A History of Exploration for Petroleum* for the American Association of Petroleum Geologists, admitted that his own work in Bahrain "consisted only in checking a map that had been prepared by 'Dusty' Rhoades of Gulf Oil a year or so earlier" on work already done by Frank Holmes on the Bahrain field.)

Lombardi's plan had failed but Francis Loomis wasn't giving up. He took the extraordinary step of asking Edmund Janson of E&GSynd, as a personal favor, to call on Anglo-Persian's John Cadman in London to sound out the possibility of Standard Oil of California getting into Iraq, independently of Iraq Petroleum.

Not surprisingly, Janson reported no joy from this meeting. Cadman told him, Janson said, that he could not comment on what areas of Iraq might be outside Iraq Petroleum's concession because he "did not have a map of the old Turkish Petroleum Company's concession."

Cadman said the Iraq government was "debarred" from dealing with any areas until Iraq Petroleum had "definitely settled" its selections. He warned Janson it was not "the slightest good" approaching Nuri Pasha, the Iraq prime minister and an old friend of Frank Holmes who was currently in London, "until this question has been settled."

Cadman did not miss the opportunity to also claim meaningfully to Janson that "Bahrain, Kuwait, Al Hasa, the Neutral Zone, and 'an area to the south in Arabia' are in 'Anglo-Persian's sphere of influence.'" Janson wrote soothingly to Loomis, promising that Holmes would be able to suggest the "best way of obtaining concessions in Iraq when Nuri Pasha arrives back home." Holmes would know exactly what areas to try for, Janson said, because apart from his own knowledge and observations, Holmes had a copy "of the original Turkish Concession that he got from the Admiralty Office."[8]

PHILBY CONVERTS TO WAHHABI ISLAM

As the Standard Oil men were departing the area, Philby was in Mecca converting to Islam under the sponsorship of Abdul Aziz bin Saud. In 1928 he wrote bin Saud asking him to be his sponsor. Distracted by the Ikhwan rebellion, bin Saud replied that he was too busy. Philby wrote again in 1930, and this time the king agreed. Philby could have converted at any time (Islam does not require a sponsor for converts). Because he combined the vow of faith with the ceremonies of the minor pilgrimage, Philby imagined even his conversion to be unique. "I must surely be the first Muslim to have performed the *Tawaf* (circling of the Kaaba) before ever saying a single prayer," he trilled.

The king was aware that many doubted Philby's sincerity. When Philby arrived from Mecca to the court at Taif, bin Saud gave a sort of pep talk to the assembly around him. He repeated almost word for word the claim to his obligation that Philby himself had laid on bin Saud. He said that Philby "had sacrificed a great deal for Arabia, had disagreed with his own government and had resigned his post at Amman because the British would not back me, bin Saud."

The king said that Philby had never taken a salary from him but, prudently perhaps, stated that he would not be given a government position and added that a theologian would be engaged to give Philby religious instruction.

Bin Saud bestowed the name of Abdulla on Philby. Within hours the joke was going around the merchants, who believed Philby had converted only for the advantage of readier access to the king, that instead of "Abdulla, slave of God," Philby should be called "Abd'al Qirsh, slave of pennies." Philby's wife, Dora, was also not convinced. "I see what a difference it is going to make to you, being able to get around the whole country," she wrote. Curiously, Philby replied to Dora: "I still hope to end my days as a member of the British Parliament."

The conversion to Wahhabi Islam made Philby an international celebrity. He rode the wave, writing articles for the British, Egyptian, and international press. Philby soon had access to the king's palace in Mecca where he attended the three daily court sessions in which the king received guests, dispensed justice, ordained punishment, and read the Scriptures. In early 1931, bin Saud would give him a house in Mecca and a slave girl, Mariam, as a concubine. Philby described her to his long-suffering wife in England as "not beautiful by any means but young and shapely enough." Enthusiastically, he would inform his partners in Sharqieh: "My coming to Mecca has entirely changed our position."

NO PROGRESS ON KUWAIT

Frank Holmes had moved on to finalize the Kuwait oil concession and complete E&GSynd's agreement with Gulf Oil. The shaikh of Kuwait was happy to have Holmes back and to hear of progress on Bahrain. Holmes presented, with a copy to the political agent, the draft agreement drawn up by Gulf Oil while he was in New York that "embodied the ideas of Mr. Wallace" and reported that Ahmad was "quite willing to give us the concession."

Shaikh Ahmad was keen to have the investment in Kuwait that development of the concession would bring. Growth of the Japanese cultured pearl industry was badly affecting Kuwait's own pearl diving, and bin Saud's ongoing boycott was stifling trade. In addition, the shaikh of Kuwait's personal income from his property and date gardens in Basra was now subject to taxation from the Government of Iraq.

Although the British had promised him at the time of the Saudi-Iraq Treaty of Friendship that this would not occur, the Iraq government had withdrawn his tax exemption. The British government had not protested.

Holmes wrote to his wife: "Town quiet, the people are away in Fao and Basra attending to their date harvest. I live in a house right on the seashore, boat building is the industry, there are four being built right now. The Political Agent has gone to India until the really hot weather passes and will not be back until the end of September."

Holmes seemed to be feeling the heat, the strain, and the loneliness, as he wrote wistfully to Dorothy: "I have been working out here for five years. During this time, I have only had about ten months with you, that is rather a little, ten months out of 60 months."

Holmes's tranquil contemplation of the Arab boats, the graceful dhows, on the shore near his house was rudely interrupted by a letter from the resident. In it, Lt. Col. Hugh Biscoe said that another version of the British nationality clause than that used in Bahrain must be inserted in the draft document for the Kuwait concession before he would allow Shaikh Ahmad to even discuss the subject.

The new exclusive British nationality clause dictated that not only control but the capital of the concessionaire company must remain purely British, and that transference of any concession rights could only be effected to another purely British company, and only with the specific approval of the British government.

The new definition was moving back to the concept of the 1913 "convention," which the viceroy of India had noted "gave monopoly rights to Great Britain against foreigners and . . . gave us control over selection of even British Companies."

The definition in the Bahrain concession was "a British company registered in Great Britain or a British Colony" without specific mention of the source of capital. This had allowed Standard Oil of California to register the Bahrain Petroleum Company in Canada as the nominees of the Bahrain concession, even though Canada allowed total repatriation of profit to a parent company.

In effect all profit from Bahrain received in Canada could be repatriated to Standard Oil of California in San Francisco. The Gulf Oil Corporation had planned to follow the same mechanism for the Kuwait concession.

Holmes decided to do something about the long months away from his wife. On August 23, 1930, while Philby was in Mecca converting to Islam, Frank Holmes retuned to London to once again do battle with British officialdom.[9]

NOTES

1. John Bagot Glubb, *The Changing Scenes of Life: An Autobiography* (London: Quartet Books, 1983), p. 87, "Bin Saud could not control his own followers"; Christine Moss Helms, *The Cohesion of Saudi Arabia: Evolution of Political Identity* (London: Croom Helm, 1981), p. 235, "ceased to exist as a political factor" and p. 228, citing February 28, 1928, high commissioner, Iraq to secretary of state for the Colonies, "the whole desert . . . will become untenable"; Benjamin Shwadran, *The Middle East: Oil and the Great Powers* (London: Atlantic, 1956), p. 247, details the BOD approach and the original "selection" requirement; on p. 248, Shwadran states, "the British had never meant to abide by this system."

2. Hafiz Wahba, *Arabian Days* (London: A. Barker, 1964), p. 138, "no communication between Nejd and Mecca"; Holmes Papers, December 2, 1925, Craufurd to Janson (forwarded Holmes in Bahrain), "Holmes to get from bin Saud some idea of his definite requirements for wireless communications . . . we can get the contract with reasonable advantages accruing to us if we take on the job properly." See Helms, *Cohesion of Saudi Arabia*, p. 252 (after the 1925 capture of the Hijaz holy cities), the use or installation of such modern innovations as telegraph, telephone, cars, and airplanes was delayed or halted entirely because of Ikhwan opposition. Even the bicycle was viewed as the "vehicle of the Devil." Elizabeth Monroe, *Philby of Arabia* (London: Faber and Faber, 1973), p. 158, quoting Philby, "the King was favourably considering this idea, one which was all my own." A 17,000-sterling contract with Marconi was signed in 1930; Philby Papers,

box XXX-7, August 2, 1929, report of a meeting in London among Cree, Philby, and Dundas arranged by C. F. Mayer, vice president of Standard Oil Company of New York. Standard Oil New York was part of the American group in Turkish Petroleum. In 1931 they became Socony-Vacuum; and November 26, 1929, Cree to Philby, "any chance Al Hasa concession being available?"

3. Helms, *Cohesion of Saudi Arabia*, p. 241, "Both Faisal of Iraq and Ahmad of Kuwait sent the rebels money, food, arms and ammunition, camels and horses"; Glubb, *Changing Scenes of Life*, p. 91, "It was agreed . . . the rebel tribes would be evicted from Kuwait and returned to bin Saud"; Kenneth Williams, *Ibn Sa'ud, The Puritan King of Arabia* (London: J. Cape, 1933), p. 235, "the value of arms and ammunition supplied by the Government of India was 31,500 sterling . . . provision has been made for the British taxpayer to repay to the Government of India two-thirds of the amount"; Helms, p. 271, citing January 2, 1930, minute, India Office (Laithwaite), "to some extent hemmed in"; for this episode see Helms, pp. 250–74, Robert Lacey, *The Kingdom* (London: Hutchinson, 1981), pp. 204–14, David Howarth, *The Desert King* (London: William Collins, 1965), pp. 152–64. On pp. 162–63 Howarth says, "for the first time in the long history of the Mutair tribe the herd of 'sacred' black camels which had been the symbol of their pride were scattered . . . (after their defeat bin Saud) seized the sacred camel herd. It still exists today among the royal herds of the House of Saud"; Wahba, *Arabian Days*, p. 143, "friendship and kindness."

4. Shwadran, *The Middle East*, p. 251, the first pipeline was a parallel twelve-inch line from Kirkuk to Haditha a distance of 150 miles, where it branched off in two directions, one line going to Haifa (Palestine) for a distance of 470 miles and one to Tripoli (Lebanon) a distance of 380 miles; construction was completed in 1934. Anglo-Persian and Dutch Shell received their share of Iraq oil at Haifa, the Americans and French received theirs at Tripoli. The hostility of the Iraqis, displayed by the 1920 Iraq rebellion, had prompted the British to manipulate Faisal bin Husain al Hashem (son of Sharif Husain of Mecca) into position as king of Iraq in August 1921, by this means they hoped to continue "indirect" rule. However, King Faisal I, and the emerging Iraqi nationalists, were committed to achieving full sovereignty. See Philip K. Hitti, *History of the Arabs from the Earliest Times to the Present*, 10th ed. (London: Macmillan, 1970), pp. 752–53, Arthur Goldschmidt Jr., *A Concise History of the Middle East*, 3rd ed. (Boulder: Westview Press, 1988), pp. 202, 276; Glubb, *Changing Scenes of Life*, pp. 91–92, "each monarch believed he had gained precedence"; Administrative Report for Bahrain 1930 (Prior), "the tenuity of our hold on Bahrain." Bin Saud left Ras Tanura on February 20, and arrived in Bahrain on February 25, after the conclusion of the *Lupin* conference.

5. Elizabeth Monroe, *Philby of Arabia*, p. 159, February 14, 1930, Philby to his wife, Dora, "If I had the job I should of course run the show"; and pp. 159–61, for Philby's interferences, meetings with Zionist leaders and grand proposals to personally solve the Arab-Jewish problems of Palestine; Philby Papers, box XXVI-2, Crane Letters; November 11, 1929, Philby to Twitchell, and December 11, 1929, Twitchell to Philby, and December 27 and March 11, 1929, Philby to Charles Crane. See Monroe, p. 155, the Ford Agency was not secured until 1934, and p. 157, "cars were forbidden to all but the royal house until 1934." Philby Papers, box XVI-2, June 19, 1930, Philby to Charles Crane, "the King is particularly keen . . . on agriculture."

6. Chevron Archives, box 120791, May 2, 1930, Lombardi to Loomis, "how far (the

geologist) can go . . ."; box 120797, April 7, 1930, Bill Taylor at Hotel Metropole, London, to William McLaughlin (Mac) Standard Oil San Francisco and April 30, 1930, Bill Taylor at the Carlton Hotel Baghdad to McLaughlin "treated royally and told nothing"; box 0120796, May 26, 1930, Dr. Arthur Wade, consulting petroleum and mining geologist, London, to Francis Loomis at the Hotel Plaze-Athene in Paris, "hard to find anyone with any experience of that part of the world today."

7. Chevron Archives, box 120797, May 17, 1930, Fred A. Davis on Bahrain to San Francisco and May 30, 1930, Bill Taylor on Bahrain to San Francisco, "Holmes could go farther toward obtaining concessions, both in Arabia and Iraq, than any other European or American here"; box 120791, June 6, 1930, Lombardi to Loomis, "do we have any call on Holmes's services . . . outside Bahrain"; box 120791, June 7 and June 11, 1930, Bill Taylor to San Francisco, "we thought we would get away (from Bahrain) in two to three weeks" and June 10, 1930, Lombardi to Loomis, memorandum of telephone conversation with William Wallace, "visit those localities that are geologically important in connection with Bahrain" and instruct Holmes "to interview bin Saud and to get permission for Standard Oil of California to visit the mainland."

8. Chevron Archives, box 0120796, August 12, 1930, Fred Davis in Baghdad to San Francisco, "I had very ambitious hopes of coming back from over here with a great fund of general information on oil"; Standard Oil of California company magazine, *Among Ourselves*, December 1930, "We Travel East of Suez" by Taylor and Davis; see also Administrative Report Bahrain 1930 (Prior), "Major Holmes arrived on April 15, and was followed by two American geologists, Taylor and Davies, a month later. They left again on July 21st and without making any definite statement, for obvious reasons, but are understood to have been optimistic in the possibilities of finding oil at Bahrain." American Heritage Center (International Archive of Economic Geology), University of Wyoming, Edgar Wesley Owen Papers, box 1a, October 11, 1974, Fred A. Davis to G. Gish, "only in checking a map"; Chevron Archives, box 0120798, August 14, 1930, "Private and Confidential," Edmund Janson to Francis Loomis, Standard Oil of California, "Holmes will suggest best way of obtaining concessions in Iraq."

9. Monroe, *Philby of Arabia*, pp. 165–69, for events leading up to Philby's conversation in August 1930, reactions, and his situation immediately afterward, p. 171, "young and shapely enough"; Holmes Papers, August 10, 1930, Holmes in Kuwait to Dorothy at MillHill, "that is rather a little, ten months out of 60 months. . . ."

CHAPTER 16

KUWAIT OUT—
BAHRAIN IN

Kuwait became a casualty of Britain's aiding of bin Saud's putdown of the Ikhwan rebellion that raged across the area where construction of the Iraq oil pipeline was planned. As British attention focused on Iraq, and bin Saud's independence was recognized as a factor in ensuring the security of the pipeline, Kuwait was viewed as a bargaining chip to be played either with bin Saud or with Iraq, as the occasion demanded.

The Persian claim to sovereignty over Bahrain at the League of Nations put a brake on Britain simply imposing its "assumed" rights over the Bahrain Islands. Moreover, Britain was aware that refusal to allow American investment in development of the Bahrain concession rested on very dubious legal grounds, grounds that would not stand up to international scrutiny. In addition, Britain needed to present itself as ensuring the "prosperity of Bahrain's Arab inhabitants" as it claimed to be doing in its submission to the League of Nations.

But there was no international spotlight on Kuwait. The only interest in Kuwait was coming from its covetous neighbors, Abdul Aziz bin Saud and Iraq. Bin Saud had displayed superiority by achieving sovereignty over Kuwait's traditional tribes and desert frontiers at the 1922 Ujair Conference. Kuwait was suffering from the continued trade blockade, which the political agent described as demonstrating "bin Saud's avowed objective that still appears to be to force Kuwait to acknowledge him as suzerain, through the simple process of strangulation."

British officials administering Iraq were also covetously eyeing Kuwait. The possibility of allowing Kuwait to quietly pass to Iraq had been under discussion since the end of the Ikhwan rebellion. Just a month before Holmes's visit to Shaikh Ahmad, the acting high commissioner of Iraq seriously proposed to London that the British government "should in fact assist any tendency there may be in the direction of encouraging the absorption of Kuwait by Iraq."[1]

The whole discussion of the status of Kuwait highlighted the tension between the Colonial Office holding overall responsibility for political policy, including toward Iraq and Saudi Arabia on the one hand, and, on the other, the authority of the Government of India administering the shaikhdoms of the Persian Gulf. The prospect of Kuwait being taken outraged the Government of India. Having been deprived of the goal of gaining Mesopotamia as a colony for India, they were not about to stand by and see their post–World War I compensation package— authority over the shaikhdoms of the Persian Gulf—picked off one by one.

HAROLD DICKSON RETURNS

The political agent in Kuwait was now Col. Harold Dickson, finally rehabilitated from India through the good offices of friends, promoted from major to colonel, and posted to Kuwait after twelve months as secretary to the resident in Bushire. Kuwait was considered a backwater. The arrival of Dickson and his family in May 1929 brought the number of European residents to eleven. He experienced a flurry of excitement while the British were cracking down on the Ikhwan rebels on Kuwait's borders and a moment of glory when he accompanied the political resident, Lt. Col. Hugh Vincent Biscoe, to oversee the rebel leaders' surrender to bin Saud.

To his intense irritation, Dickson had not been invited to "represent" the shaikh of Kuwait at the British-sponsored conference onboard HMS *Lupin* as his predecessor Major More had done at the Ujair Conference. Since the *Lupin* meeting between bin Saud and King Faisal that marked the end of the Ikhwan rebellion, Kuwait had sunk into insignificance, so much so that its absorption into Iraq could be openly advocated.[2]

Dickson viewed the prospect of an oil concession in Kuwait as his chance to move center stage. In 1923 when he thought he was going to be attached to bin Saud, Dickson had drawn castles in the air for his superiors of a powerful British High Commission overseeing Bahrain, Kuwait, and the Trucial States, with its headquarters in Riyadh, and himself as high commissioner. Once again, Dickson set about inflating the importance of his post and, by no means coincidentally, his own importance. Now he sketched a scene in which the British would militarily occupy Kuwait and turn it into "the Gibraltar of the Gulf."

Dickson's idea was that "especially after 1932 when Iraq will obtain her independence" Kuwait under British military occupation could be used to dominate the Gulf, protect the eastern air route, and provide a base from which the British could "threaten" bin Saud's territory and Iraq and Persia.

From Kuwait, Dickson said, Britain would be able to "strike at the flank" of either bin Saud's territory or Iraq should they be at war with each other, "and it is to our advantage to come to the assistance of one or the other." Dickson claimed he could handle the ruler of Kuwait in such a way that "he would welcome" being under British military occupation.

He forwarded his missive to the resident, Biscoe, laying out his detailed scheme for military occupation that would transform Kuwait into the Gibraltar of the Gulf, accompanied by a personal letter. In this he enthusiastically agreed with Biscoe's own objections to the development proposals contained in Holmes's concession.

"The points you have also noticed and mentioned in your letters to London . . . his ambitious plans for railways, telegraphs, canals, ports, etc. Fancy if it all came true and we found them all in American hands! It would be an intolerable position," Dickson cried. The resident had earlier told the Colonial Office that if Holmes's syndicate succeeded in developing roads, ports, lighthouses, telegraphs, and so on, "the shaikh would become a puppet in the hands of the Americans and Britain would, to a large extent, lose the influence it had exercised. Its whole position in the Persian Gulf would be threatened."

GIBRALTAR OF THE GULF

Dickson now fancied himself as something of an expert on oil. "Between you and me," he confided to Biscoe, "I firmly believe Holmes thinks there is oil close to and in the immediate vicinity of the town. His water bore of 700 feet near the shaikh's palace, I believe, has proved a lot to Holmes. From there he proposes to do the usual triangular boring of wells. If I am correct and he sees success at two points of the triangle then one brings him almost dead onto the old British leased area which we had for a naval coal depot and the other is conveniently close to the south of it."

Dickson's imaginary Gibraltar of the Gulf was already a reality in his own mind as he warned: "In any case if he failed here and succeeded elsewhere we shall see both the old naval base and the suggested cantonment area covered with barracks, oil tanks, railways, telephones, dirt, etc. etc., in a very short time. It would be good-bye to our ever getting a decent harbour and landing place for our flying boats, ships, depots and warehouses, not to mention the most pleasant parts of Kuwait for our personnel, officers and men."

On receipt of Dickson's package, the resident was inspired to write again, this time addressing his "further comments" directly to Lord Passfield, now the colonial secretary, with a copy to the Government of India. He said it was "most undesirable" that the shaikh of Kuwait should grant his concession to American interests. "Although they have obtained a concession in Bahrain," he said, "it would seem desirable, if possible, to exclude them from other places in the Persian Gulf."

Dickson wanted a watertight "British Nationality Clause" and denial to Holmes of "certain areas" already included in the concession because these would be required by Dickson's imagined military occupation force. The British government should "instruct" Holmes to make these "corrections," Dickson said.

If not, when Holmes restarted the discussions and the shaikh of Kuwait asked for Dickson's advice, he said, he would "require Shaikh Ahmad to get Holmes to insert our clauses."

Biscoe's recommendation to Lord Passfield was more simple. Just keep insisting on a really tough "British Nationality Clause," he advised, and "Holmes and his syndicate will abandon the attempt to obtain a concession and then there will be no need to consider the matter further."

In view of the fact that the British might wish to use Kuwait to "threaten" neighboring countries, the resident repeated his advice that "on strategic grounds it is desirable to maintain the independence of Kuwait" and not allow its absorption into Iraq.

This opinion, he told Lord Passfield, was shared by "the Air Ministry and possibly the Admiralty." Dickson concurred wholeheartedly, saying, "particularly as Iraq is now going from us."

Encroachment by Iraq could not be contemplated. Moreover, the political officers saw the possibility of Frank Holmes and his American partners gaining a foothold in Kuwait as a further threat to their hegemony. In the guise of objecting to American involvement, the officers were in fact arguing for a tightening of the Government of India's grip on Kuwait. Consultations were arranged between the India Office, the Colonial Office, and the Foreign Office. The Government of India won the day. As Holmes was returning to London once again to discuss the Kuwait concession with the Colonial Office, the instruction was being circulated that there would be no change in policy toward Kuwait.

Iraq would not be encouraged to absorb Kuwait. It would be maintained as before, as an unofficial "protected state" administered by the Government of India, through the political resident and the political agent. Conveniently, it would also serve as a "buffer" between bin Saud and Iraq. There would be no difficulty in maintaining this position, the officials stated, because they could easily "command obedience" from the shaikh of Kuwait—as long as he was kept constantly aware that protection from the territorial ambitions of both Iraq and bin Saud depended solely on British goodwill. Cheerfully, they concluded, "this fortunate situation may be expected to continue."[3]

THE PETROLEUM DEPARTMENT ERROR

Holmes and Janson took the Kuwait concession documentation along to the Colonial Office. They pointed out that the shaikh of Kuwait was keen to finalize and also pointed out that the documentation included, exactly, the conditions that had been approved just months before as the nationality clause in the Bahrain concession.

Neither Holmes nor Janson was aware of the decision just circulated that confirmed the Indian government, and its local officials, in their authority over

Kuwait. Perhaps because of the success of the Bahrain negotiations conducted with the Colonial Office, Holmes and Janson had forgotten the extent of the paranoia of the officials of the Government of India in the Persian Gulf. They were about to be sharply reminded as the Colonial Office replied that before the matter could be discussed, comment needed to be obtained from "interested departments" and the responsible officers of the Indian Political Service in the Persian Gulf. There was plenty of comment to go around.

Biscoe's abhorrence at the possibility of interference in the authority of the Government of India over the Persian Gulf, as he believed would occur with the entry of Americans, coalesced into a personal animosity toward Frank Holmes. He seemed to think if Frank Holmes could be contained, the Americans would be curtailed along with him.

He drew the colonial secretary's attention to the fact that "Major Holmes proposes to be chief local representative at Kuwait as well as at Bahrain. I do not see how this is possible." He pressed his case, saying: "The intention of appointing him as Chief Local Representative at Bahrain was to have someone, more or less permanently on the spot, with whom any questions that arose could be at once discussed. It would seem impossible for Major Holmes to fulfil this function in two separate places."

He said he had gained the impression from Holmes that it was "by no means his intention to reside permanently at Bahrain." Biscoe wanted the Colonial Office to institute "some arrangement under which Holmes should reside for some stipulated period, say nine months in each year, at Bahrain."

The India Office took up this theme. The Colonial Office, which had insisted on Holmes's appointment in the first place, was cool to the request. Their thoughts were, they said, "not to attach too much importance to the Resident's objections to Holmes being the local representative in Kuwait as well as in Bahrain in the event of a concession in Kuwait being granted to the present applicants. We feel that, if he spent most of the year in either one place or the other, he could probably fulfil his duties satisfactorily."[4]

The Petroleum Department of the Board of Trade joined the fray. After all, oil concessions should be their business more than that of the Government of India or the Colonial Office, although they seemed to have a very poor grasp indeed of the facts.

They ran through what they thought was the background and noted: "At one time Anglo-Persian were interested in Bahrain and Kuwait but it is understood that their geologists have now condemned practically the whole of the mainland of Arabia." From correspondence in their files, they noted that the Iraq Petroleum Company was not inclined toward consideration of either Bahrain or Kuwait.

The Petroleum Department now set in motion a serious error that would color official opinion for months to come. They did not seem aware of the existence of the Gulf Oil Corporation and simply assumed that all Holmes's interests were destined to be transferred to the Standard Oil Company of California, as had happened with Bahrain. They commented: "If the other concessions in which E&GSynd are interested on the Arabian coast are completed and transferred, the Standard Oil Company of California through their fully-owned Bahrain Petroleum Company will control a very considerable area in Arabia on the shores of the Persian Gulf."

This might not be all bad, they surmised, because "the investigations of Anglo-Persian have indicated that the prospects of finding oil are not very favourable." Encouraging the American-owned Bahrain Petroleum Company to undertake "as much active prospecting as they are prepared to do" presented the tempting possibility of seeing the Americans waste their time and money. Dealing specifically with the extension of the Bahrain concession area requested by Holmes, the Petroleum Department wanted Standard Oil of California told that "this will be considered when they can show they have carried out satisfactory development work on the property they already have." The Petroleum Department considered this to be a most satisfactory arrangement because, if the extension was withheld, "it may then be possible to secure some share to British companies if oil is discovered in commercial quantities in Bahrain."[5]

ANGLO-PERSIAN OFFERED KUWAIT

Having dealt with Bahrain, the Petroleum Department said it hoped Anglo-Persian could be persuaded to take on Kuwait. They wrote to Sir John Cadman, telling him that E&GSynd had applied to close an oil concession with the shaikh of Kuwait and added, mistakenly, "and to transfer their rights to a Canadian Company in which the bulk of the capital will be held by Standard Oil of California." As they already had Bahrain, they said, "before recommending the grant of any further areas to American interests we should like to know definitely whether there is any prospect of British interests being disposed to assist in prospecting for oil in Kuwait."

They asked Cadman: "Do you consider there is any chance of either Anglo-Persian or Iraq Petroleum being prepared to undertake exploration for oil in this area? Or have your geological advisers considered this district and turned it down on the ground that prospects are too unfavourable to make it worth carrying out any test drilling?"

They added: "The Americans appear to be willing to risk some money and we do not like to see any area which offers any promise going entirely into American hands." For the Petroleum Department of the British government, they showed remarkable ignorance when they inquired of Cadman: "Perhaps you could say whether Standard Oil of California is in any way linked with the Standard Oil companies who at present have shareholdings in the Iraq Petroleum Company?"

Cadman seemed to be almost as badly informed as was the British Petroleum Department. His latest information was, he said, "Frank Holmes left
Kuwait in 1928 after the breakdown of negotiations with the Shaikh." Cadman
claimed that Anglo-Persian had been interested in Bahrain and Kuwait but had
been a victim of Colonial Office's "unwillingness to recommend modification of
terms" offered by rival E&GSynd "such as to suit our views . . . including lower
royalties." He said that Holmes and his syndicate had been "ready to accept obligations to which we objected, and to outbid us on other points."

Employing Anglo-Persian's habitual line of slanderously discrediting its
competitors, Cadman charged that "E&GSynd's objective was not our objective.
What they wanted were concessions with which to traffic. Ours was to obtain a
concession for honest development should oil be found to exist." He warned the
Petroleum Department: "They appear to have got away with the Bahrain concession. I do not doubt their capacity, provided they bid sufficiently high, to do the
same in Kuwait."

John Cadman knew the value of discovering the detail of the competition's
business. He said that for Bahrain and Kuwait "very little detailed geological
work has been done and reports leave little room for optimism" and that the
search for oil was "a very long shot" on which expenditure would be futile.

But, before he could "answer specifically your question as to whether
Anglo-Persian is prepared to reconsider its own position in this matter," Cadman
said he wanted to know "the terms that E&GSynd is prepared to offer." And
while they were at it, Cadman asked, "also let me know the terms on which the
concession in Bahrain was actually concluded." The Petroleum Department
obligingly forwarded this confidential information.

Like the Petroleum Department, Dickson was hoping Anglo-Persian or the
Anglo-Persian-dominated Iraq Petroleum Company would step in and save the
day by taking on the Kuwait concession. Any such hope was well and truly
dashed. After receiving the confidential information about Holmes's agreements,
Anglo-Persian advised the Petroleum Department that they had decided "not to
take any further steps in the matter." Cadman said the geological information he
possessed on the prospects for Kuwait was that "the country appears to be
entirely covered with alluvium and rocks of recent geological age, which would
involve drilling of a very speculative type."[6]

DICKSON SPIES FOR WILSON

Dickson was particularly disappointed because he had submitted the results of
his "prowling around the state trying to find out what has made Holmes so keen
on getting an oil concession from the Shaikh of Kuwait" to Arnold Wilson at
Anglo-Persian in London. He had urged the resident to also send a copy to the

executives at Anglo-Persian. Dickson's sleuthing consisted of an attempt to visit all the places where he thought Holmes had shown interest.

Holmes had been to the Burgan Hills and examined the springs, Dickson said, so he went there. He claimed to have found "a most interesting spring of pure bitumen or tar oozing out of the ground in the crater like depression . . . though the area is now heavily covered with a deposit of sand. I brought back samples." Holmes had also examined Wara Hill, ten miles north of Burgan, so Dickson also went there and brought back a "piece of stone covered with Bitumen deposit." In Bahra, an area due north of Kuwait Town, where Holmes had sent his assistant engineer/driller, Dickson said he came across "two obviously recently excavated holes in the ground out of which crude oil was slowly oozing."

It was significant, Dickson said, that "the so called water bore that Holmes put down near the shaikh's palace lay exactly on the straight line drawn between the oil spring mentioned above and the Burgan hill to the south." Dickson said "an informant" had told him that from this well "samples of some kind of black deposit of the consistency of putty, and strongly impregnated with crude oil were brought to the surface." He said that Holmes's assistant engineer had told someone, who had told him, that "the oil bearing zone moved in a westerly direction from Anglo-Persian's main fields at Maidan-i-Naphtun and that the most likely place to find it in Kuwait territory was towards the south west."

With his newfound expertise in oil geology, Dickson told Arnold Wilson and the political resident: "It is quite patent to me that there are strong 'surface' oil indications in Kuwait. But whether the 'srata' lies the right way is another matter: Holmes quite obviously thinks it does—or he would not be so anxious to get his agreement through."[7]

After receiving the Petroleum Department's error confusing the Gulf Oil Corporation with Standard Oil of California, a secret memorandum from the Political Department of the Colonial Office was circulated. It noted: "The Standard Oil Company of California will ultimately come to control a very considerable area in Arabia if the other concessions in which E&GSynd are interested on the Arabian coast (a concession in Kuwait and concessions which may properly be regarded as having lapsed in Al Hasa and the Neutral Zone) are completed, or revived, and transferred to the Bahrain Petroleum Company." It warned: "The difficulty of dealing with a wealthy and powerful corporation such as Standard Oil of California, if it once establishes itself firmly on the Arabian coast, is a strong argument for giving as little away as possible." A comment on the memo noted: "The Anglo-Persian Oil Company does not think the probability of oil in Arabia is very good. This is satisfactory from the political point of view."[8]

TAKING THE SHAIKH'S NAME IN VAIN

The resident, Hugh Biscoe, made a personal visit to Kuwait. Dickson's plan to "require Shaikh Ahmad to get Holmes to insert our clauses" when he asked the political agent for advice had now transformed into a claim by Dickson that Ahmad was "extremely emphatic" that he personally required an appropriate "British Nationality Clause."

Repeatedly assured by Dickson that this was so, Biscoe confidently reported to Lord Passfield that "under no circumstances whatsoever" would the shaikh of Kuwait grant a concession to any concern that "was not entirely and altogether British." He said that "it is clear he will not grant the concession to Holmes and his syndicate if there is any risk of it being transferred by them to a foreign concern." Relying on Dickson, Biscoe said that Shaikh Ahmad was "frankly alarmed at the idea of a large and powerful syndicate being established in his territory."

Biscoe was so sure Shaikh Ahmad would "not commit himself to any conditions that might be detrimental to British interests" that he advised there was no problem in allowing Frank Holmes to discuss the concession on the understanding that "any agreement arrived at would be subject to the approval of His Majesty's government." Biscoe confidently predicted: "I personally consider it unlikely that any agreement will be arrived at, unless of course, E&GSynd definitely sever their connection with their American associates and considerably modify the terms contained in their documentation."

This report from the political resident, supported by statements from Dickson, now formed the basis of insistence that it was the shaikh of Kuwait himself who was demanding the insertion of an exclusive British nationality clause. Prompted by the India Office, the Colonial Office wrote to E&GSynd:

> It has now been ascertained that the Shaikh of Kuwait has definitely refused to grant a concession to any concern that is not entirely British. He will insist on the inclusion of a "British Control" clause in any concession that he might grant in respect of his territory. In the light of this information, your syndicate's application has been carefully considered, in consultation with the other interested Departments and the Government of India. We now inform you that His Majesty's government are not prepared to advise the Shaikh to reconsider his attitude in this matter.[9]

While the Colonial Office acceded to Government of India prodding, it does not appear to have been wholeheartedly in agreement. Holmes had been in Bahrain and Iraq. When he returned again to London and expressed his surprise—and frank disbelief—that this was Shaikh Ahmad's attitude, the Colonial Office suggested that perhaps he could be induced to "change his mind." They suggested Holmes return to Kuwait.

To Dickson's disgust, British officials in Baghdad "endorsed Holmes's passport for as many journeys as he chooses to make to Kuwait during the year." Dickson couldn't stand it. Holmes, he reported to Biscoe, arrived in Kuwait on April 16, 1931, "with a flourish of trumpets informing the shaikh by wire from different places en route that he was coming." Dickson was not mollified by Holmes's explanation that he had come to Kuwait to close down his water boring plant and ship it to Bahrain. "In reality the plant is an oil boring machine," Dickson hissed to Biscoe. He was furious at Ahmad's obvious delight at seeing Holmes again and his praise for the generosity of a "most unexpected" ten thousand rupees that Holmes brought him as "compensation for not finding water."

Almost in passing, Ahmad told Dickson that he now had two draft oil agreements from Holmes to consider "but had not had time to see them." Dickson could only contain himself for a week before asking Ahmad to let him read the papers. As he had suspected, he wrote to Biscoe, neither set of documents contained an exclusive British nationality clause. The newer version did have a clause similar to Bahrain's, but Dickson decreed this as "in no way meeting the case. It is to my mind even more objectionable than the one before, in the 1930 documentation."

While Dickson continued to claim, as did Biscoe, that Shaikh Ahmad would "not look at" any agreement unless it contained an exclusive British nationality clause, he does seem to have had some inkling that he was deluding himself. He was due to go on leave, and wrote the resident that he was "filled with concern" at the prospect of Holmes being in Kuwait while he was away. Dickson said Holmes would "probably start a fresh 'attack' on the shaikh over his precious oil concession." He was sure Holmes hoped to "push his latest draft though by judicious palm-oiling of certain high dignitaries here, after I am gone." Lamely, Dickson added, "I do not, of course, believe the shaikh will succumb to such an attack. I have far too much confidence in his sense."

Holmes stayed three days. Before departing for Bahrain he informed Dickson that he was coming back in about fifteen days and would "see what he could do about creating a special nationality clause of his own that would suit any qualms the shaikh might have." Knowing the claim that Shaikh Ahmad was demanding an exclusive British nationality clause was emanating from him, and through him from Biscoe, Dickson alleged, "Holmes intention must be to persuade the shaikh to waive this demand 'by hook or by crook.'"

Dickson was beside himself. "I confess I cannot understand the man's goings on and consider him dangerous," he told Biscoe. It was unforgivable, Dickson said, that "he seems to take pleasure in running down all British government officials and especially Political Officers of the Government of India." According to a hostile Dickson, Holmes implied that his offers to the Arabs were being obstructed and he "tries to make out we are all rogues and rascals trying to prevent honest men doing business with, and bringing grist to the pockets of, the poor Arabs."[10]

CRANE RETURNS TO JEDDA

The American philanthropist Charles Crane had again visited Jedda, from February 25 to March 3. This time, primed by Philby on the economic benefits that might be elicited from this multimillionaire with his bathroom-fitting fortune, Abdul Aziz bin Saud himself played host.[11]

Crane was feted as guest of honor at dinner parties at the palace and a two-hundred-person banquet mounted by the Jedda Municipality. "Every conceivable topic of economic interest was freely discussed," Philby reported, adding that the Jedda Municipality had identified the primary need of "providing the town with a piped water supply." The Municipality was not disappointed. Before departure, Crane said he would instruct the American engineer he had sent to Yemen, Karl Twitchell, to visit on the way home and advise on the question of underground water in the neighborhood of Jedda. Crane was enthusiastic about a search for water as he sincerely believed that agricultural development was the necessary first step for any country attempting to improve the lot of its citizens.

He also believed it was absolutely imperative to institute a program of free education. Crane had educated a number of Arab youths including Rashid Zok, "a young Muslim of a good Syrian family" whom Crane had sponsored to the United States where he trained in modern methods of dry agriculture. Zok had studied six years in America at Crane's expense and was presently working for him in Yemen. Crane said he would also send Zok to Jedda. On his departure, bin Saud showered Crane with gifts including assorted Persian carpets, daggers, and swords. Crane accepted two mares from the six Arab pedigree horses offered.

Within days, Crane cabled bin Saud. "Greetings, I am taking the liberty of sending Twitchell to test drill for water (and gold gravel) subject to your approval but entirely at my expense." Philby had tried to channel the discussions into persuading bin Saud that awarding concessions might be a good thing. In response to Philby's not-so-veiled references, Crane added: "So, I am asking you defer any general development program for approximately six months because I do not wish to be connected in any way with any subsequent commercial applications. I am writing you fully. Twitchell will arrive October with an Empire drill. Salaams. Charles Crane."[12]

START UP IN BAHRAIN

In Bahrain, Holmes was flush with success. From Kuwait, via Basra, he had arrived with five Americans in tow from Standard Oil of California. Led by Edward Alan Skinner, this first group included a rig builder, a driller, and an accountant. Five more Americans soon followed. About a thousand tons of equipment arrived in a succession of steamers that were unloaded by a crane capable of lifting five tons. Holmes had constructed the crane earlier in the year at the end of the Bahrain pier. He had put together a construction gang from Iraq and had hired Iraqi rig hands and Sikhs to run the powerhouse. Some four hundred men could eventually be employed, he told the Bahrain political agent.

Holmes had got to this point almost through sheer willpower. At Standard Oil's San Francisco headquarters, Lombardi was holding out. He wanted to be sure he could get his team from Bahrain to Holmes's other concessions. In January he wrote: "Have been awaiting permission to visit the mainland before selecting the machinery for drilling on Bahrain. See my letters of November 6th and December 30th." He told Holmes he planned to "send a man to Bahrain . . . just as soon as we are sure he can get on the mainland."

Standard Oil of California was doggedly maintaining the fiction that they wanted to examine the mainland purely for its geological implications on Bahrain. As it was Gulf Oil that held the option with E&GSynd on the old Al Hasa concession through its second agreement of 1927, the Mainland Agreement, Standard Oil of California had no claim to either this concession or to Holmes's time and effort in relation to Al Hasa or Kuwait.

In fact, their interest in these concessions was a direct contravention of their understanding with the Gulf Oil Corporation. Standard Oil of California's contract was solely for the Bahrain concession. The one-third of Holmes's salary they contributed was to cover his expertise in development of the Bahrain concession only. Nevertheless, Lombardi of Standard Oil heeded Holmes's many messages that the shaikh of Bahrain was "impatient to have work under way at once," although, Lombardi noted, "our contracts do not require this." He wondered if "the drilling of a water well and some odd jobs this year" would satisfy Shaikh Hamad of Bahrain. But Holmes persisted and he had pulled it off. Now, at last, he had something to show.

Even the Bahrain political agent was impressed and reported:

> The camp is fitted with everything that practical comfort could suggest. Refrigeration, electric light and heaters, full length baths, hot and cold showers and water borne sanitation are apparently a *sine qua non* for Americans on "location" although they would be considered luxuries by most British Officials in India. The organisation of the oil company is remarkable. Of the immense quantity of stores required nothing was mislaid and everything arrived on time. With the exception of the Nissin huts which are made in England, and with which they are dissatisfied, all their requirements from machinery to lead pencils have been imported from the United States. Except for Major Holmes all the operating staff are Americans and the rig hands are mostly Iraqis. In fact, except for the Sikhs in the Powerhouse, all are non-British subjects. The American employees that have been sent so far proved to be of a good type and have given little trouble.[13]

With Bahrain moving along at a steady clip, Holmes headed back to Kuwait at the end of June. He needed something in writing to take back to the Colonial Office with which to demonstrate that it was not the shaikh of Kuwait insisting on an exclusive British nationality clause as claimed by the political officers of the Government of India. He asked Shaikh Ahmad to give him something "emphasising your willingness and approval of my application for an oil concession . . . as you recently promised me during our verbal discussions."

On July 2, 1931, Shaikh Ahmad addressed a letter to Holmes that read: "The draft of the conditions that you presented for an oil concession in the territories of Kuwait has been considered by us. We have mentioned to you in the course of our conversations the clauses which His Majesty's government has indicated to you that their incorporation in an oil concession is necessary." Ever mindful that the independence of his country depended on his "obedience" to the British, the shaikh inserted a careful statement: "Because I believe that His Majesty's government is my own sincere government, which is always evoking its care to the welfare of my country and the safeguard of my rights, we must not ignore what it considers useful to us and to our country."

Having got that out of the way, Shaikh Ahmad's final paragraph seemed to meet Holmes's needs: "If you and your company agree with His Majesty's government on the said clauses mentioned to you, and it allows you to omit them, then we shall have another opportunity of discussing matters with you."

Before delivering this letter to Holmes, the shaikh prudently sent it for vetting by the acting political agent who pronounced it just fine. "You ask for my opinion on your letter," he wrote the shaikh of Kuwait. "Your Excellency has explained in your letter to Major Holmes, in language so simple and to the point, that I could not add anything to make it better."

Breaking his journey in Baghdad in order to get the letter officially translated by his Arab legal office, Holmes returned to London armed with the evidence that the Colonial Office needed to face down the claims of the political officers of the Government of India.[14]

A SPOKE IN DICKSON'S WHEEL

Before Holmes even left Baghdad, Anglo-Persian had a copy of the shaikh of Kuwait's letter. A copy also preceded him to the Colonial Office, and to the Government of India, accompanied by a "confidential" message from Maj. T. C. Fowle, the acting political resident. Although Fowle was merely standing in while Biscoe was on vacation, he recognized the significance of this letter. He commented, "the attitude hitherto maintained" has been that the shaikh of Kuwait "insisted on the inclusion of the 'British Control' clause and that HMG were not prepared to advise him to reconsider this attitude." The shaikh has "retreated from the position he has hitherto maintained," Fowle said, "and has now thrown on HMG the onus of insistence on the inclusion of a 'British Control' clause." Perceptively, Fowle concluded: "It will probably be difficult therefore for His Majesty's government to maintain vis-a-vis Holmes's syndicate the attitude hitherto adopted."

E&GSynd requested the Colonial Office to notify the resident that the concession could now be granted. The shaikh of Kuwait's letter, they pointed out, "clearly establishes his willingness to omit an exclusive 'British Nationality' clause from the concession that he is willing to grant E&GSynd, provided HMG will agree to omission of same." They stated that Shaikh Ahmad understood the concession was to be developed through American interests and that, in the matter of protective clauses, conditions similar to those approved by the Colonial Office for the Bahrain Concession would be included. They passed on Holmes's message that Shaikh Ahmad was trying to meet the "pressing demands of his people" to develop "without delay" the natural resources of Kuwait because of "the severe trade depression and unemployment."[15]

Both Dickson and Biscoe were home on leave. The Colonial Office called them in. Dickson went for bluster. He assured the meeting that the shaikh's written word should be discounted and that he would surely be "vexed" if he knew of "the interpretation" now being put on it. He said that he, Dickson, knew what the shaikh of Kuwait was really trying to say. Dickson said it was obvious the shaikh had just wanted to get rid of Holmes. What the shaikh of Kuwait had really meant, Dickson claimed, was not what he had written in black and white, but, "I am sick of being bothered by you, so go and talk to the British government, and if you get them to agree to anything, come back and we will see about it." Anyway, there was no real problem Dickson later guaranteed the India Office, because "the shaikh will certainly be prepared to eat any words that he may have uttered . . . *if we want him to do so.*"

Anglo-Persian was watching this new development with interest. They were not as confident as Dickson that there would be no difficulty bringing Shaikh Ahmad to heel. They saw the possibility that, despite the Government of India's best efforts to the contrary, Holmes might actually get the Kuwait concession. A stalling action seemed called for. So soon after their latest dismissal of Kuwait as a place where the search for oil would be "a very long shot" on which expenditure would be futile, they needed a good excuse. They approached the Petroleum Department, saying: "Recent borings on the Persian coast south of Bushire give us some reason to think that prospects of the existence of petroleum in Kuwait territory are perhaps less remote than have hitherto appeared to be the case."

What they wanted was permission to send a "small party of geologists" to Kuwait to "make a thorough examination of the surface geology." This reconnaissance, they pointed out, "would take several months," and they assumed the shaikh of Kuwait would "give us every facility without requiring any payment from us." Obligingly, the Petroleum Department advised the Colonial Office that "Anglo-Persian should be given every encouragement and all facilities that can reasonably be afforded them."

Dickson was overjoyed and wrote to Arnold Wilson. "You can rely on me to try and get the Shaikh of Kuwait to hold up things where Holmes and his crowd are concerned pending your people coming to a decision. Between ourselves I feel confident of doing this," he assured Wilson. Further pledging to use his official position in support of a purely commercial company, Dickson confided to Wilson: "Let me know if ever I can be of further assistance. I am very imperialistic and cannot stand the thought of an American concern getting in here."[16]

DICKSON "MAKES IT CLEAR"

When he returned to Kuwait, Dickson set about attempting to validate his misleading claim on which the resident had relied, and subsequently found himself embarrassed at the Colonial Office meeting. Everything he had surmised was indeed correct, he assured Biscoe. He said that Holmes had worried Shaikh Ahmad "so often and so long" that "in order to get rid of him gracefully" he had written the letter. Ahmad had told him, Dickson claimed, that "Arab politeness" forbade him "turning Holmes down."

He explained, "Although one cannot entirely excuse Shaikh Ahmad of weakness and vacillation in this matter, he really is a simple Bedouin at heart, and there is some excuse for his having succumbed to the continuous and direct attack of a man of Holmes's calibre." Dickson understood all this, he said, because "knowing the shaikh, as I do, he is easily swayed and not possessed of much strength of character." He said that Shaikh Ahmad's "simple faith in us is pathetically sincere and he can hardly be expected to appreciate the finer points of our diplomacy which apparently requires that he, the shaikh himself, should insist on the necessity of an exclusive 'British Nationality' clause." However, it would be "far too risky to leave it at this." It must be made quite clear to the shaikh of Kuwait, Dickson advised, that "we *expect* him to insist on the insertion of an exclusive 'British Nationality' clause."

It grated badly with Dickson that Holmes did not afford him the pomp and prestige he believed to be his right. For Biscoe's edification, Dickson pontificated on the character of Frank Holmes. "Major Holmes is rather an unscrupulous person with persistency his strong suite. He has an unpleasant habit of ignoring the local Political Agent and trying to short circuit him on every occasion possible. This more American than English individual, and the cavalier way in which he has always treated this agency, are objectionable to say the least. It is still more objectionable to see the way in which he has settled down and taken a house in Kuwait, as if he already had definite rights in this principality," he complained.

"Perhaps his most unpleasant habit is his habit of running down British Offi-cials in general and in particular of holding up the Officials of the Government of India to ridicule. If he says the things he does in my presence, one can only suppose he is not very careful of what he says in front of the Shaikh or the Arabs of the town." Dickson was dreaming of having this disrespectful individual removed from his fiefdom.

He advised Biscoe: "The time may come when the question of granting Major Holmes a British visa for Kuwait may have to be reconsidered. In dealing with adventurers of the Holmes type, a sort of 'Monroe' doctrine for the Persian Gulf would seem to be the sort of thing that is required."[17]

Dickson heard that Holmes was on his way to Kuwait "and has probably now reached Baghdad." He marched off to see Shaikh Ahmad for the purpose of "making it quite clear to him." He told Ahmad he was sympathetic because he knew it was not his fault. It was obvious that Holmes was too smart for him. After all, Holmes had waited until he, Dickson, was away to "some how or the other get around the shaikh" and in doing so had also "cleverly bluffed" the acting political agent to "acquiesce in the wording of the shaikh's letter."

Dickson said that Biscoe had three times written to the High Government assuring them that Shaikh Ahmad had "in most vigorous terms" declared that he would never grant any concession to Holmes without an exclusive British nation-ality clause. He warned Ahmad that his "apparent *volte face*" was annoying the High Government. He told him that "it had suited their purpose that the shaikh himself should be the person to tell Holmes that an exclusive 'British Nationality' clause *must* be inserted." He said that the shaikh now having "thrown the burden of saying this on the High Government . . . puts them in an awkward position."

Dickson pushed on to a not-so-veiled threat. He referred to the "efforts and intrigues" of Shaikh Ahmad's neighbors. He reminded him that bin Saud's trade blockade and Iraq's tax on his date gardens were matters still unresolved. Mean-ingfully, Dickson said, "Therefore it would be the greatest pity if the shaikh, in any way, now alienated the High Government's sympathy and support by opposing their well known wishes and doing anything foolish in the matter of Holmes and his syndicate."

But all was not lost, Dickson told Shaikh Ahmad. He could "clear up all the misunderstanding and suspicion in the minds of everyone concerned." All he had to do was tell Holmes immediately that he would have nothing to do with any con-cession unless the exclusive British nationality clause was inserted. Speed was essential, Dickson said, because Biscoe would arrive shortly and "his first thing will be to ask the shaikh about his apparent change of mind." Dickson said it was so important "in his own interests" for Shaikh Ahmad to "clear" himself in the res-ident's eyes that he must take immediate action, before the resident arrived.

Dickson said he would help Ahmad by writing two letters. The first would be from him, "requesting the shaikh's assurance in writing that he understood the

High Government's desires and intentions in the matter of any concession granted to Holmes and his syndicate and that he would definitely insist on an exclusive 'British Nationality' clause being inserted." Ahmad was to reply with the second letter, to be written by Dickson "in suitable terms." Alternatively, the shaikh could wait for Biscoe and "verbally give him this same assurance." He must do one or the other, Dickson told the ruler of Kuwait, in order to "recover all lost ground and obtain the High Government's support and approval. Only in this way would he be able to clear up the misconception that had been created."

Anxious to rehabilitate his own reputation, Dickson claimed in a "Very Confidential" report to Biscoe that Shaikh Ahmad had told him it had all been a "stupid misunderstanding" and his letter had been "a mistake." The shaikh "had really wanted to gain time," Dickson told Biscoe, but he "has in no way altered his views about the exclusive 'British Nationality' clause." Dickson claimed Ahmad had told him that "his honour and the interests of his State were beyond purchase" and that if Holmes did not agree to insert the clause "he would ask Major Holmes to leave his State." Completely carried away by his own version of events, Dickson did not seem at all suspicious of Ahmad's refusal of the two ready-written letters. The shaikh of Kuwait said he preferred to wait until the resident arrived.[18]

NOTES

1. Administrative Report for Kuwait, 1931 (Dickson), "to force Kuwait to acknowledge (bin Saud) as suzerain"; see, for example, India Office Library, R/15/1/638 D73, "Confidential," political resident (Biscoe) to Lord Passfield, colonial secretary; "In my despatch of April 30, 1930, I expressed the view that on strategic grounds it was desirable to maintain the independence of Kuwait." But, this view was opposed from Iraq, see Yossef Bilovich, "Great Power and Corporate Rivalry in Kuwait, 1912–1934: A Study in International Politics" (PhD diss., University of London, 1982), p. 185, citing Colonial Office memo of August 1930 that quotes July 24, 1930, acting high commissioner Iraq, "encouraging the absorption of Kuwait by Iraq."

2. Violet Dickson, *Forty Years in Kuwait* (London: Allen and Unwin, 1971), p. 66, Violet says her husband was given the job of secretary to the political resident in the Persian Gulf in 1928 through "the good offices" of his then employer, the Maharajah of Bikaner; Zahra Freeth, *Kuwait Was My Home* (London: Allen and Unwin, 1956), p. 11, "in 1929 . . . Europeans and Americans numbered only eleven." Zahra Freeth is Dickson's daughter.

3. IOL/R/15/5/238, vol. 111, September 22, 1930, "Secret," political agent Kuwait (Dickson) to political resident (Biscoe) contains Dickson's plans for military occupation of Kuwait and "Gibraltar of the Gulf" scheme; IOL/R/15/1/638 D73, September 25, 1930, "Personal," political agent Kuwait (Dickson) to resident (Biscoe)," require Shaikh Ahmad to get Holmes to insert our clauses"; and September 30, 1930, confidential, resident (Biscoe) to Lord Passfield, colonial secretary, "just keep insisting" and "on strategic grounds . . . maintain the independence of Kuwait . . ."; Bilovich, "Great Power and Cor-

porate Rivalry in Kuwait," p. 185, quoting Colonial Office memo of August 1930, "this fortunate situation."

4. IOL/R/15/1/638 D73, October 6, 1930, confidential, resident (Biscoe) to colonial secretary (Lord Passfield) cc Government of India, political agent Kuwait, in which the resident advocates that Holmes be confined to Bahrain, "say nine months in each year"; IOL/L/PS/10.10R 13188.9, vol. 993, November 12, 1930, "Private," K. W. Blaxter at the Colonial Office to J. G. Laithwaite at the India Office, "not to attach too much importance to the Resident's objections."

5. IOL/L/PS/10/R1388.9, vol. 993, October 11, 1930, Petroleum Department/ Mines Department (F. C. Starling) to undersecretary, Colonial Office, "Standard Oil of California, through the Bahrain Petroleum Company, would control a very considerable area in Arabia on the shores of the Persian Gulf"; note that Dickson did recognize that there were two different American companies involved, but appears to have chosen not to correct the Petroleum Department, or other officials, see Administrative Report Kuwait 1930, "Holmes left for UK 20/8 via Iraq. He stated before he left that he hoped to arrange things to the satisfaction of HMG when he got to London, though he admitted the American 'Gulf' group, headed by Mr. Mellon, would be disappointed at the turn of events (Nationality Clause)."

6. Archibald H. T. Chisholm, *The First Kuwait Oil Concession Agreement: A Record of the Negotiations, 1911–1934* (London: F. Cass, 1975), p. 127, October 18, 1930, H. W. Cole, Petroleum Department to Sir John Cadman, "like to know definitely whether there is any prospect of British interests being disposed to assist in prospecting for oil in Kuwait" and October 21, 1930, Sir John Cadman to H. W. Cole, Petroleum Department, "also let me know the terms on which the concession in Bahrain was actually concluded"; IOL/R/15/5/238, vol. 111, "Confidential," Colonial Office (J. E. W. Flood) to political resident (Biscoe), "No objection was felt to supplying Sir John with a copy of the Bahrain concession . . . and this was done. He was also advised that the proposed Kuwait concession was on generally similar lines, with variations as to payment." Chisholm, p. 128, January 20, 1931, Petroleum Department, to A. C. Hearn of Anglo-Persian, "not take any further steps in the matter" and "drilling of a very speculative type."

7. IOL/R/15/1/638 D73, January 12, 1931, "Confidential," political agent Kuwait (Dickson) to political resident (Biscoe) is Dickson's report of going "everywhere Holmes has been." The copy to Arnold Wilson is IOL/R/15/5/238, vol. 111, September 25, 1931, Dickson to Wilson with a further reference by Dickson to having forwarded this report to Wilson in IOL/R/15/5/239, vol. IV.

8. IOL/L/PS/10.10R 1388.9, vol. 993, October 29, 1930, minute, paper, "Secret," Political Department and also IOL/R15/5/238, vol. 111, Colonial Office memo (J. E. W. Flood), "Standard Oil of California will ultimately come to control a very considerable area in Arabia", see also Bilovich, "Great Power and Corporate Rivalry in Kuwait," p. 173, citing Colonial Office minute of November 6, 1930, "satisfactory from the political point of view."

9. IOL/R/15/1/638 D73, November 9, 1930, "Confidential," political resident (Biscoe) to Lord Passfield, colonial secretary "extremely emphatic" and "frankly alarmed" and "unlikely any agreement will be arrived at"; Chisholm, *First Kuwait Oil Concession Agreement*, p. 17, January 31, 1931, Colonial Office to E&GSynd, "HMG not prepared to advise the shaikh to reconsider his attitude in this matter."

10. Note that Chisholm, *First Kuwait Oil Concession Agreement*, p. 18, states

Biscoe returned on April 28, 1931, to advise the shaikh of Kuwait not to grant an oil concession to Holmes's company, but Shaikh Ahmad had replied that "he had promised a concession to Major Holmes and desired to carry out his promise." If this is so, the resident seems not to have informed the British officials. IOL/R/15/1/638 D73, May 21, 1931, and IOL/R/15/5/238, vol. 111, May 25, 1931, both letters, "Confidential," political agent Kuwait (Dickson) to political resident (Biscoe), "Holmes intention must be to persuade the sheikh . . . by hook or by crook" and "tries to make out we are all rogues and rascals."

11. Mostly discounted by serious scholars is Philby's tale of a 1931 car ride with bin Saud during which Philby, apparently, raised the possibility of concessions. According to Philby he quoted the Koran and remarked that bin Saud's people were "like folk asleep on buried treasure." Philby says bin Saud beseeched him, "Oh Philby, if anyone would offer me a million pounds I would give him all the concessions he wanted." Philby says he replied that Charles Crane, who had "helped" the Yemen, was currently in Cairo and could also "help" bin Saud. "He is a very rich man with important contacts in the American industrial community." Philby says he "guaranteed" to bring Crane to bin Saud. Philby first tells this story in *Arabian Days, An Autobiography* (London: R. Hale, 1948), then extends it in every following book. Philby's biographer, Elizabeth Monroe, *Philby of Arabia* (London: Faber and Faber, 1973), p. 203, doubts it and remarks, "whether the invitation was prompted by Philby, as he claims . . ."; Robert Lacey, *The Kingdom* (London: Hutchinson, 1981), p. 226, says Crane was visiting Shaikh Fawzan al Sabik, bin Saud's representative in Cairo, seeking to purchase a pedigree Arab horse to mate with "Saudia," bin Saud's gift in 1926. Lacey says after Al Sabik telegraphed Mecca bin Saud issued an invitation to Crane. In the Charles Crane Papers, his Arab traveling companion and interpreter, George Antonious, who kept a diary of this 1931 visit writes, "bin Saud received us the same day we arrived, February 25th. Crane said how anxious he was to meet bin Saud these last years. On his return from China, he had found bin Saud's invitation and had gladly availed himself of it."

12. H. St. J. B. Philby, *Arabian Jubilee* (London: Hale, 1952), p. 164, "a project for supplying Jedda Municipality with a piped water supply"; note Charles Crane Papers, Antonious diary, p. 17, Antonious comments acidly on Philby's presence, "As he styles himself since his recent conversion 'Shaikh Abdulla Philby.'" Note that in his book and writings on his time in the Yemen and Saudi Arabia, Karl Twitchell never once credits Rashid Zok. Twitchell Papers (uncataloged), cable from Charles Crane to bin Saud, March 1931, "sending Twitchell to test drill for water."

13. Administrative Report for Bahrain for the Year 1931 (Prior), "Some 400 men could eventually be employed"; Chevron Archives, box 120797, January 12, 1931, Lombardi in San Francisco to Frank Holmes in London, "would some odd jobs this year satisfy the Shaikh"; Administrative Report for Bahrain, 1931 (Prior), "would be considered luxuries by most British Officials in India."

14. IOL/R/15/5/239, vol. IV, June 28, 1931, Holmes in Kuwait to shaikh of Kuwait, "emphasising your willingness and approval of my application for an oil concession"; and July 2, 1931, shaikh of Kuwait to Frank Holmes in Kuwait, "if you and your company agree with HMG on the said clauses mentioned to you, and it allows you to omit them . . ."; and IOL/R/15/1/639 D78, July 2, 1931, "Confidential Letter" from acting political Agent (A. L. Greenway) to shaikh of Kuwait, "language so simple and to the point, that I could not add anything to make it better."

15. Chisholm, *First Kuwait Oil Concession Agreement*, p. 18, a copy of the shaikh's letter was forwarded from Anglo-Persian's Kuwait agent to the general manager in Abadan and on to their London office; IOL/R/15/5/239, vol. IV, July 7, 1931, acting resident (Fowle) to Lord Passfield, colonial secretary, copies to Government of India and political agent Kuwait. Fowle had also reported on June 18 that Holmes had recently visited Farsi Island and "appears to be taking an interest in that group of islands. . . . The ownership of this group of islands . . . is described as being undetermined" and enquiries made by the high commissioner Iraq in 1928 failed to trace the history of an Arab claim to the slands"; IOL/R/15/5/239, vol. IV, August 4, 1931, E&GSynd to Colonial Office enclosed with Colonial Office covering letter of August 12, 1931, to Government of India, "willing to omit a British Nationality Clause" and "pressing demands of his people."

16. Bilovich, "Great Power and Corporate Rivalry in Kuwait," p. 182, citing August 10, 1931, Colonial Office minute of the interview with Dickson forwarded to Foreign Office and India Office and forwarded by India Office to political resident, "shaikh will certainly be prepared to eat any words that he may have uttered . . . if we wish him to do so"; IOL/R/15/1/621 F79, August 25, 1931, Anglo-Persian London (A. C. Hearn) to the director, Petroleum Department, Mines Department, Board of Trade, "reconnaissance . . . would take several months" and Bilovich, p. 190, citing September 10, 1931, Petroleum Department to Colonial Office, "Anglo-Persian should be given every encouragement and all facilities." IOL/R/15/5/239, vol. IV, September 25, 1931, "Personal," Dickson in Kuwait to Arnold Wilson, London, "rely on me to get Shaikh of Kuwait to hold up things where Holmes is concerned."

17. IOL/R/15/5/239, vol. IV, October 3, 1931, "Confidential," political agent Kuwait (Dickson) to resident (Biscoe), "the Shaikh of Kuwait really is a simple Bedouin" and "he is easily swayed and not possessed of much strength of character"; St. Antony's College, Dickson Papers, box 3, file 5, October 3, 1931, "Personal," Dickson to Biscoe, "Holmes is an unscrupulous person" and "question of granting Holmes a British visa for Kuwait may have to be reconsidered."

18. IOL/R/15/5/239, vol. IV, October 10, 1931, political agent Kuwait (Dickson) to resident (Biscoe), "Very Confidential" report on Major Holmes's activities/ notes on a general conversation held between the political agent and the shaikh from Kuwait, "if Holmes does not agree to insert the clause the Shaikh of Kuwait will ask Major Holmes to leave his State."

CHAPTER 17

KING SOLOMON'S MINES

Karl Twitchell landed in Jedda in April 1931, well ahead of Crane's estimate of October. The work of installing a mail-order bridge in Yemen, across a wadi about one hundred miles from Hodeida, was proving more difficult than expected. When he received the message from Crane, Twitchell promptly abandoned the bridge to his foreman. He and his wife caught the next boat to Arabia. In Jedda they moved into Philby's house. There was lots of room because Philby was spending most of his time in Mecca, with Mariam, in the house given him by bin Saud following his conversion to Islam.

Crane's understanding with bin Saud was that the man for whom he was meeting all expenses, Karl Twitchell, should search specifically for water, particularly along the pilgrim routes where the danger of dying of thirst was as ever-present as it had been in the days of the Prophet. The high-minded Charles Crane hoped to offer some relief to the many thousands of pious men and women from every corner of the world who made their way across the desert to the sacred city. But neither bin Saud nor Karl Twitchell was nearly as interested as was the charitable Crane in spending time and money seeking water for the benefit of foreign pilgrims or the benefit of the Municipality of Jedda.

Bin Saud was skeptical of there being oil in his country; he had said as much to Frank Holmes when he gave him the Al Hasa concession in 1923. In his memoirs, Philby would note: "In 1923, bin Saud was happy to receive rent for the Al Hasa concession while the experts went about the dreary task of discovering that Arabia had no oil." Even in later years, Philby said, "bin Saud did not himself believe there was oil in his country" and he "remained pessimistic about the prospects of oil being found in his land until his scepticism was blown sky high by the great 'gusher' of March 1938."

GOLD IN SAUDI ARABIA

Bin Saud did, however, believe he had other resources. Shortly after his conquest of Jedda, he awarded a banking monopoly to the Dutch and asked the Dutch government to send him specialists to prospect for minerals. D. van der Meulen, the Dutch minister in Jedda, recorded that in 1928: "Dutch geologists produced an extensive plan for a geological survey of the whole country that would take time and cost much money but would at least ensure that nothing would be overlooked."

But the stipulation that funds to cover the cost should be lodged in the Dutch Bank in Jedda before the work started was unacceptable to bin Saud's government and caused the plan to be rejected. Van der Meulen said that after 1928 "other small nations were invited and produced less elaborate and so less expensive projects," although none were implemented.

In his memoir *The Wells of Ibn Sa'ud*, van der Meulen wrote: "Every Arab of education knew the ancient authors and even the holy books of Jews and Christians mentioned Arabia as a land of gold, incense and myrrh." He said bin Saud and his ministers believed "Western science should be able to trace the veins of gold and to exploit them on modern lines." When he arrived in bin Saud's country, van der Meulen noted observantly, "Karl Twitchell gradually came to realise that he had found the chance of a lifetime. He was convinced that there must be gold in Arabia for he knew his Bible and believed what was written in it."[1]

Not only his Bible. Twitchell had spent many lively evenings in Yemen debating the probable location of King Solomon's Mines with Holmes's colleague, the eccentric Commander Craufurd, who was convinced the mines were in Yemen. Both Twitchell and Craufurd could quote passages "alluding to gold" from the Old Testament. Twitchell knew them by heart. Both men had pored over Sir Richard Burton's classics, particularly *The Gold-Mines of Midian*. Twitchell was particularly taken with Rider Haggard, who turned out some forty popular books including the early 1900 *King Solomon's Mines* that Twitchell referred to quite seriously as a work "on the presumed whereabouts of these mines."

In the early 1950s Twitchell would write an article for an American publication called *The Explorer* under the title "King Solomon's Mines?" He opened by saying, "Nearly everyone has heard fabulous stories of the vast wealth of King Solomon and the mines which furnished the gold for his temple in Jerusalem. . . . The location varies from Rhodesia to the Sudan and Arabia." He would tell his readers that "In the first books of the Old Testament there are allusions to gold. The phrase in Genesis is 'there is gold in the land of Havila, and that gold is good.' On a biblical map Havila is shown to be northern Hijaz, Saudi Arabia. An interesting fact is the record that the Midianites paid tribute in gold to Jerusalem. The Land of Midian lies to the north of the Wadi Hamdth, in northern Hijaz." Twitchell added his own interesting opinion that "undoubtedly" at the mines "most of the labour consisted of slaves, most likely men captured in raids and wars."[2]

Twitchell was equally romantic about his first glimpse of Abdul Aziz bin Saud. "I was to meet the King at the religious boundary beyond which infidels cannot go towards Mecca. When I got there a great tent had been erected and I was asked by one of bin Saud's soldiers to await the King there." Breathlessly, Twitchell recalled the "great sight" of bin Saud's arrival. "He came down in a great caravan of motor cars—he himself had a very big Packard on each side of which was a big Nejd soldier with brilliant flowing robes and flowing headdress flying in the sunlight—it was a wonderful sight. The king got out, greeted me very cordially, and we went into the tent and had quite a feast."

Crane's gift was now providing an almost heaven-sent opportunity for both bin Saud and Twitchell to explore under the guise of seeking water with which to alleviate the pilgrims' thirst. Bin Saud "instructed his people to give me all possible help for travelling around the Hijaz . . . to investigate water possibilities," Twitchell recorded enigmatically. He was given the services of a "most efficient interpreter and secretary" (who would go on to become Saudi Arabia's director of mines and public works). He notes, almost in passing, that along the way he "inspected some ancient workings, mine ruins and tailings." When he returned to Jedda from this first trip, he told bin Saud's astute minister of finance, Abdulla Sulaiman Al Hamdan, that there was little hope for artesian water. "But there might be minerals of commercial value in the country," he said. He told Suleiman that he would need a team of mining engineers "with facilities to travel over the country, take samples, and make the necessary surveys and reports."

Curiously, an article in *Um El Qurra* (Mother of Towns), the official Mecca newspaper, did not seem to share Twitchell's gloomy picture of water possibilities and also seemed to think he had discovered a "petroleum mine." The article said:

Mr. Twitchell, the American specialist in mining and water, and Khalid Al Waleed, Government Deputy, have returned from their trip along the northern coast after spending three weeks in studying and investigating, covering a distance of 3,500 kilometres. We have been informed that from their investigations it appears the water available in this area is not less in quantity than that between Wadi Fatima and Jedda and that this water is very near to the surface so that the maximum depth does not exceed ten meters. The water being so easily reached in this area there is no need for artesian boring. They have also struck a great gold mine on the boundaries near Wedj containing many gold veins. Also they have found near Wedj signs of an old lead mine. During their investigations and studies they have struck a great petroleum mine between Al Lubana and El Muwaileh in an area of not less than 350 kilometres in length. This is what we have learned in this connection.[3]

Twitchell did say, in an unpublished manuscript titled "Summary of Gifts of Charles Crane, administered by K. S. Twitchell" that while traveling north down the coast from Jedda "at Nuwaits there were some inactive oil seeps but the area was so limited, being too close to the volcanos to make it, or any of the coast, attractive for commercial oil possibilities." This was the area where Philby's original partners, Midian Ltd., had attempted to resume oil and mineral concessions granted by the Ottoman government while Sharif Husain was ruler of Mecca. It had fallen through because of complications involving Anglo-Persian and its subsidiary the D'Arcy Exploration Company over succession to Turkish assets.

Abdulla Suleiman did not have the finances to provide Twitchell with a team of mining engineers, so he did the next best thing. Armed with a "pamphlet" written by Twitchell describing what to look for that would indicate "evidence of ancient mines," Suleiman himself trekked across the country.

He told Twitchell that "everywhere he stopped he explained to the shaikhs, the villagers, and the Bedouins what minerals and ancient mine workings would look like and ordered that he be informed as soon as such objects and conditions were found."

Mr. and Mrs. Twitchell went back to installing the bridge in Yemen. Perhaps their attention was no longer focused on this project. Van der Meulen reported that he "met Twitchell's assistant trying to rebuild an iron bridge that had been severely damaged by floods. Alas, the bridge was never completed. Poor indeed were the results of Twitchell's efforts in the Yemen."[4]

CRANE'S EXTENSION

By the time the Twitchells returned to Jedda on their way back to New York, Suleiman had "two nuggets of gold found near Taif by a Turkish prospector." Twitchell confirmed it was "native gold" but advised there was "little chance of developing any property of commercial size" there, although he thought there may prove to be "small areas, beneficial to bin Saud's government or local groups." Twitchell put a proposal "in order to make a competent report on gold properties that might interest foreign capital."

Abdulla Suleiman managed to scrape together seven hundred sterling for Twitchell to "cover the initial travelling and other expenses of the engineers I was to engage on their behalf." Back in New York, reporting to Charles Crane, Twitchell strongly lobbied for "drilling and testing of ground water in the vicinity of Jedda" and slipped in, almost incidentally, "examination of mineral properties."

Generously, Crane said he would pay for a further six months in bin Saud's territory and Yemen. "But," Crane said, "because of the depression" this would be the last expedition he would finance for Twitchell. With Crane's funding behind him for the next six months, Twitchell, with his wife and one assistant, returned to Jedda in October 1931 just as spudding of the first oil well was getting under way in Bahrain.

This was an exciting moment for the Bahrainis. The political agent recorded that the ruler, Shaikh Hamad Al Khalifa, "worked the drill for its first few blows.

The rig quivered under the impacts and His Excellency glanced at it and quipped 'the machine is drunk!'"

Frank Holmes was elated. He predicted: "So we are on the way to prove if there be oil or not. I have much faith we will have luck and strike a good flow."

In what could be viewed as a spiteful response, Anglo-Persian, the sole supplier in Bahrain of petrol and kerosene, raised their prices to more than double that of Persia and 50 percent above the price in England. Charles Prior, the Bahrain political agent, wrote to Harold Dickson, his counterpart in Kuwait, saying: "I talked to Anglo-Persian's chief distributor and found him quite impenitent. In fact, he seemed to think it right that Bahrain should have to pay more for its petrol than does the rest of the world." As Russian steamers were "not allowed" to call at Bahrain, Prior asked Dickson "if Russian petrol is available at Kuwait, will it be possible to send it down by launch" in order to "break this ring" of Anglo-Persian monopoly. Unfortunately, Dickson's reply is not recorded.[5]

Meanwhile, Abdul Aziz bin Saud was once more militarily active with another fight on his hands. Although he had in 1926 "adopted" the Assir as his "protectorate," Imam Yahya had sent a Yemeni force to move into the richly fertile inland oasis of Wadi Najran, located within the Assir province and the key to the route into Arabia for Yemen's coffee trade.

Bin Saud would not tolerate this challenge to what he perceived to be the status quo and was busy organizing the dispatch of his warriors to see off the Yemenis from what had been and still was, officially at least, their own country. The Imam Yahya's troops would not be able to withstand bin Saud's better-equipped warriors and in early 1932 would pull back, well away from the unde-marcated frontier of the Assir province. Bin Saud would send a deputation to Sana'a to negotiate a settlement with an unwilling Imam Yahya. The talks would drag on for more than three years.

While not putting any effort into the pilgrim route, to meet Charles Crane's desire to guarantee a supply of water, for which he was paying after all, Twitchell did work in Jedda at "rehabilitating the city's water supply." This was the system originally built by the Turks some sixty years earlier. As he had done in Yemen, he installed an American windmill. This one raised an average of forty-gallons per minute into the old Turkish water tunnel from where it flowed seven miles to Jedda. Twitchell reported to Crane that the windmill "makes an appreciable addition to the city water supply."

Twitchell's assistant engineer, working with an Arab team, "drilled, pitted and sluiced" for almost six months in the Taif area. Unfortunately for Twitchell "nothing developed that would be profitable." Nevertheless, he assured Abdulla Suleiman that the possibilities of gold, and particularly the gold samples Suleiman had brought him, gave every cause for continuing optimism and "a gold production will be of the greatest value to your country." Twitchell promised Suleiman: "It is quite possible I may be able to help you in arranging financing."[6]

KUWAIT STANDS FIRM

The ruler of Kuwait, Shaikh Ahmad Al Sabah, meanwhile, was not behaving in the "easily swayed" and weak-charactered manner that Harold Dickson had so confidently predicted. He was not responding to pressure from the political officers by "eating his words" as Dickson had promised. As Karl Twitchell was advising Abdulla Suleiman he could "arrange financing" for exploring the ancient mines in bin Saud's country, Shaikh Ahmad in Kuwait was assuring Frank Holmes that he stood firmly by the content of his letter of four months earlier, even though Dickson "had done nothing but badger" him.

The political resident, Hugh Biscoe, arrived in Kuwait soon after Holmes and spent five days adding his considerable power to Dickson's badgering. Once again, Biscoe regurgitated Dickson's version wholesale to the colonial secretary, Lord Passfield. Biscoe reported that Frank Holmes had "for some time been pestering" the shaikh, who, "anxious to get rid of him," gave him the letter. He said that Ahmad's undertaking "certainly does not commit the shaikh to anything definite but is in that evasive style in which the Arab mind seems to delight."

Biscoe told Lord Passfield he had personally "reminded Shaikh Ahmad that he had all the advantages of an independent ruler, since we do not interfere in the internal affairs of his State." He said he was "quite convinced" Ahmad "really does not desire the presence of an American company in his State and will not willingly agree to any abrogation of the exclusive 'British Nationality' clause." Biscoe was so sure of this that he advised "there is absolutely no need for us to bring any pressure to bear upon him in this matter."

He said that the shaikh of Kuwait was "very grateful indeed" to learn that the Anglo-Persian Oil Company was contemplating another geological survey. Biscoe said he hoped "Anglo-Persian will not delay in carrying this out." If the survey was successful, he said, "and results in a definite application by Anglo-Persian for a concession, it will relieve us from the somewhat delicate position in which we have been placed by the activities of Major Holmes on behalf of American interests."

As soon as the resident left Kuwait, Shaikh Ahmad called Holmes in and told him that despite the "many talks" during which Biscoe had tried to persuade him to withdraw his letter, he had not done so. Mullah Saleh told Holmes that Shaikh Ahmad had indeed stood firm. At one stage he had forcefully told the resident "where does it state in any Treaty between the Kuwait government and the British government that the Shaikh is not permitted to write to persons like Holmes regarding concessions."

Saleh also reported that Ahmad had told Biscoe and Dickson that his people believed Anglo-Persian was primarily seeking political power and that he did not want to be made "the cockpit of troubles" between E&GSynd and the British authorities. He told them the matter of the exclusive British nationality clause was "between Holmes and the British government." Shaikh Ahmad told Biscoe and Dickson that he would not rescind the letter and, furthermore, if the High Government advised the political officers that opposition was withdrawn he was immediately "willing to grant the Kuwait concession to Major Holmes and his company."

The fact that Anglo-Persian was not "contemplating" any firm offer for a concession but only another geological survey—this would be the seventh—did not impress Shaikh Ahmad. Despite the six previous negative surveys and Anglo-Persian's long insistence that Kuwait was oil dry, the Kuwaitis had always believed they had oil. Yet, twenty-five years would pass between 1913 when Ahmad's grandfather, Shaikh Mubarak, willingly agreed to the "exclusively British" compact signed with Sir Percy Cox and the first Kuwait oil flowing in October 1938. Shaikh Ahmad was anxious to develop the oil he was sure his country possessed. He had visited Bahrain twice during the year, once by seaplane. "It was a stab to my heart when I observed the oil work at Bahrain and nothing here," he wrote to Holmes.[7]

THE STATE DEPARTMENT—ROUND TWO

Frank Holmes was as fed up as the shaikh of Kuwait. He wrote his London colleagues that the only way to resolve the situation was to press the Colonial Office for a "definite statement as to their intentions." He considered that: "We have been patient and considerate to a fault. We are now justified in making use of every weapon, diplomatic or otherwise, that we possess." It had been "proven" to the Colonial Office that the shaikh was willing to grant the concession to E&GSynd without an exclusive British nationality clause. As this was the case, Holmes concluded: "It is the Colonial Office, and not the shaikh, who is the stumbling block. The shaikh and his people are anxious to have exploration work commenced at the earliest opportunity."

The mention of "diplomatic means" acted like a starting pistol on the Americans at the Gulf Oil Corporation. They truly believed that it was American diplomatic representations that had forged the breakthrough and subsequent relaxing of the terms of the nationality clause for the Bahrain concession. They may not even have been aware of Persia's claim to sovereignty over Bahrain and the implications of Britain's submissions to the League of Nations. In their patriotic view, American diplomacy had bent the British to its will.

They reasoned if they had done it once they could do it again. Within days of receiving a copy of Holmes's letter, Gulf Oil's executives were in the State Department in Washington once more pouring out a story of woe with the Government of India, the 51 percent British government–owned Anglo-Persian Oil Company, and the Colonial Office.

Gulf Oil's William Wallace and his assistant Guy Stevens tackled Washington while Maj. Harry Davis used the London contacts he had developed at the American Embassy since arriving there as Gulf Oil's representative. Both Wallace and Davis knew the routine. They had played this game before in 1929 over the Bahrain concession. By December 1931 the State Department was instructing the American Embassy in London to take up the matter with the Foreign Office.[8]

Gulf Oil's complaint delivered by the American Embassy came out of the blue for the Foreign Office. They were annoyed. The Foreign Office had no responsibility for the Persian Gulf. Nor did they have anything to do with oil. In their opinion, any matter related to the Arab shaikhdoms of the Persian Gulf could only be of minuscule importance in their global responsibilities. Yet, they were currently involved in two issues for which they certainly needed to maintain the utmost goodwill of the United States. On the Foreign Office agenda was Britain's request, on the grounds of the Depression, to postpone repayment of the substantial WWI debt owed to the Americans. The Foreign Office also wanted US cooperation in restraining Japanese expansion in the Far East, particularly China.

Consequently, the Foreign Office did not want the State Department or the American Embassy offside over an area as peripheral as the Arab shaikhdoms in the Persian Gulf. They examined the complaint and concluded that it would be impossible to "justify, to any impartial tribunal, insistence upon stricter conditions in the case of Kuwait than in the case of Bahrain." This adjudication from the Foreign Office ran like a brushfire through the Government of India, the India Office, the Petroleum Department, the Admiralty, the Air Ministry, and the Colonial Office. As it began to dawn on the Foreign Office just how many departments were involved, they bravely took on the task of mediating a solution.

THE FOREIGN OFFICE MEDIATES

The various departments did seem to agree on what should happen. The best thing to do was nothing, they assured the Foreign Office. A conference on January 1, 1932, advocated that, since Anglo-Persian had begun its seventh survey on December 15, any reply to the Americans should simply be avoided until such time as Anglo-Persian reached a decision on whether or not they were interested in Kuwait.

The Foreign Office was under pressure, it had to get this out of the way and get on with more important matters, such as putting off repayment of the WWI debt. A tactic was adopted aimed at bringing each department, one by one, around to the Foreign Office point of view.

The Colonial Office moved into opposition to the Foreign Office by insisting that Britain "ought not to allow United States oil interests into Kuwait" and called in the Admiralty to support this stand. The Admiralty came in with flags flying and went further. They objected strenuously to "what appears to be an assertion that Kuwait is independent." Kuwait, the Admiralty declared, "only has some degree of self government—but not independence in the usual meaning of the term."

The India Office was told it was no longer credible "to shelter behind the shaikh." The viceroy of India came to the defense of the India Office and went back to the turn of the century to find his justification in imperial tradition. He stated that the "policy of His Majesty's government in the Persian Gulf region was authoritatively defined by the Secretary of State for Foreign Affairs, Lord Curzon, speaking in the House of Lords on May 5th 1903." He did concede that "it was perhaps a mistake of tactics to attempt to raise the nationality bar over the Kuwait agreement" and "in view of the shaikh's tortuous attitude it is, anyhow, now no longer possible to pretend the objection comes from him."

The viceroy thought the Americans should be made to understand that Britain's main concern was not so much the matter of nationality but to "secure best terms possible" for the shaikh of Kuwait and other Arab rulers "whom we regard as our wards." He saw no problem with saying it straight as he reminded his colleagues: "Our opposition to admission of foreign interests hitherto has been, at least in part, based on anticipation that foreigners would prove less amenable to our guidance than British concerns."

Biscoe and Dickson were right in the middle of the scrum. Biscoe shot off a hot letter to the Government of India in which he castigated Frank Holmes for not only seeking permission to explore and prospect for oil "but to construct roads, ports, lighthouses, telegraph, telephones and railways." The result of this, he shrilly warned, would be that "in a short time, the Arab shaikhs would look to foreign interests, who would doubtless not stint money, rather than to our government as the dispenser of favours. We would, to a large extent, lose the influence that we now exercise. The whole position that we have laboriously built up would be threatened."

In a personal note to Dickson, he complained that it was "the shaikh's stupid letter to Holmes that has fairly put us in the soup." He observed that the Foreign Office seemed "very sensitive on the subject of the Americans" and reminded Dickson that "we are never anxious to define or emphasise our 'special position' in the Persian Gulf."[9]

Dickson, meanwhile, was doing his best to spur on Anglo-Persian by keeping up a steady stream of information and advice to Arnold Wilson. He informed on Holmes's every move and urged Wilson to do more "to get the shaikh on Anglo-Persian's side. Your people would do well to start ingratiating themselves with Shaikh Ahmad." Holmes spends money on local charities, he said, as well as "keeping the shaikh happy with small gifts, such as cigarette boxes, silver trinkets, etc. He has, of course, given him several expensive gifts in the past such as a six cylinder Sunbeam car, Rs10000 in cash, and now, I believe, an eight cylinder Buick is on order." As a starter, Dickson told Wilson, "you could not do better than suggest a good 12 inch searchlight, with Dynamo, for the shaikh's yacht, *Kuwait*. I know he is very keen on one."

For his part, Holmes was keeping the shaikh of Kuwait informed of progress in Bahrain. He sent a thank-you note for the "fine lot of birds" Ahmad had sent onboard as Holmes sailed for Bahrain and said on arrival he had given many to Shaikh Hamad. "The work at the borehole has progressed wonderfully, the depth reached is 835 feet. A series of bands of oil-bearing shales have been encountered in drilling. These showings of oil are very encouraging and are favourable indications of oil being struck at a lower depth," he wrote. He confided to Shaikh Ahmad: "I wish I was boring in Kuwait territory for you, as I feel it would result in great benefit to you and your people."[10]

DICKSON: WHAT HOLMES KNOWS

Dickson was determined to discover what it was that made Holmes so certain there was oil in Kuwait, so certain that he would endure the extreme opposition of the Government of India, and the slander and insults of that government's officials in the Gulf, in order to obtain the concession. Dickson was genuinely intrigued but also intent on secretly forwarding whatever information he could gather to Holmes's implacable competitor, the Anglo-Persian Oil Company.

There was a triumphant tone to Dickson's "Confidential" report, complete with three sketch maps, headed "Views of Major Frank Holmes of the Eastern & General Syndicate on the possibility of 'oil' being found in Kuwait, Bahrain and the Gulf generally." He prefaced the report by warning that Holmes had "emphasised that certain experts would by no means agree with everything he says." Dickson's "confidential" report read:

> In the first place Major Holmes gave it as his considered opinion that there was a close connection between the "oil" and the numerous fresh water springs (all of which were warm, some also very hot) which were to be found everywhere in the sea off the North-East coast of Arabia, on the islands of Bahrain, on the mainland of Hasa both in the vicinity of Qatif and up the whole length of the Wadi al Miyah (valley of waters), which series also obviously stretched up through Kuwait (where they weakened) and the country south of the Euphrates right up to the vicinity of Nejef (as instance the wells of Shagra Abu Ghar, Ruhba, etc.).
>
> This water, Major Holmes was of the opinion, came from the Persian mountains and not from the highlands of Jejaz and Nejd as hitherto supposed, and passing under the Persian Gulf and the lower end of Iraq at a great depth (as instance the heat) was eventually forced up again on the line described in the preceding paragraph by a sort of fault in the earth's surface which took the form, to use an easily understandable simile, of a great rocky wall or cliff, which came up from a great depth and ran in the North, West and South-easterly direction following the general line of the Persian Gulf and the valley of the Euphrates.
>
> The same "cliff" formation was in the same way trying to help the "oil" to the surface, as witness the "bitumen" seepages which are to be found stretching

from Hit on the Euphrates, past Tel al Magaiyir (Ur), Burgan in Kuwait territory, also Qatif and Bahrain. This line which, he said, followed the same source
as the water springs mentioned above, Major Holmes was of the opinion,
formed the Southern edge of the "oil" bearing zone or line, and it was close to
this line that the underground cliff so to speak ran.

There was another probable series of water and oil surface springs, said
Major Holmes, which took the centre line of the Persian Gulf and passing
through Halul Island, Arabi and Farsi Islands, Qarw Island, Mudaira on the
north side of the bay of Kuwait and Jebel Sanem, proceeded in a north-westerly
direction and parallel to the line mentioned above. At all the places mentioned
except Jebel Sanam, traces of oil or bitumen springs are, he said, known to exist.

The third great oil bearing line, said Major Holmes, was of course the well-
known one that passed through Kirkuk, Shustar region and north of the Tigris
and Persian Gulf and went in a southeasterly direction. This line, as far as Major
Holmes could see, probably also took a parallel course to his mid-Persian Gulf
and Iraq line, and his Arabian coast and Euphrates line.

With the above data to go on Major Holmes was of the opinion that chances
of striking oil either in Bahrain or at Al Hasa or in Kuwait were very bright. In
spite of the fact that oil experts always pretended that surface oil indications
never really meant anything, he, Major Holmes, said that he had yet to find an
oil geologist who did not carefully follow such indications, or an oil company
which did not find its eventual "spouter" in the vicinity of such indications.

As regards the formation of Jebel Durkhan at Bahrain, where he is now
boring, Major Holmes volunteered the statement that never in his long and
varied experience had he found such a perfect example of an oiliferous cone—
everything including general strata and formation of the under surface rocks,
etc. etc., promised success, he said.

His "bore" had gone down 800 feet or so and already "oily shales" were
being extracted. At Kuwait there was, he said, a surface ooze of oil both at
Bahra (close to Mudaira) and also in the sea close to Ras al Abid (visible at low
tide) near the southern boundary of Kuwait. These are apart from the extensive
bitumen deposits at Jebel Burgan on the southern border of Kuwait. He said all
of these indicated every likelihood of oil being found if bored for.

The attached sketch map, the original in the rough of which I persuaded
Major Holmes to draw for me, will probably show better than I have been able
to explain above, the theories held by Major Holmes.

Dickson set out again to see for himself what it was that Holmes knew. He made
"a careful search for the supposed oil seepage which Holmes said was visible at
low tide, but I was unable to find any trace of oil or bitumen. I also made careful
inquiries from an old Arab official who knows the coast line there very well
indeed, and he could give me no information of value," he reported. To the resident, he admitted: "It is possible that Major Holmes, with his expert eyes, is able
to see what I can not."[11]

ECONOMIC DISTRESS

While Kuwait waited for the US State Department, the Foreign Office, the Government of India, the Petroleum Department, the Admiralty, the Air Ministry, and the Colonial Office to resolve their differences so they could reap some financial benefit from awarding the Kuwait oil concession, Dickson's 1931 Administrative Report was a model of benevolent paternalism. He wrote:

> The relations between His Majesty's Political Agent Kuwait and the Ruler have continued to be of a satisfactory nature throughout the year. It is doubtful whether they have ever been on a more pleasant or more friendly footing. The policy of the Political Agent has been, and is, to interfere as little as he possibly can with the internal administration of the State and in return the Ruler trusts him and consults him in most things, especially in matters connected with his foreign relations. This is, in my opinion, as near the ideal as possible for the Ruler is flattered and pleased to think that he is allowed to "run his own show"without irritating interference from the Political Agent.
>
> The Ruler's foreign policy, on the other hand, is a different category. It is here that watchful control is necessary, for a mistake at any time could easily be made which might have wide repercussions and which eventually might land HMG, as the protecting power, in an awkward situation which she might find it difficult to recede from to say the least.

Dickson struck a compassionate chord as he continued:

> The closeness of bin Saud's blockade, coupled with the complete failure of the 1931 pearl season for the third year in succession, has reduced Kuwait to such a state of poverty and economic distress that it has required all the skill and patience of the Ruler to prevent the hot-headed members of his family from attempting retaliatory measures in the form of intriguing with the border tribes and his leading merchants from leaving Kuwait and transferring their business elsewhere.
>
> The Ruler's relief measures and free distribution of food has been appreciated by the poorer and hungry elements in the Town. He is literally playing the role of father to his people and by regularly going about among them and by adopting a frugal and unostentatious way of living, and by cutting down his own household expenses to a paltry Rs1500 per month, has set an example to all in the way they should meet their difficulties.
>
> In the matter of the blockade, the Ruler has steadfastly placed his reliance on the promise made by His Majesty's government during the Ikhwan rebellion of 1929–30 which was to the effect that in return for the Shaikh's loyal cooperation the British government would do its utmost by diplomatic pressure, or other means, to get the blockade raised.

Dickson's crocodile tears for the "poor and hungry" people of Kuwait are hard to accept when he personally was expending every effort to deny these same people the means of relief through the granting of the oil concession.

As Biscoe had recently reminded the Government of India, the proposed investment Holmes was offering was so open-ended that it included possible "roads, ports, canals, lighthouses, telegraphs, telephones and railways." The economic boost that development of an oil concession generated, even in its early stages, was now obvious from the activity in Bahrain. As he had proudly reminded Arnold Wilson, Dickson was "very imperialistic" with a viewpoint blinkered by a lifetime with the Government of India. He was unable to see beyond empire, and perhaps the exercise of personal power, to truly empathize with the daily suffering of the Kuwaitis, the solution to which was in his hands.

ANGLO-PERSIAN REFUSES KUWAIT

Although the Americans had not received an official reply, compared to the usual bureaucratic standards of the day the Foreign Office was making rapid progress. By March 1932 it was more or less accepted that the fiction of the shaikh of Kuwait demanding an exclusive British nationality clause could not be maintained and consequently its insertion per se could no longer be insisted upon. It now remained to reconcile the conditions that the various departments were demanding must be inserted in the Kuwait concession in lieu.

While the Foreign Office maneuvered, Andrew Mellon, the seventy-seven-year-old senior member of the Gulf Oil Corporation founding family, arrived to take up his post as United States ambassador to Britain, appointed by President Herbert Hoover. From Bahrain, Holmes confided to his wife: "I do not think they are progressing very well with Kuwait in London. They keep asking me to worry the Shaikh of Kuwait. But I know if London does not agree, I can do nothing here."

Finally, on April 9, the Foreign Office replied formally to the American Embassy that the British government would no longer object if E&GSynd renewed its application to the shaikh of Kuwait for an oil concession, to be subsequently transferred to the Gulf Oil Corporation. Two days later, the Foreign Office called in Sir John Cadman to inform him of this decision transmitted to the Americans.

Cadman stated that, according to the latest geological information, some oil might exist in Kuwait but it would be of a very heavy nature, which, in the long run, would be of no interest to his company. From Anglo-Persian's point of view, Cadman told the Foreign Office, "the Americans are welcome to what they can find there."[12]

The "latest geological information" that Cadman referred to came from Anglo-Persian's just-completed seventh survey of Kuwait. The geologists were Peter Cox, who would later become a director of Kuwait Petroleum Company, and A. H. Taitt. Although the two men had gone beyond surface surveying to actual test drilling, their survey had done no better than the previous six. "We went there without great hopes of finding anything very attractive," Cox would later tell the American Association of Petroleum Geologists. "Our geophysicists were not optimistic."

Cox and Taitt had needed to fit their surveying and drilling into the time left over from their other task, that of impressing the natives. Dickson took them in hand from the moment they arrived and arranged a program of activities and splashy entertainments designed to dazzle. Holmes's colleague, Muhammad Yateem, who had been in Kuwait for several months, reported: "They are following the advice of the Political Agent and they are entertaining and making friends very liberally. They have full powers to invite, or ask, or discuss, Anglo-Persian with anyone they think fit." Yateem said Dickson was "working in the background with great zeal."

The geologists were spending money lavishly and hiring men and equipment locally. All of this was going down well, reported Yateem, because "the people are in need of work and don't care from who it comes." He said the men from Anglo-Persian had a monoplane with which they were "causing a big show of their greatness." He said they had taken Shaikh Ahmad and some of his important friends "for a spin" in the plane.

Yateem said Mullah Saleh had told him that the shaikh of Kuwait had given permission for the survey, that unexpectedly included test drilling, because "he does not want to do anything to cause Dickson to suspect that he is against the Anglo-Persian Oil Company." Yateem gave Holmes heart by telling him "when Ahmad receives your letters, he talks about your success in Bahrain."

Commenting that Dickson was "literally acting as the agent of the Anglo-Persian Oil Company," Holmes passed Yateem's news to his colleagues in London. He reminded them that "the shaikh, and his family, are very much afraid of the Anglo-Persian Oil Company." From experience right across the Arab shaikhdoms, Holmes said, "they know exactly what the company stands for as to their own future."

When they were recalled in early April, Cox and Taitt's report concluded, as Sir John Cadman had told the Foreign Office, that "Kuwait's prospects appeared to be the possibility, even probability, of finding only very heavy oil accumulations." Anglo-Persian's general manager at Abadan wrote, thanking Dickson profusely for his "cooperation, assistance and hospitality in our work," and gave him the news. "As a result of our investigations at Kuwait, we have decided to abandon operations in the area and to take no steps at present to acquire a concession from the Shaikh of Kuwait," he said. When Biscoe learned of this, he commented philosophically: "The only consolation is that if Anglo-Persian does not anticipate finding oil there, no one else is likely to do so."

SCARE TACTICS

Dickson was devastated. He declared, "I see Holmes's and his friend Andrew Mellon's hands in this." Holmes, said Dickson, "is, of course, continuing his blandishments from Bahrain and writes almost weekly personal letters to the shaikh. Last week he sent him a second sample of *so called* 'oil' extracted from his well at Bahrain and expressed the hope that Shaikh Ahmad would soon allow him to pull up a similar sample in Kuwait."

Dickson predicted darkly that the "backdown" on the exclusive British nationality clause would be seen by the shaikh of Kuwait as "another instance of weakness and surrender on our part and give further support to the statements of anti-British propagandists that we are on the run." Dickson was certain it was all Holmes's fault. He fumed to the resident: "Holmes is a renegade and, I consider, one of the worst anti-British propagandists in the Persian Gulf. He, more than anyone else, stands for and advises the Shaikh of Kuwait to kick against our control and independently strike out a line for himself."[13]

To Dickson's chagrin, Holmes was soon back in Kuwait. "He arrived yesterday," Dickson wrote Biscoe. "He was welcomed on board by one of the shaikh's representatives and a special launch was sent out to bring him ashore. Holmes dined alone with Shaikh Ahmad that same evening." Dickson couldn't bear the fact that Holmes had a personal relationship with the shaikh of Kuwait and with many Kuwaitis. "One of the difficulties I am up against where Holmes is concerned," he complained to Biscoe, "is that he sees the shaikh privately whenever he chooses, and it is impossible to know what conversation takes place."

Dickson thought he had a card to play. In a long letter to Biscoe he said he had noticed that the cable to Holmes—which Dickson had, as usual, read before delivery—mentioned that the Americans at the Gulf Oil Corporation were the first to know of the Foreign Office communique. Gulf Oil had been informed by the State Department. Dickson thought this could be "counted on to alarm the shaikh" if it was made to look like American involvement at the government-to-government level.

When Holmes next saw Ahmad, he had been thoroughly spooked. He was gripped by the idea that the High Government had agreed to the transfer to an American company "without first consulting him." In a state of high anxiety he conjectured that this "indicated a change of disposition on the part of the High government towards him." Holmes reported that Ahmad was totally preoccupied with the thought that he had lost the backing of the High Government, which, as he had been told so many times, was all that stood between him and the ambitions of Iraq and bin Saud to absorb his country.

What it might mean to be absorbed by bin Saud could be read in the Bahrain political agent's memorable report on the "outstanding point of interest" of his own visit to Hofuf with Biscoe and Dickson. He remarked on "the apparent friendliness of bin Saud" to the British visitors "contrasting with the sour looks of his followers." He recorded: "The party left Hofuf on a characteristic note of savagery, seeing a hand and a foot of two men nailed over the town gate as we went out. The men had been caught stealing camels and owed their lives to the clemency of bin Saud, bin Jiluwi the redoubtable Amir of Hasa having wished to put them to death."

Shaikh Ahmad had just returned from a visit to Riyadh. The two rulers had pledged eternal loyalty to each other. Yet, bin Saud had not made any move to lift his crippling trade embargo. Shaikh Ahmad had told Dickson he was struck with how aged and ill bin Saud appeared. Dickson suggested this should be a worry for Ahmad because, if anything happened to bin Saud, he could be facing an even worse situation. He could be facing a new leadership on his border, perhaps even an Ikhwan resurgence, possibly involving folk like the amir of Hasa, the unforgiving bin Jiluwi. Dickson intimated this might not be a good time for Ahmad to be dumped by the British.

Ahmad was so flustered that he "put all the responsibility on me," Holmes wrote his London colleagues of his extraordinary meeting at the palace. Ahmad accused Holmes of "wilfully misleading him" in the use of his letter. He claimed that alterations to contract clauses had been made "behind my back." He charged that "neither the British government, nor the Americans, have any concern or any right to conclude anything, except through me."[14]

Ahmad's tantrum did not last. Dickson's machinations became suspect when he popped up again to announce the "good news" that Anglo-Persian wanted to reopen negotiations. The "good news" was contained in a personal letter from Arnold Wilson to Dickson, which asked whether Shaikh Ahmad would be prepared to give Anglo-Persian a two-year "exclusive option to prospect and survey Kuwait territory." The suggested payment for this two-year exclusivity was two thousand sterling.

This strange offer made Holmes and the Americans look truly serious, not to mention generous, in comparison. Shaikh Ahmad responded to Dickson's "good news" by saying he had been severely criticized for having allowed Anglo-Persian geologists into Kuwait when another company's proposal was before him. He said he had permitted it only because the High Government had "supported and blessed" the request. However, he told Dickson sharply, there was no way he was going to let them back in again to "sink shafts and dig holes and generally lay bare the possibilities of my country."

If Anglo-Persian was interested, he said, "they should come forward, like Holmes has done, and make a straight business offer and take their chance whether there is oil or not." Ahmad said that Holmes had never deviated from his original offer to directly take up a concession of financial benefit to Kuwait.

"The going is difficult," Dickson reported back to Anglo-Persian. "The shaikh did not jump at Wilson's proposal." Nevertheless he reassured them, "I am closely watching Holmes." Anglo-Persian should recognize, he instructed, that "Frank Holmes is quite positive that Kuwait is the centre of, or near, an immense oil field. He doesn't want to make any examination or survey of the territory, beyond what he has already seen and done, and is ready to risk heavy money for an immediate concession."

He urged the company to do the same, adding, "Like Holmes, I am convinced there is oil here and if you people don't move Holmes and his Americans will cut you out. Holmes is one of the cleverest men and smartest diplomat in the Gulf. Don't you make any mistake about that."

Even Biscoe, the political resident, was slowly losing faith in Anglo-Persian. He again visited Kuwait "to discuss the oil question" with Shaikh Ahmad. Displaying his impatience with the Anglo-Persian Oil Company, Biscoe reported to the colonial secretary: "On the one hand we have Frank Holmes, who presents the shaikh with considerable cash presents, motor cars, wireless, etc., and asks for a definite concession and is obviously anxious to get one, and, judging from Bahrain, to exploit it when he has got it. On the other hand," Biscoe continued, "we have Anglo-Persian who have for years shilly-shallied. And now they first of all inform the shaikh that they do not want a concession and then in a very informal manner say that they will give 2,000 sterling for an exclusive option to prospect and survey, although no request is made for a formal concession. In these circumstances it is pretty obvious Anglo-Persian does not mean business."[15]

NOTES

1. Karl S. Twitchell, interview by Paul C. Merritt, Twitchell Papers, Society of Mining Engineering, September 1965, pp. 78–83, mentions living in Philby's house, "the house in Jedda was also partly occupied by St John Philby—whose son (Kim) was recently branded a traitor." And also the mail-order bridge, "I bought a bridge, a 120 foot span to install . . ." and "when I received Crane's cable I left my assistant, Harry Ballard, with the bridge while I went up to Jedda with my wife"; H. St. J. B. Philby, *Arabian Oil Ventures* (Washington, DC: Middle East Institute, 1964), p. 66, "happy to receive rent for the Al Hasa concession while the experts went about the dreary task of discovering that Arabia had no oil . . . until his scepticism was blown sky high by the great 'gusher' of March 1938" and p. 133, "bin Saud . . . did not himself believe that there was oil in his country." Note: This certainly contradicts Philby's claim, in the same book, that Twitchell had produced evidence of the presence of oil. D. van der Meulen, *The Wells of Ibn Sa'ud* (London: Murray, 1957), pp. 135–37, "Dutch geologists produced an extensive plan for a geological survey of the whole country" and p. 136, "Twitchell gradually came to realise that he had found the chance of a lifetime . . . he knew his Bible and believed what was written in it."

2. H. Rider Haggard, *King Solomon's Mines* (London: Cassell 1966; repr. as Nelson Classic); Twitchell Papers, original typed manuscript headed King Solomon's Mine? "Havila, in the Northern Hijaz mentioned in Genesis as the land of gold"; Richard Francis Burton, *The Gold-Mines of Midian* (London: Kegan Paul, 1878), *The Ruined Mid-*

ianite Cities (London: Kegan Paul, 1878), and *The Land of Midian* (London: Kegan Paul, 1879). Burton's ostensible reason for journeys to northwest Arabia was to prospect for gold and other economic minerals, his real purpose was to "study and survey the region culturally and 'scientifically'"; Twitchell Papers, Merritt interview with Twitchell, pp. 78–83, "Rider Haggard on the presumed whereabouts of these mines."

3. The "most efficient" interpreter/secretary was Jajib Salla, two other local assistants were Habid Bey Gargoni and Nazib; K. S. Twitchell with Edward J. Jurji, *Saudi Arabia, with an Account of the Development of Its Natural Resources* (Princeton, NJ: Princeton University Press, 1947), pp. 139–52, contains Twitchell's version of his part in identifying gold mines and assisting Standard Oil of California to obtain an oil concession. "Bin Saud instructed his people to give me all possible help" and "need a team of mining engineers"; Twitchell Papers, "Translated extract from Umm El Gurra, dated May 15, 1931, the official newspaper published in Mecca, Hedjaz, Arabia."

4. Twitchell Papers, original typed manuscript, signed by Twitchell, titled "Explorer's Log—King Solomon's Mine." The manuscript is Twitchell's story of his original examination of the ancient gold mine Mahad Dahbab (Cradle of Gold) and its subsequent development by Twitchell's syndicate. Suleiman told Twitchell that "everywhere he stopped he explained . . ."; van der Meulen, *Wells of Ibn Sa'ud*, p. 136, "the bridge was never completed . . . poor indeed were the results of Twitchell's efforts in the Yemen"; Twitchell and Jurji, *Saudi Arabia, with an Account of the Development*, p. 143, "a competent report on gold properties that might interest foreign capital" and "Crane said, because of the depression this would be the last expedition he would finance for Twitchell."

5. Bahrain's Well No. 1 was spudded in October 16, 1931; Administrative Report for Bahrain, 1931 (Prior), "Shaikh Hamad worked the drill for its first few blows"; Angela Clarke, *Bahrain Oil and Development, 1929–1989* (London: Immel Publishing, 1991), p. 127, citing November 2, 1931, letter from Holmes to Ward, "So we are on the way to prove if there be oil or not"; IOL/R/15/5/239, vol. IV, October 1, 1931, political agent Bahrain (Prior) to political agent Kuwait (Dickson), "seemed to think it right that Bahrain should have to pay more for its petrol than does the rest of the world."

6. Twitchell Papers; this entry in Twitchell's diary dated October 20, 1931, "(I told Suleiman) I may be able to help you in arranging financing . . ." together with his earlier advice to Suleiman that he could make a competent report on gold properties "that might interest foreign capital," is of paramount importance. Twitchell later made much of a claim, see *Saudi Arabia, with an Account of the Development*, p. 147, that it was bin Saud who requested him to "try and find capital to carry out the development previously discussed" that Twitchell identifies as "mines, oil and roads." Despite his two previous offers to "find capital," Twitchell says, on p. 147, he responded to bin Saud's request by protesting he was an engineer "and not a promoter." In subsequent publications the story becomes Twitchell being the first person to recognize the oil potential of Al Hasa and that, when he told bin Saud about Al Hasa oil possibilities, bin Saud asked Twitchell to find the capital *specifically* to develop *oil* in his territories. This story has been enthusiastically propagated by Standard Oil of California and its successors in Aramco. Note: Twitchell never corrected this distortion. In fact, it is possible to trace the shift through the years in interviews with Twitchell, and speeches and articles by him, away from his original single-minded concentration on finding gold in Saudi Arabia to a claim that he was the first person to recognize the oil possibilities of Saudi Arabia.

7. Archibald H. T. Chisholm, *The First Kuwait Oil Concession Agreement: A Record of the Negotiations, 1911–1934* (London: F. Cass, 1975), p. 19, "nothing but badger him for days"; IOL/R/15/5/239, vol. IV, November 3, 1931, political resident (Biscoe) to colonial secretary (Lord Passfield), "relieve us from the somewhat delicate position in which we have been placed by the activities of Major Holmes on behalf of American interests"; Chevron Archives, box 0426154, November 8, 1931, Frank Holmes to E&GSynd, "immediately willing to grant the Kuwait concession to Major Holmes and his company"; Administrative Report Bahrain 1931 (Prior), "Shaikh of Kuwait . . . two visits to Bahrain"; Chisholm, p. 19, "stab to my heart when I observed the oil work at Bahrain and nothing here."

8. Chisholm, *First Kuwait Oil Concession Agreement*, p. 130, citing November 1, 1931, Holmes in Kuwait to E&GSynd, "the Colonial Office and not the shaikh is the stumbling block," pp. 130–41, reprints all related American State Department and Embassy correspondence and Foreign Office replies. Yossef Bilovich, "Great Power and Corporate Rivalry in Kuwait, 1912–1934: A Study in International Politics" (PhD diss., University of London, 1982), pp. 193–230, examines this State Department/Gulf Oil/Kuwait episode in detail.

9. See the traffic following Foreign Office statement, "impossible to justify to any impartial tribunal, stricter conditions in Kuwait than Bahrain" including IOL/R/15/5/239, vol. IV, February 3, 1932, India Office to Government of India, and February 5, 1932, political resident (Biscoe) to political agent Kuwait (Dickson), and February 6, 1932, political resident (Biscoe) to Government of India, and February 25, 1932, viceroy of India to India Office; IOL/R/15/1/639, D78, March 15, 1932, "Secret," Admiralty memorandum; PRO/FO371/16001, March 16, 1932, Admiralty to Foreign Office; IOL/R/15/1/639, D78, March 19, 1932, Air Ministry to Foreign Office; Bilovich, "Great Power and Corporate Rivalry in Kuwait, pp. 205–206, "we are never anxious to define or emphasise our 'special position' in the Persian Gulf."

10. IOL/R/15/5/239, vol. IV, November 4, 1931, "Confidential," Dickson in Kuwait to Arnold Wilson in London, "get the shaikh on Anglo-Persian's side"; and November 15, 1931, Holmes in Bahrain to shaikh of Kuwait, "I wish I was boring in Kuwait territory . . . I feel that it would result in great benefit to you and your people."

11. IOL/R/15/5/239, vol. IV, December 5, 1931, confidential, political agent Kuwait (Dickson) to resident (Biscoe). Unfortunately, the sketch maps are not with this file. See also March 3, 1932, confidential, Arnold Wilson, general manager, Anglo-Persian Oil Company to Dickson thanking him for this report, "which contains much of interest"; IOL/R/15/5/239, vol. IV, December 12, 1931, confidential, political agent Kuwait (Dickson) to the political resident (Biscoe), "with his expert eyes."

12. Andrew W. Mellon took up this position in February 1932, he personally retained substantial financial interests in Gulf Oil Corporation; Holmes Papers March 26, 1932, Holmes in Bahrain to Dorothy, "if London does not agree, I can do nothing here"; Bilovich, "Great Power and Corporate Rivalry in Kuwait," p. 218, "Americans welcome to what they can find there."

13. Chevron Archives, box 0426154, February 25, 1932, Muhammad Yateem in Kuwait to Holmes in Bahrain and March 5, 1932, Holmes in Bahrain to E&GSynd, "Dickson literally acting as the agent of the Anglo-Persian Oil Company" see also Bilovich, "Great Power and Corporate Rivalry in Kuwait," p. 274, "Dickson was known to be a staunch supporter of Anglo-Persian and was soon identified as its Agent in

Kuwait"; IOL/R/15/5/239, vol. IV, April 13, 1932, general manager Anglo-Persian Abadan (Elkington) to political agent Kuwait (Dickson), "we have decided to abandon operations in the area"; IOL/R/15/1/621 F79, April 22, 1932, political resident (Biscoe) to Anglo-Persian Abadan, "if you do not anticipate finding oil there, no one else is likely to do so"; IOL/R/15/1/639 D78, April 14, 1932, political agent Kuwait (Dickson) to political resident (Biscoe), "Holmes advises the Shaikh of Kuwait to kick against our control and independently strike out a line for himself."

14. IOL/R/15/5/239, vol. IV, April 28, 1932, "Confidential," political agent Kuwait (Dickson) to political resident (Biscoe), "Holmes sees the shaikh privately whenever he chooses and it is impossible to know what conversation takes place;" IOL/R/15/5/239, vol. IV, May 1, 1932, political agent Kuwait (Dickson) to political resident (Biscoe) "counted on to alarm the shaikh"; Administrative Report for Bahrain, 1932 (Prior), "a hand and a foot of two men nailed over the town gate," intriguingly the report refers to a January visit to Bahrain by Shaikh Ahmad, Biscoe, "and members of Britain's Tehran Legation . . . a somewhat embarrassing visit"; Daniel Yergin, *The Prize: Epic Quest for Oil, Money, and Power* (New York: Simon & Schuster, 1991), p. 288, citing April 6, 1932, political agent Kuwait (Dickson) notes on Shaikh Ahmad's trip to Riyadh, "how aged and ill bin Saud appeared"; Chisholm, *First Kuwait Oil Concession Agreement*, p. 142, May 4, 1932, Holmes in Kuwait to E&GSynd, "the shaikh put all the responsibility on me." Note that Chisholm, p. 21, remarks that "this episode was the only one of its kind in the long history of Holmes's relationship with Shaikh Ahmad, from 1923 until Holmes's death in 1947."

15. IOL/R/15/5/239, vol. IV, April 29, 1932, "Personal," Arnold Wilson to Dickson, "two-year exclusive prospecting licence"; and May 16, 1932, political agent Kuwait (Dickson) to political resident (Biscoe), "'told the shaikh I was bringing the good news" and "not going to let them in again to sink shafts and dig holes" and "should come forward like Holmes has done"; and May 21, 1932, "Strictly Confidential," political agent Kuwait (Dickson) to Charles Myles, Anglo-Persian Abadan, "if you people don't move Holmes and his Americans will cut you out." On the same date Dickson wrote Myles that the shaikh was thrilled at the thought of getting the searchlight for his yacht. "Your people have made a good move . . . keep it up"; IOL/R/15/5/239, vol. IV, May 26, 1932, "Confidential," political resident (Biscoe) to Colonial Office, "it is pretty obvious Anglo-Persian does not mean business."

CHAPTER 18

ALL EYES ON AL HASA

Harry Philby had become annoyed at bin Saud's reluctance to set him up for the trek across the Rub al-Khali that would enable him to claim the title of "desert explorer." Even though bin Saud was obviously preoccupied with chasing the Yemenis out of their own Assir province, Philby's annoyance became disenchantment and spilled over into criticism. At the end of 1931 he wrote his wife in England, "bin Saud has disappointed me for the time being. At the actual present moment his show is worse than Sharif Husain's was, which is enough said. . . ."

His estimation of the situation shot up considerably in mid-December when he received word that bin Saud had given permission and would meet all the costs including guides and provisions. He could set out on his exploration of the Empty Quarter. He wrote Dora: "After this trip I shall never want anything more from the Arabs."

Philby had observed Karl Twitchell's progress with interest and had written T. D. Cree about the American's romantic obsession with finding gold. Cree considered this a quaint pursuit and replied, "The result of his expedition will be most interesting. I am not hopeful but one never knows. Gold is about the only metal that is worth mining at the moment although of course silver is much better again."

Bin Saud's minister of finance, Abdulla Suleiman, felt the Hijaz had had quite enough attention. He thought Twitchell should now move to bin Saud's "real" country, Nejd, and the province of Al Hasa. Of course, as he was still being funded by Charles Crane, he should look for water . . . and "minerals." Twitchell left his wife to get on with the search for water around Jedda. By herself, she competently and efficiently ran the Empire drill. Without any problems she directed a crew of thirty tough Bedouin workmen.

PHILBY AND TWITCHELL

Twitchell and Philby met up at bin Saud's hunting camp, about fifty miles north of Riyadh. Curiously, although they shared the same house in Jedda, Twitchell would claim this was where he first met Philby. The two met "again" here according to Philby. He recorded in *Arabian Oil Ventures*: "It was not until December that I saw Twitchell again at the King's camp in the Muzailij area near Riyadh. He had been asked by Abdulla Suleiman to visit the Al Hasa province to study its famous springs and report on ways and means of expanding irrigation and agriculture. . . . It should be noted that the prospect of oil being found in the Persian Gulf region had virtually been written off with the collapse of E&GSynd's concession."[1]

After Philby and Twitchell surreptitiously joined forces to obtain Frank Holmes's Al Hasa concession for Standard Oil of California, Philby would claim that Twitchell had discovered "three oil seepages" during this trip across Al Hasa. In 1964 Twitchell would admit to the Middle East Institute that his claim to having identified the oil of Saudi Arabia might not be true, particularly as Aramco had now publicly stated: "No oil seepages in Al Hasa have ever been noted by Aramco geologists."

Twitchell would admit: "I do not remember ever having seen oil seeps in Al Hasa and while my memory is far from perfect, especially after some thirty odd years, I do not recall ever telling Mr. Philby so. I do remember seeing seeps along the Red Sea Coast. Perhaps I mentioned these to Mr. Philby and was misunderstood to have been talking of seeps in Al Hasa." Until that admission, however, Twitchell would not correct Philby's claims nor would he be at all averse to letting it be thought it was him, rather than Frank Holmes, who identified the fabulously rich Al Hasa oil field of Saudi Arabia.

When Twitchell joined Philby at the king's hunting camp, he had nothing to report. He told bin Saud and Suleiman that he had not found "anything worthwhile" on the way from Jedda to Riyadh.

Bin Saud imagined that Twitchell putting in a road and installing a mail-order bridge in Yemen and Holmes's engineering work on a bridge and roads in Bahrain must mean that both men possessed the same sort of skills. He remembered Holmes's recommendation that Ras Tanura was the place where he should try to develop his much-desired harbor and port.

According to Twitchell, bin Saud asked him to go to Ras Tanura and investigate the possibility of a harbor. Twitchell said that bin Saud suggested he should also drop across to Bahrain "where there is an oil company drilling and tell me what you think about it."

In his memoirs, published in 1947, Twitchell avoids the point of whether or not bin Saud told him the oil company was working on Frank Holmes's concession and that Holmes had similarly identified oil prospects in Al Hasa in 1923. He does claim that after he returned from visiting the oil site at Bahrain, bin Saud told him "he had a former arrangement with a foreign oil company concerning Al Hasa but as they had not lived up to their terms of agreement he wanted nothing more whatsoever to do with them."

Twitchell's single oblique reference to Frank Holmes and E&GSynd must be taken with a grain of salt. As will be seen, when Twitchell arrived back in Saudi Arabia twelve months later as the representative of Standard Oil of California charged with obtaining Holmes's Al Hasa concession, far from "wanting nothing more whatsoever to do with them," bin Saud would personally call immediately for Holmes to come.

Twitchell and Philby left bin Saud's hunting camp together. Philby reported they traveled as far as Hofuf, "which was to be the starting point of my own venture of crossing the Empty Quarter, and there we parted."

Unlike Holmes, Karl Twitchell was not inspired to visions of a modern harbor development by the view of Ras Tanura. All he could see when he got there, he reported, was "nothing but a dead porpoise."[2]

TWITCHELL IN BAHRAIN

Twitchell duly arrived in Bahrain on January 7, 1932, and failed utterly to impress his fellow American, Edward Skinner, Standard Oil of California's manager at Bahrain Petroleum Company.

Skinner was a self-taught oilman who had worked his way up from the mailroom through jobs as a roustabout and driller. The rough-and-ready Skinner was particularly put off by Twitchell's affectation of sweeping about the Bahrain camp dressed in full, flowing Nejdi Arab robes, complete with *ghutra* and *ighal*.

The Bahrainis, Iraqis, and Sikhs working on the site also thought this a very odd way for a European to behave. Holmes's reaction is not recorded. He was out of Bahrain during Twitchell's theatrical appearance.

Skinner quickly came to the conclusion that this strange American swishing around in his fancy dress outfit didn't know anything about oil. Skinner wondered what he was doing there. Twitchell soon confided that he needed to make a report for bin Saud and Abdulla Suleiman. In all innocence he asked Skinner whether, in his opinion, there might be any oil in bin Saud's territories. Perhaps intent on deflating Twitchell's not inconsiderable ego, Skinner sarcastically replied that the only way to find out was to let some "skilled" geologist take a look, meaning someone other than Twitchell.

But then the penny dropped. It suddenly hit Skinner that Twitchell had arrived in Bahrain from the Eastern Province of the mainland, *from Al Hasa*. This was exactly where Standard Oil of California had been trying to get to since Lombardi's 1927 sighting of Frank Holmes's map and observations on Al Hasa in the Gulf Oil Corporation's office.

Skinner cabled his head office to pass on the news of Twitchell's visit. He told San Francisco that Frank Holmes was completely "tied up" trying to get the Kuwait concession and furthering the drilling in Bahrain. Furthermore, Skinner said Holmes had told him "in conversation" that, for the moment, he could not "undertake" to get a permit for Standard Oil of California to mount a geological survey of Al Hasa. Get busy, Skinner advised San Francisco, get in touch with Twitchell and maybe we can get to bin Saud and into Al Hasa through him.[3]

After three days in Bahrain, Twitchell headed back to bin Saud's party that was now at Riyadh. According to Twitchell, he told bin Saud exactly what Skinner had told him, that "the only way to find out if he had oil was to let some skilled geologists take a look." He says bin Saud thought this was excellent advice and asked Twitchell to "get him an oil geologist" and asked who and what would be necessary to test for oil in his country.

Not only did Twitchell not know the answers, his interest was in gold. He was impatient at this diverting of bin Saud's attention. The end of Crane's six-month extension was now very close.

Twitchell said he told bin Saud to wait and see whether Bahrain was successful or not "because it entails a great deal of money to get equipment and competent men." He said bin Saud was "disappointed, but said he would wait." Disappointment would seem to be an inappropriate reaction from the man who, according to both Holmes and Philby, did not believe there *was* any oil in his country.[4]

MAHAD DHAHAB—CRADLE OF GOLD

At Riyadh, Abdulla Suleiman had something to show Twitchell. The Bedouins had followed his instructions and they had come through. Twitchell described the scene. "On the Persian rug in front of me, the contents of several gunny sacks were emptied. They were mineral samples. The white quartz glistened and made an impressive contrast with the golden yellow of the copper-iron-sulphides, the lighter yellow of the iron sulphides, the green of the copper carbonates and the red of the iron oxides." Twitchell asked Suleiman where he had gotten them. Suleiman replied that "a Bedouin chief had told him of a large hill with many holes in it, small black glassy (slag) heaps, many small ruined walls, and many grinding stones on the large flat field." The location was between Mecca and Medina.

With a twelve-soldier escort and an assortment of tents, spare parts, tools, and live sheep and goats, Twitchell set off from Jedda in a convoy of six cars and trucks. Twitchell, his wife, an interpreter, a cook, and a houseboy traveled in a model-T Ford. They followed the route of the pilgrim caravans along the Red Sea for two hundred thirty miles, then went north and crossed the derelict Hijaz Railway. East of Medina they bumped their way across the ancient lava flow. After six days and five hundred miles they arrived at a mountain "scarred with the gashes of the ancient miners." Twitchell had found his gold mine. The Arabs called it *Mahad Dhahab*, the Cradle of Gold.

Twitchell spent ten rapturous days taking samples of the tailings, digging holes to get samples at depth and breaking off pieces of ore where he could reach. When they returned to Jedda, he sent his samples for assaying to London and New York. While he waited for the results, Twitchell and his wife wound up Crane's programs in the Yemen. Twitchell knew he was in a tough spot financially. On March 5 he noted despondently in his diary, "I must pay myself for this work (*Mahad Dhahab*) from the Hijaz government because Mr. Crane's gift has terminated."

Returning again to Jedda, this time "at bin Saud's expense," Twitchell was thrilled with the assay reports that indicated ore of commercial value. But he was completely dashed when Suleiman told him bin Saud's government had no money. The worldwide depression had caused the number of pilgrims to fall to a trickle, and the pilgrimage was almost the only revenue they could count on.

Suleiman had not forgotten that Twitchell once stated he "could make a competent report on gold properties that might interest foreign capital." Nor had he forgotten that Twitchell had told him, "I may be able to help you in arranging financing." Not unreasonably, bin Saud and Suleiman suggested that now might be the time. Suleiman said that bin Saud thought Twitchell should "find capital to carry out the development previously discussed." Bin Saud's idea of develop-ment was not just Twitchell's gold mine; it was roads, a harbor, reliable water, electricity supplies, and mineral resources other than gold. For bin Saud, any thought of oil was a long way behind.

Twitchell claimed that he, like a reluctant bride, protested that he "was an engineer not a promoter." Be that as it may, at his insistence he was issued with a letter "authorising and requesting" him "to undertake this project" of finding finance for national developments on behalf of bin Saud's government. At the end of May 1932, armed with bin Saud's blessings in writing, Twitchell departed for New York, via London, to find investors for his gold mine, and, maybe, for the development program that bin Saud had in mind.[5]

In the last week of May 1932, as Twitchell and his wife sailed away from Jedda, the first oil flowed on the Arabian side of the Persian Gulf, from Frank Holmes's well in Bahrain.

BAHRAIN STRIKES OIL

Arnold Wilson followed his request for a two-year exclusive prospecting license over Kuwait, for two thousand sterling, by writing to thank Harold Dickson, the political agent, for his representations to the shaikh of Kuwait on behalf of the Anglo-Persian Oil Company. With typical Wilson pomposity, he declared: "The latest information that has reached us from Bahrain as to the progress of

Holmes's well there is by no means favourable. I have little doubt that it will shortly be abandoned."

Before this letter reached Dickson, a cable arrived from Hugh Biscoe, the political resident. It read: "Political Agent Bahrain reports that the Bahrain Petroleum Company have struck oil and that Well No. 1 is making over 70 tons a day."

Charles Belgrave, the adviser in Bahrain and a strong supporter of Frank Holmes, recorded in his memoirs how this magnificent event was downplayed by the men at Anglo-Persian and the political officers of the Government of India. "At the end of May 1932, oil was found in the first well which was drilled, but the flow was small and there was not much gas," he wrote. "So instead of being pleased, as we were, the experts talked gloomily about there being insufficient oil for commercial exploitation and not enough gas to push the oil out of the ground. But the great thing was that oil had been found in Bahrain."

Wilson's response was typical. "In Bahrain," he wrote Dickson, "they have struck a little oil, rather less than 2000 feet below sea level. But there is not enough pressure to bring it to the surface and it appears to be heavy oil at that. This increases the prospects of unfavourable results at Kuwait, and makes us correspondingly the less inclined to leap into negotiations without looking. But I need say no more as you will be in touch with our people at Abadan."

Ten days after the Bahrain well came in, despite the unequivocal statement to the Foreign Office that they had no interest whatsoever in Kuwait, Anglo-Persian sent their people from Abadan to repeat Wilson's offer for a low-rent exclusive prospecting license.

In contrast to Frank Holmes's unwavering friendly relations with Shaikh Ahmad maintained over nine years, this was the first occasion since Arnold Wilson's visit in 1923 that any Anglo-Persian official had even bothered to visit the shaikh of Kuwait. They had been quite content to leave protection of their interests in the willing hands of the political officers of the Government of India.

Although Abadan's deputy general manager was accompanied by a persuasive Dickson, the shaikh's reception was "extremely frigid." Again he rejected Anglo-Persian's proposition. Again he said that he would consider nothing less than a proper draft concession, on the same lines as that submitted by Holmes.[6]

"SAFEGUARDS"

The political resident, Biscoe, returned with a vengeance to his campaign aimed at restricting Holmes to Bahrain for nine months of the year. He pulled the director of the Geological Survey of India into his scheme and was soon writing the colonial secretary that this illustrious official was suggesting "that HMG should object to Major Frank Holmes being Chief Local Representative in Kuwait as well as Bahrain." Feigning innocence, Biscoe claimed: "I entirely agree with this view."

Biscoe went on to explain: "The company have apparently struck oil in Bahrain and it is probable that they will in consequence in the near future considerably extend their operations and increase the number of their employees. If so, there will be many matters of detail arising for discussion between the company and the Bahrain government and it is essential the Chief Local Representative of the company should be permanently located at Bahrain." Sounding reasonable, he continued: "At present Major Holmes only visits periodically and it is obvious that he cannot attend to the work in Bahrain if he also occupies a similar position in Kuwait." But he could not help himself as he added: "I may mention that I do not consider Major Holmes a particularly desirable person. I should be glad to see him replaced by someone else in Bahrain when occasion offers. In any case, however, he should not be the local representative in Kuwait."

Biscoe claimed it was the director of the Geological Survey of India who was not happy with "the extent of protection of British interests" contained in Frank Holmes's Kuwait concession. "Therefore it should be rejected," Biscoe decreed, "and an entirely fresh concession should be drafted . . . and the safeguards desired to be incorporated should be discussed with Holmes's syndicate in London."

Biscoe felt it had to be done in London as discussions "out here" were "very unsatisfactory" because "the shaikh is a child in the hands of Major Holmes."

The "safeguards" now being proposed for the Kuwait concession by the panicked political officers extended to the Government of India's right to exert its jurisdiction over foreigners, including power over individual Americans. The political resident claimed: "So far as subjects of non-Moslem powers are concerned, the right of jurisdiction ceded us by the Shaikh of Kuwait is no less than that ceded by the Shaikh of Bahrain. If in Bahrain we exercise such jurisdiction without the prior consent of Foreign Powers, as we do, it seems illogical to seek such consent in the case of Kuwait." Furthermore, Biscoe continued: "It seems to me that an attempt to secure such formal consent from, say the Government of USA, in regard to Kuwait would lead that government to question our right to act without it, as in the case of Bahrain."

The resident pushed home his argument by reminding his colleagues that it was the Foreign Office, which was now giving them grief, that had itself suggested in 1912 "the form" currently in use. The form, Biscoe pointed out, "is to ignore the necessity for the consent of Foreign Powers."

DISCONTENT IN PERSIA

Holmes quickly followed through on his Bahrain success. He sent the young Husain Yateem, nephew of his longtime colleague Muhammad Yateem, to call on the the shaikh of Qatar to inquire if he was willing to grant his concession to Holmes. Anglo-Persian's eighteen-month exclusive right to "reconnaisance" of

Qatar, obtained in March 1926, had passed without any activity being initiated, then or since.

Now the Admiralty went into a spin. They thought up a devious possibility. "No opportunity of applying the proposed Kuwait safeguards to the Bahrain concession should be missed. We suggest a favourable opportunity will arise if Holmes's syndicate applies for a concession in Qatar," they wrote the colonial secretary. "The Qatar peninsula is so close to Bahrain (Bahrain to Doha is about 90 miles) that we believe we can contend that the Qatar concession is, in effect, an extension of the Bahrain concession. Therefore, we can insist that as a precondition of advising Shaikh Abdullah Al Thani to grant the Qatar concession—the original Bahrain concession must first be amended to include the safeguards as proposed for Kuwait."

The Admiralty warned that if Holmes and his syndicate obtained the Qatar concession, "the Bahrain concession may gradually be merged into a much larger undertaking, the whole of which will be under the control of American interests." It was most important, they advised, "that no step should be neglected which might enable the Anglo-Persian Oil Company to bring Qatar within their sphere of control."[7]

At the end of June, Anglo-Persian chairman John Cadman issued instructions to prepare a concession, similar to Holmes's, that would combine exploration, prospecting, and mining license into one agreement to be presented to the shaikh of Kuwait. Although Anglo-Persian knew exactly what was in Holmes's draft—because Dickson had passed them a copy—they still offered lower royalties, the same as they were paying in Persia and Iraq.

Anglo-Persian's strange behavior seems to have had more to do with Persia than with Kuwait. For more than two years, Persia had been agitating for a fair deal from the Anglo-Persian Oil Company. The Persians now clearly understood that a wealthy and powerful industry had been developed in their country in which they had no real share and little financial reward. The Persian press consistently depicted the Anglo-Persian concession as serving the interests of imperialists and capitalists. Newspaper reports portrayed the granting of the 1901 original D'Arcy concession as one perpetrated by ignorant officials bribed by unscrupulous foreigners.

Persia was demanding a shareholding in the company, a percentage of the profits from all the company's operations—not just those in Persia—and elimination of all "free" allowances to both Anglo-Persian and its personnel. Now in June 1932, the annual statement for 1931 had just been completed. The Persian government was protesting the paltry amount of royalty allocated and refused to accept. Things were not looking good.[8]

By feigning interest in Kuwait and lodging an offer, albeit one unlikely to be accepted because it contained terms lower than those offered by Holmes, the Anglo-Persian Oil Company could play up the appearance of having an alternative to its Persian fields and by this means shake the confidence of the Persian protests. This tactic was having some effect in Persia where officials began to attribute Anglo-Persian's "bad faith" in refusing to seriously renegotiate to an intention to transfer future development activities "elsewhere."

WOOING E&GSYND

Standard Oil of California had been seeking a way into bin Saud's territory, and Holmes's Al Hasa concession, since Lombardi viewed Holmes's maps and documents in the Gulf Oil Corporation's office.

This was now urgent because, in response to a request from E&GSynd, the Gulf Oil Corporation had released any claim it had on the Al Hasa and the Neutral Zone concessions under the Mainland Agreement of November 1927. This meant that the Eastern & General Syndicate, represented by Frank Holmes, was free to pursue revalidation of the Al Hasa concession on its own behalf. This was why Holmes had told Skinner in conversation in Bahrain that he would not "undertake" to get Standard Oil of California a permit for a geological survey in Al Hasa.

The Gulf Oil Corporation was happy. They believed, mistakenly, as it turned out, that pressure from the State Department had now removed all obstacles between them and the Kuwait concession. Magnanimously, they suggested to E&GSynd: "We do not know whether Standard Oil of California would be interested in the concessions on the Al Hasa and Neutral Zone territories. But we believe they might well be inasmuch as they already have personnel and equipment nearby and these areas would possibly prove a valuable supplement to the Bahrain territory. We suggest you take this up with them."

Whether or not E&GSynd intended to follow this suggestion, they didn't have time because Francis Loomis of Standard Oil of California promptly arrived in London. His purpose was to initiate discussion with the syndicate's directors aimed at reaching a joint venture arrangement for the Al Hasa and Neutral Zone concessions on which the Gulf Oil Corporation had released its claim. Loomis was pursuing these discussions when the presence in London of Harry Philby caught his attention. Philby was enjoying celebrity status as an explorer of bin Saud's deserts.

WOOING PHILBY

When he set out on his crossing of the Empty Quarter, Philby advised his financially desperate wife, Dora, to borrow money from their lawyer with which to keep herself and the children. On his return to Mecca he informed her, "I am a pauper . . . not a bean in the till." He instructed Dora to negotiate a publisher's advance on a book and an advance on newspaper articles with the *Times*. In

London now, Philby was busy trying to capitalize on his recent adventure through speaking engagements. The American consul wrote on Loomis's behalf: "Please allow me to introduce to you the Honourable Francis Loomis, formerly Under Secretary of State of the United States. . . . Mr. Loomis has been impressed with your work in the desert of Arabia and would like to meet you. I am quite certain that the meeting would be mutually agreeable."

On July 11, Philby and Loomis met for the first time, over lunch at Simpsons in the Strand. Loomis was traveling on to The Hague later that same day but would be back in two or three weeks. He told Philby he was "particularly anxious to know whether it would be possible to obtain an oil concession in bin Saud's territory." Philby recorded modestly that he "responded positively," declaring "he would be glad to help in any scheme which would contribute to the prosperity of Arabia."

Two weeks later, on July 27, Karl Twitchell, on his way to New York to find backers for his gold mine and armed with a commission to generate financing for bin Saud's development projects, also met Philby in London. They lunched at the less expensive Holborn restaurant.[9]

WOOING TWITCHELL

Arriving in New York in August, Twitchell set out on a round of visits to mining companies he hoped would finance development of *Mahad Dhahab*, the gold mine he believed to be that of King Solomon. Eleven companies turned him down. They said that with their operations in India, Africa, and Asia, if there was anything of value in Arabia as Twitchell seemed to think, then it was the British who should be interested.

In trying to explain how he came to be hired by Standard Oil of California while on a trusted mission to raise development funds for bin Saud, Twitchell would later give many different versions of what happened next (see note 10). Coincidental or not, Francis Loomis met Philby again in London on September 3, this time they lunched at the Mayfair. He returned to New York the next day, and Standard Oil of California immediately began to woo Twitchell. He was invited to meet a Standard Oil geologist and, shortly after, Francis Loomis himself made the trip out to Virginia where Twitchell was staying.[10]

Loomis arranged for Twitchell to meet M. E. Lombardi in New York. Lombardi asked Twitchell if he was able to get a prospecting permit for Standard Oil of California's geologists to examine the area of Holmes's Al Hasa concession. No point in being satisfied with just a prospecting permit, Twitchell replied grandly and boasted that he could do much better than that. He assured Lombardi: "It will take no more time for me to get the concession than it will to secure a permit for a geological survey." After all the years of casting covetous eyes at Holmes's Al Hasa concession, this was surely what Lombardi of Standard Oil of California wanted to hear.

Lombardi was so impressed that he recommended to his colleagues that, with the assistance of Karl Twitchell, negotiations for the Al Hasa oil concession should be initiated immediately.[11]

SAUDI ARABIA

While Loomis was on his way to Virginia, Abdul Aziz bin Saud was proclaiming the unification of the two territories of Hijaz and Nejd into one country.

In September 1932 he declared he was naming this new country after his own family. From now on it would be known as "Saudi" Arabia. The new kingdom of Saudi Arabia would fly the green flag of Islam, embellished with two crossed swords and a palm tree. Wahhabi puritanism frowned on music. The Arabia of the Al Sauds would not have a national anthem until 1946 when King Farouk's bandmaster would quickly compose one with which to welcome Abdul Aziz bin Saud on an official visit to Egypt.[12]

TWITCHELL GETS THE JOB

Francis Loomis did not agree with Lombardi's recommendation that Twitchell be hired immediately. Personally, he was very taken with Philby. Perhaps a little dazzled by the star status Philby was enjoying in London, Loomis had believed Philby's own estimate of himself and his self-acclaimed "important position" vis-à-vis bin Saud.

Loomis had also believed the cruel stories fed to him by Philby about Frank Holmes. Philby had roundly denigrated Holmes whom he saw, correctly, as the main obstacle to his obtaining from bin Saud, on behalf of Standard Oil of California, the very concession he had been trying to get for years on the repeated urging of Stuart Morgan.

Loomis cabled Philby in London asking if on his return to Jedda he "would be good enough, on behalf of our company" to inform bin Saud that Standard Oil of California intended to "formally request permission to make a careful geological survey of Al Hasa and the Neutral Zone." If the results proved favorable, the company would then want to "enter into a practical working contract for development of petroleum." Loomis thoughtfully offered to mail a check to cover the cost of cables.

He followed up with a second cable, dated November 2 and now addressed to Philby in Jedda, emphasizing that the company wanted "exclusive rights" to examine Al Hasa and neutral territories lying between Kuwait and the Qatar

Peninsula. He repeated that, if geological results were favorable, the company would "contemplate a concession for exploration for petroleum to be followed by a lease for producing petroleum." Philby, who was traveling overland to Jedda, did not receive the cables for some weeks.

In keeping with his Establishment background, Loomis liked to pursue his business in defensible stages. He called on the State Department's Division of Near Eastern Affairs. Despite its impressive title, the division was staffed by a single individual. Loomis inquired whether the British had any control over Al Hasa, "a portion of bin Saud's realm bordering on the Persian Gulf," and whether there would be any objection to negotiating directly with bin Saud with "a view to obtaining a concession to prospect for oil in that region." He was told that "as far as the State Department is aware the British have no control over Al Hasa and bin Saud would appear to be entirely free to grant a concession in that area should he please to do so."

Covering himself for future difficulties with E&GSynd—with whom he was still discussing arrangements for a joint venture—Francis Loomis played down the fact that it was Frank Holmes who had identified the oil of both Bahrain and Al Hasa, had marked out the fields, and had personally obtained both concessions. Misleadingly, Loomis told the official at the Division of Near Eastern Affairs: "Major Frank Holmes, who *assisted in obtaining* the oil concession now held by Standard Oil of California in the Bahrain Islands, has offered to assist the company in obtaining the desired concession in Al Hasa."

Revealing how much he had been swayed by Philby's maliciousness during their long London lunches, Loomis went on to say he "did not repose complete confidence in Major Holmes and did not believe he has the influence with bin Saud that he pretends to have." He plowed ahead with a completely untrue statement that could only have come from Philby, saying: "On one occasion when Holmes went to Nejd to see bin Saud about this matter bin Saud refused to receive him."

Seemingly unaware of the successive waves of Ikhwan rebellion that had convulsed Saudi Arabia, Loomis parroted what was Philby's own argument for why he should be appointed Standard Oil of California's representative to bin Saud. Loomis repeated Philby's unfounded charge that bin Saud's supposed refusal to receive Holmes "results from the fact that Holmes had earlier negotiated with bin Saud on behalf of E&GSynd and the promises then given were not fulfilled."

Now he got down to the nitty-gritty. He said he had two persons in mind as possible "negotiating agents" for Standard Oil of California, "St. John Philby and Mr. K. S. Twitchell." He said he knew Philby "very well" and was impressed by him. Curiously, considering their several meetings in Virginia, he said he did not know Twitchell personally, "although my company has been in correspondence with him."

The sole officer of the Division of Near Eastern Affairs did know Twitchell. He had, he told Loomis, "had several conferences with Twitchell and he has made a very favourable impression." The division gave its advice. "Everything else being equal," the officer said, "it will be desirable to entrust any eventual negotiations that Standard Oil of California might have with bin Saud to an American, rather than to a foreigner." He told Loomis that he did not want to give the impression that he was "vouching" for Twitchell or that he did not "have confidence" in Philby.

Displaying a chauvinism that would have made the British proud, the director of the Division of Near Eastern Affairs declared: "In matters of this kind where concessions are being sought by an American company, the American nationality of the negotiator is of no little importance."[13]

Twitchell was in.

PHILBY INFORMS ANGLO-PERSIAN

Philby finally replied to Loomis's cables. He went straight for the jugular and advised Loomis to make a firm offer based on "annual rent, in advance of 5,000 sterling in gold, government to have 30% of all net profits and a loan, in gold, of 100,000 sterling recoverable from payments due." Philby peremptorily instructed Loomis to "telegraph immediately acceptance in principle, or precise counter offer on these three points . . . or send representative to continue discussions, with *my* assistance."

Loomis was taken aback. His reply on November 28, 1932, said that he understood, obviously from Twitchell although he did not say so, that bin Saud "is favourably inclined" to Standard Oil of California and "its intention to first undertake a geological survey." Somewhat alarmed, he told Philby to "hold matters in abeyance" because Lombardi would soon arrive in London and "communicate with you further."

Twitchell's breezy confidence that he could get the concession "with ease" looked very good indeed in comparison to Philby's demanding response. Twitchell knew when he was onto a good thing. He drove a hard bargain.

Twitchell's contract with Standard Oil of California included a salary of $1,000 per month, for a minimum of three months, to be extended "as necessary." There would be an immediate deposit of $2,000 and expenses paid monthly. If, "through Twitchell's efforts," the company was successful in securing an oil concession and proceeded on to test drill, he would receive a bonus of $15,000 within thirty days of the well being spudded. If the company did find "commercial production of oil," Twitchell's additional bonus would be $50,000.

To allow room for developing his *Mahad Dhahab* gold mine, and anything else that might come his way, the contract allowed for Twitchell to "do work for other companies in other lines of business while in Arabia." Monthly expenses were to be apportioned to each company "according to the amount of time spent for each." Twitchell was to leave New York in January 1933 for London where he was to meet Lombardi and team up with Lloyd Nelson Hamilton, a practiced negotiator in Standard Oil of California's legal and contract department.

Philby might have seen an opportunity to get paid from more than one source for doing exactly the same thing. Or perhaps he was unhappy at the less than enthusiastic reply from Loomis telling him to "hold matters in abeyance." Whatever may have been his motivation, Philby contacted the Anglo-Persian Oil Company.

Philby boasted to Anglo-Persian's chief geologist, G. M. Lees, about the "high level approach" from Standard Oil of California and the expensive lunches in London with Francis Loomis, a former undersecretary at the State Department. He proudly advised the management of the Anglo-Persian Oil Company that he had been "approached by an American concern to apply to bin Saud's government for a concession in the Al Hasa province."

Saying "of course, this information is strictly private between you and me," Philby gave the terms he had dictated to Loomis. He told Anglo-Persian that if they wanted the Al Hasa concession they should appoint him immediately "to place a definite offer before bin Saud's government."

Standard Oil of California now had several horses in the same race. They were in the concluding stages of the agreement for a joint venture with the directors of E&GSynd and Holmes. Harry Philby was convinced Standard Oil of California had an agreement with him. Karl Twitchell was packing his bags to fulfill the contract he had with Standard Oil of California. While no party knew of Standard Oil of California's arrangements with any other party, all were aimed at the same objective—securing Frank Holmes's Al Hasa concession.[14]

NOTES

1. Elizabeth Monroe, *Philby of Arabia* (London: Faber and Faber, 1973), pp. 176–77, "his show is worse than Sharif Husains's" and "never want anything more from the Arabs"; Philby Papers, box, XXX-7, November 19, 1931, Cree to Philby, "result of Twitchell's expedition will be most interesting"; H. St. J. B. Philby, *Arabian Oil Ventures* (Washington, DC: Middle East Institute, 1964), p. 76, "I saw Twitchell again at the King's camp . . . near Riyadh. . . . [W]hen Twitchell resumed his journey . . . I accompanied him to Hofuf, which was to be the starting point of my own venture. And there we parted, he to wander in Al Hasa, and incidentally to find encouraging signs of oil in the hills of Dhahran."

2. Philby's book, *Arabian Oil Ventures*, published posthumously in 1964 by the Middle East Institute, was underwritten by Aramco, which consequently obtained proofing and editing rights. George Rentz, manager local government relations at Dhahran for Aramco, did the supervision and galley proofing and wrote a prepublicity article for *Middle East Journal*. John M. Curry of Aramco's law department requested "more deletions." See Philby Papers, box XLII.3, April 18, 1962, Gary Owens of Aramco's Washington office to Bill Sands at the Middle East Institute, "there are three references to oil seepages in Al Hasa all attributable to Karl Twitchell. No oil seepages have ever been

noted by Aramco geologists ... but there are certain factions critical of American oil efforts in Saudi Arabia who like to say that the Americans knew there was oil there all the time and took advantage of the bare foot boys." Philby's three references are (1) "and there we parted, he to wander in the Hasa, and incidentally to find encouraging signs of oil in the hills of Dhahran." (2) In a quoted letter from Philby to Loomis, he says, "I discovered a Miocene deposit ... between Selwa and Jabri, this probably underlies a part of Al Hasa, along the coast of which between Ras Tanura and Jubail Twitchell ... discovered substantial seepages." (3) "Twitchell was asked to visit Al Hasa in connection with its water problems and ... he found favourable indications of petroleum." The editors contacted Twitchell for clarification and the following editor's note appears on p. 83, "No oil seepages ever have been noted by geologists of the Arabian American Oil Company (Aramco)." Twitchell, in a recent private communication, states: "I do not remember ever having seen oil seeps in Al Hasa ... I do not recall ever telling Philby so ..."; note, also, in Twitchell's Papers, this interesting entry in a March 6, 1940, letter from Twitchell to bin Saud, "you may remember that after my first trip to El Hasa I reported that I considered that the Dahana had favourable structures (for oil) ..."; Twitchell Papers, Merritt interview, p. 81, "an oil company drilling and tell me what you think about it" and "nothing but a dead porpoise"; K. S. Twitchell with Edward J. Jurji, *Saudi Arabia, with an Account of the Development of Its Natural Resources* (Princeton, NJ: Princeton University Press, 1947), p. 217, "former arrangement with a foreign oil company concerning Al Hasa but as they had not lived up to their terms of agreement he wanted nothing more whatsoever to do with them." This supposed quote from bin Saud in reference to Holmes and his syndicate, coming after the fact of Twitchell's employment by Standard Oil of California to obtain Holmes's Al Hasa concession, is certainly suspect. It is not mentioned again in any other of Twitchell's articles, speeches, or interviews.

3. E. A. Skinner, interview by JWS, Chevron Archives, box 120791, in-house memorandum, August 15, 1955, "dressed in full flowing Nejdi Arab robes" and "didn't know anything about oil" and "cabled San Francisco"; see also Twitchell Papers, Merritt interview, p. 81; Twitchell refers to wearing "Arab costume" in Bahrain and says, "the American in charge was not at all cordial at first."

4. Twitchell Papers, Merritt interview, p. 81, "bin Saud asked Twitchell to get him an oil geologist." Twitchell's own version is somewhat suspect. Coming after the fact of Twitchell's employment by Standard Oil of California to obtain Holmes's concession, it can be viewed as self-serving. In this 1965 interview he says he told bin Saud if Bahrain did find oil, "you would probably have it in your country and could get better terms if you wanted anyone to help you." And he claims bin Saud specified "he wanted only American capital because he had seen British go into foreign countries to do pioneering and then make that country a colony."

5. Twitchell Papers, personal diary, March 5, 1932, "I must pay myself for this work ... Crane's gift has terminated"; Twitchell Papers, original typed manuscript for Saudi Arabia, p. 54, "at bin Saud's expense." Twitchell and Jurji, *Saudi Arabia*, pp. 147–48. This "authorisation" letter of bin Saud's does not seem to have survived. It is my belief that the letter made no mention of oil. Twitchell was not seeking investment in oil; he was seeking investors in gold mining, see Merritt interview, p. 82, "on the basis of that assay report I had to find capital willing to develop it (*Mahad Dhahab*). I approached about eleven New York companies. . . ." Note that in Twitchell and Jurji, *Saudi Arabia*, p. 145, Twitchell, after

claiming he advised bin Saud to wait the oil result in Bahrain, inserts a sentence, "Furthermore, it seemed quite possible that American capital might be found to undertake the great expense of oil development in Hasa." And Philby, *Arabian Oil Ventures*, p. 77, says, "Twitchell went off to America to try to sell Arabian oil to its industrialists." Both these statements are suspect and more likely inspired by the later actual events involving Philby and Twitchell. Moreover, Philby had outlined bin Saud's development plans to Crane in May 1930; these were already drawn up long before Twitchell's trip to Al Hasa and did not include oil. See also Twitchell and Jurji, p. 148, Twitchell tried unsuccessfully to interest Charles Crane in investing in his gold mine; oil was never mentioned.

6. IOL/R/15/5/239, vol. IV, May 30, 1932, Wilson to Dickson, "shortly be abandoned" and June 4, 1932, telegram, resident (Biscoe) to political agent Kuwait (Dickson). Oil was actually struck in Bahrain on the night of May 31, 1932. The well was tested next day, June 1, and proved to be in commercial quantities; Charles Belgrave, *Personal Column* (Beirut: Librarie du Liban, 1960), pp. 82–83, "instead of being pleased, as we were. . . . But the great thing was that oil had been found in Bahrain"; IOL/R/15/5/239, vol. IV, May 30, 1932, Wilson to Dickson, "increases the prospects of unfavourable results at Kuwait"; Archibald H. T. Chisholm, *The First Kuwait Oil Concession Agreement: A Record of the Negotiations, 1911–1934* (London: F. Cass, 1975), p. 25, "prospecting licence only," Yossef Bilovich, "Great Power and Corporate Rivalry in Kuwait, 1912–1934: A Study in International Politics" (PhD diss., University of London, 1982), pp. 235–39, "extremely frigid."

7. PRO/FO/371/16002, June 10, 1932, political resident (Biscoe) to colonial secretary and IOL/R/15/5/875, vol. 394, June 11, 1932, "Confidential," political resident (Biscoe) to colonial secretary, "it should be rejected" and "the Shaikh of Kuwait is a child in the hands of Major Holmes"; IOL/R/15/1/639 D78, June 17, 1932, political resident (Biscoe) memorandum on HMG's right of jurisdiction over foreigners in Kuwait and Bahrain, "ignore the necessity for the consent of Foreign Powers"; IOL/R/15/5/875, vol. 394, July 28, 1932, "Important and Confidential," Admiralty to colonial secretary, "we can contend that the Qatar concession is, in effect, an extension of the Bahrain concession" and "no step should be neglected which might enable Anglo-Persian to bring Qatar within their sphere of control." Note that Holmes and Dorothy had taken Husain as a child to England. He had lived with Dorothy at the farm and attended school in Brighton, all at Holmes's expense. In my interviews with Husain Yateem in Bahrain in 1988, he recounted for me this visit to Qatar.

8. The growing Persian disaffection is noted in R. W. Ferrier, *The History of the British Petroleum Company: the Developing Years, 1901–1932* (Cambridge: Cambridge University Press, 1982), pp. 600–607, and Benjamin Shwadran, *The Middle East: Oil and the Great Powers* (London: Atlantic, 1956), p. 42. On August 7, 1931, Cadman declared Anglo-Persian would no longer contemplate a revision of the concession because "the demands of the Persian government were greatly in excess of anything which the company could accept." Discussion continued on the basis for calculating the 16 percent net profit to which the Persian government was entitled. Persia claimed it was owed money from previous years. Ferrier, p. 618, says that Anglo-Persian viewed Persia's demands as threatening "future developments elsewhere," Shwadran, p. 42, says Persia's "excessive demand" was for a guaranteed annual income of 2,700,000 sterling; Shwadran, p. 42, "Persian government protested."

9. Chevron Archives, box 0426154, April 15, 1932, Leovy of (Eastern) Gulf Oil to E&GSynd confirming, "E&Gsynd will be free to negotiate for and secure, on its own behalf, independently of the Gulf Oil Corp, concessions on the Hasa and Neutral Zone territories"; Elizabeth Monroe, *Philby of Arabia*, pp. 186–87, "borrow from the lawyer" and "I am a pauper"; Thomas E. Ward, *Negotiations for Oil Concessions in Bahrain, El Hasa (Saudi Arabia), the Neutral Zone, Qatar and Kuwait* (New York: privately published, 1965), p. 205, quotes the letter and cites an interview with Feuille in 1933, "as retold by Feuille in 1933" for the American Consul's introduction to Philby; Philby, *Arabian Oil Ventures*, p. 78, for this first lunch with Loomis "glad to help"; Twitchell Papers, diary entry, July 27, 1932, "lunched with Philby at Holburn Restaurant."

10. Twitchell gives a number of versions of how he came to be employed by Standard Oil of California. In *Saudi Arabia*, pp. 148–50, he says that through a friend he was introduced first to Texas Oil, who suggested he try Near East Development and Standard Oil of California. He says "by chance" he went to Stuart Morgan of Near East Development, the body that represented the American group in the Iraq Petroleum Company. On p. 149 he says Stuart Morgan assured him that the Iraq Petroleum Company "was the most powerful group of oil companies in the world. Although they had been caught napping while the Bahrain oil fields were negotiated for by another organisation, it would not permit itself to do so again." If this was Morgan's assurance, then it is hard to see why, as Twitchell claims, he immediately went to Guy Stevens at the Gulf Oil Corporation, which "turned him down." He says "after this disappointment" he was contacted by Corriell, the New York representative of Standard Oil of California, followed by meetings with Loomis and Lombardi. Twitchell gives another version in the Merritt interview, p. 82, in which he says that after being turned down by the mining companies, "before leaving New York" he called on the Texas Company, "which could not undertake a foreign project" and says, "a few days later a representative of the Standard Oil Company of California contacted me." In this interview, he makes the highly unlikely claim that "after studying my notes on Saudi Arabia . . . Standard Oil of California . . . suggested I contact Stuart Morgan." He goes on to say that "shortly" after sending him off to Stuart Morgan, "Standard Oil of California decided to go ahead with the Arabian oil venture on condition that I act for them in negotiations with the Saudi Arabian government." In the manuscript for the *Explorer*, Twitchell does not even mention any discussions with the oil companies. He says only that "all principle mining companies and groups" turned his gold mine down and suggested he tried London. "This was done and after several disappointments the Saudi Arabian Mining Syndicate Ltd was formed composed of representatives of nearly all the most prominent mining groups in London." Philby, *Arabian Oil Ventures*, p. 78, for the Mayfair lunch with Loomis; Twitchell and Jurji, *Saudi Arabia*, p. 150, for the New York meeting with Standard Oil of California geologist, see also Chevron Archives, box 120791, in-house memorandum of interview with E. A. Skinner, August 15, 1955, "Twitchell's first meeting at Standard Oil of California was with geologist H. J. Hawley. . . . Hawley's statement to me was that Twitchell did not know there was more than one Standard Oil Company and thought when talking to Near East Development that he had canvassed Standard Oil of California" and "Skinner was not familiar with the peregrinations described in Twitchell's book"; Twitchell Papers, original typed manuscript for *Saudi Arabia*, p. 57, for Francis Loomis traveling to visit Twitchell in Virginia.

11. Ward, *Negotiations for Oil Concessions*, p. 205, citing an interview and lunch

with Standard Oil of California's legal adviser, Judge Feuille, retold by Feuille in 1933, for Loomis arranging the meeting between Lombardi and Twitchell and quoting Twitchell assuring Lombardi: "It will take no more time for me to get a concession than it will to secure a permit for a geological survey."

12. Robert Lacey, *The Kingdom* (London: Hutchinson, 1981), p. 280, "(the Saudi national anthem) a ditty knocked out on the spur of the moment by the bandmaster of King Farouk of Egypt. . . ."

13. Philby, *Arabian Oil Ventures*, p. 79, for the two telegrams Loomis to Philby; Abu Dhabi Documentation Center, United Arab Emirates, copy of the Records of the US Department of State Relating to Internal Affairs of Saudi Arabia 1930–44, Film T1179/2, December 1, 1932, Division of Near Eastern Affairs, Department of State, Conversation which took place November 16, 1932, with Mr. Francis Loomis regarding the interest of Standard Oil Company of California in El Hasa, "After getting the issue of Al Hasa out of the way, Loomis went on to complain of the British 'placing petty obstacles' in the way of his company since they had discovered oil in Bahrain and of the difficulties with the British being encountered by Gulf Oil in Kuwait."

14. Philby, *Arabian Oil Ventures*, pp. 80–81, for this exchange of cables between Philby and Loomis and "I understand bin Saud is favourably inclined towards Standard Oil of California"; Twitchell Papers, January 10, 1933, Terms and Conditions, Standard Oil Company of California and K. S. Twitchell (signed New York between Lombardi and Twitchell); Philby, pp. 80–82, "if Anglo-Persian interested in Al Hasa . . . should immediately appoint me"; Philby would claim that he orchestrated the entire sequence. See H. St. J. B. Philby, *Arabian Days, An Autobiography* (London: R. Hale, 1948), p. 295, "On arrival at Jedda I found a telegram waiting from Loomis, one of the Vice-Presidents of the Standard Oil of California, whom I had met on several occasions during the summer to discuss the prospects of an oil concession in Sa'udi Arabia. . . . Loomis confirmed his unabated interest in the matter (a concession) and formally authorised me to ascertain the conditions on which the King might be prepared to grant a concession. After discussions with the latter and his Finance Minister I was able to telegraph a reply, to which the prompt response was the arrival in January 1933 of Lloyd N Hamilton accompanied by Karl Twitchell as technical adviser to begin negotiations for Standard Oil of California."

CHAPTER 19
INTRIGUE AND ILLUSION

As Karl Twitchell was crossing the Atlantic on his quest for an American mining company to finance development of the *Mahad Dhahab* gold mine, Harold Dickson and Hugh Biscoe were onboard HMS *Bideford* in the waters of the Persian Gulf. They were traveling together for the purpose of persuading the shaikh of Sharjah to sign an agreement allowing Imperial Airways to use his territory as a stopover on long-haul flights. On July 17, 1932, while Dickson slept, his fellow pro-Anglo-Persian conspirator and main anti-Holmes, anti-American ally, political resident Hugh Vincent Biscoe, died of a heart attack.

In sight of Sharjah, Dickson and the ship's crew appeared ondeck in full dress uniform. To the sound of a mournful tune from the ship's band, and an appropriate volley to see Biscoe off, they buried at sea the chief representative of His Britannic Majesty in the Persian Gulf. Relentlessly, Dickson then continued on in pursuit of his duty. After what he had just witnessed, the shaikh of Sharjah did not voice his objections to passenger aircraft full of foreigners landing in his country. He signed.

In August, Lt. Col. Trenchard Craven Fowle took up the appointment of political resident. He knew the issues, having covered for Biscoe's six-month leave the year before. Fowle was a career officer with the Indian Political Service. His previous appointment, as Major Fowle, was as political agent in Muscat. As a mere captain in 1916, he had filled a six-month vacancy as political agent, Bahrain.

Dickson's first intimation that Fowle might not be on the same wavelength as Biscoe had come fairly quickly. Holmes was on a trip to London when Anglo-Persian's draft concession, ordered by John Cadman, arrived for the shaikh of Kuwait. Dickson wrote to Fowle, saying "a good result" would be obtained if he authorized Dickson "to tell Shaikh Ahmad that you (the Resident) approve of Anglo-Persian's draft . . . and have no objection to his opening formal discussions with the Anglo-Persian Oil Company." Instead, Fowle wrote the Colonial Office, raising "Britain's moral obligation to advise the shaikh." As this was the case, he suggested, the government's "unbiased views" of both concessions should be promptly drawn up.

On his way back to Kuwait in September, Frank Holmes wrote his wife from Baghdad: "I arrived here at noon yesterday and received a telegram from Muhammad Yateem, instigated by the Shaikh of Kuwait, asking me to stay and wait for his arrival by air on Saturday. Yateem arrived from Basra bringing me good news regarding the attitude of the shaikh and, more importantly, that the people of Kuwait are sticking strongly for me."[1]

ANDREW MELLON TAKES ACTION

Although the Gulf Oil executives believed there was a strong diplomatic offensive running on their behalf, little was actually going on. At the beginning of September the American Embassy in London, under Ambassador Andrew Mellon, sent a note to the Foreign Office listing Gulf Oil's complaints and reminding the Foreign Office of its promise that American interests would be placed in "as favourable a position as British ones" in having the draft concessions considered by the shaikh of Kuwait. The Foreign Office assured the Americans that the offer from the Gulf Oil Corporation and that of the Anglo-Persian Oil Company were in the process of being examined by the various departments of the British government "on their merits" purely for the purpose of ascertaining which "suited the shaikh's best interests commercially."

Nothing happened. The "various departments of the British government" continued their barnyard squabbling. With a substantial private interest in the Gulf Oil Corporation and therefore in the Kuwait concession, Ambassador Mellon finally snapped. Without advising the State Department he called personally on Sir Robert Vansittart, permanent undersecretary at the Foreign Office, first on October 17 and again two weeks later.

The urbane Vansittart was mortally embarrassed. He fired off a sharply worded memo "to all departments concerned" stating that he would "no longer submit to being placed in this intolerable position." He would not, he said, "accept any further excuse whatsoever for any further delay." He was particularly scathing of the Anglo-Persian Oil Company, which, he said, "had thrown away their chance six years previously when they could have bought out the Eastern & General Syndicate and now, apparently, wish to pop in on a change of mind."

Shocked into action, before the week was out the Petroleum Department was circulating a memorandum in which they concluded that Anglo-Persian's offer was the most advantageous to the shaikh of Kuwait's commercial interests. The comparison was remarkably creative when reporting on Anglo-Persian's lower royalty and refusal to pay customs duty. They blithely advised that the lower royalty "will have the greater inducement to produce oil" while "too high a royalty will retard production." Anglo-Persian's royalty, it was noted, "is approximately" the same as that in Iraq.

A copy of the comparison reached the State Department via Ambassador Mellon after the shaikh of Kuwait passed a copy to Holmes. The State Department was not impressed. They noted: "Not one favourable comment is made with respect to the E&GSynd draft. On the contrary, some of E&GSynd's offers are twisted and misrepresented in such a manner as to make them appear unattractive even when they are obviously more advantageous to the shaikh than are the terms of Anglo-Persian's draft."

Anglo-Persian's royalty was 2.10 rupees per ton. E&GSynd's was 3.8. E&GSynd offered 1 percent custom duty on all oil produced, calculated on the value of the oil at the wells, while Anglo-Persian offered "no similar provision." The Petroleum Department's memorandum had concluded that this was not a good idea anyway, as "it is not usual to impress a customs duty in addition to a royalty."[2]

The Foreign Office assured Ambassador Andrew Mellon that the comparison would be forwarded with instructions to the shaikh of Kuwait that he was free to choose whichever concession he wished. This inspired another round of bickering as the Petroleum Department, the Admiralty, the India Office, and the Colonial Office lobbied the Foreign Office demanding that the shaikh of Kuwait be "advised" to select Anglo-Persian. Dickson, for one, warned shrilly, "HMG *must* bring a certain amount of direct pressure on the shaikh. Otherwise, the concession will be lost to Holmes."

By December 13, with still no movement toward the Kuwait concession, Ambassador Mellon would be back in the Foreign Office, protesting strongly. When news of Mellon's independent and somewhat undiplomatic actions on behalf of "his company" reached the State Department, he would be politely requested to leave the matter in their hands.

KUWAIT ECONOMIC STRESS

As had happened in Bahrain, events in Persia now impacted on the Kuwait concession. Like Kuwait, Persia was in desperate need of money. The full force of her frustration now turned on the foreign entity in her midst. The shah of Persia personally announced, at a cabinet meeting on November 26, 1932, the outright cancellation of the concession held by the Anglo-Persian Oil Company from the Persian government.

The Shah of Persia charged that the Anglo-Persian Oil Company had never allowed Persia to inspect expenditures in order to safeguard the supposed 16 percent of net profit, that the company had never submitted to the Persian government any detailed account or other evidence of its expenditures or the expenditure of all its subsidiaries, that it had refused to pay Persia a share of the royalty of subsidiary companies, and that it had refused to pay income tax (introduced in 1930). Furthermore, Persia stated it could not accept that, according to Anglo-Persian, all it was owed from the 1931 net profit of some 3 million sterling was 307,000.

Kuwait was in dire economic distress for the second year in a row. Dickson's report for 1932 was even more heartrending than that of 1931. In Kuwait, he reported,

> suffering and acute want among the lower classes of the town is a new and pathetic feature . . . gangs of beggars are roaming the town . . . some 2,000 starving Persian refugees arrived, driven across to the Arab coast from their own country by hunger . . . they filled the streets and byways imploring all and sundry to assist them. They were followed by a wave of Persian fishermen. The Ruler decreed all Persians without visible means of support be repatriated to their own country . . . (Kuwait) sent them in dhows and dropped them on the left bank of the Shatt Al Arab . . . some 2000 Kuwaitis died from smallpox in the town alone in the four months between July and October.

Yet in the same report, Dickson could callously comment that "oil negotiations are not progressing in sympathy with British interests" and warn of the possibility of Shaikh Ahmad "falling a too easy victim to a powerful American Oil Corporation." Apart from this, Dickson cheerfully reported, because he did not "obviously" interfere in internal administration "the shaikh is flattered and pleased to think that he is being allowed to run his own show in internal matters. He is wise enough to appreciate that the Protecting Power should have a say in his dealings with his neighbours (particularly bin Saud and Iraq)."

There was, he said, "one notable exception" to this rosy picture. This was "the negotiations of the American oil group, represented by Major Frank Holmes."

Dickson chafed at the intimate friendship between Holmes and Shaikh Ahmad. Since his days with Arnold Wilson in the Mesopotamia Administration Dickson had boasted of having "a wonderful hold" over Arabs. He had claimed to Biscoe, the India Office, and the Colonial Office that he and Ahmad were so close that he knew better what he was thinking than did Ahmad himself. Yet, here was Ahmad obviously preferring to be in the company of Holmes than Dickson. He sought to explain this, saying: "The shaikh thinks it is unwise to say too much to me, the representative of His Majesty's government in Kuwait, who must perforce favour one party because it is a British concern and so tender biased advice." With superb aplomb considering his indefatigable activities on behalf of the Anglo-Persian Oil Company, Dickson declared: "I have taken some pains to point out to the shaikh the fallacy of such an argument."[3]

ACTIVITY IN BAHRAIN

Bahrain, meantime, was a hive of activity. The political agent's report for 1932 recorded that "the Bahrain Petroleum Company struck oil on June 1st and before shutting off the well obtained a flow of 70 tons a day. That rig was dismantled and re-erected some two miles away on a fresh site where drilling commenced on August 1st. A second rig was brought from America and erected on yet a third site. The second well has reached a depth of over 1500 feet and traces of oil have been found."

The report continued: "There is little doubt the company is beginning to exert a considerable local influence. This will undoubtedly increase, and combined with the American Mission, represents a most unfortunate intrusion of foreign influence into Bahrain. They have hired all their workers in Iraq or locally and have not given them formal agreements."

Standard Oil of California, through its wholly owned subsidy Bahrain Petroleum Company, was far from being a benevolent employer. Its laborers worked a seven-day week and this, combined with the lack of transport, meant Bahraini employees only got home to see their families once in eight weeks, longer if they came from outlying islands.

The Americans were housed in Nissin portables with, as the political agent had reported the year before, "everything that practical comfort could suggest and refrigeration, electric light and heaters, full length baths and hot and cold showers." The Bahrainis, Iraqis, and Indians, by now totaling some three hundred fifty men, lived in *barasiti* huts, a fragile and leaky shelter made from local palm leaves.

The political agent was concerned that this influx of men and industry would deplete the water supply so hard-won by Frank Holmes. "Water is still being wasted in incredible quantities and this, Bahrain's more valuable asset, is being frittered away as though it were inexhaustible," he wrote. "Towards the end of the year Major Holmes, the pioneer of artesian wells in Bahrain, addressed a long letter to Shaikh Hamad pointing out the danger of squandering so valuable an asset and stating that certain artesian wells have now tapped the lowest water holding strata. Shaikh Hamad has promised to take action."[4]

STRIKE TWO IN BAHRAIN

Frank Holmes had probably the greatest Christmas present of his life when, on the morning of December 25, 1932, Bahrain's second oil well came in, with a rush.

Holmes excitedly sent for Charles Belgrave and his wife to share the moment. They were attending the church service at the American Mission. They raced off to Holmes's site, arriving at the spectacle in their very best clothes. "When Marjorie and I reached the well, which was in the foothills near Jebel

Durkhan, we saw great ponds of black oil and black rivulets flowing down the wadis," Belgrave recorded. "Oil, and what looked like smoke but which was in fact gas, spouted gustily from the drilling rig and all the machinery and the men who were working were dripping with oil. It was impossible to tell which of them were Americans and which were Arabs. It was not a pretty sight but it was an exciting moment for me," he wrote in his memoirs. "I could see, without any doubt, that there was an oil field in Bahrain."

Belgrave had been supportive of Frank Holmes since he discovered they were "on the same wavelength" the night he arrived to take up his post as adviser. With some satisfaction, Belgrave observed: "It was a great day for Major Frank Holmes, who now saw the visible proof of what he had always believed."

Charles and Marjorie Belgrave turned their children's Christmas party at their house that evening into "a very gay celebration of Bahrain's first real oil well." Frank Holmes was the guest of honor.[5]

Word of the Bahrain strike spread like wildfire up and down the Gulf. Across the shaikhdoms the Arabs dubbed Frank Holmes "*Abu Al Naft*—the Father of Oil." All the Arabs knew Standard Oil of California had bought the concession that Holmes had identified and mapped in 1922. And they knew that Holmes had never given up on that concession; nor had he given up on his other finds in Al Hasa, Kuwait, and the Neutral Zone. Not for a moment did they believe the stories that now erupted in the Western media about the Americans discovering oil in Arabia. Right across the Persian Gulf everybody knew it was the New Zealander Frank Holmes—*Abu Al Naft*—who had found their oil.

The 1932 oil strike in Bahrain (and then in Kuwait, Saudi Arabia, and the Neutral Zone, in areas originally delineated by Frank Holmes a decade earlier) left many highly rewarded people very embarrassed indeed. The men from Anglo-Persian had been dogmatic in their insistence that there was no oil in Arabia. They were now considerably annoyed at having been shown up by someone they considered a rank outsider and well below their own caliber. A number of myths would subsequently be fostered in an effort to gloss over the fact that, in the conviction there was no oil in Arabia, the established experts were so wrong, so often, and for so long. Meantime, the Arabs of the Persian Gulf would continue to praise Holmes—*Abu Al Naft*—as the discoverer of the Arabian oil fields.

By January 1933 when the dispute between the Persian government and the Anglo-Persian Oil Company went before the League of Nations, the British were publicly threatening Persia that they would "take measures" to "preserve" Anglo-Persian's rights. The Persians could not miss the implications of the British warships that appeared in the Gulf.

Arguing on behalf of "a British company," the British government dismissed much of Persia's claims as "ancient history" and argued that "exploitation of oil had brought nothing but good to Persia, as evidenced by the total of 11,000,000 sterling," paid over some thirty years, "in royalties to the Persian government." The Persian government pointed out, among other issues, that "as to the 11,000,000 sterling, which it had received in revenue, customs duties alone for the period 1901–1932 would have amounted to almost 20,000,000 sterling, had the company paid them." The League of Nations hearing would be suspended in February to allow for an attempt at negotiations between the two parties in Geneva and Paris. By the end of February, on the grounds that the delights of Paris were a distraction, the negotiations would be moved to Tehran.[6]

PHILBY GETS HIS CONTRACT

On January 13, 1933, while the Persian government was fighting before the League of Nations for a better deal from the Anglo-Persian Oil Company, Karl Twitchell with his wife and son sailed from New York to London. They were met by Standard Oil of California's experienced lawyer and negotiator, Lloyd Hamilton, also accompanied by his wife. Although Philby did not know Twitchell had been hired by Standard Oil of California, in Jedda he heard that Twitchell was returning "on behalf of some American group who appear to be interested both in the oil and gold possibilities of bin Saud's country."

Philby promptly re-presented himself to Francis Loomis in a letter dated January 16. He said he "quite understood that the speculative nature of the Hasa terrain deters your friends from immediate acceptance of the Government's conditions." He drew Loomis's attention to "the revolt" of the Persian government against the terms of their concession held by the Anglo-Persian Oil Company. In comparison, he spoke in glowing terms about what he called the "lavish arrangements" made by the Iraq Petroleum Company with the Iraq government. He raised the spectre of "British official pressure" by saying this was the way Anglo-Persian's prospecting license in Qatar had been obtained on "easy terms."

Unaware that Twitchell had been hired by Standard Oil of California, Philby claimed, quite untruthfully, that during Twitchell's tour of the Eastern Province he had "discovered substantial oil seepages." He advised Loomis: "The fact that Twitchell is now returning for some American group should be seen as proof that, doubtless, he has been able to give his people some idea of what they may expect to find." Again untruthfully, he said that the Saudi government "is making a vigorous attempt to place its concessions on the market" and urged Loomis to "lose no time in arriving at a decision in this matter." The only person trying to get the concession on the market was Philby and by "this matter" he meant the necessity of employing him to obtain the concession from Abdul Aziz bin Saud.

Not receiving any reply, Philby cabled again ten days later. A diplomatic response came from M. E. Lombardi in London. Revealing that Twitchell was now associated with Standard Oil of California, but not informing Philby of the goal of that association, Lombardi said that Hamilton and Twitchell were due to arrive in Jedda on February 15 "for the purpose of conferring with you and discussing with you the Arabian government terms for a proposed arrangement. I hope they can have your continued support."

The Anglo-Persian Oil Company was also expecting the continued support of Philby. They had received his offer, saying that if they wanted the Al Hasa concession they should immediately appoint him "to place a definite offer before bin Saud's government." They replied that, in spite of their current concession difficulties with Persia (or perhaps because of), "we are definitely in the bidding for Al Hasa."

The reply came from Anglo-Persian's chief geologist, G. M. Lees, who pointed out that, because of bin Saud's long-standing antagonism toward the company, they could not "act directly." But, he said, Anglo-Persian had "of course, a certain influence in the councils of the Iraq Petroleum Company." He said that Philby would hear from the Iraq Petroleum Company, who would send someone to "talk things over on the spot." They probably wouldn't make an offer "anywhere near what you say the Saudi government is expecting." Nevertheless, Lees suggested, perhaps "a more modest, and more equitable, offer might be found acceptable."

Philby wrote Lees that high-level executives of Standard Oil of California, "the people who have been in correspondence with me," were due to arrive shortly. In a masterstroke he directed Lee's attention to what he should be thinking about. Assuming an air of innocence, he told Lees he wondered why the Americans would want his assistance, *"unless it is to prevent my helping some other party."* Leaving Lees to contemplate that, Philby reminded him that he "was not pledged to any party" and was "available" to Iraq Petroleum/Anglo-Persian.

He received a letter from Twitchell next, but this didn't give him "any details regarding the object of his impending visit." This was quickly followed by another letter from Lombardi who seemed to have been well briefed on the extent of Philby's ego. He was at pains to flatter. He was sorry he wasn't coming, Lombardi said, as he would like "to meet and confer with you." Francis Loomis was "fortunate enough to meet you last summer," he wrote, "and you were good enough to agree to aid us to approach bin Saud. We much appreciate your efforts on our behalf."

Lombardi said that Twitchell would "consult" with Philby, and with Philby's "goodwill and advice and influence at court" Standard Oil of California looked forward to a conclusion satisfactory to both his company and Saudi Arabia. Lombardi said he hoped to have, in the near future, "the pleasure of making your acquaintance and expressing thanks and appreciation for your efforts on our behalf." Unaware that Philby himself had planted this particular rumor with Loomis, Lombardi confided, "According to reports we have, and which seem reliable, the Anglo-Persian Oil Company is now endeavouring to secure an agreement covering Al Hasa."

During their London lunches, Philby had represented himself to Francis Loomis as utterly devoted to bin Saud and purely, altruistically even, concerned with "helping in any scheme which would contribute to the prosperity of Arabia." While it sounded good, this wasn't at all what Philby really had in mind. He was furious now he realized that Twitchell had landed a paid "engagement" to work for Standard Oil of California to obtain the concession in Saudi Arabia. "The company had not given the slightest hint that they ever contemplated the making of comparable arrangements with me," he fumed.

While he was seething over this situation, Anglo-Persian wrote informing him the Iraq Petroleum Company would be sending Stephen Helmsley Longrigg as their negotiator "unless you will act for us. In which case he will acquaint you with our terms and leave negotiations with you."[7]

Philby had built his position inside Saudi Arabia by promoting his criticism of British government policy in the Middle East. In addition, he was well aware of bin Saud's long-standing personal hostility toward the Anglo-Persian Oil Company, now exacerbated by events in Persia. There was no way he could afford to be seen as involved with Anglo-Persian by becoming their official negotiator. (In this he was right. When the concession was signed with the Americans, it contained a clause prohibiting the transfer of rights and obligations, without the consent of the Saudi government. Exactly as bin Saud had required of Frank Holmes in 1923 in order to specifically exclude the Anglo-Persian Oil Company.)

Send Longrigg, he told Anglo-Persian, and he, Philby, would work "behind the scenes." Longrigg and Philby were old friends from the Mesopotamia Administration. In 1920 they had worked together at the Interior Ministry. Longrigg had been political officer at Kut before replacing Harold Dickson as political officer at Hilla. Like so many of Arnold Wilson's "Young Men," Longrigg had followed Wilson into the Anglo-Persian Oil Company.[8]

Philby struck much the same deal with Standard Oil of California. Within forty-eight hours of the arrival of Hamilton and Twitchell, Philby had what he thought was a pretty good deal, better than Twitchell's, to "work on behalf of Standard Oil of California."

Philby would receive a salary of $1,000 a month, but guaranteed for a minimum six months rather than Twitchell's three months, $10,000 when the concession was signed, $25,000 when commercial exploitation began plus fifty cents per ton exported until a second $25,000 was reached. Philby convinced Hamilton it would be better if this arrangement was not made known, most particularly to the Saudi Arabian officials. Repeating exactly what he had told Anglo-Persian, Philby said that the best working arrangement would be for Hamilton "to conduct negotiations with the government, with Twitchell in attendance, while I remain in the background." Philby would, he said, provide "information and advice."

Mr. and Mrs. Hamilton and Mr. and Mrs. Twitchell moved in with Mr. and Mrs. Philby in the Jedda house. Dora Philby was visiting at the time. Her relief at the possibility of now meeting her loans, household debts, and school fees was palpable. She generously noted of a husband who had never before cared about his family's finances: "Jack is a different man. He was beginning to get very jumpy about the money problems."

Philby would be extraordinarily successful in keeping secret from the Saudi Arabians his lucrative arrangement with Standard Oil of California. He would record that Yousuf Yassin, a primary adviser to bin Saud, accused him years later saying: "If I had known, at the time, that you were actively working for Standard Oil of California, I would have done my best to block the grant of the concession to them." It has been claimed that not even Twitchell knew Philby was being paid by Standard Oil of California.[9]

Harry Philby had a natural bent for intrigue (which his son obviously inherited. During the cold war, Kim Philby became chief of anti-Soviet counterespionage in British intelligence while actively spying for the Soviets). In the three and a half months of negotiation that followed, Philby would mislead everyone. Bin Saud would assure Philby that he was "confident you will protect our interests, both economic and political, just as you would protect your own personal interests."

Longrigg, the Iraq Petroleum/Anglo-Persian negotiator, with encouragement from G. M. Lees, would believe Philby was working on his behalf. Sir Andrew Ryan, British minister at Jedda—the man who got the job that Philby had so badly wanted—would believe that because Philby was looking out solely for British interests, the concession would go to Iraq Petroleum/Anglo-Persian.[10]

The only person who did not trust Philby was Frank Holmes, whose work in identifying and mapping the oil field of Al Hasa Philby was now setting out to profit from. The discovery in Bahrain had overturned all previous geological theories that there was no oil in Arabia. It vindicated Holmes's unique skills and abilities and greatly enhanced his status and reputation. The Arabs of the Persian Gulf were feting him as *Abu Al Naft*—Father of Oil. Despite the concerted effort he would put into undermining and denigrating Frank Holmes to all parties, to Philby's fury Holmes would arrive in Jedda in April.

ARCHIBALD CHISHOLM

While Philby was intriguing in Saudi Arabia, Dickson in Kuwait was continuing to malign Holmes with a barrage of wild stories and accusations. He reported that Shaikh Ahmad no longer asked for advice or confided in him, a development that Dickson attributed not to his own behavior but to Frank Holmes. He claimed this situation had come about because Holmes depicted the political officers of the Government of India as "retrograde sort of people, bent on keeping local Rulers in leading strings."

Dickson said in Holmes's company, "The Shaikh of Kuwait seems to become the victim of some sort of hypnotic influence." He claimed that "this rather disturbing state of affairs is due to the ideas with which Major Holmes fills the shaikh's mind." It bothered him that "Shaikh Ahmad is a constant visitor to Major Holmes's house." Dickson frequently warned his colleagues: "Major Holmes is a remarkably clever man and knows the psychology of the Arabs better than anyone I know of today." He compared the shaikh of Kuwait with the "clever" Holmes by describing Ahmad as "simple minded and possessed of no more than the suspicious cunning inherent in every Bedouin."

Both companies now had negotiators in Kuwait. Frank Holmes was acting on behalf of the Gulf Oil Corporation. The Anglo-Persian Oil Company sent the thirty-year-old, tall, slim, and monocled Archibald Chisholm. Educated at Oxford, descended from Lord Byron, with family connections to Sir John Cadman's clan, Chisholm was an aristocrat. Following a stint as a journalist (his father was editor of the *Times* and chairman of the Athenaeum Club, Arnold Wilson's London haunt), he had been hired by Cadman in 1927 and sent to Persia. In an unlikely pairing with the elegant Archibald Chisholm, Anglo-Persian sent William (Hajji Abdulla) Williamson as "general assistant and interpreter." This was the same Williamson, the not-so-ex-spy, sent to Kuwait in 1925 charged with discrediting Holmes and bribing Shaikh Ahmad's advisers.[11]

Dealing now with two entirely different sets of negotiations, the shaikh of Kuwait called in reinforcement. To advise him on the merits of the two proposals, he sent for his Armenian lawyer, J. Gabriel, who handled Ahmad's legal affairs from his practice in Basra. Gabriel offered a creative solution. He advised Ahmad to divide his territory into two concessions. In his opinion, he said, Ahmad could then get more in total payment while also being able to satisfy both the Americans and the British. The shaikh of Kuwait was still mulling this over when the first of his international visitors arrived. Now that Imperial Airways had transformed travel to Arabia into a gentleman's pastime, Edmund Janson of E&GSynd was persuaded to visit Bahrain and Kuwait. M. E. Lombardi of Standard Oil of California also made the trip.

THE PLOT UNRAVELS

Lombardi, with Hamilton (but not Twitchell), had continued the discussion in London with Edmund Janson aimed at reaching a joint venture arrangement by putting Standard Oil of California in the position previously held by Gulf Oil under the 1927 Mainland Agreement covering Al Hasa and the Neutral Zone. Lombardi and Hamilton concealed from Janson, Holmes, and the directors of E&GSynd the fact that they were, right then, on the way to Saudi Arabia to obtain Holmes's Al Hasa concession with Twitchell's assistance.

When Edmund Janson suggested in the most friendly manner that they all meet up in Bahrain, Lombardi could hardly refuse, at least not if he wanted to maintain the pretense.

This was why he had had to pass up the opportunity, as he had written Philby, "of the pleasure of making your acquaintance and expressing thanks and appreciation for your efforts on our behalf." Lombardi would later admit: "After a short stay in London, I set out alone for Bahrain Island while Hamilton (and Twitchell), *unknown to E&GSynd*, left for Jedda to negotiate the Al Hasa lease."

Janson explained it in a letter to Ward, who was most anxious to know when he would begin receiving his share of the Bahrain royalties.

> On February 25th I flew out to Arabia to have a look at Bahrain and see if I could do anything with the Shaikh of Kuwait. I met Michael Lombardi in London before he went out and also met him again out in Bahrain. There are two wells down to oil in Bahrain and they are both Gushers, but neither of them is in the oil strata. The proposal is to put No. 3 well through the strata and see what is its depth. A better opinion of what the capacity of the oilfield may be will then be formed. It will probably be another six weeks or two months before this is done. The rub seems to be that the formation is rather too flat. Anyway, I believe Mr. Lombardi is now employed in seeing where he can market the oil, so it looks as if they expect to be in a position to do so before long. I should think it will be quite two years before we shall draw any royalty.

Dickson knew in advance about the visitors. In mid-February he advised the resident that "Mr. Janson and Mr. Lombardi, a Director of the Standard Oil Company, will pass through Kuwait by air en route to Bahrain on March 2nd. Major Holmes proposes to accompany them from here and all three will return to Kuwait on March 9th and stay with the shaikh. After that Janson's and Lombardi's movements are uncertain. This news was given me yesterday by the shaikh and confirmed by Holmes."[12]

When Janson, and Holmes with his wife, passed through Kuwait on the way to Bahrain, Shaikh Ahmad put forward Gabriel's "two concession" suggestion. He explained the idea was that one concession, of twelve hundred square miles, would "be reserved for British interests." The other, of four hundred square miles, would be "reserved for American interests." Holmes argued that this didn't look very fair. By the time they returned from Bahrain on March 14, Ahmad had amended the idea to dividing the original area equally between British and American interests. Obligingly, Holmes agreed to try and make it work in the way Ahmad wanted and began to draw up the appropriate documentation.

Dickson wrote pompously to tell Fowle, the resident, that Janson and Holmes had paid homage through "the usual official calls on the British Agency . . . but Mr. Lombardi did not think fit to do so." Dickson had taken to writing a fortnightly "Report on the Oil Situation." In these documents circulated to the resident, the Government of India, and the India Office, Frank Holmes's every move was relayed and his supposed intent interpreted in the worst possible light. In this lengthy report, Dickson said he believed Ahmad's idea of splitting the concession came "from Baghdad." He couldn't be sure because, as he complained, "It is unfortunate that the shaikh prefers at this juncture to keep me in the dark as to his real intentions. I put this down to the influence of Major Holmes and the interested advice he continually receives from that person." If the idea didn't come from Baghdad, Dickson said, it might be "a result of Shaikh Ahmad's recent meeting with bin Saud in the desert, which from later corroborative evidence I am now almost certain took place."

As Dickson and Fowle suspected, the shaikh of Kuwait had indeed met "secretly" with bin Saud, in other words, without telling the British. In fact, they met twice, on February 20 and 25, in the southern desert. Holmes's colleague, Muhammad Yateem, was with Shaikh Ahmad's party on both occasions. He had been spending most of his time with either the shaikh or his own close friend, Mullah Saleh, Ahmad's principal adviser, particularly since the Anglo-Persian geologists' dedicated socializing efforts of the year before. Yateem's mere presence at the desert meetings incensed Dickson and aroused his deepest suspicion; his malice extended to everyone around Holmes. Dickson had taken to referring to Muhammad Yateem as "Holmes's jackal."[13]

Unfortunately for Lombardi, the Saudis had been keen to tell the Kuwaitis about the Americans who had arrived in Jedda and were trying to obtain Holmes's Al Hasa oil concession. Lombardi's secret was out. When the Kuwaitis returned from the desert, they immediately passed on the news. From the Kuwaitis, Holmes and Janson now knew that when Hamilton left London he had not returned to San Francisco as they were led to believe. Hamilton had gone straight to Jedda where he still was, with Karl Twitchell, trying to get the Al Hasa concession—Holmes's concession—behind the back of their supposed partners, the Eastern & General Syndicate.

LOMBARDI FLEES

Lombardi returned from Bahrain to Kuwait happily anticipating a colorful Oriental experience staying with Holmes and Janson as Shaikh Ahmad's guest in Dasman Palace. Within hours, a panicked telegram was winging to San Francisco. "Janson and Holmes propose breaking relations with Standard Oil of California," Lombardi announced. "They are strongly opposing us in Arabia and withdrawing support for the Bahrain Petroleum Company because Standard Oil of California approached bin Saud for the Al Hasa concession through others."

Nicely shifting the blame, Lombardi said that he had understood "Francis Loomis notified E&GSynd last year in London that we would proceed through others after two months." He said that Holmes insisted there had never been any reference to Standard Oil of California going after the Al Hasa concession as such. What had been raised, Holmes stressed, was discussion only of the possibility of a permit for "geological work for the benefit of the Bahrain Petroleum Company." Lombardi told San Francisco he had sighted E&GSynd's cables to Gulf Oil on the matter and "the situation is serious." Judging discretion to be the better part of valor, Lombardi advised his office, "wire me in Cairo" and quickly departed Kuwait.[14]

On the day Lombardi was fleeing to Cairo, Abdul Aziz bin Saud was urgently cabling Holmes in Kuwait. Explaining that his Al Hasa concession was now the subject of offers, bin Saud urged Holmes: "Come at once to Jedda." As usual, Dickson had a copy of this cable before it was delivered to Holmes. He reported to the resident and followed up with a dispatch to the intelligence officer at Basra. "On March 18th bin Saud wired to Holmes to come to Jedda at once if he wished to be a bidder for the Nejd oil concession (Al Hasa)."

Bouncing blame was the order of the day at Standard Oil of California. Loomis claimed: "I distinctly told Janson last summer that, if they did not do so, we proposed establishing contact with bin Saud at the end of October." Right up to the very moment they were exposed through the Saudi-Kuwaiti meeting in the desert, Standard Oil of California had continued discussions with E&GSynd aimed at a joint venture for Al Hasa and the Neutral Zone. Moreover, had anyone checked Loomis's diary it would be obvious that the double-cross began long before the end of October 1932.

Loomis had put the prospect to Philby over lunch, first on July 11, then on September 3. His cable asking Philby to inform bin Saud of their "formal request" for a survey followed by a "practical working contract" for the Al Hasa concession was dated October 5. The wooing of Karl Twitchell had begun in September. Francis Loomis had made the trip to visit Twitchell in Virginia in the middle of that month.

The fact was that Standard Oil of California had led E&GSynd to believe they were negotiating in good faith for a joint arrangement covering Al Hasa and the Neutral Zone. They had gone to some length to hide their agreements with Philby and Twitchell and to keep E&GSynd from discovering that when Lloyd Hamilton left the talks in London, he was not going back to San Francisco but on to Jedda.

The company's legal advisers were called in for an emergency session. The issue to be considered was whether Standard Oil of California was under a legal obligation to E&GSynd in the matter of negotiating for the Al Hasa and Neutral Zone concessions. The company's executives were comfortably maintaining that "Janson and Holmes are bluffing . . . we do not need to take the matter very seriously." A lengthy legal opinion was drawn up on the basis of the correspondence and documentation. By March 22, the lawyers were urgently advising, "Postpone any action until a full discussion can take place."

"Major Holmes and Mr. Janson have suddenly decided to fly tomorrow to Cairo and have a further conference with Mr. Lombardi of the Standard Oil," Dickson flagged in an eight-page fortnightly report, dated March 22. The sudden departure from Kuwait of M. E. Lombardi followed by the equally sudden departure of Edmund Janson, Frank Holmes, and Muhammad Yateem left Dickson dying of curiosity. "I cannot say, at present, whether they are trying to drag Lombardi into the Kuwait business or whether their conversations are connected with bin Saud and Nejdi oil," he surmised. "I rather fancy that the latter is the correct answer and that Major Holmes will try and go on from Cairo to Jedda."[15]

NOTES

1. See Violet Dickson, *Forty Years in Kuwait* (London: Allen and Unwin, 1971), p. 115, for Biscoe's burial at sea; Penelope Tuson, *The Records of the British Residency and Agencies in the Persian Gulf* (London: India Office Library and Records, 1979), pp. 184–85, for Fowle's background, he would remain as resident until 1939; Yossef Bilovich, "Great Power and Corporate Rivalry in Kuwait, 1912–1934: A Study in International Politics" (PhD diss., University of London, 1982), p. 241, citing August 23, 1932, political agent Kuwait (Dickson) to political resident (Fowle), "tell Shaikh Ahmad that you (the Resident) approve of Anglo-Persian's draft" and August 25, 1932, political resident (Fowle) to Colonial Office, "our unbiased views of both concessions should be promptly drawn up"; Holmes Papers, September 1, 1932, Holmes at the Riverfront Hotel Baghdad to Dorothy at MillHill, "the shaikh and, more importantly, the people of Kuwait are sticking strongly for me."

2. Bilovich, "Great Power and Corporate Rivalry in Kuwait," pp. 245–49, "Ambassador Andrew Mellon . . . without the knowledge of his State Department called on the Foreign Office"; Chevron Archives, box 0426154, February 8, 1933, Holmes in Kuwait to E&GSynd, enclosing a copy of the British government "comparison" passed to him by the shaikh of Kuwait. Bilovich, pp. 250–53, gives the State Department reaction "not one favourable comment is made. . . ."

3. R. W. Ferrier, *The History of the British Petroleum Company: The Developing Years, 1901–1932* (Cambridge: Cambridge University Press, 1982), p. 628, and Benjamin Shwadran, *The Middle East: Oil and the Great Powers* (London: Atlantic, 1956), p. 45, detail the charges and statements of the shah of Persia and the unilateral cancellation of Anglo-Persian's concession; Administrative Report Kuwait, 1932 (Dickson), "suffering and acute want" and "one notable exception."

4. Administrative Report Bahrain, 1932 (Prior), "influence will undoubtedly increase and combined with the American Mission represents a most unfortunate intrusion of foreign influence into Bahrain." The report also notes "a Mr. K. S. Twitchell an American Prospector believed to be in the pay of Charles Crane the American bathroom millionaire, arrived from Hasa on January 7th." On November 28, 1932, Lt. Col. Percy Gordon Loch took over as political agent, Bahrain. Angela Clarke, *Bahrain Oil and Development, 1929–1989* (London: Immel Publishing, 1991), pp. 133–36, describes the early working and living conditions at Bapco.

5. Charles Belgrave, *Personal Column* (Beirut: Librarie du Liban, 1960), p. 83, "It was a great day for Major Frank Holmes, who now saw the visible proof of what he had always believed."

6. The prime myth holds that the Anglo-Persian Oil Company (in 1935 renamed Anglo-Iranian Oil Company and, in 1955, British Petroleum) and its geologists did know there was oil on the Arabian side of the Persian Gulf, but either contrived a strategy of publicly stating there was not in order to deter competition from what they considered to be their turf, or employed tactics aimed at delaying competitors until they themselves were ready to develop these fields; Shwadran, *The Middle East*, pp. 44–50, details this episode including, "British warships did appear in the Persian Gulf."

7. H. St. J. B. Philby, *Arabian Oil Ventures* (Washington, DC: Middle East Institute, 1964), pp. 80–85, for this various correspondence between Philby, Loomis, Lombardi, Twitchell, and Lees of Anglo-Persian.

8. Ibid., p. 78, "I told (Loomis) I would be glad to help in any scheme which would contribute to the prosperity of Arabia"; p. 86, "(never) contemplated the making of comparable arrangements with me" and "sending Longrigg . . . unless you will act for us"; Abu Dhabi Documentation Center, copy of Records of the US State Department Relating to Internal Affairs of Saudi Arabia, 1930–44, Film-T1179/2, August 6, 1930, from US Consul Alexander Sloan, Baghdad to Department of State, "King bin Saud will not permit the Anglo-Persian Oil Company to obtain any concession within the boundaries of Nejd. Furthermore (Bill Taylor, Standard Oil of California geologist visiting Bahrain) was informed that the King had stated that if he gave an oil concession to any company and later found that that company was affiliated in any way with Anglo-Persian or had sold out to any company affiliated with Anglo-Persian he would immediately cancel that concession"; Shwadran, *The Middle East*, p. 293, for the clause prohibiting transfer of rights and obligations.

9. Philby, *Arabian Oil Ventures*, pp. 127–130, "conduct negotiations with the government, with Twitchell in attendance, while I remain in the background"; Shwadran, *The Middle East*, p. 290, points out that Twitchell, in Saudi Arabia, never mentions any role by Philby in helping Standard Oil of California obtain the concession. He notes that Aramco's first official publication, Roy Lebkicher's "Aramco and World Oil," in *Arabian American Oil Company Handbook for American Employees* 1 (New York: Russell F. Moore, 1950), confirms Philby's role but minimizes Twitchell's; Elizabeth Monroe, *Philby of Arabia*, p. 204, for the terms of Philby's contract with Standard Oil of California and Dora Philby writing, "Jack is a different man . . ."; Wallace Stegner, *Discovery! The Search for Arabian Oil*, abridged for *Aramco World* (Beirut, Lebanon: Middle East Export Press, 1971), claims not even Twitchell knew Philby was being paid by Standard Oil of California.

10. There are many books on Kim Philby, see, for example, Bruce Page, David Leitch, and Phillip Knightly, *Philby, The Spy Who Betrayed a Generation* (London: Deutsch, 1968), and Kim Philby's *My Silent War* (London: MacGibbon and Kee, 1968); Philby, *Arabian Oil Ventures*, p. 81, "the King took me aside one day and expressed his conviction that I would give priority to the interests of his government," and p. 89, "you will protect our interests . . . as you would protect your own personal interests."

11. IOL/R/15/1/623, vols. 82 and 86, February 15, 1933, political agent Kuwait (Dickson) to political resident (Fowle), "some sort of hypnotic influence" and "no more than the suspicious cunning inherent in every Nejdi or Bedouin Shaikh"; Ralph Hewins, *A*

Golden Dream: The Miracle of Kuwait (London: W. H. Allen London, 1963), p. 217, "tall, slim and monocled" and Chisholm's own description of his background, pp. 109, 159.

12. Chevron Archives, box 0120814, interview with Mr. Lombardi re: Socal-Antitrust Dec. 18, 1953: (note: As could be expected, there are many factual errors and misleading implications in the documents prepared by Standard Oil of California for the antitrust hearings), "On the afternoon of December 10, 1953, James O'Brien, John Sonnett and the writer (D. G. McInerney) met M. E. Lombardi at his home also attached M. E. Lombardi Confidential Memorandum of September 1949." In these documents Lombardi claims he "was greatly surprised to find Janson on the plane from Baghdad to Bahrain." The document says, "Janson who was a son-in-law of E&GSynd owner, was to trail Lombardi all the rest of the time he was in the Middle East!" Lombardi also claims in these documents, "After looking over the Bahrain concession, at Major Holmes request I stopped off in the city of Kuwait to meet the Shaikh of that province. Mr. Janson also stopped off. It is all very obvious now that Holmes wanted to get me away by myself for a while so that he could pump me for information concerning our intentions in the Middle East—particularly with regard to Saudi Arabia." Note that Lombardi's claims in these documents are proved untrue by IOL/R/15/1/623, vols. 82 and 86, February 19, 1933, political agent Kuwait (Dickson) to political resident (Fowle), "Mr. Janson and Mr. Lombardi, a Director of the Standard Oil Company, will pass through Kuwait by air en route to Bahrain on March 2nd . . . and all three will return to Kuwait and stay with the Shaikh on March 9th . . ."; see also Chevron Archives, box 0120814, Lombardi confidential memorandum of September 1949, "I set out alone for Bahrain Island while Hamilton (and Twitchell), unknown to E&GSynd, left for Jedda to negotiate the Al Hasa lease"; Ward Papers, box 1, March 31, 1933, Janson to Ward, "I met Lombardi in London before he went out and also met him again out in Bahrain."

13. IOL/R/15/1/623, vols. 82 and 86, March 22, 1933, secret, political agent Kuwait (Dickson) to resident (Fowle) headed "Oil Activities of Major Holmes" details the idea of splitting the concession, Holmes's objection, "he demurred for two days" and the decision to make the split equal, "Holmes had agreed to try and carry out the shaikh's wish"; and March 17, 1933, "confidential," resident (Fowle) to Colonial Office, "Dickson reports two journeys of the Shaikh of Kuwait across his southern boundary, ostensibly for hunting, in company among others of Holmes 'jackal,' Muhammad Yateem" and March 22, 1933, "secret" political agent Kuwait (Dickson) to political resident (Fowle), "Ahmad's recent meeting with bin Saud in the desert, which from later corroborative evidence I am now almost certain took place."

14. Clarke, *Bahrain Oil and Development*, p. 138, reprints this March 17, 1933, cable, Lombardi in Kuwait to San Francisco, "Janson and Holmes propose breaking relations with Standard Oil of California . . . situation serious." Without any experience of the area, Lombardi had no idea how his secret had come out. See Chevron Archives, box 0120814, Lombardi confidential memorandum of September 1949, in this document Lombardi says he believed "E&GSynd had broken our cable code and all my wires to Hamilton were known to them. . . . So the Kuwait discussions broke up in a row."

15. IOL/R/15/1/623, vols. 82 and 86, April 17, 1933, political agent Kuwait (Dickson) to resident (Fowle) and April 17, 1933, Dickson to special service officer Basra (Intelligence) and April 22, 1933, "Report on Oil Activities," "Bin Saud wired to Holmes to come to Jedda at once," and "Holmes and Janson have suddenly decided to fly

tomorrow to Cairo." Again on April 27, Dickson reminded the resident: "Holmes received a telegram from bin Saud on March 18, inviting him to Jedda." In his April 27 report Dickson lamented that "nothing of an unusual nature occurred during this fortnight" because Major Holmes was out of the area, in Cairo and Jedda; Clarke, *Bahrain Oil and Development*, pp. 138–39, "Janson and Holmes are bluffing" and "postpone any action."

CHAPTER 20

AN UNCLEAN SUBJECT

As if the affable Frank Holmes was the vanguard of the Apocalypse, the resident sent out an alert to Jedda, India, and London, warning: "It is more than likely that Holmes intends visiting Abdul Aziz bin Saud in Riyadh, via Jedda, after this conference in Cairo."

The prospect of Frank Holmes arriving in Saudi Arabia reduced Standard Oil of California's three representatives there to shadows of their former confident selves. Lloyd Hamilton rushed to tell Philby that he had heard from Lombardi, "who had apparently met Holmes in Bahrain," that Holmes had "made arrangements to meet bin Saud." Not very convincingly, Hamilton confided to Philby that he was "not particularly frightened about competition from that source—especially since my conversations with you about Major Frank Holmes."

Twitchell went into a tizzy when he heard Holmes was coming. He warned Philby: "Holmes has left by air for Jedda. He might go via Cairo and Khartoum and then to Port Sudan to catch the steamer for Jedda." Like Hamilton, Twitchell tried to shore up his courage, telling Philby: "I think you and I do not fear him as a serious factor."

Twitchell reported that Holmes had telegraphed bin Saud asking him not to conclude any negotiations until he arrived, and as a result the Saudis "now want to prolong negotiations." Twitchell was worried that Holmes had "impressed Lombardi," and said, "if he meets Lombardi and Hamilton in Cairo, I imagine there will be some fireworks." In his reply, Philby worried that Holmes may have beaten him at his own game by secretly coming to an arrangement to work with Lombardi and Hamilton "and has telegraphed to the king *with their approval* to delay a decision until he appears on the scene."

Philby was unnerved. He believed Holmes was coming to "act" for Standard Oil of California. If so he could kiss good-bye the "windfall" that he was counting on to solve the problem of "bills for Kim at Cambridge and three daughters at first class schools." Because, as Philby knew better than anyone and had mentioned to Lombardi, Holmes had "a past record in connection with this very Al Hasa concession." He wrote again to Lombardi: "I cannot help feeling a little uneasy . . . now that you have secured the services of Holmes."[1]

COMPROMISE IN CAIRO

On March 23, Lloyd Hamilton was called from Jedda to join Lombardi, Janson, Holmes, and Yateem in Cairo to try to comply with instructions issued by the lawyers retained by Standard Oil of California and thrash out a working arrangement.[2]

From Cairo, Lombardi wrote Philby on April 3: "As regards the Neutral Zone, it seems best for the moment to allow the Eastern & General Syndicate, represented in Arabia by Major Frank Holmes, to attempt to bring together the Saudi Arabian government and the Shaikh of Kuwait so that a contract, satisfactory to both, may be obtained." He added: "Major Holmes, however, is not representing us in our attempt to get a contract for Al Hasa; but his interests are in accord with our attempt to get such a concession." Philby recorded that he was "astonished at this statement of policy." From Philby's point of view, Holmes was the competitor who had to be overcome at all costs. He found it disconcerting that Lombardi regarded Holmes as a colleague.

Moreover, it was Lloyd Hamilton who had personally added "oil rights" in the Neutral Zone to those of Al Hasa in Standard Oil of California's proposal. He had included this extension soon after arriving from London and the discussions there with Janson, Holmes, and Lombardi. Hamilton had remarked to Philby: "Because of its relatively small area, the Neutral Zone may appear to be unimportant in our negotiations, but my principals do not regard it so." (The Saudis had presented a counterclause that stated "the Zone shall not be exploited without the further consent of the Saudi Arabian government.")

In his letter from Cairo, Lombardi explained further: "In the present circumstances I believe it would be unwise for me to go to Jedda. Mr. Hamilton will return if, within a reasonable time, you think the situation is ripe for another try. In the meantime I hope you will keep us fully advised of developments and let us know what you think of the situation."

Philby did not know of the legal pickle Standard Oil of California had landed in through committing to a joint venture with E&GSynd, and gaining valuable information during the discussions. His contacts had been with Francis Loomis and it was Loomis he had convinced of the necessity of hiring him. Lombardi had hired Twitchell and was now speaking of Frank Holmes in very amenable terms. Philby mistakenly concluded: "Evidently something had happened in San Francisco. It may well have been that Mr. Lombardi's influence had, indeed, resulted in a sort of *volte-face* on the part of Standard Oil of California."[3]

PHILBY'S MISINFORMATION

Philby was in a spin. He had put a lot of effort into castigating Frank Holmes. He had convinced Loomis, Hamilton, Twitchell, the British minister at Jedda Sir Andrew Ryan, and the people at Iraq Petroleum/Anglo-Persian that Frank Holmes was persona non grata. He had told them all, quite untruthfully, that bin Saud was implacably hostile toward Holmes because the original Al Hasa concession had not been developed. The 1923 Al Hasa concession held by Holmes had never been officially cancelled. It certainly had lapsed. But E&GSynd had paid over some sixteen thousand sterling to bin Saud before enforced inactivity resulting from bin Saud's own campaigns followed by the waves of the Ikhwan revolt that swept the area until the British military actions of late 1929.

This didn't stop Philby telling everyone, again quite untruthfully, that in 1928 bin Saud had "warned" Frank Holmes and E&GSynd to pay "back rent" and when this was "not forthcoming" had "angrily cancelled" the concession. (The rent was three thousand sterling each year. The concession was signed in May 1923. Having paid sixteen thousand sterling, E&GSynd would have been in credit in 1928.)[4]

Philby painted Holmes as "no more than an adventurer" and wildly embellished his story that bin Saud had suffered such a "grievous disappointment in the matter of the Al Hasa concession" that since then he had carried a "powerful personal grudge" against Frank Holmes. Now Philby was faced with the fact that, far from being hostile, bin Saud had personally called Holmes to come to him, a fact that was confirmed by bin Saud himself when Philby next saw him in Mecca. How was Philby going to explain this?

Putting his trust in Philby, Ryan the British minister in Jedda, had assured his London colleagues that British interests would come out on top in the awarding of bin Saud's concession. "Philby will be my informant," he confided. The aristocratic Ryan had been with the embassy in Constantinople both before and after World War I; it was he who had facilitated the concession on Mesopotamia's oil in 1912–14 to the British-dominated Turkish Petroleum rather than the contending Germans.

Andrew Ryan swallowed completely Philby's vicious stories about Frank Holmes. In mid-March he had written to the Foreign Office about "information of mine that Holmes is *persona ingrata* in bin Saud's eyes." He said: "This is because of the non-payment of certain dues claimed by the king when the former prospecting concession petered out." It was a shock for him to receive the resident's warning that Holmes was coming to Jedda.

Ryan was so taken in by Philby that he doubted this was possible. He reminded his colleagues: "Holmes has hitherto been believed to be distasteful to bin Saud." A flummoxed Andrew Ryan was soon appealing for guidance to the Foreign Office. On April 8 he cabled, "Telegrams have come for Holmes 'care of

Legation.' How should I treat him officially if he arrives? Personal relations will probably be unavoidable."

The Foreign Office reply came on April 11, headed "Treatment to be Accorded Major Holmes." The cable read: "Personal relations should be limited to normal courtesies. Holmes's visit is for the purpose of furthering American interests and he is entitled to no official assistance. Moreover, it would be undesirable that he should give the Saudis the impression that the Legation is behind him. You may certainly, if you think fit, take him to task for having his correspondence sent to the Legation without your permission." By the time Ryan received this instruction, Frank Holmes had already arrived in Jedda.

On April 9, looking for all the world like a group of best friends arriving for a seaside vacation, Mr. and Mrs. Frank Holmes and Mr. and Mrs. Lloyd Hamilton disembarked at Jedda from the Egyptian steamer SS *Taif*.[5]

COLONIAL OFFICE BAULKS

The Colonial Office had had enough of the whole subject of the Persian Gulf. In particular, they were exasperated with the quarrels over oil. Administration of the Arab shaikhdoms had previously been left entirely in the hands of the Government of India, with referrals to the Colonial Office only in matters of political policy, such as the Persian claim to sovereignty over Bahrain.

Now the Colonial Office experience was that the matter of oil was dragging them ever deeper into a quagmire of interdepartmental quarrels, plots, and counterplots, and activities that by no stretch of the imagination could be characterized as aiming for the best interests of either the people or the rulers of the individual countries of the Persian Gulf. The Colonial Office wanted out. In high dudgeon they recorded that, in their view, "oil is an unclean subject." Expressing extreme irritation they declared: "We are not in control. We are not interested. And the Middle East Department is not staffed to deal with it."[6]

Colonial Office interests would soon be formally passed to the India Office, thereby removing any objective oversight of the Government of India's administration of the shaikhdoms of the Persian Gulf. The Colonial Office may have had cause to be pleased with their timing. When they took off for Cairo, Holmes and Janson left two issues bubbling in Kuwait. The first was their tacit agreement to discuss Shaikh Ahmad's idea of a split concession.

The second was an invitation from Edmund Janson for the shaikh of Kuwait to join him, at Janson's expense, for the summer at his London and country homes and sailing on his yacht. Janson had enjoyed himself immensely in Kuwait, as he recorded: "I stayed with Shaikh Ahmad in his palace and got on very friendly terms with him. I have invited him to come and stay with me." The thought of the shaikh of Kuwait actually taking up this invitation filled Dickson with anguish.

Writing to the resident, Dickson put forward the highly unlikely claim that the shaikh of Kuwait had admitted to Janson that he could not budge, could not even accept a personal invitation, until he asked the British. Dickson asserted that Shaikh Ahmed had told Janson that "he could in no way accept until His Majesty's government had been consulted and given their permission because he is under Great Britain's protection and therefore under definite obligations as regards leaving his State."

But Dickson could not shake off a suspicion that Shaikh Ahmad might want to go. "It is clearly our *duty* to prevent the shaikh accepting a private invitation of this sort," he harried the resident.

In case he didn't quite get the idea, Dickson thoughtfully included the letter he had written for the resident to sign. Addressed to Shaikh Ahmad it said, in part: "While a Ruler of a State often pays short visits to an oil field, or a factory, belonging to a company for the purpose of inspection, it is unusual for a Ruler to visit for the purpose of pleasure, or a change of air, a country where the company has no works of its own to show him." If Ahmad did accept Janson's invitation, Dickson's letter warned: "Your Excellency cannot expect to be received officially by His Majesty's government with the honours customary on an official visit." Lieutenant Colonel Fowle declined to send Dickson's letter. In his opinion, he wrote the Government of India's secretary for foreign affairs: "The days have definitely gone by when we can keep the more important Arab Rulers of the Gulf in leading strings."[7]

GULF OIL VERSUS CADMAN

The discovery of oil at Bahrain had severely embarrassed Anglo-Persian's chairman, Sir John Cadman, in June; questions had even been raised in Parliament. Just five months later, Persia's cancellation of Anglo-Persian's concession had added to his discomfort. Cadman was now a driven man.

He was bailed up at a London lunch by the chairman of the Gulf Oil Corporation, Col. J. F. Drake, who accused him of "hampering the endeavours" of Gulf Oil even though Anglo-Persian had been offered Kuwait and refused it. Cadman denied that his company had ever refused to go into Kuwait. In November when Cadman delivered a speech titled "Petroleum Policy" at the American Petroleum Institute Convention in Houston, Texas, Gulf Oil bailed him up again. In this speech Cadman foreshadowed what would be the great price-fixing scandal of 1951 in which seven international oil companies—the "Seven Sisters"—combined to artificially maintain high oil prices.

To the 1933 American Petroleum Institute Convention, Cadman advocated the multiple benefits the oil companies could gain through forming an international "oil cartel" designed to lock out competition and tighten supply through the allocation of regional oil production quotas. It was ten days after this speech that the shah of Persia canceled Anglo-Persian's concession.

Cadman said: "Consumption everywhere has decreased. Competition is too keen and prices contain no margin for gain. One possible step towards rehabilitation of trade is evidently the readjustment of supply to demand and the prevention of excessive competition by allotting to each country a quota which it will undertake not to exceed."[8]

This time Cadman was intercepted by Gulf Oil's vice president, F. A. Leovy, who was still smarting from his company's mistake in giving up Holmes's concession in Bahrain. At the convention, he told Thomas Ward that "the loss of Bahrain by Gulf Oil was the greatest regret of his life." He was not going to make the same mistake twice. Leovy thrust under John Cadman's nose evidence that, contrary to Cadman's denial, Anglo-Persian had indeed refused to consider the Kuwait concession. It ranged from the 1923 approach to Anglo-Persian by the London office of E&GSynd and Arnold Wilson's 1926 rebuff, when he famously declared, "only Albania is hopeful," to Cadman's own April 1932 statement to the Foreign Office that "the Americans are welcome to what they can find there."

Leovy repeated the charge that the Anglo-Persian Oil Company had no right to block Gulf Oil's efforts since, from the early 1920s, they had shown nothing but disdain for Kuwait. On the spot Cadman personally created the myth that Anglo-Persian would maintain forever after rather than admit its technical failures and Holmes's geological skill.[9]

When E&GSynd had first come to them, Cadman said, they did not want the concession "at that time." He said Anglo-Persian had continued to believe they could permit Kuwait "to lie fallow" and probably make a much better trade with the shaikh of Kuwait at "some later date." He said the company thought it was "good trading tactics to advise Frank Holmes that they were not interested." Cadman maintained that Anglo-Persian did not think "for a moment" that anybody else might come in and work with E&GSynd to develop the concession. This was a remarkably unconvincing excuse from Cadman as Anglo-Persian had already experienced precisely this scenario unfold in Bahrain, which they had also turned down.

Now Cadman strongly advised Gulf Oil to agree to a partnership with Anglo-Persian along the lines just detailed in his "Petroleum Policy" speech. While warning darkly that he "would use every power in his means to do so," he told Leovy he was now "determined" to obtain the Kuwait concession. Offended, but not intimidated, Leovy replied sharply that he did not think his colleagues would be interested in any arrangement with the Anglo-Persian Oil Company.[10]

Cadman was pleased with his quick response and felt he had achieved a bluff. He let the matter drop. But, by February 1933 when the League of Nations' hearing over the canceled concession was suspended in favor of negotiations with Persia, John Cadman turned to Andrew Mellon, whose term as American ambassador to London was due to expire at the end of March.

The day before he was to leave for Tehran, where the negotiations had been moved away from the distractions of Paris, Cadman reported progress to his fellow directors and the Foreign Office. Andrew Mellon, he said, was fully informed and "all out" for "lending a hand" to E&GSynd, which Cadman still referred to as "the Edmund Davis group." He said he had pointed out that Anglo-Persian and Gulf Oil competing would lead to a form of "costly bidding" and would "raise the price" of the Kuwait concession. He had told Andrew Mellon, he said, that it was the desire of Anglo-Persian to make the "most friendly over-tures" to Gulf Oil in order to bring the two companies together to "exploit this region on a 50-50 basis."

Cadman's definition of "50-50" was that the Anglo-Persian Oil Company would appoint the chairman, furnish the staff, and "work the concession at such a rate as would be compatible with world market conditions." In other words, "tightening supply through production quotas," as he had advocated in his speech to the American Petroleum Institute. Cadman was spelling out to Gulf Oil exactly what the shaikh of Kuwait had long suspected—that, unlike Frank Holmes, Anglo-Persian had no intention of honestly exploring for and/or developing oil in Kuwait. Cadman severely criticized what he called Gulf Oil's use of a third party in its negotiations, meaning Holmes and E&GSynd, and in man-to-man fashion told Andrew Mellon that "we are both capable of handling this matter without outside assistance."

Cadman followed up his approach to Andrew Mellon by cabling Anglo-Per-sian's New York agent. He had "spoken to" Andrew Mellon, he said, and the New York Agent should now "get Gulf Oil to withdraw." He said this would leave Anglo-Persian "free to negotiate on the basis of a 50-50 partnership." The New York agent duly set about following orders.

Cadman reminded Gulf Oil that "my company very definitely regard Kuwait as within their natural and particular sphere of influence and will not passively look on while any other company, with no reasonable interest whatsoever in that area, are endeavouring to secure an interest there."

He dropped in personally to tell Leovy that Cadman had "spoken" to Andrew Mellon. Leovy replied brusquely that he and Mellon had worked together "for years" and Mellon would never interfere in a matter he was han-dling and, therefore, "you can take it that Gulf Oil's policy will not be changed." When the now-former ambassador Mellon arrived back in the United States, Anglo-Persian's American agent reported that New York oil circles considered "Andy Mellon has returned determined to keep his hands on Kuwait."

John Cadman instructed Anglo-Persian's Abadan office to "put all pressure at Kuwait." He said the goal was to scare Gulf Oil into thinking Anglo-Persian was in "dead earnest to secure the concession—and will go to any lengths to do so." In Kuwait, Archibald Chisholm and Hajji Abdulla Williamson redoubled their efforts.[11]

CADMAN IN KUWAIT

Cadman's tactics were well understood by the Foreign Office. They immediately recognized the game as one of playing Kuwait off against Persia, and they were all for it. They concluded that Cadman should not go directly to Tehran, but should detour to Kuwait and make the shaikh an improved offer. There would be advantages, the Foreign Office said, in not "showing any undue haste to reach Tehran" and "keeping the Persians guessing" as to what he might be doing in Kuwait. The Foreign Office said the Persians might even think that Cadman "was negotiating another concession on which to fall back if his negotiations for renewal of the Persian concession did not go well."

Cadman didn't need the Foreign Office's advice; this was what he had intended to do anyway. Apart from the possible effect on the Persians, with the British government's blessing, it would certainly fit in with his plan to make Gulf Oil believe he was in "dead earnest" to secure the Kuwait concession.[12]

It was a high-powered delegation that appeared in Shaikh Ahmad's office on March 25, 1933. John Cadman was accompanied by Anglo-Persian's general manager, deputy general manager, the company's lawyer, Archibald Chisholm, the ubiquitous Hajji Abdulla Williamson, and a bilingual secretary. To his chagrin, Dickson was not asked to attend.

After praising the role of the British government as the majority shareholder in the Anglo-Persian Oil Company, which he described as "the largest British oil producing company in the world," Cadman got straight into slandering E&GSynd. Their only intention, he shamelessly avowed to Shaikh Ahmad, was to obtain the concession in order to sell it. He played fast and loose with the truth as he claimed Holmes's syndicate had recently approached his company "on the grounds as they alleged" that they knew of the shaikh's aversion to Anglo-Persian. He said they had "offered to obtain the concession *on behalf of* the Anglo-Persian Oil Company." He told Shaikh Ahmad he had "indignantly" refused this offer because he preferred to deal direct with Ahmad, "so you will, personally, receive the full payments."

He scornfully declared that Holmes and his syndicate were "merely brokers." Imperiously he said that it was not fair to place him, Sir John Cadman, who was the chairman of a great company that itself possessed the finances to exploit the concession, "in a position of bidding against a broker."

As to Holmes's association with Americans, he sniffed, he, Cadman, was in a far better position to invoke American participation "than E&GSynd could ever be." He said he was on "intimate terms of friendship" with many of the leaders of the American oil industry. Possibly referring to the acrimonious exchanges with the executives of Gulf Oil, Cadman told Shaikh Ahmad that he had "discussed Kuwait oil affairs with some of these leaders" during a recent visit to the United States. He declared he was "currently in discussion" with the Gulf Oil

Corporation for a joint partnership, a partnership that would "dispose of" Holmes and his syndicate "forever."

Even so, Cadman said, his strong advice was that the shaikh of Kuwait should not seek international partnerships. He told Ahmad he had only to look at how easily Anglo-Persian had developed the Persian concession and compare this to the "constant delays and difficulties" experienced by the Iraq Petroleum Company, which had international partners.

The shaikh of Kuwait listened patiently, and didn't buy a word of it. The nationality of a developing company was "a matter of indifference" to him, he said, as long as it met the payments stipulated in any agreement. Nor would he agree with Cadman that Anglo-Persian's operations being so close to his own territory was an advantage. Ahmad said he didn't care where their operations were so long as the developing company eventually selected met its payments and fulfilled the contract obligations. He also would not agree that an intermediary, what Cadman sneeringly referred to as "a broker," would reduce the amount he received for the concession. He said he would give the concession to the highest bidder and it would be up to him to be satisfied he was making the best deal.

Shaikh Ahmad followed through by saying he had just received an updated offer from Holmes, and it was better than Anglo-Persian's. He said if Cadman now wanted to raise his bid, he would take it onboard—and then give Frank Holmes another opportunity to revise his. It was possible he might do this for some time, he suggested, accepting bids and counterbids until he decided definitely to tell one party or the other that the concession would be granted to them.

Without batting an eye, Shaikh Ahmad turned down Cadman's offer to immediately lift the upfront payment from sixty-five thousand to two hundred thousand rupees and agree to scale up the annual rent "after declaration of commercial production." However, if Ahmad signed with him this very minute, Cadman promised, after commercial production was declared he would "double" the signature payment, making it four hundred thousand rupees, and the third year's rent. He stressed, however, that this generous offer would be withdrawn if Ahmad didn't sign right now. Cadman's "bonus offer" failed to excite the shaikh of Kuwait who remained calm and serene. He replied that he had "given his promise" to Frank Holmes that he would not close with Anglo-Persian without giving Holmes an opportunity to revise his offer.

Sir John Cadman, chairman of "the largest British oil producing company in the world," did not take this rebuff at all well. Before the meeting closed, he was threatening the ruler of Kuwait that if he gave his concession to the Americans, the Anglo-Persian Oil Company would be "compelled" to declare an "oil war" that would prevent Kuwait ever developing any oil potential it might have.[13]

ANDREW RYAN'S REPORTS

From the moment Frank Holmes stepped ashore in Jedda from the SS *Taif*, in the company of his wife, Muhammad Yateem, and their new American friends, Lloyd Hamilton and his wife, Airy, a dozen eyes were on him.

Philby and Twitchell were watching like hawks. Sir Andrew Ryan at the British Legation was gathering every rumor, every innuendo, every passing comment. He penned these into amusing reports for the Foreign Office which circulated them in turn to the Colonial Office, the India Office, the Petroleum Department, and the Admiralty. Ryan's droll reports were further embellished along the way with individual contributions, opinions, and conjecture. In Kuwait, Dickson impatiently grilled any arrival from the mainland for news and gossip of Holmes's activities.

Ryan was thriving on the unusual excitement. Even before Hamilton's trip to Cairo and the arrival of Frank Holmes, he had artistically sketched a scenario for the enjoyment of his desk-bound colleagues in London. "The stage is set," he wrote creatively. "The *dramatis persona* is an avid Abdulla Suleiman who thinks of oil in Al Hasa as already a marketable commodity. Twitchell and Hamilton are featuring for Standard Oil of California. A nice fellow (Major?) Longrigg and his Syrian Muslim assistant, a young Mr. Ahmad Mudarris educated at Oxford, are on for the Iraq Petroleum Company."

"It may amuse you to know," he continued, "that all the persons named above, except of course Abdulla Suleiman, sat at my wife's reception recently and that all the representatives from abroad are now in the same hotel, a Government institution principally for high class Pilgrims. It has been suggested that what we really need now is a separate hotel for concession hunters!" Ten days later Ryan reported that Hamilton had left "temporarily" for Egypt and Longrigg had gone to "Port Sudan for a few days." Unkindly, Philby said Longrigg had gone "to buy a hat."[14]

(In writing from Arabia in a fashion designed to amuse his deskbound colleagues, Ryan was following a style favored by many Foreign Office officials in exotic places but most particularly by his predecessor, Sir Reader Bullard, also ex-Constantinople. In his highly respected 1938 work on early Arab nationalism, *The Arab Awakening*, George Antonious observed that Britain's withdrawal of support for Sharif Husain of Mecca was brought about because, in part, "the fashion had started of circulating funny stories about the old man's idiosyncrasies, some of which were undoubtedly laughable, and as the stories went round they begat others and created a demand for more, as funny stories will. A time came when the British Agent's [Bullard] political reports on the situation in the Hijaz would arrive packed with material for official laughter, to be circulated more widely than usual in Whitehall on account of their comic value. Sharif Husain became a laughing stock.")[15]

CONSPIRACY THEORIES

With everybody speculating on his movements and intentions, Holmes used the next three days to talk with Abdulla Suleiman. He explained the understanding reached in Cairo with Standard Oil of California. He said the Americans were serious about spending money on his Al Hasa field. He said he would step aside from the negotiations so that development could begin as soon as possible. He asked Suleiman to tell Abdul Aziz bin Saud, who was still in Mecca presiding over the close of the pilgrimage, that there was no longer any obligation between them in regard to the Al Hasa concession.

He went on to explain to Suleiman that the Neutral Zone was excluded from the understanding he had reached with the Americans of Standard Oil of California. He asked Suleiman not to include the Neutral Zone in any concession currently under discussion. (Both parties honored this request with neither the Saudis nor Standard Oil of California broaching the subject of the Neutral Zone.) Holmes left with Suleiman a package of documents addressed personally to bin Saud. Everybody wanted to know what was in the package.

In Kuwait, Dickson was pumping "an important individual residing at the court of His Majesty King bin Saud" and had his own version of what was going on in Jedda. His convoluted theory began with the shaikh of Kuwait's "secret" meeting in the desert with bin Saud. Dickson believed "the intention of this meeting was to prepare the ground for a submission by Frank Holmes of a proposal for the Neutral Zone."

Chafing from Shaikh Ahmad's rebuff of John Cadman, Dickson cast the shaikh of Kuwait as the villain. Dickson said the meeting in the desert was part of a plot to ensure Frank Holmes got the Kuwait concession. He said the plan was "originated" by Shaikh Ahmad who "suggested to Holmes" that he should obtain the Neutral Zone concession from bin Saud and then bin Saud should "appear to press" Ahmad to grant the Kuwait concession to Holmes.

Dickson continued to lay out how this imagined conspiracy would work, saying: "The Shaikh of Kuwait believes that with bin Saud supporting him he need not fear His Majesty's government's subsequent annoyance if he agreed to this form of pressure." He said that Major Holmes was "shrewd" enough to know that if he could get bin Saud to give him the Neutral Zone, the shaikh of Kuwait's consent would follow "as a matter of course." Dickson conjectured wildly that this would happen "because bin Saud has only to voice a wish to Shaikh Ahmad that he should give his approval and Ahmad will fall into line." Dickson declared it was "obvious" that it could all be "arranged" so as to appear that bin Saud was urging the shaikh of Kuwait to give the Kuwait concession to Holmes, rather than to Anglo-Persian. He said the excuse would be "that it would be impossible to have one oil company working in the Kuwait Neutral Zone and another in Kuwait territory."[16]

For his eyes and ears, Andrew Ryan was still relying on Philby. He repeated the story Philby was now giving out in explanation of how it was that bin Saud had personally invited Frank Holmes to Jedda, despite Philby's much promoted assertion that Holmes was persona non grata.

"Regarding the movement of Holmes," Ryan reported to London, "I gather from a very confidential conversation with Philby that the king still has a low opinion of Frank Holmes but acceded to Holmes's suggestion of a visit in order to enlarge the field of competitors." He said: "Holmes represents himself as being tied to Standard Oil of California only in respect of concessions already obtained in the Persian Gulf and of which you doubtless have full particulars." Ryan confided that Philby had told him of a conversation in which Holmes revealed "that his immediate purpose is to get the consent of bin Saud so far as he is most interested with the Shaikh of Kuwait for the concession in the Neutral Zone. He has a gentleman's agreement with Standard Oil of California not to queer their pitch in Al Hasa, provided that they leave him a free hand in regard to the Neutral Zone."

From Philby, Ryan believed he knew the content of the document package Holmes had addressed personally to Abdul Aziz bin Saud. He said: "Holmes has evolved a scheme for the Al Hasa area, the principle of which would be to divide the whole area into, say, sixteen blocks which would be taken up sectionally on contribution of proportionate payments approximating to the Saudi government's demands. Philby thinks some such scheme might be acceptable to the government but does not believe Holmes has at present serious backing for it."

The Foreign Office circulated Ryan's report in a memo headed "Oil Negotiations in Saudi Arabia." At the various stops along the way, handwritten comments were added. The Colonial Office noted: "Holmes seems to be putting his eggs into as many baskets as possible, by working out a scheme which now enables Standard Oil of California to meet amicably with bin Saud, and by securing for his syndicate a Neutral Zone concession which will be a consolation prize if they fail to obtain the Kuwait concession and a profitable addition if they do obtain it." The India Office scrawled acidly: "It is to the good that bin Saud should be showing signs of suspicion towards Major Holmes who seems to be the 'jackal' for every non-British interest."[17]

PHILBY INTRIGUES

Before sailing off to Port Sudan to "buy a hat," Stephen Longrigg had been poorly advised. As soon as he arrived in Jedda, Philby had called on him and continued to drop by. Convinced that Philby was working behind the scenes for him, Longrigg listened to what his visitor had to say—then made an offer so low that it went a long way toward assisting Philby in his disguised lobbying of Standard Oil of California to the Saudi Arabians. In his memoir, Philby would brag about how he gained information each time he invited Longrigg to dinner and how he immediately passed on to Lloyd Hamilton the detail of Longrigg's discussions with Abdulla Suleiman.

In his "very confidential conversation" with Andrew Ryan, Philby detailed a conversation he had supposedly had with Holmes. As reported by Ryan, that conversation centered on Holmes's "gentleman's agreement with Standard Oil of California" and his scheme to divide the Al Hasa concession into sixteen blocks. Ryan's account of this Philby-Holmes conversation is at odds with the one later described by Philby in *Arabian Oil Ventures*, his book on the negotiations for Al Hasa, published in 1964, well after Holmes's death in 1947.

Curiously, Philby does not give any hint of a supposed 1933 confrontation with Holmes in Jedda in his *Arabian Jubilee*, published in 1954 when many of the participants were still living, including Holmes's wife, Dorothy, who was with her husband in Jedda.[18]

Be that as it may, in this colorful account Philby claims Holmes's "first call" in Jedda, straight off the SS *Taif*, was to him; conveniently, it seems, without a witness. Philby wrote:

> Without any beating about the bush, he opened his guns on me "What is all this nonsense about, Philby? Here we are, three competitors for the right of exploring the country in search of the oil, which will make it rich, if it exists. Can't they see that the best way of ensuring its discovery is to get us all on the job together, instead of setting us all at each other's throats.
>
> "Obviously, their best course is to divide the whole territory concerned into a series of triple strips; giving each of us one strip in each district, without payment except for escorts, protection, etc., to operate for the ultimate benefit of the Government itself, which would get better terms in the end, if the oil were proved to exist.
>
> "Certainly no one is going to offer them large sums of money, just for the privilege of providing the Government with valuable knowledge of the resources of the country."

And, Philby says, "much else in the same strain."

Had it ever happened, such a conversation would be very out of character for Holmes, in particular that he would doubt the very existence of oil and expect the Arabs to come onboard "without payment." Frank Holmes had aroused great resentment among the oil companies by his habit of always moving straight to a concession rather than the safer exploration license. He was ostracized in part because he upset the status quo by always offering the highest terms. In addition, his offers had always reflected his own implicit belief that each country should have shareholding in the operating companies, which Holmes's concessions were the first to offer.

Yet, Philby forges on, saying he "listened very patiently" before pointing out that the Saudi Arabian government "is in desperate need of money" and the concession will go to the highest bidder "provided his bid is near the government's target of 100,000 sterling in gold." Philby says Holmes "scoffed at the very idea as absurd."

Philby claims that he attacked Holmes, telling him he had had "this very concession" ten years ago "on very easy terms" and that "it was not the Saudi government's fault that he had failed to make good with it." Had Philby ever had this conversation, there is no doubt Holmes would have reminded him of the scarcely forgettable Ikhwan rebellion that took place in the very area where the Al Hasa concession was located.

Philby plows ahead, claiming that he challenged Holmes with: "And by the by, I seem to remember that you still owe the government something, a matter of about 6,000 sterling in gold, on account of the rental of the area for the three years during which you held the concession rights but did no exploring. I rather think that Abdulla Suleiman will insist on your paying that outstanding debt before being permitted to negotiate at all in the present case. You have given them reason to doubt whether you will, or can, fulfil your engagements." Philby alleges: "That shaft went home!"

Frank Holmes does not seem to have remembered any in-depth conversations, or confrontation, with Philby. The Europeans in Jedda while Holmes and Dorothy were there included the Hamiltons, the Twitchells, and the Philbys. This created a rare opportunity for lively social occasions. Perhaps Andrew Ryan's quip about the need for a separate hotel for concession hunters might not have been too far off the mark. When Holmes arrived back in Kuwait, he told Dickson that in Jedda he had met up with Hamilton, Twitchell, Longrigg, and Philby and had also come across "a Swiss and an Italian" interested in the oil concession. According to Dickson, Holmes said that he did not think much of Longrigg but was "much impressed with Philby—but more especially his talented wife!"[19]

Frank Holmes, Dorothy, and Muhammad Yateem left Jedda on April 12. They were seen off by a grateful Lloyd Hamilton, relieved that Holmes had now cleared the way for Standard Oil of California with the Saudi officials. To Philby's annoyance, Hamilton sent him a message putting off their appointment because "I am busy getting the Major off on the *Taif* this morning."

In his *Arabian Oil Ventures*, Philby claims his challenge hastened Frank Holmes's departure. He writes: "Next day, after my meeting with Holmes, the SS *Taif* was due at Jedda from Port Sudan on its way back to Suez; and Major Holmes was on the passenger list." Philby does not seem aware of Holmes's discussions over three days with Abdulla Suleiman. He goes on to say: "Holmes had apparently changed his mind about seeing Abdulla Suleiman and the King. . . . Thus one of the three players had thrown in his hand and only Longrigg was left to fight it out with Hamilton."

The display of close friendly relations between Hamilton and Holmes seems to have puzzled Philby. He adds: "Naturally, I gave Hamilton a full account of my interview with Holmes. But I always felt there had been some understanding between them for the purchase of the concession in the event of Holmes being able to secure it on better terms than those proposed by the Saudi government to Hamilton."

While Philby did what he felt he had to do in order to ensure his own private interest, including his atrocious attempts to destroy Frank Holmes's reputation both personal and professional, he does seem to have had a grudging admiration for him. In his memoirs, Philby wrote: "It must be admitted that Frank Holmes had an extraordinary flair. The gallant Major, a man of considerable personal charm, certainly achieved some striking successes. This tough and patient New Zealander achieved a proud record in his lifetime. His name will always be associated with the development of the Arabian oilfields." But, says Philby, "I never thought he had a chance of securing the Al Hasa concession for the second time, for a song."[20]

For all his skill at intrigue and misdirection, Philby never knew of a private meeting between Hamilton and Abdul Aziz bin Saud that took place soon after Holmes's departure. Arriving in Jedda from Mecca, the king discreetly called Hamilton to his Kazma palace on April 20. Bin Saud spoke of Standard Oil of California's activities in Bahrain, which Frank Holmes had told him about. He implied that Hamilton should now feel confident about getting the Al Hasa concession. Hamilton did not tell anybody, particularly Philby, of this meeting. He soon received a letter from Philby saying bin Saud was "putting pressure" on his ministers and things were looking "propitious." Naturally, Philby took credit for this state of affairs.

Dickson reported from Kuwait that Holmes had told him Abdul Aziz bin Saud would not make a decision on the Al Hasa concession "until the end of May." Dickson astutely observed that Frank Holmes was far from being persona non grata with bin Saud. Dickson was still pursuing his conspiracy theory as he warned his colleagues that bin Saud was about to force Shaikh Ahmad to give the Kuwait concession to Holmes. Ever the interpreter of the "Arab Mind," Dickson explained: "To the Arab world this would mean that Kuwait was under bin Saud's political control . . . such a move would not necessarily be unpleasing to the Shaikh of Kuwait either, even though he might make some sacrifice financially."[21]

AL HASA SIGNED

Sure enough, just as Holmes told Dickson, bin Saud did not make his decision until the end of May. But by then, Standard Oil of California was the only contender left standing.

Perhaps pleased that the concession had gone to the Americans, and to one of the Standard Oil companies at that, Stuart Morgan at the Iraq Petroleum Company must nevertheless have been a bitterly disappointed man. Since Frank Holmes had first appeared in his New York office in 1926, and aroused his interest in the oil resources of Arabia, Morgan had been maneuvering and manipulating for this very Al Hasa concession. He had exhorted his colleagues at Iraq Petroleum/Anglo-Persian telling them "the Al Hasa concession would be theirs within 48 hours" if they would just authorize Longrigg to "plunk down 50,000 sterling." They told him he was "far too hasty."

Longrigg would later remark that "the chief mistake I made was that I did not believe there was paying oil in Al Hasa." Rather than listening to Stuart Morgan, Iraq Petroleum/Anglo-Persian was swayed by the negative reports of its geologists. Longrigg was instructed that "a proved field is a very different thing from one yet to be explored." He was told to push for an exploration license, for a low monthly fee, rather than an immediate concession. Perhaps Iraq Petroleum/Anglo-Persian thought they didn't have to exert much effort because, as they believed, Philby was working for them. Longrigg was so trusting that at one stage he asked Philby to take over the negotiations on his behalf while he went back to check on the construction of the Iraq oil pipeline.

Andrew Ryan believed this and more; he thought Philby was furthering the interests of the British per se by working for the Iraq Petroleum/Anglo-Persian bid.

On May 30 Andrew Ryan reported with some consternation: "Representative of Iraq Petroleum Company left for London on May 24th at a moment's notice." By the time Ryan learned of Longrigg's departure, the deed was already done. The Al Hasa concession was signed with Standard Oil of California on May 29, 1933. Philby soon came by to gloat. With malicious glee Philby records how he enjoyed his revenge against the man who got the job of British minister to Saudi Arabia—the job that Philby believed should have been his. The job about which he had told his wife, "I should of course run the show. And expect the Foreign Office to leave me to do so."

In *Arabian Oil Ventures* Philby describes his visit to Andrew Ryan, saying: "We talked about everything under the Arabian sun; but it was only when I got up to take my leave that I said to him: 'I suppose you have heard that the Americans have got the concession.'" Philby almost glows as he chirps: "He was thunderstruck and his face darkened with anger and disappointment." Never once admitting that he misled Ryan from the very beginning into thinking that through him the British would get the concession, he writes with ill-disguised triumph: "He had been certain that his influence behind the scenes . . . would have turned the scales in favour of the British competitor. Our final leave taking was somewhat strained."

Philby dismisses Ryan with a final insult, saying: "He was indeed the 'Last of the Dragomans,' bred in the school of traditional dominance in the eastern world, while I was surely one of the first champions of eastern emancipation from all foreign controls."[22]

A BARGAIN BUY

If anyone at Iraq Petroleum/Anglo-Persian had acted on Stuart Morgan's advice and "plunked down 50,000 sterling" they might indeed have got the Al Hasa concession in forty-eight hours. In the end it was a cheap concession. Had he not been involved, Philby might even have described the price as "a song." Despite their later claims to have been exceedingly generous with the Saudi Arabians, the Americans at Standard Oil of California paid less for Al Hasa in 1933 than Holmes had done in 1923.

The financial detail of Standard Oil of California's agreement, signed on May 29, 1933, provided for a lump sum upfront of $173,000, composed of $150,000 (thirty thousand sterling in gold) together with the first-year rent of $25,000 (five thousand sterling in gold). However, the $150,000 was not a payment, it was a loan, to be repaid from future royalties. The $25,000 (five thousand sterling) first-year rent was, in fact, the only actual payment. Everything else was in the form of loans with clauses in the contract dictating how these loans were to be repaid. After eighteen months the company would advance a second loan of $100,000.

When oil was discovered, and produced in commercial quantities, the company would make a third loan of $500,000. The royalty was four shillings a ton (matching the new royalty now obtained by Persia). The company would pay neither import nor export customs duty and was exempt from all direct or indirect taxation. If the company organized its own subordinate company for the purpose of exploiting the concession, and if it offered shares in the subordinate company for sale to the general public, Saudi Arabian subjects would be given the opportunity to subscribe a minimum of 20 percent of such shares.

Standard Oil of California's 1933 agreement can be seen to be far from generous when contrasted with Holmes's of a decade earlier. In 1923 Holmes's terms were an upfront payment of $30,000, as a direct payment, not a loan. Annual payments were $45,000 against Standard Oil of California's 1933 rent of $20,000. Holmes's annual payment was made up of rent at $15,000 per annum plus payment each six months of $15,000, designated "security protection" provided by bin Saud. Under Holmes's terms, the Saudi Arabian government held full participation in the development, from the beginning, in the form of a 20 percent shareholding and the right to take up, for cash, an additional 20 percent; if bin Saud exercised this right he would hold 40 percent of the company. The company would pay customs duty and, after the commencement of commercial production, would be liable for local taxes.[23]

THE WASH-UP

Apart from the bonuses in his contract, Philby was rewarded with a retainer of one thousand sterling a year, which he kept until the outbreak of World War II. Philby says he was paid "for my contacts . . . without having any very burdensome duties to perform." Perhaps Standard Oil of California viewed it as insurance.

Twitchell got his bonuses and, after oil was discovered, five hundred shares in Standard Oil of California. Fountain pen sets were distributed to the various Saudi officials connected with the negotiations, and the sole interpreter/secretary, Najib Salha, got a briefcase. Twitchell also stepped forward to claim "a reward from the Saudi Arabian government, as they had promised." Twitchell argued that the concession signed by the Americans fulfilled his mission of "finding capital to carry out development." He never did reveal the amount of this "reward." Twitchell left shortly after for London to resume the search for investors in *Mahad Dhahab*, the ancient gold mine that he was convinced was that of King Solomon. He would succeed in beginning production there in 1937 and extract $32 million worth of gold, silver, and copper before closing in 1954.[24]

Standard Oil of California created the wholly owned Californian Arabian Standard Oil Company (Casoc) organized under the laws of Delaware with a capitalization of $700,000 to which the concession was assigned at the end of 1933. As this was not a public shareholding company, no participation was offered to Saudi Arabia. The name was later changed to the Arabian American Oil Company (Aramco).

That the concession was underpriced can be judged by the fact that just three years later, in December 1936 (before oil was proven in March 1938), the Texas Oil Company bought a half share in the Saudi Arabian concession, for which they paid Standard Oil of California $3 million in cash and undertook to pay a further $18 million out of the oil produced in Saudi Arabia. Little wonder an Aramco official would comment some years later about "certain factions critical of American oil efforts in Saudi Arabia" who charged that Standard Oil of California "took advantage of the bare foot boys" in its purchase of Holmes's lapsed Al Hasa concession.[25]

That Standard Oil of California's concession was Frank Holmes's original Al Hasa concession was never in doubt. It was Holmes's identification of this field as being rich with oil that first aroused the interest of Standard Oil of California. The dimension of the field, as originally mapped by Holmes, was not altered for the resale. And Philby, secretly working for Standard Oil of California, had a copy of Holmes's original map and dimensions received from Stuart Morgan. The concession obtained by Standard Oil of California in 1933 was Holmes's original of 1923, with the addition of options for other areas.

Dickson knew it and reported to the resident: "The Hasa Concession, in respect of area especially, is virtually the same as was given to Major Holmes in

1923." Philby would admit it in his 1950 *Arabian Jubilee*, describing Holmes's 1923 concession as "roughly the same as that now being operated in Eastern Arabia by the Arabian American Oil Company."

And Fred Davis, the Standard Oil of California geologist who worked with Holmes on the Bahrain concession in 1930, and went on to become chairman of the Aramco board, admitted in 1974 that the Al Hasa concession, as identified and mapped by Holmes, "covered almost the entire area we had there at the start *and is where all our oil has been found*."[26]

Unlike Bahrain, where Holmes did all the preliminary work on his own concession and where oil was found less than eight months after the spudding of the first well, it would take Standard Oil of California five years, until March 1938, to unlock the secrets of Al Hasa that Holmes had spotted a decade previously. Standard Oil geologists "drilled seven holes in Arabia and found no oil . . . these seven wells cost $7 million," an Aramco geologist would record. When M. E. Lombardi was asked in later years whether he "had trouble getting the money" to continue the drilling program due to the series of dry holes, he would say he had not. In what must be taken as a tribute to Holmes, even if unintended, Lombardi explained: "Because of the discovery in Bahrain, we were sure there was oil in Al Hasa."[27]

NOTES

1. IOL/R/15/2/421.10/5, March 22, 1933, political resident (Fowle) to Colonial Office copied to HMG's minister, Jedda (Ryan) and to Government of India, "Holmes intends visiting bin Saud in Riyadh, via Jedda, after . . . Cairo"; H. St. J. B. Philby, *Arabian Oil Ventures* (Washington, DC: Middle East Institute, 1964), p. 94, "I am not particularly frightened, especially since my conversations with you about the Major," p. 96, "I think you and I do not fear him"; p. 97, "telegraphed the king *with their approval* to delay a decision"; p. 88, "the prospect of this windfall . . . eased the problem of bills for a son at Cambridge and three daughters at first class schools"; p. 104, "past record with this very Al Hasa concession" and "now that you have secured the services of Holmes."

2. For the Senate investigations, Lloyd Hamilton would later give an extraordinary version of this whole Cairo episode. See Chevron Archives, box 0120814, in-house interview with Lloyd Hamilton written in connection with Senate investigation dated April 2, 1947: "Hamilton carried on negotiations for a concession arrangement on the easterly portion of Saudi Arabia with Shaikh Abdulla Suleiman Minister of Finance and other representatives of the Saudi government until the middle of March 1933 when negotiations threatened to break down due to excessive demands on the part of the government. Hamilton went to Cairo late in March to meet Lombardi for a conference on the policy to be followed in further discussions with the government. He was with Lombardi in Cairo until April 5th and arrived back in Jedda on April 9th 1933."

3. Philby, *Arabian Oil Ventures*, pp. 94–95, "Holmes's interests are in accord with our attempt to get such a concession" and "astonished" and "my principals do not regard

it so" and "*volte-face*"; note on p. 97 Philby claims he received a cable from the American lawyers asking him "to confirm or refute Holmes's claim to have been invited by bin Saud to visit him, some time during the period 1930–32. Naturally, I had no specific information on this point, although I had no doubt that, if Holmes had proposed such a visit, bin Saud would have agreed to it." Such a reply would appear to be very out of character for Philby considering the energy he put into widely circulating his story that Holmes had been persona non grata to bin Saud since 1928.

4. Ibid., pp. 103–104, in a letter to Loomis, "Evidently Lombardi has been impressed by Holmes, it is of course entirely your business whom you employ to work for you . . . I believe Holmes to be a *persona non grata* with the king"; ILO/R/15/1/641/ 86.IV, February 2, 1934, India Office (Laithwaite) note of interview with Mr. Janson and Major Holmes, "It was important for E&GSynd to recoup themselves for their previous losses in respect of the Hasa concession which amounted to some 16,000 sterling paid over to bin Saud, over a period of years, without any return"; Benjamin Shwadran, *The Middle East: Oil and the Great Powers* (London: Atlantic, 1956), p. 287, citing Philby, repeats this story of bin Saud supposedly angrily canceling the concession. Philby tells this story in *Arabian Oil Ventures* and also in *Arabian Jubilee* (London: Hale, 1952), pp. 69, 178. Shwadran's account of the Saudi Arabian negotiations is based entirely on Philby's *Arabian Oil Ventures*. Unfortunately, even now many authors do rely without question on Philby's account, see, for example, Anthony Cave Brown, *Oil, God, and Gold: The Story of Aramco and the Saudi Kings* (Boston: Houghton Mifflin, 1999).

5. Philby, *Arabian Oil Ventures*, p. 103, "there were rumours about that Holmes was about to butt into the Al Hasa business . . . when I saw the king he told me that Holmes was indeed coming to see him at Jedda in connection with this oil business"; IOL/R/15/2/421.10/5, March 15, 1933, British minister to Jedda (Ryan) to Foreign Office, Ryan's four-page report on the situation in regard to the oil negotiations, "information of mine that Holmes is *persona ingrata* in bin Saud's eyes"; and March 26, 1933, Ryan to Foreign Office, "Holmes has hitherto been believed to be distasteful to bin Saud"; and April 8, 1933, Ryan to Foreign Office (copy to India), "How should I treat Holmes officially if he arrives?"; and April 11, 1933, Foreign Office to Ryan (copy to India), Subject: Treatment to Be Accorded to Major Holmes, "should be limited to normal courtesies. Holmes's visit is for the purpose of furthering American interests and he is entitled to no official assistance"; Philby, *Arabian Oil Ventures*, is mistaken in giving the date as the "morning of April 10th"; date of arrival in Jedda was April 9. Philby comments Holmes arrived "curiously enough in the same steamer as Hamilton . . . who had now been away from the scene of action for nearly three precious weeks."

6. Yossef Bilovich, "Great Power and Corporate Rivalry in Kuwait, 1912–1934: A Study in International Politics" (PhD diss., University of London, 1982), p. 256, citing February 20, 1933, minute of the Colonial Office, "We are not in control. We are not interested. And the Middle East Department is not staffed to deal with it (oil)."

7. IOL/R/15/1/623, vols. 82 and 86, February 19, 1933, and March 22, 1933, political agent Kuwait (Dickson) to political resident (Fowle), "It is clearly our duty to prevent the shaikh accepting a private invitation of this sort" and March 31, 1933, "Confidential," political resident (Fowle) to foreign secretary, Government of India cc Colonial Office, "days have definitely gone when we can keep the more important Arab Rulers of the Gulf in leading strings." Nevertheless, he did write a gentle personal note.

8. Thomas E. Ward, *Negotiations for Oil Concessions in Bahrain, El Hasa (Saudi Arabia), the Neutral Zone, Qatar and Kuwait* (New York: privately published, 1965), p. 224, for detail of Cadman's November 1932 "Petroleum Policy" speech. Anthony Sampson, *The Seven Sisters: The Great Oil Companies and the World They Made* (London: Hodder and Stoughton, 1975), pp. 137–40, "The US Federal Trade Commission investigated the foreign agreements of the oil companies, the report released in August 1951 charged the seven companies controlled all the principal oil producing areas outside the US, all foreign refineries, patents and refining technology, that they divided the world markets between them, and shared pipelines and tankers throughout the world—and that they maintained artificially high prices for oil. The report said the companies were 'engaged in a criminal conspiracy for the purpose of predatory exploitation.'" The price fixing and exploitative actions revealed involved five American companies—Standard Oil of California, Texas Oil, Standard Oil of New Jersey, Socony Vacuum (formerly Standard Oil of New York), and the Gulf Oil Corporation—together with British Petroleum (formerly the Anglo-Persian Oil Company) and the Dutch Shell groups. The tactics of juggling quotas and production restrictions, first mooted by Sir John Cadman in 1932 and perfected by the international petroleum cartel that became known as the Seven Sisters, was the model on which OPEC would later draw.

9. Ward, *Negotiations for Oil Concessions*, p. 224, "the loss of Bahrain by Gulf Oil was the greatest regret of his life"; see chap. 5 of this book. In 1923 in order to appear to meet Sir Percy Cox's instructions that they seek a joint venture, E&GSynd approached Anglo-Persian with, it must be said, no serious intention of pursuing Cox's vision, and IOL/R/15/5/238, vol. 111, April 12, 1926, Wilson in London to political agent Kuwait (More), "Holmes's London principals have been bothering us for the last few months trying to sell us the Kuwait concession before they have got it! And all the other concessions too . . . but we are not having any," and Bilovich, "Great Power and Corporate Rivalry in Kuwait," p. 218, April 1932, John Cadman to the Foreign Office, "Americans welcome to what they can find there."

10. Chevron Archives, box 0426154, Principal Documents in Regard to Kuwait Organization (1951), pt. 11, Background of the Kuwait Oil Concession, Its 50-50 Ownership by the D'Arcy Exploration Company (Anglo-Iranian Oil Company Ltd.) and Gulf Exploration Company (Gulf Oil Corporation) through the Kuwait Oil Company, and the Restrictions of the Agreement of December 14, 1933, p. 18, Cadman saying, "at the time the Kuwait concession was taken up with Anglo-Persian by Holmes . . . lie fallow." Gulf Oil recognized Cadman's move to initiate an oil cartel, a 1932 memorandum from Gulf Oil's William Wallace in this file observed, "Anglo-Persian has recently completed its combination with the Dutch Shell for the protection of their present 'as is' position . . . they are frankly afraid this situation will be disturbed and possibly upset if we are left at liberty to proceed in Kuwait without any let or hindrance"; see also Archibald H. T. Chisholm, *The First Kuwait Oil Concession Agreement: A Record of the Negotiations, 1911–1934* (London: F. Cass, 1975), p. 176, March 16, 1933, Department of State Division of Near East Affairs memorandum regarding Cadman's conversation with Gulf Oil, "use every power in his means to do so." See also Ward Papers, box 2, September 12, 1933, Madgwick to Ward, "Beeby Thompson tells me Cadman is wild at having missed Bahrain. His geologists turned it down."

11. Chisholm, *First Kuwait Oil Concession Agreement*, p. 176, March 1, 1933, John

Cadman to APOC's Sir John Lloyd, "(I told Mellon) we are both capable of handling this matter without outside assistance" and p. 175, "Andy Mellon has returned determined to keep his hands on Kuwait." On pp. 175–76, Chisholm reprints the March 16, 1933, memorandum, Department of State Division of Near East Affairs recounting John Cadman's approaches and conversations with Gulf Oil; Chevron Archives, box 0426154, Principal Documents in Regard to Kuwait Organization (1951), pt. 11, Background of the Kuwait Oil Concession, Its 50-50 Ownership by the D'Arcy Exploration Company (Anglo-Iranian Oil Company Ltd.) and Gulf Exploration Company (Gulf Oil Corporation) through the Kuwait Oil Company, and the Restrictions of the Agreement of December 14, 1933, p. 17, Basil R. Jackson, American representative of Anglo-Persian, to William Wallace of Gulf Oil, "will not passively look on while any other company, with no reasonable interest whatsoever in that area. . . ."

12. Bilovich, "Great Power and Corporate Rivalry in Kuwait," pp. 278–80, citing March 16, 1933, minute by Foreign Office, (Cadman will) "keep the Persians guessing" as to what he might be doing in Kuwait.

13. Chisholm, *First Kuwait Oil Concession Agreement*, pp. 32–33, 177–78, reprints in full the Anglo-Persian minutes of Sir John Cadman's interview with Shaikh Ahmad on March 25, 1933; see also Chisholm, p. 180, May 17, 1933, Frank Holmes to E&GSynd, "Anglo-Persian has told Ahmad that the Gulf Oil Corporation approached Cadman with a view to combining forces in obtaining and working the Kuwait concession"; see also Bilovich, "Great Power and Corporate Rivalry in Kuwait," p. 281. Note that a myth has been perpetuated that Holmes made a "dawn" visit to Shaikh Ahmad before Cadman arrived in order to extract the promise that he could top any offer Cadman made. As Holmes was in Cairo on March 25, this is obviously impossible. Ward gives this dramatic story on p. 227 and seems to have got it from Dickson whom he interviewed in 1955, see Ward, *Negotiations for Oil Concessions*, 2nd appendix, p. 255(a). This myth has been made much of in subsequent publications.

14. IOL/R/15/2/421.10/5, March 15, 1933, British Legation Jedda (Ryan) to Foreign Office (C. F. A. Warner), "the stage is set . . . the *dramatis persona* is . . ." and March 27, 1933, British Legation Jedda (Ryan) to secretary of state for Foreign Affairs London, cc resident Bushire, "Port Sudan for a few days"; Philby, *Arabian Oil Ventures*, p. 102, "Longrigg departed on a flying visit to Port Sudan to buy a hat."

15. See George Antonious, *The Arab Awakening* (London: H. Hamilton, 1938), Antonious was secretary/translator to Charles Crane during his 1923 visit to Sharif Husain in Jedda and again in December 1926, when Crane was hosted by bin Saud's son, Faisal, in Jedda; and E. C. Hodgkin, ed., *Two Kings in Arabia: Letters from Jeddah, 1923–5 and 1936–9* (Reading, UK: Ithaca Press, 1993). Bullard was consul to Jedda 1923–25, and would replace Andrew Ryan as minister to Jedda 1936–39, before becoming ambassador to Tehran.

16. IOL/R/15/1/623, vols. 82 and 86, April 27, 1933, political agent Kuwait (Dickson) to political resident (Fowle) Holmes, "had not attempted in any way to get an oil concession out of bin Saud, for the whole or part of, Saudi Arabia, but was after 'something else' which he could not divulge to me." Holmes "had not seen the king but had seen Abdulla Suleiman, staying three days in Jedda during which he submitted his official proposals for the 'thing' he wanted, in writing, and addressed these to the King." This long report also details Dickson's conspiracy theory, "it would be impossible to have one oil company working in the Kuwait Neutral Zone and another in Kuwait territory."

17. Philby, *Arabian Oil Ventures*, pp. 105–20, has no compunction in explaining how he immediately passed all detail of Longrigg's official discussions on to Hamilton; PRO/FO 371/16871, April 11, 1933, "No Distribution" British Legation Jedda (Ryan) to Foreign Office, "a very confidential conversation with Philby"; and April 27, 1933, Foreign Office memorandum, "Oil Negotiations in Saudi Arabia" circulated Ryan's report, and "Holmes seems to be putting his eggs into as many baskets as possible," and "good also that bin Saud should be showing signs of suspicion towards Major Holmes who seems to be the 'jackal' for every non-British interest."

18. *Arabian Oil Ventures* was published by the Arabian American Oil Company (Aramco) in 1964, after Philby's own death in 1960. While Philby several times refers to the Al Hasa concession, and to Frank Holmes, in the 1954 *Arabian Jubilee*, there is no mention of any conversation or confrontation.

19. Note that Philby, *Arabian Oil Ventures*, p. 62, had no reluctance in revealing that he had Holmes's map and documentation related to Al Hasa (passed to him by Stuart Morgan via T. D. Cree), "the actual document signed by Holmes and bin Saud has never been published; but its general character and content seem clear enough from copies in my possession of the original draft brought out to Arabia by Holmes." Furthermore, in the editor's note, p. 66, it is pointed out that Philby's incorrect estimate that Holmes's annual rent was 2,000 sterling is based on this early draft, the amount was later raised to 3,000 sterling in gold, plus payment each six months of a further 3,000 designated "protection fee." Unfortunately, Philby's error on Holmes's 1923 terms and conditions is continued in the literature until today; IOL/R/15/1/623, vols. 82 and 86, April 27, 1933, political agent Kuwait (Dickson) to political resident (Fowle) quoting Holmes, "much impressed with Philby—but more especially his talented wife!"

20. Philby, *Arabian Oil Ventures*, pp. 98–99, note that the dates do not support Philby's claims. Holmes arrived in Jedda on April 9 and departed April 12, Philby says he was Holmes's "first call" but gives the date of the conversation as April 10. Had Philby scared him off "next day" as he says, Holmes would have departed April 10 or perhaps April 11, but not April 12. On p. 128, Philby refers to "the *flight* of Holmes from Jedda"; p. 99, "I am busy getting the Major off on the Taif this morning" and "I always felt that there had been some understanding between Hamilton and Holmes" and "for a song"; pp. 54, 98, 99, "achieved a proud record in his lifetime" and "his name will always be associated with the development of the Arabian oil fields." On p. 54 Philby says, "the first of Holmes ventures (Al Hasa), covering the greatest potential area of production, came *unaccountably* to grief."

21. Wallace Stegner, *Discovery! The Search for Arabian Oil*, abridged for *Aramco World* (Beirut, Lebanon: Middle East Export Press, 1971), p. 19, "after that interview with the king, about which he told no one" and "Philby's letter . . . things looked propitious"; IOL/R/15/1/623, vols. 82 and 86, April 27, 1933, political agent Kuwait (Dickson) to political resident (Fowle). Dickson repeats his conspiracy scenario, with more detail, in IOL/R/15/5/242, vol. II, June 7, 1933, Kuwait political agent (Dickson) to political resident (Fowle).

22. Shwadran, *The Middle East*, p. 291, "the General Manager of Iraq Petroleum reported in 1935 . . . authorise Longrigg to plunk down 50,000 sterling"; Brown, *Oil, God, and Gold*, p. 46, "I did not believe there was paying oil in Al Hasa"; IOL/R/15/5/242, vol. VII, May 30, 1933, British minister at Jedda (Ryan) to Foreign Office,

"Representative of Iraq Petroleum Company left for London on May 24th at a moment's notice." Philby, *Arabian Oil Ventures*, pp. 125–26, "He was thunderstruck and his face darkened with anger and disappointment" and "he was the last . . . while I was surely the first." *The Last of the Dragomans: Sir Andrew Ryan, 1876–1949* (London: G. Bles, 1951) was the title of Sir Andrew Ryan's biography, edited by Sir Reader Bullard, and not surprising containing a large selection of Ryan's reports and letters from his foreign postings.

23. Financial details of the Standard Oil of California 1933 purchase is in Chevron Archives, box 120797, minutes of board of directors meeting, December 4, 1933; Shwadran, *The Middle East*, pp. 292–93; and in Daniel Yergin, *The Prize: Epic Quest for Oil, Money, and Power* (New York: Simon and Schuster, 1991), p. 291. Philby, *Arabian Oil Ventures*, p. 124, gives the date of signature as May 10, 1933, this is certainly an error; Ralph Hewins, *Mr. Five Per Cent: The Story of Calouste Gulbenkian* (New York: Rinehart, 1958), p. 222, referring to bin Saud's original request for a down payment of $800,000 (100,000 sterling) comments bin Saud was "whittled down" and calls the final terms "derisory"; Holmes's financial details are in the Frank Holmes Personal Papers that includes the original lease made in the name of "Abdul Aziz bin Abdul Rahman bin Faisal bin Saud, Sultan of Nejd and its dependencies and Major Frank Holmes of 20 Cecil Street SW London in his capacity as the true and lawful attorney of the Eastern & General Syndicate."

24. Philby, *Arabian Oil Ventures*, p. 130, "a retainer of 1000 sterling per annum"; Twitchell Papers, typewritten manuscript for Saudi Arabia, p. 58, "reward from the Saudi Arabian government, as they had promised"; Twitchell's diaries, entry for January 14, 1939, "cable arrived saying Standard Oil of California had deposited $24,278,89 FOR ME!!! [*sic*] Asked if I want to purchase 450 instead of 225 shares in Standard Oil of California. I replied yes"; Twitchell diary entry for September 15, 1942, shows him then holding 1,175 shares in Standard Oil of California; they paid thirty-five cents dividend; typewritten manuscript for Saudi Arabia, p. 58, "fountain pens . . . a briefcase"; Twitchell Papers, Merritt interview, p. 83, "I terminated my services with Standard Oil of California and went to London (to find investors)."

25. Shwadran, *The Middle East*, p. 295, for financial detail of the 1936 deal; Philby Papers, box XLII-3, April 18, 1962, re (posthumous) publication of Philby's "Arabian Oil Ventures," Gary Owen of Arabian American Oil Company, Washington, to Bill Sands, "there are certain factions critical of American oil efforts in Saudi Arabia who like to say that the Americans knew there was oil there all the time and took advantage of the bare foot boys."

26. Philby, *Arabian Oil Ventures*, p. 62, "the same as that now being operated"; IOL/R/15/5/242, vol. VII, June 7, 1933, political agent Kuwait (Dickson) to resident (Fowle), "same as was given to Holmes in 1923"; Owen Papers, box 1a; October 11, 1974, Fred A. Davis to G. Gish, "is where all our oil has been found."

27. The Bahrain Well No. 1 was spudded October 16, 1931, this well "came in" on June 2, 1932; Chevron Archives, box 120797, in-house interview with G. G. Gaylord April 25, 1958: Gaylord was sent to Al Hasa in 1936, "seven holes and no oil . . . cost US$7 million"; and in-house interview with M. E. Lombardi, September 9, 1955, "because of the discovery in Bahrain," the interviewer attached this comment to his transcript, "This does not support Aramco publicity that attributes unusual courage to the company in pouring money into Arabia in depression years."

CHAPTER 21

PRESSURE POINTS

W hile Andrew Ryan's dramatis personae were playing out their various roles in the Jedda negotiations, in London the Colonial Office was transferring its political responsibilities for the Persian Gulf over to the India Office. The Colonial Office, the Foreign Office, the India Office, the Admiralty, and the Petroleum Department gathered for two conference sessions on April 26 and May 3, 1933. The political resident, Lt. Col. Trenchard Fowle, was instructed to come to London and join them. The goal was to draw up definitive policy and guidelines on oil concessions and development to be implemented by the India Office and the Government of India acting through its political officers in the Persian Gulf.

THE AGENDA

When the meeting convened, the item at the top of the list was the rebuff handed out by the shaikh of Kuwait to Anglo-Persian's Sir John Cadman—despite his munificent offer and contrary to the advice contained in the Petroleum Department's "comparison" of both offers. Without any evidence, the participants at the meeting surmised that the rebuff could only mean that the shaikh of Kuwait was secretly receiving a personal commission from Frank Holmes and his syndicate, "which he would not have to pass on to his family."

The meeting participants were highly suspicious of Holmes's tacit agreement to Shaikh Ahmad's proposal that he split Kuwait into two concessions. This was a problem. They couldn't intervene on the grounds that Kuwait was too small for two concessions because they knew "British interests" (Anglo-Persian) were currently attempting to "acquire a footing in Bahrain," which was tiny in comparison to Kuwait.

The participants at the meeting had "reason to believe" that during his explorations for water, Holmes might have discovered "that one area has more oil bearing possibilities than the other." They strongly suspected Holmes of angling so that "the part to be allotted to the British might be comparatively unremunerative." They might have been right. Holmes had selected the southern half of the split concession; it would prove to be far richer in oil than the northern half. Moreover, the southern section adjoined the Neutral Zone, which would result in Holmes having a large area if he succeeded in concluding a joint concession with the shaikh of Kuwait and bin Saud.[1]

Although the participants of the meetings did not know it, by May 4 when the second session gathered in London, Holmes had already persuaded Shaikh Ahmad away from the idea of a split concession. He had countered Cadman's "bonus offer" with a higher one of his own and added free supplies of petrol and kerosene for Ahmad and his family. Holmes's outbidding of Cadman was contingent on Ahmad's returning to a concession covering the whole of Kuwait territory, the same as Cadman had asked for. Ahmad had dropped the whole idea of splitting the concession.

Qatar was on the meeting's agenda. With pleasure the gathering noted that Anglo-Persian had recently secured, with the active assistance of several of the participants present, a low-rent exclusive option to pursue geological surveys for two years. The meeting decreed that Shaikh Abdullah Al Thani of Qatar should be further locked in by being "reminded" of his obligation under the 1916 treaty not to grant any concessions in his territory without the approval of His Majesty's government. The Petroleum Department was told to inform the Anglo-Persian Oil Company they could be confident this restriction on the shaikh of Qatar would be rigidly applied.

Al Hasa came up for discussion. Perhaps reassured by Sir Andrew Ryan's reports that with Philby's help the British would certainly get this concession, the meeting thought Al Hasa was not of immediate concern. They noted merely "various interests were currently negotiating" and "apparently Major Holmes is taking a part."

As to the Kuwait Neutral Zone, they weren't sure whether or not Holmes had had this concession in 1923, or if he had, whether or not it had lapsed. The meeting decided it probably had lapsed. The shaikh of Kuwait had been told that in awarding his concession "the ultimate decision rested with him." Despite this, the meeting participants now agreed that, like the shaikh of Qatar, the shaikh of Kuwait would be told he could not grant any concession in the Neutral Zone without first obtaining the approval of His Majesty's government. This instruction should be issued quickly, the group agreed, in case the shaikh of Kuwait under "a misapprehension"—or even "wilfully"—acted in regard to the Neutral Zone "without reference to His Majesty's government." A note of caution was raised from one participant who said, "our right of control in the Neutral Zone is less than in Kuwait," which might make such an action difficult to justify. The meeting decided the Shaikh of Kuwait should be instructed orally rather than in writing.

On the subject of the concession in "Kuwait proper" the group spent a lot of time debating just how far they could go in insisting Shaikh Ahmad accept "our advice" and choose Anglo-Persian, without arousing the ire of the Americans. He had, after all, flatly turned down the advice contained in the "comparison." They decided that "tendering of advice" could not become the subject of arbitration or legal proceedings so long as they could make it appear "bona fide" and could mount a "reasonable defence" that they had not violated the pledge given to the Americans. If they handled it this way, there would be no need to "take too seriously any protests we might receive from the United States government." One participant added enthusiastically that as long as the advice "appeared" sincerely given on "the merits of the case" then they were free to exert "all legitimate pressure on the shaikh."

Knowing the shaikh of Kuwait's well-founded fear of the territorial ambitions of bin Saud and Iraq, the meeting participants agreed that "legitimate pressure" should begin with a warning from the resident that by "disregarding our advice" the shaikh was "endangering his own interests." With no hint of sarcasm, Fowle agreed that Shaikh Ahmad would probably "accept advice tendered in this fashion."

The meeting covered Shaikh Ahmad's disappointment at not being "allowed" to take up Edmund Janson's jaunty invitation to visit his London and country houses and sail on his yacht. The group instructed the resident that while he shouldn't actually commit to anything, he could "intimate orally" to the shaikh of Kuwait that when the oil question was settled, he might "possibly" receive an invitation to visit England as the guest of His Majesty's government.

Discussion of Bahrain followed. The Bahrain Petroleum Company insisted it was entitled, under the terms of the original agreement, to a twelve-month extension of its prospecting license. In return for now selecting two hundred thousand instead of one hundred thousand acres, the company was proposing to increase the annual payment and make a substantial loan to the shaikh of Bahrain to be repaid over five years from future royalties. As the participants knew, Anglo-Persian was playing a "spoiler" role by lodging an offer to the shaikh of Bahrain for the additional area Bapco now wanted. The shaikh of Bahrain was resisting Anglo-Persian's offer. Although the Bahrain Petroleum Company was making it clear that their immediate concern was the extension of the prospecting license, the meeting members saw an opportunity to push them into accepting the "amendments" they "desired" to see introduced, retrospectively, into the original agreement. The desired amendments included the right of the British government to preempt the oil of Bahrain. So that the oil would be suitable for use by the British navy, another amendment would ensure that refining, according to British navy specifications, was done on the spot in Bahrain, or at least in British territory.

It was obvious the India Office was now in charge when a well-practiced Government of India tactic was adopted. The group decided that a letter to Bahrain Petroleum Company would be written *"as if it is* from the Shaikh of

Bahrain." Nobody thought it necessary to inform the shaikh of Bahrain what was being undertaken in his name. The letter would state that in return for extending the prospecting license, the company should make the loan to the shaikh as they had planned to do in return for the additional area. The letter, written in the shaikh's name, would also tell the Bahrain Petroleum Company they should agree to the "amendments" to the concession as "suggested."

The one issue left for the group to discuss was Frank Holmes.

Everybody agreed that Holmes's "activities" got in the way of the Government of India's control. They agreed he would also be a hindrance to smooth implementation of the decisions now being reached. The idea put forth by Biscoe, the former resident, before he was so splendidly buried at sea off the coast of Sharjah, was enthusiastically revived. Almost everybody was keen on the campaign as pursued by Biscoe and now agreed that Holmes ought to spend the bulk of the year, they thought ten months, confined to Bahrain.

The sole dissenting voice came from the Colonial Office. They pointed out that in 1930 "Holmes's appointment was, in fact, made at the insistence of His Majesty's government." They repeated several times that "it was His Majesty's government, and not the company, who wanted Major Holmes appointed." Too bad, the other participants responded in chorus. The Bahrain Petroleum Company would now be told that "Major Holmes activities are *not* regarded with favour by His Majesty's government. It is essential that a representative of the company should be continually resident in Bahrain. Perhaps another representative should be appointed in Holmes's place."

It was thought that either the manager of the Eastern Bank or the manager of the Mesopotamia-Persia Corporation, both British subjects and with British companies, would make "quite a satisfactory" local representative for the Bahrain Petroleum Company. The former Bahrain political agent (Charles Prior), who was attending the meeting, predicted that "if Major Holmes is required to reside in Bahrain for, say ten months in every year, he will probably resign the appointment, if only for reasons of health." The meeting members thought this was an excellent prophecy. If so, they agreed, "insistence on such a period of residence might well be the easiest way out of our difficulties."

The participants had so indulged themselves in extravagant spleen venting that they ran out of time. Without a decision on how to actually proceed in order to achieve the goal of getting rid of that thorn in their side, the New Zealander Frank Holmes, the meeting adjourned.[2]

PRESSURE MOUNTS ON AHMAD

The resident left London to pursue his mission of warning Shaikh Ahmad that by "disregarding" advice he was "endangering his own interests." He arrived in

Kuwait on May 14, just three days after a visit from Anglo-Persian's Abadan general manager. The general manager and Chisholm had again pressed Ahmad to agree to the terms offered by Sir John Cadman. Like the participants at the London meeting, the two men concluded that Shaikh Ahmad's refusal of Cadman's offer must be based on the prospect of a substantial "secret personal bribe."

Cleverly, they thought, they asked Ahmad if he had any "special commitment" to Frank Holmes. Otherwise, they hinted, how else could it be possible that the shaikh could "regard E&GSynd as Anglo-Persian's equal in every way." Ahmad didn't think his visitors were clever. He was deeply stung by the obvious inference that he was personally corrupt. Haughtily, he replied that he did not allow his personal friendship with Holmes to interfere with the business of state. He told the Anglo-Persian representatives that he was absolutely impartial in making this business decision and considered only what he and his advisers thought to be in the best interests of his state and his people.

When he took his turn with Shaikh Ahmad, the resident began by instructing him that the British government now required to "vet" any new offers, particularly those drawn up by Holmes for the possible split concessions. Ahmad did not tell him the split concession idea had been dropped as the resident went on to tell him he was not free to deal with the Neutral Zone without first gaining permission from the British government. He was also informed that he was "committed" to "abide" by the advice of the High Government.

As outlined at the London meeting, the resident left the shaikh of Kuwait in no doubt that British government's "support" for his country in relation to its neighbors would be jeopardized should he choose the wrong contender. Not very subtly, the resident said that in a matter of such importance to his state and people it was "desirable" to consider the "undoubted strength of Anglo-Persian," particularly as John Cadman had even offered to "bring in American partners, if the shaikh wanted."

It was all too much for the shaikh of Kuwait. He quickly understood that the supposedly free hand, the promise that in awarding his concession the "ultimate decision" would be his alone, had evaporated, probably forever. Harried into an untenable position, he took the only way out. He closed down the negotiations altogether. He had decided, he said, that to continue and still more to conclude his oil negotiations at this point "was not in the interests of the State." He told Muhammad Yateem to inform Holmes who was in Bahrain and personally told Archibald Chisholm that he was suspending all negotiations "until such time as he was prepared to reopen them."

But Holmes turned up on May 17. The ever-vigilant Dickson reported that he was met by Muhammad Yateem, "who remained closeted in a cabin with his chief for over an hour on the steamer before landing." There was much for Yateem to cover. Later that day Holmes cabled his London colleagues: "Mullah Saleh told me that Anglo-Persian told Shaikh Ahmad that we, E&GSynd, had approached them in London trying to dispose of our interests to them. In addition, Anglo-Persian told Shaikh Ahmad that the Gulf Oil Corporation had also approached them asking to 'combine interests' in order to obtain and work the Kuwait concession." Not surprisingly, Holmes reported that the shaikh of Kuwait "is angry and distrustful of me because of these alleged secret deals." Holmes was as confused as Ahmad as to what might be going on in London. Just in case, he warned: "Please keep away from Anglo-Persian in London and do not complicate matters by even seeing, let alone discussing terms of disposing of Kuwait, over which we hold no title."

Shaikh Ahmad of Kuwait was under immense pressure, Holmes told London, because "the Resident threatened that the British government would cease to support him and his country if he does not give his concession to the Anglo-Persian Oil Company."

Holmes was not the only one who knew of the threats being made to Ahmad. Archibald Chisholm also noted that "the shaikh's sudden decision to suspend the negotiations was probably due to a realisation that he might be putting at risk the confidence in himself of the British protectors of his State and regime in a way which he could ill afford. Particularly in view of the difficult state of his relations with his neighbours Saudi Arabia, Iraq and Persia, who do not even acknowledge the existence of his State."[3]

Over the next few days, Holmes and Ahmad met with what Chisholm described as "unusual frequency." Dickson charged along to tackle Holmes. He reported to the resident that Holmes denied knowing the negotiations had been suspended, "lost his head altogether," "vowed vengeance on us all in a most scandalous manner," and threatened that the US government "would enter the lists" to ensure their oil company "got a fair go." Dickson was off to the shaikh the next day to tell the story. He was a bit miffed when Ahmad mildly cracked a joke, saying, "Perhaps Holmes had been drinking Brandy."

Dickson nevertheless forged ahead in his report to the resident, claiming Ahmad agreed with him that Holmes had lied when he said he did not know about the suspension. He claimed Ahmad had criticized Holmes by saying he had made a "great and serious mistake."

Archibald Chisholm observed this scene with some amusement. He concluded that Holmes had deliberately set out to provoke Dickson in the hope he would let something slip about Anglo-Persian's machinations. As Dickson also reluctantly noted, Shaikh Ahmad was at Holmes's house that same evening for a long and friendly dinner followed by a few rounds of bridge.[4]

The shaikh of Kuwait and Frank Holmes, close personal friends for some eleven years, had more in common at this time than ever before. There was only a decade between them; in 1933 Ahmad was forty-seven years old and Holmes was fifty-eight. They shared many character traits; both were affable and enjoyed a joke, they shared a liking for good food, good coffee, and good company. Both men had carried responsibility from a young age. Holmes was not quite eighteen when he was sent to the South African goldfields and took part in the Jamieson Raids. He had experienced the Boxer Rebellion, uprisings in Mexico, earned recognition at Gallipoli, and lived through Flanders. Ahmad was an admired veteran of desert warfare and had distinguished himself at the 1920 Battle of Jahra.

The two men had first formed a bond based on the implicit belief each held that Kuwait did have substantial oil resources. Unlike bin Saud, and even the shaikh of Bahrain, Shaikh Ahmad Al Sabah had always believed there was oil in his territory.

The belief had passed to him from the earlier ruler, the revered Shaikh Mubarak Al Sabah. Sure that doing so was the initiating act to development of his country's oil, Mubarak had signed the exclusion agreement with the British during the 1913 visit to Kuwait of Admiral Slade's commission while urging them to go with his son to see "the place of oil." John Cadman had been the "oil expert" accompanying Slade on the 1913 commission arranged by Winston Churchill.

Now both Ahmad and Holmes were facing the biggest obstacle to their shared goal of liberating Kuwait's oil riches. Under the threat of being left naked before the territorial predations of his neighbours, Ahmad was under pressure from the British government to grant the concession to the Anglo-Persian Oil Company, which he doubted had any serious intention of faithfully looking for oil, much less working the concession. With some justification Ahmad believed Anglo-Persian wanted the Kuwait concession only for the purpose of curtailing its development in order to protect its interests in Persia and Iraq.

Apart from the Government of India pushing to have him removed from Bahrain and his position there as chief local representative, Holmes was facing a concerted effort by Anglo-Persian to entirely dislodge both him, and E&GSynd from their relationship with the Gulf Oil Corporation and, consequently, with Shaikh Ahmad and Kuwait.

PRESSURE MOUNTS ON GULF OIL

It was true that the Gulf Oil Corporation was in discussions with the Anglo-Persian Oil Company. William Wallace had gone to London to take over from his assistant, Guy Stevens, in talks with Anglo-Persian's top men, including chairman John Cadman and deputy chairman William Fraser, both men flush with the success of dealing with the Imperial Government of Persia.

The announcement had come on April 29, 1933, that the Anglo-Persian Oil Company had signed a new agreement with Persia. The report of the League of Nations noted: "The dispute between His Majesty's government in the United Kingdom and the Imperial Government of Persia, which was brought before it last December, is now settled." While Persia had improved some of its conditions, most particularly a minimum income guarantee and the right to have a representative view selected documents and observe specific board meetings, it was by no means a clear victory for the Persians. The unresolved issues would simmer for years and finally culminate in 1951 with the acrimonious nationalization of Iran's (the renamed Persia) oil industry.

At the time Ahmad and Holmes were summing up their situation in Kuwait, William Wallace was cabling his New York colleagues, "Sir John Cadman and Sir William Fraser of Anglo-Persian insist we must now agree definitely to future policy with respect to a joint operation. They state frankly we will not get a concession over their opposition."

Wallace reported that "while they have no objection to us as a joint partner, they are unwilling to assume any set development program." Equally importantly, Wallace informed his colleagues, "under no circumstances will they deal with E&GSynd." Nor would Anglo-Persian recognize E&GSynd's overriding royalty contained in its agreement with Gulf Oil, he reported.

That was it. The ultimatum was on the table. Gulf Oil must break its contract with Anglo-Persian's long-time nemesis, E&GSynd, and join forces with the Anglo-Persian Oil Company or any prospects for the Gulf Oil Corporation in Kuwait would be crushed.

When Wallace returned to New York, he wrote a memorandum for his executive colleagues of what had transpired during three meetings with Cadman and Fraser at Anglo-Persian. He clearly understood: "They will oppose any combination that does not have as its primary objective the development of Kuwait *only* at such a rate and at such times as would meet the needs of Anglo-Persian."

So much for Kuwait and what Dickson had described as the "suffering and acute want of the poor and hungry people of Kuwait." Anglo-Persian's intention was precisely as the shaikh of Kuwait had always suspected.

Wallace's memorandum continued:

> From the talks I have had with Cadman and Fraser, I get the very distinct understanding they are afraid that if we obtain any sizeable production in the Kuwait territory, and are left free to produce same, that we might, in our effort to market such oil, encroach upon their territory. The Anglo-Persian Oil Company has recently completed its combination with the Dutch Shell for the protection of their present "as is" position and I judge that the consummation of this deal involved a great deal of work, study and careful thought. Therefore, they are frankly afraid this situation will be disturbed and possibly upset if we are left at liberty to proceed in Kuwait without any let or hindrance.

Gulf Oil concluded that while the shaikh of Kuwait would not "willingly" grant his concession to the 51 percent British government–owned Anglo-Persian Oil Company, the British government would prevent him giving it to the Americans, even if they set up a British or Canadian company to receive and operate the concession. Despite Gulf Oil's constant entreaties and active cooperation the US government had proved unable, or unwilling, to exert the level of pressure required on the British government. The State Department had declared itself fully satisfied with the Foreign Office undertaking that the decision on awarding the concession would be that of the shaikh of Kuwait alone.

To William Wallace, it was now glaringly obvious that although the shaikh of Kuwait would not grant his oil concession to Anglo-Persian, he was effectively being prevented from granting it to the Gulf Oil Corporation. Wallace and his executive colleagues at the Gulf Oil Corporation decided to throw in their lot with Anglo-Persian—and ditch their partners of six years, the Eastern & General Syndicate.[5]

THE LONDON MEETING RESUMES

The meeting of the Colonial Office, the Foreign Office, the India Office, the Admiralty, the Petroleum Department, and the political resident reconvened in London on July 26 to review implementation of the decisions reached in the earlier meeting on guidelines to be followed by the Government of India acting through its political officers on oil matters in the Persian Gulf.

Qatar did not need attention. Anglo-Persian was currently "putting a proposal" to Shaikh Abdullah Al Thani of Qatar, and the Petroleum Department declared there would be a "positive outcome" because they had the matter well in hand.

As the matter of a split concession in Kuwait was no longer an issue since Holmes had persuaded Shaikh Ahmad away from his proposal to divide the concession, the meeting's first discussion centered on the Kuwait Neutral Zone. Although barely credible in any logical sense, Dickson's hysterical outline of a conspiracy between Frank Holmes and Abdul Aziz bin Saud aimed at gaining political supremacy over Kuwait had found fertile ground with the Government of India.

A response perhaps not so surprising in view of the fact that the Government of India and its political officers were themselves well-practiced exponents of the art of conspiracy. Dickson's series of feverish reports had warned that bin Saud was planning to "get the Americans, through Holmes, into the Neutral Zone and then using this as a lever get them into Kuwait as well, which would give bin Saud the ascendancy he has for so long sought over the Shaikh of Kuwait."

Dickson's suspicion was increased because bin Saud had recently said he was "anxious to compose his differences with all his neighbors." Dickson declared proof of the alleged conspiracy was a letter from bin Saud to Shaikh Ahmad "propounding certain proposals of a commercial nature." He vowed this meant that bin Saud was "definitely going to enter the lists in the matter of Kuwait Oil." Dickson predicted shrilly that "the secret transference of our political ascendancy over Kuwait to bin Saud, thus leaving the British out in the cold." Caught up in Dickson's paranoia, the resident agreed. He later reported that bin Saud's letter to the shaikh of Kuwait "did not, after all, mention oil as Dickson assumed."[6]

REVERSING COX'S HANDIWORK

Having unceremoniously removed the Neutral Zone from Kuwait at Percy Cox's 1922 Ujair Conference and awarded equal rights in that territory to bin Saud together with sovereignty over the tribes, the British now looked for a way they could put it back again.

The India Office had taken on this challenge.

Circulated to those attending the meeting, the India Office report began by giving the background as: "The status of Kuwait was the subject of considerable discussion in the period before the war until an agreement was finally reached and a Convention signed between Britain and Turkey in July 1913." Neglecting to mention that this convention was never ratified, the report said that under this agreement "the territory of Kuwait was recognised as an autonomous *qada* (sub-district) of the Ottoman Empire and the Ottoman flag was to be flown by the Shaikh; he had the right to insert a distinctive emblem in the corner. The shaikh was to continue to be a Turkish *Qaimaqam* (governor of a sub district) and his successors were to be appointed to a similar position by the Ottoman government."

The India Office continued, saying that at the time, the limit of the shaikh of Kuwait's administrative autonomy was defined on a map by two circles. The area within the inner red circle was not in contention. His authority was less clear within the outer green circle although it was recognized that the tribes within this area were dependent on him. The convention decreed that the Shaikh of Kuwait would levy tribute from these tribes and perform the administrative duties of a Turkish *qaimaqam* in both regions. Turkey undertook to abstain entirely from any interference.

The report said this position continued until the December 1922 Ujair Conference when it was "decided" that the area that became known as the Kuwait Neutral Zone would be recognized as common to the neighboring states of Nejd and Kuwait. Creatively, the India Office report concluded that Nejd and Kuwait were to enjoy equal rights in this area only *"until such time as a fresh agreement should be arrived at between them regarding it, through the good offices of His Majesty's government."*

The India Office's imaginative interpretation was a radical departure from all previous statements on this subject. It was, in fact, a 180-degree turnaround. "It is fairly clear," the India Office said, "that in the pre-war period the area now included in the Neutral Zone was predominantly under Kuwait control. It was quite definitely not under the control of Abdul Aziz bin Saud, whose active interest in it dates from a much later period."

The India Office produced its trump card. They decreed: "It would, we think, be reasonable to contend in all the circumstances that, for the purposes of international oil agreements affecting 'Kuwait' or 'the Sultanate of Kuwait,' the Neutral Zone can properly be regarded as covered by either of these terms." In this way the meeting participants intended to restore the Neutral Zone to Kuwait, but only for the "purposes of international oil agreements."[7]

The participants were satisfied they had produced grounds on which they could block any move by bin Saud, as predicted by Dickson. Now they could turn their entire attention toward getting rid of Frank Holmes.

TARGETING HOLMES

Referring mysteriously to "action outlined in recent telegraphic correspondence," the meeting members drafted a letter that would be addressed to the London office of the Bahrain Petroleum Company. It said, "it had now become necessary" to approach the company because of (a) "Frank Holmes absence from Bahrain," (b) his "Breach of Agreement by approaching the shaikh directly," and the rather vague charge of (c) "an inaccurate account of an interview with the shaikh."

At the same time, the political agent in Bahrain would be instructed to "obtain" the shaikh of Bahrain's "concurrence" to a second letter to be sent to the company's Bahrain office "as if it is from the shaikh." The participants were reassured that the wording of the letter purportedly from the shaikh of Bahrain had already been agreed upon, well before the meeting, "by the interested departments." The interested parties did not include Shaikh Hamad of Bahrain.

Gleeful at the prospect, the group went on to discuss who would be Frank Holmes's replacement. The manager of the Eastern Bank was dropped because of "administrative difficulties." It was thought that Bahrain Petroleum Company's own field manager would do "faute de mieux" (for the moment). This was the rough-and-ready Edward Skinner who had been so put off by Karl Twitchell's swishing around his oil site dressed in flowing Arab robes.

Both the Colonial Office and the Foreign Office appealed for common sense rather than sheer spite. Appointing Skinner, they pointed out, "would be a reversal of HMG policy." Skinner was American. They reminded the meeting participants that the whole idea of the chief local representative was based on the necessity of his being a British subject. They were promptly howled down. Announcing that they "would prefer not to take any action in this matter," the Colonial Office bowed gracefully out of the Persian Gulf.

REMOVING HOLMES

Following the shaikh of Kuwait's decision to suspend negotiations, Holmes spent June and July in London. He returned to Bahrain in August—just in time to be fired.

The directive came in an August 25 "secret" telegram from the India Office to the Government of India and the political resident in the Gulf. They reported that after being subjected to "oral representations," the Bahrain Petroleum Company had "decided" to replace Holmes as chief local representative. Notification "from the company" goes to Holmes "by this week's air mail," they trumpeted.

The "oral representations" linked Bahrain Petroleum Company's application for an extension of the prospecting license to the demand that Frank Holmes be relieved of his position as chief local representative.

The company had clearly seen the linkage. They cabled the San Francisco head office of Standard Oil of California, saying: "While reserving all our rights . . . we have tentatively arranged with the India Office that Major Frank Holmes be permitted to retire . . . thereby avoiding hiatus." Tellingly, the India Office directive revealed that "the company will not renew application for extension of its prospecting licence, for the moment, but propose to do so once Holmes has 'handed over.'"

The Bahrain Petroleum Company had submitted a written declaration attesting that Holmes had been in Bahrain "at all the times required." The India Office had rejected this and countered by saying that in the one year and eleven months since Bahrain had struck oil, Holmes had only spent six months and fourteen days in Bahrain.

M. E. Lombardi at Standard Oil of California, owners of the Bahrain Petroleum Company, did not take kindly to being blackmailed by the India Office. He said he did not know of any "overt act" that Holmes had committed that would require his resignation.

Lombardi said he could not see any problems related to Frank Holmes. The sole "difference of opinion" he knew of had been between E&GSynd and Standard Oil of California only, he said, and had been related purely to corporate matters around the Al Hasa concession. Nevertheless, his pragmatic opinion was that "there is little option now to asking Homes to resign. But it should be definitely, and obviously, shown to be forced by the India Office and not by Standard Oil of California."

The India Office had no intention of showing its hand. Claiming it was a suggestion from the Bahrain Petroleum Company, the India Office advised that "as the company's decision is likely to be very ill taken by Holmes, who may cause embarrassment locally by making accusations against British Officials, etc., it would be well, if possible, that the Political Agent should avoid Holmes until the effect of the air letter has subsided." The shaikh of Bahrain had also been bullied into submission. He told Holmes that the political officers "attacked him with sail and oars in action."

Holmes tendered his official resignation on September 14, 1933. On receipt, the Bahrain Petroleum Company's London office commented laconically, "No doubt the voice of the British government will now play an all important part." The request for an extension of the prospecting license was resubmitted and duly granted. With the "concurrence of His Majesty's government," Holmes was replaced by the American Edward Skinner. The political officers lost no time in telling all the shaikhs of the Persian Gulf that Frank Holmes had been "exiled" and "would not be permitted to return."

Anglo-Persian's negotiator in Kuwait, Archibald Chisholm, was quick to follow up on the perceived advantage. He quickly visited the shaikh of Kuwait to inquire if, "now that Major Holmes is apparently unemployed," the shaikh would soon be reopening his negotiations with the Anglo-Persian Oil Company. Shaikh Ahmad replied that he would not be taking any action "for the present" and refused to agree with Chisholm that Holmes would not return. Ahmad sent Mullah Saleh to spend the stopover at Kuwait airport, in deep discussion with Holmes, as he passed through on September 21 on his way to London.[8]

As Frank Holmes and Shaikh Ahmad now knew, the Gulf Oil Corporation was about to join forces with the Anglo-Persian Oil Company. Holmes and Ahmad had spent hours talking about the ramifications. They had agreed Ahmad would not make a move until Holmes had had a chance to see what he could sort out.

Even when the India Office forced Holmes out of Bahrain, Ahmad still had confidence that he would find a way to protect him and his concession. As Chisholm perceptively understood, Shaikh Ahmad "was content to bide his time . . . due to his faith in his old friend Holmes's willingness, and ability, to get him the best possible terms for his concession, in any circumstances."[9]

Holmes also knew that Gulf Oil, on orders from Anglo-Persian's John Cadman, needed to rid itself of all contractual obligations with their partners the Eastern & General Syndicate. In London, Holmes, Janson, and the Gulf Oil people would set about breaking up what had been a mostly harmonious relationship.

GULF OIL SPLIT FROM E&GSYND

The Gulf Oil Corporation was not about to let go a bird in the hand on the mere say-so of the Anglo-Persian Oil Company. The first agreement of dissolution with E&GSynd, dated November 6, 1933, was cautiously drawn up as an option. The option delivered to Gulf Oil the complete transfer of Eastern & General Syndicate's entire rights, title, and interests in any Kuwait concession. Gulf Oil was to pay E&GSynd a consideration of thirty-six thousand sterling (in the event Anglo-Persian contributed half) for agreeing to free them to join forces with

Anglo-Persian and giving up their right to an "overriding royalty" in any oil found in Kuwait. With the option in hand, Gulf Oil moved on to serious discussion with Anglo-Persian.

The first point was relatively simple. The Kuwait oil concession would be acquired jointly, by both companies acting through the Kuwait Oil Company, a British company to be formed with shares held equally between the Anglo-Persian Oil Company and the Gulf Oil Corporation. The next point, insisted on by Anglo-Persian, was far from straightforward. John Cadman and William Fraser's requirement was for a "private preliminary agreement" that would restrict the Kuwait Oil Company from "ever injuring" Anglo-Persian's markets "wherever situated, at any time, either directly or indirectly."

The agreement specified that Kuwait Oil Company orders would be supplied from Anglo-Persian's own operations. Only if more were needed would the oil of Kuwait be put into the market. Clause 8 specified: "Anglo-Persian to supply the Gulf Oil Corporation's requirements from Persia and/or Iraq in lieu of Gulf requiring the company to produce oil or additional oil in Kuwait." The restriction was further defined: "Provided Anglo-Persian is in position conveniently to furnish such supply—of which Anglo-Persian will be the sole judge—it will supply the Gulf Oil Corporation with any quantity of crude . . . at a price and on conditions to be discussed and settled by mutual agreement."

Gulf Oil was confronted with a stark choice. They could accept Anglo-Persian dictating where, and in what quantities, they could market any oil found in Kuwait. Or they could refuse to accept this condition—and give up any chance of ever obtaining an interest in the oil prospects of Kuwait. Rather than be completely shut out, the Gulf Oil Corporation agreed. Anglo-Persian further insisted that this "private preliminary agreement" signed on December 14, 1933, was to remain secret between the two companies. The shaikh of Kuwait should never be told.[10]

There was one point on which Gulf Oil could not give in. Holmes, Janson, and E&GSynd refused to sign the Release & Quit option that Gulf Oil needed unless this point was first agreed upon. Frank Holmes had promised to do everything he could to protect the interests of Shaikh Ahmad of Kuwait. Ahmad had confidently predicted to Chisholm that Holmes would return. He had told his friends and advisers that Frank Holmes would not "throw me to the wolves." Ahmad had stated that he was certain Holmes would "preserve me both politically and commercially."

Holmes had kept his house in Kuwait. His cook and servants had returned there in November, following the drawing up of the Release & Quit option with Gulf Oil. This had sparked a burst of rumor and prompted an agitated Dickson to write to the resident, "it is common property in Kuwait" that Major Holmes will shortly arrive "to start fresh negotiations for an oil concession with the shaikh."

The clause that mattered in the Release & Quit documentation stated: "Should Gulf Oil request the assistance of E&GSynd, and Major Frank Holmes, to obtain the Kuwait Concession, they shall use their best efforts to comply therewith. Any expense connected with such assistance will be to Gulf Oil's exclusive charge. If so requested, syndicate and Frank Holmes will endeavour to obtain the Kuwait concession on such terms and conditions as Gulf Oil may indicate." (Prudently, another clause stated that if Holmes and E&GSynd obtained the concession they would "promptly assign and transfer said concession to Gulf Oil." Holmes and E&GSynd were forbidden to undertake "at any time or in any manner" to obtain any concession in Kuwait for themselves or any other party, or to assist any other party.)

Because Holmes, Janson, and E&GSynd refused to release them otherwise, Gulf Oil could not budge in insisting to its new partners, the Anglo-Persian Oil Company, that in the negotiations for the Kuwait oil concession that would now be conducted on behalf of the new Kuwait Oil Company, Gulf Oil's negotiator would be Maj. Frank Holmes. They pointed to the clauses in the Release & Quit agreement with E&GSynd and claimed that Holmes would work for Gulf Oil as "an independent consultant."[11]

Having gained everything else they wanted, believing they held the upper hand anyway and that negotiation of the terms and conditions of the concession was now merely pro forma, Anglo-Persian agreed. Knowing how the Government of India and the political officers felt about Frank Holmes—a marred reputation to which Anglo-Persian had considerably contributed—Sir William Fraser dropped around for a courtesy call at the India Office.

INDIA OFFICE IN CHARGE

The first thing the two gentlemen at the India Office impressed upon Fraser was that, although on oil matters they needed to consult with the Government of India, the Admiralty, and the Petroleum Department, it was the India Office that was now very definitely in charge in the Persian Gulf. The Colonial Office no longer had a role. They told Fraser that he could not take it for granted that his American-British combine would automatically be allowed the Kuwait concession. They told him that now everybody must "consult effectively with the India Office before any decisions are made affecting the Arab shaikhdoms."

Having firmly established the power balance, they brought to Fraser's attention "the political difficulties that might arise in connection with the operations" unless all personnel were British. They suggested they might put this forward as one of the prerequisite conditions of any agreement. Fraser replied that he thought there would be "considerable difficulty with the Americans in securing their formal agreement to solely British personnel."

When Fraser moved on to the real purpose of his visit—the matter of Frank Holmes negotiating the Kuwait concession on behalf of Gulf Oil—the men at the

India Office went into shock. After years of trying, the political officers, the Government of India, and the India Office had only just succeeded in manipulating Holmes out of the Persian Gulf, and they had done it, in part, on Anglo-Persian's behalf. They couldn't believe what they were now hearing.

Fraser adopted a conciliatory tone. He realized there were "certain difficulties in connection with Major Holmes," he said. "But," he asked, "were these really serious?" (Anglo-Persian should have been in a good position to judge how serious were the difficulties. After all, they had been responsible for manufacturing many of them.) Holmes was a British subject, Fraser said, and he did have great local influence and knowledge. Moreover, he assured, he would personally "keep Holmes under control" and would remove him "without hesitation" if he "misbehaved."

Stunned, the men at the India Office responded vigorously that the difficulties were "quite definitely serious" and reeled through a litany of their grievances. Top of the list was that Holmes had attempted to "short circuit" Harold Dickson, the political agent in Kuwait. They said it was the very fact of Holmes's great knowledge and influence that made for "uneasy relations" with the Government of India's political officers.

Fraser asked if the India Office would "rather see an American used as a negotiator on behalf of the American company than Major Holmes" because he was pretty sure this would be Gulf Oil's counterdemand. The India Office men were aghast. As their memorandum noted, "It was agreed that this question must be considered further." Before Fraser left, the India Office decreed "after considerable discussion" that Dickson, the political agent at Kuwait, "must be present at all discussions between the representatives of the company and the shaikh of Kuwait."[12]

INDIA OFFICE CONDITIONS

Matters moved up a notch the next day when John Cadman and William Fraser called on the head of the India Office, Secretary of State to India Louis Kershaw.

Claiming that he had "absolutely no desire" to discriminate against the Americans, Kershaw also opened the discussions by pushing the necessity for all-British personnel in Kuwait. He said the "peculiar conditions" in the Gulf shaikhdoms "necessitate caution in regard to the admission of Europeans and Americans."

"Great caution is necessary in the case of non-British subjects," he intoned, "because these might not understand as well as British subjects the desirability of following the advice of the political authorities. And also because, if they should get into trouble with the local or neighbouring populations, the case might involve international complications instead of being merely a domestic matter for the British government." Kershaw thought there was "no special reason" to suppose the Americans would want to employ their own people. He imagined they would prefer "English" personnel because "they are cheaper and can be readily provided by Anglo-Persian."

Kershaw told Cadman and Fraser how the whole matter was going to be handled. There would be two agreements, he said. The first would be between the company and the shaikh of Kuwait and would cover commercial matters. The second would be between the company and His Majesty's government and would cover "political questions." Having just completed their own secret cartel agreement with the Gulf Oil Corporation, neither Cadman nor Fraser could have any objections to that.

Cadman, who had personal experience of Harold Dickson, did not want the political agent present at the negotiations with the shaikh of Kuwait. He marshaled a convincing argument. Now that the two companies had combined, he said, the shaikh of Kuwait "will have to put up with less attractive financial terms." He told Kershaw that if the political agent was present when the lower terms and conditions were made known, "the shaikh might attribute the less attractive nature of the offers, which he will now receive to the intervention of the Political Agent."

Cadman moved on to Holmes. He said he "fully appreciated the difficulties regarding Holmes resulting from his conduct at Bahrain." He said he was aware "as is generally known in the Gulf that Major Holmes had been 'removed' from his post at Bahrain Petroleum Company . . . and is not persona grata with the local political authorities." He said he was also aware of Holmes's "capacity for mischief."

Nevertheless, Cadman said, from the point of view of success of the negotiations it would be better to "accept Gulf Oil's proposal rather than ask them to employ someone else, presumably an American." He went on to say if Holmes were "excluded at this stage his capacity and opportunities for intrigue might be much more dangerous than if he were employed as negotiator under the eye of Anglo-Persian." Cadman assured Kershaw that "the real conduct of the negotiations will be in the hands of the Anglo-Persian negotiator. Major Holmes's role will be quite secondary."

Kershaw's memorandum of the meeting with Cadman and Fraser noted: "It was agreed that it would be better to take Major Holmes along with us in a position in which he could be easily supervised, rather than exclude him and drive him into hostility."

Within days, William Fraser introduced William Wallace's assistant, Guy Stevens of the Gulf Oil Corporation, to the India Office. "Mr. Stevens made considerable difficulty about the nationality of the operating company's personnel," the India Office reported, "it was clear that he had expected to employ a considerable number of Americans." But worse than that, the men at the India Office noted, neither Stevens nor Fraser "said anything at all about Major Holmes's position in the negotiations being in any way secondary. The whole implication of their remarks was that the two negotiators would be on an equal footing." The India Office dashed off a letter to Fraser reminding him of his promise to immediately "withdraw" Holmes in the "event of any trouble arising."[13]

The matter was settled. Frank Holmes would return on behalf of the Gulf Oil Corporation to negotiate the oil concession with the shaikh of Kuwait. To protect its interests, the Anglo-Persian Oil Company would retain its previous negotiator. In a bizarre twist, the two former rivals, Frank Holmes and Archibald Chisholm, were to become "joint negotiators" for the newly formed British-American joint venture.

On January 30, the option of dissolution between the Eastern & General Syndicate and the Gulf Oil Corporation was exercised. On February 2, 1934, the Kuwait Oil Company (KOC) was registered in London.

NOTES

1. See, for example, Yossef Bilovich, "Great Power and Corporate Rivalry in Kuwait, 1912–1934: A Study in International Politics" (PhD diss., University of London, 1982), p. 285, "Whether Holmes knew it or not the southern part of Kuwait indeed proved later to be far richer in oil than the northern part."

2. IOL/R/15/1/623, vols. 82 and 86, "Confidential" record of a meeting held at the Colonial Office on April 26, 1933, to discuss various questions relating to oil in the Persian Gulf. And IOL/R/15/5/242, vol. VII, final record of a meeting (adjourned from April 26, 1933) held at the Colonial Office on May 3, 1933, to discuss various questions relating to oil in the Persian Gulf.

3. IOL/R/15/1/623, vols. 82 and 86, May 25, 1933, political agent Kuwait (Dickson) to resident (Fowle), "remained closted in a cabin with his chief"; Archibald H. T. Chisholm, *The First Kuwait Oil Concession Agreement: A Record of the Negotiations, 1911–1934* (London: F. Cass, 1975), p. 34, the shaikh suspended the negotiations on May 14, 1933; Chisholm, p. 180, reprints the May 17, 1933, cable, Holmes to E&GSynd, "APOC told Shaikh E&GSynd approached them in London . . . (and) Gulf Oil approached them . . . to combine forces in obtaining and working the Kuwait concession. Shaikh of Kuwait angry and distrustful . . . because of these alleged secret deals. (Also) Resident threatened . . . would cease support if he did not give concession to APOC. . . . Please keep away from APOC in London"; Chisholm, p. 37, "probably due to a realisation that he might be putting at risk the confidence in himself of the British protectors of his State."

4. IOL/R/15/1/623, vols. 82 and 86, May 25, 1933, political agent Kuwait (Dickson) to political resident (Fowle): Of his wildly excited five-page typed missive, Dickson piously (and unconvincingly) says he "mentions" this story "with reluctance." He says he is repeating it "not because I wish to say anything disparaging about Major Holmes's character" but to show what he, Dickson, had to put up with. Dickson still had his information gathering intact as he refers to the content of a "very lengthy telegram costing Rs200" sent by Yateem to Holmes; Chisholm, *First Kuwait Oil Concession Agreement*, pp. 35–37, "Holmes deliberately set out to provoke Dickson."

5. Chevron Archives, box 0426154, Principal Documents in Regard to Kuwait Rerganization (1951), pt. 11, Background of the Kuwait Oil Concession, Its 50-50 Owner-

ship by the D'Arcy Exploration Company (Anglo-Iranian Oil Company Ltd.) and Gulf Exploration Company (Gulf Oil Corporation) through the Kuwait Oil Company . . . the restrictions of the agreement of December 14, 1933, p. 22, May 22, 1933, Wallace in London cable to New York, "they state frankly we will not get a concession over their opposition" and "under no circumstances will they deal with E&GSynd nor recognise the syndicate's overriding royalty," p. 23, May 31, 1933, Wallace in New York memorandum to Leovy, "they will oppose anything . . . meet the needs of Anglo-Persian"; p. 25, "Shaikh of Kuwait will not 'willingly' grant his concession to the 51% British government owned Anglo-Persian Oil Company, the British government will prevent him giving it to (us)."

6. Chisholm, *First Kuwait Oil Concession Agreement*, p. 34, "In return for his big offer, Holmes now required the whole of Kuwait territory for his concession, and not the 1600 square miles for which he had previously asked, in this respect coming into line with APOC's proposals"; IOL/R/15/5/242, vol. VII, June 7, 1933, political agent Kuwait (Dickson) to resident (Fowle), "Note on the Kuwait Oil Situation as I see it in light of British Minister at Jedda telegram, dated June 5, 1933, to Bushire/London." The Jedda telegram said that bin Saud had that day stated that he was "anxious to compose his differences with all his neighbours." And Bilovich, "Great Power and Corporate Rivalry in Kuwait," pp. 298–99.

7. IOL/R/15/5/242, vol. VII, July 14, 1933, India Office report (signed J. G. Laithwaite), "It would, we think, be reasonable to contend in all the circumstances that, for the purposes of international oil agreements affecting 'Kuwait' or 'the Sultanate of Kuwait,' the Neutral Zone can properly be regarded as covered by either of these terms."

8. IOL/R/15/5/242, vol. VII, draft of a meeting held at the Colonial Office on July 26, 1933, to discuss certain questions connected with oil in the Persian Gulf; Present: Colonial Office, Foreign Office, India Office, Admiralty, Petroleum Department; and August 25, 1933, secretary of state for India, London, to political resident (Fowle) and Government of India, Foreign and Political Department, Simla, "the Bahrain Petroleum Company has decided to replace Frank Holmes as Chief Local Representative"; Angela Clark, *Bahrain Oil and Development, 1929–1989* (London: Immel Publishing, 1991), p. 143, "attacked him with sail and oars in action" and p. 144, "but it should be definitely, and obviously, forced by the India Office" and "the voice of the British government will now play an all important part"; Chisholm, *First Kuwait Oil Concession Agreement*, p. 39, "met on the aerodrome by Mullah Saleh with whom he had a long conversation."

9. Chisholm, *First Kuwait Oil Concession Agreement*, p. 38, suspected that the shaikh and Holmes had drawn together in the face of adversity. He says Ahmad made no move to lift the suspension of the negotiations "but was content to bide his time. This can only have been due to his faith in his old friend Holmes's willingness and ability to get him the best possible terms in any circumstances."

10. Chevron Archives, box 0426154, Principal Documents in Regard to Kuwait Reorganization (1951), pt. 11, Background of the Kuwait Oil Concession, Its 50-50 Ownership by the D'Arcy Exploration Company (Anglo-Iranian Oil Company Ltd.) and Gulf Exploration Company (Gulf Oil Corporation) through the Kuwait Oil Company . . . the restrictions of the agreement of December 14, 1933; pp. 25–26. Chisholm, *First Kuwait Oil Concession Agreement*, pp. 185–89, reprints this agreement "provided Anglo-Persian is in position conveniently to furnish such supply—of which Anglo-Persian will be the sole judge—it will supply Gulf." Note that the interests of Thomas E. Ward appear to have

been overlooked when E&GSynd released all its rights to Gulf Oil, including the overriding royalty of one shilling per ton on production of 750 tons per day. See Ward, *Negotiations for Oil Concessions in Bahrain, El Hasa (Saudi Arabia), the Neutral Zone, Qatar and Kuwait* (New York: privately published, 1965), p. 241, concerning the loss of his "10% share in the overriding royalty."

11. Chisholm, *First Kuwait Oil Concession Agreement*, pp. 39–40, citing cable from Yateem in Kuwait, "not throw me to the wolves" and "Holmes would preserve me both politically and commercially"; IOL/R/15/5/242, vol. VII, November 18, 1933, Kuwait political agent (Dickson) to political resident (Fowle), "common property in Kuwait . . . Major Holmes will shortly arrive"; Ward, *Negotiations for Oil Concessions*, p. 233, reprints in its entirety this lengthy and minutely detailed agreement of dissolution between E&GSynd and Gulf Oil.

12. IOL/L/PS/12/10R.100.7-8, vol. 3808, January 3, 1934, India Office, "Confidential" note of a conversation between Sir William Fraser, deputy chairman Anglo-Persian Oil Company and the India Office's Mr. Walton and Mr. Laithwaite, "consult effectively with the India Office" and "Holmes had attempted to 'short circuit' the Political Agent in Kuwait" and "this question must be considered further."

13. ILO/R/15/1/641.86/IV, January 4, 1934, note on interview regarding Kuwait oil concession, Sir John Cadman and Sir William Fraser called on Sir Louis Kershaw, Mr. Walton and Mr. Laithwaite were also present, "Great caution is necessary in the case of non-British subjects" and "two agreements, the second between the company and His Majesty's government to cover political questions" and "real conduct of the negotiations will be in the hands of the Anglo-Persian negotiator. Major Holmes's role will be quite secondary"; IO/L/PS/12/10R.1007.8, vol. 3808, January 10, 1934, India Office (Walton) to Petroleum Department (Starling), "said anything at all about Major Holmes's position in the negotiations being in any way secondary"; ILO/R/15/1/641.86/IV, January 11, 1934, Sir William Fraser Anglo-Persian to India Office (Walton), "before Holmes goes out I shall have a personal talk with him."

CHAPTER 22

AHMAD AND HOLMES

U proar ensued when it became known that "the largest British oil producing company in the world" was assisting the Americans to enter
Kuwait. Many viewed the Anglo-Persian Oil Company's action in
joining forces with the Gulf Oil Corporation as a betrayal. This was the slippery
slope. Critics charged that with American oil companies in Iraq, Bahrain, Saudi
Arabia, and now Kuwait, the loss of British prestige and power could not be far
behind. And their very own Anglo-Persian Oil Company, with its 51 percent
British government shareholding, was contributing.

Lord John Lloyd was a particularly virulent critic, charging that oil in the
Persian Gulf should be kept "purely under British control." He had been governor of Bombay 1918–23 and British high commissioner in Egypt 1925–29, and
he was a member of Parliament with large private business interests. In foreign
affairs, he favored right-wing imperialist policies. Lloyd said the "intervention of
American interests" should be regarded with the "utmost misgiving." In a
rousing speech to the annual dinner of the prestigious Central Asian Club, Lord
Lloyd attacked the Anglo-Persian Oil Company. He said they had "missed the
opportunity" of making sure that all the oil in the Persian Gulf was under British
control. He delivered an equally stinging rebuke to the British government,
which, he said, "should have influenced the Anglo-Persian Oil Company more
strongly in this direction" and should have refused to allow American interests to
"get a foot in the Persian Gulf."

The Admiralty fired off a terse letter to Sir Louis Kershaw at the India
Office, claiming: "The concession at Bahrain has been given away without safeguards of any real value. Unless His Majesty's government insist upon them in
Kuwait and Qatar it will mean that all the oil which has hitherto been found on
the Arabian shore of the Persian Gulf has passed to partly foreign or at best international interests."

The Admiralty detailed its objections: "In 1930 when it was decided that foreign controlled companies would be permitted in future to operate in British territory it was laid down that (a) at least 50% of the oil should be refined in British territory, (b) that the refinery should be capable of producing fuel oil suitable for Admiralty use and (c) that the British government retain the right of pre-emption." The Admiralty demanded that Anglo-Persian be reminded that these conditions "must be met by any company, of whatever nationality, operating in Kuwait." Furthermore, the Admiralty said, "the refinery should not be merely in British territory, but either in Kuwait itself—which is for all practical purposes the same thing—or at some port east of Kuwait, e.g., Karachi or some other Indian port."

A scathing correspondence of the new venture passed between Anglo-Persian's general manager in Abadan, and his contact at the Petroleum Department. The general manager did not have much regard for the people of the Persian Gulf. He told the Petroleum Department: "I feel the same as you. The shaikh might not receive the Kuwait Oil Company with open arms. He might yet bring in Standard Oil of California, or others, as competitive bidders, whether through greed or a desire to remain *en rapport* with bin Saud either through genuine subservience to bin Saud or in order to play him profitably off against HMG."

He had no doubt how matters should be handled. "The next step should be taken by HMG. Both Kuwait and Qatar have openly professed the desire to have their territories tested for oil. Both rulers, who have for so long depended for their very existence on HMG, may now contemplate extorting uncommercially high terms by taking advantage of a confused political situation." There should be no deviation from the "expressed or implied obligation on British protected rulers to favour British interests," he wrote. Otherwise, he said, "they should be prepared to forego British protection." He was demanding that "HMG should now make it clear to these rulers both where they stand and where their interests lie."

The general manager was hot under the collar as he gave his opinion that Anglo-Persian's negotiators should return to Kuwait and Qatar, without the Americans. Then the political resident, Trenchard Fowle, "in his capacity as the shaikhs' guardian and guide should clearly indicate to these two shaikhs how the position stands." He thought it should be made crystal clear that "full protection" of the territory of the shaikh of Kuwait and that of the shaikh of Qatar would only be guaranteed "if the Anglo-Persian Oil Company is exploiting it under a concession."

The political resident in the Persian Gulf was also up in arms. He, too, wrote Kershaw at the India Office: "The Anglo-Persian Oil Company must be categorically informed that HMG cannot agree to the shaikh giving a concession to the joint venture unless operations are all British. The interest of the American half should be purely financial. Provided they fairly share profits they have no cause for complaint."

Resident Fowle warned: "This suggested safeguard is all the more imperative since the Anglo-Persian Oil Company, in their present negotiations with HMG, appear to have adopted distinctly evasive tactics. They apparently tried to bluff that HMG had agreed to half American half British personnel on the spot and they seem to have gone back on their assurance that Frank Holmes would occupy a secondary position in negotiations with the shaikh."

Fowle may have been regretting many of his own pro-Anglo-Persian activities as he told the India Office: "I do not wish to give the impression that I am in any way prejudiced against the Anglo-Persian Oil Company. On the contrary, I have consistently assisted them in the Gulf and my relations with their representatives are most cordial. But I cannot help feeling that, when it comes to big business, their methods are not so straight forward as one could wish."[1]

THE "POLITICAL AGREEMENT"

Louis Kershaw was not particularly perturbed by these criticisms. He was confident he had everything under control through the conditions that would be entered in the "Political Agreement" to be signed between His Majesty's government (HMG) and the Kuwait Oil Company. (A similar political agreement would be imposed on Qatar.) Kershaw wondered whether or not the shaikh of Kuwait should be informed of this "political agreement" that would effectively override any contract he drew up to govern the Kuwait oil concession. And if the shaikh of Kuwait should be told, when should he be told?

He asked Fowle for his opinion. "It seems desirable, subject to your views, that at some stage prior to conclusion of the agreement between the shaikh and Kuwait Oil Company that the shaikh should be informed of the 'Political Agreement' between the company and HMG (which will probably be ready for signature in London at an early date). This is in order that he may not later have any cause for complaint that some of the clauses in his agreement with the company are subject to another agreement of which he is unaware." An optimistic Kershaw speculated: "Presumably, the shaikh will not object to the fact that the company has agreed with HMG on matters within our province as to the conditions on which we will give consent to a concession, as required by the shaikh's Treaty arrangements with us."

Nevertheless, he conceded, "the method of approach to the shaikh may be a matter of delicacy. Please cable me your views on the tactics that should be employed with the shaikh. In particular, at what stage in his negotiations with the company, and in what manner, it will be best to tell him about the agreement between the company and HMG." Kershaw thought the "risk" had to be headed off that the shaikh of Kuwait might turn to Standard Oil of California. He advised the resident to tell the shaikh that "HMG would welcome a successful result in his negotiations with the Kuwait Oil Company." While he was doing this, Kershaw suggested, the resident could drop a little hint that HMG would consent to the grant of the concession to Kuwait Oil "subject to certain conditions regarding purely political matters to which the company has already agreed."

That the Political Agreement was of far greater import than the resident's little hint was to imply did not escape Kershaw. He told the resident: "It may be necessary by some means to provide expressly that the Kuwait Oil Company's agreement with the shaikh is subject to the terms of its parallel 'Political Agreement' with HMG. It is a question which is being examined from the legal point of view as to whether it is desirable that a clause to this effect is inserted in the agreement between the company and the shaikh."

Kershaw said he would be glad to know the resident's views "as to the shaikh's probable reaction to insertion of such a clause during the negotiations at Kuwait. This would probably make it necessary for us to communicate to the shaikh the content of the parallel 'Political Agreement,' or at least the substance. But still it will be a question of what stage in his negotiations with the company this should be done."

Headed "India Office—Strictly Confidential" the Political Agreement between His Majesty's government in the UK and the Kuwait Oil Company was signed on March 5, 1934. The shaikh of Kuwait was not informed. Nor were the Kuwait Oil Company's own negotiators, Frank Holmes and Archibald Chisholm, told of the content of the Political Agreement.

The Political Agreement dictated that "regardless of anything agreed between the shaikh and the company," no obligations or benefits could be transferred without prior consent in writing of HMG and never to any concern with less than 50 percent of the capital and voting power held by British subjects. The majority of employees must be British or subjects of the shaikh, and "regardless of any agreement with the shaikh" the import of foreign labor must be subject to the approval of the political agent.

A chief local representative would be appointed by HMG and only permitted to deal with the shaikh through the political agent. The company was obliged to pay "due deference" to the "advice" of the political agent and the political resident. Also, "regardless of any agreement with the shaikh," the company could not use or occupy any sites that "may have been selected" by His Majesty's government for defense purposes.

The company was to construct a refinery according to HMG's directions as to "suitable type and capacity." In a national emergency or war—the existence of which HMG would be the sole judge—His Majesty's government had the right to preempt all the oil produced in Kuwait and additionally to require the company to produce the maximum possible of fuel oil, to British Admiralty specifications, and deliver it to a place determined by HMG. The price to be paid for such oil so commandeered would be decided solely by HMG. Finally, His Majesty's government would be "at liberty" to take control of the works, plant, and premises of the Kuwait Oil Company at any time. Compensation for such confiscation would be paid only for any resulting "loss or damage."[2]

HOLMES AND CHISHOLM: JOINT NEGOTIATORS

Chisholm and Holmes worked surprisingly well together. The tension that did exist came from Chisholm's "interpreter/assistant"—Anglo-Persian's resident agent provocateur Hajji Abdulla Williamson. Shaikh Ahmad took mischievous delight in discomforting Williamson. Ahmad once invited Chisholm and Williamson to dinner. To Chisholm's surprise, "the only other guest was Muhammad Yateem whom Ahmad knew was extremely disliked by Williamson."

In the final stages of the negotiations, Williamson would be replaced. "The reason for this," Chisholm would explain rather naively, "was that Williamson's antipathy for Holmes's assistant, Yateem, and to Holmes himself, had by then become so strong that it became impossible to retain him. Holmes and Yateem reciprocated the antipathy felt by Williamson. It was connected with some bitter business antagonism that dated back to 1922. As far as Muhammad Yateem was concerned, his convivial habits were appreciated by Shaikh Ahmad."

The good humor of the negotiations, and the negotiators, may have been contributed to by the absence of Harold Dickson at the sessions, a concession that John Cadman had personally extracted from Louis Kershaw at the India Office. The negotiations with the shaikh were interspersed with enjoyable lunches, dinners, and rounds of bridge, the game Holmes had taught Ahmad. Although Frank Holmes's cook was reputed to be the best in the Gulf, Ahmad, Holmes, and Chisholm took turns playing host. Muhammad Yateem joined the evenings. Hajji Abdulla Williamson was not invited.

The resident had dropped his little hint that there would be some form of agreement between HMG and the Kuwait Oil Company designed to "safeguard" British interests. But as both the resident and the political agent assured him that this matter "would not concern him," Shaikh Ahmad did not expect anything more onerous than the five conditions in the British nationality clause as imposed on Bahrain. Neither did Holmes, who, although entrusted with negotiating all the terms and conditions of the concession, was not informed by either Gulf Oil or the Kuwait Oil Company of the detail of this agreement. Although he was Anglo-Persian's own man, Archibald Chisholm wasn't told either.

Consequently, neither Holmes nor Chisholm attached much importance to the instruction they received in early March to insert two new clauses into the draft agreement they were discussing with Ahmad. The first clause referred to the means of arbitration if the company "failed to observe any of the terms of the Political Agreement." The second declared that if any of the terms of the concession agreement reached with the shaikh "be inconsistent or in conflict with the Political Agreement signed between the company and HMG," the shaikh's agreement would be "subordinate to and controlled by the terms of the Political Agreement."

Holmes and Chisholm didn't even bother talking about these two clauses with Shaikh Ahmad. They agreed it was obvious they couldn't ask him to sanction their insertion until he had at least seen the actual Political Agreement. He wouldn't get to see that for some time.

COMPLICATIONS

The India Office had canvassed the "interested departments" for views as to when the shaikh of Kuwait should be told about the content of the Political Agreement. All agreed the Political Agreement should be held back.

The Admiralty led the pack, saying the shaikh should not be informed until he "has undertaken to give the concession to the Kuwait Oil Company." The Admiralty was adamant that secrecy must be maintained until such time as the deed was done. "If the Standard Oil Company of California got hold of the 'Political Agreement,' as might very well happen if it is communicated to the shaikh of Kuwait, they might use it to advance a claim to the concession," they warned. Furthermore, the Admiralty reminded the India Office, "we do not want the terms of this agreement to be made more public than can be helped." Dickson received the Political Agreement on March 24 but did not communicate it to Shaikh Ahmad, or to Holmes and Chisholm.[3]

At the beginning of April another development took everyone by surprise. Abdul Aziz bin Saud had tired of the talks in Sana'a going nowhere with a hostile Imam Yahya. A settlement of the Assir-Saudi border dispute was no closer now than it had been in 1931 when bin Saud had forced the retreat of Yemeni forces. In March 1934 bin Saud dispatched his son Faisal to launch a full-scale attack across the border into Yemen. The chief port of Hodeida was soon captured and command of the road to Sana'a was secured.

Military activities were again putting bin Saud under financial pressure. He owed ten thousand sterling to a merchant from Syria, Abdul Ghani Al Ydlibi, now in business in Manchester, England. To settle this debt and so keep open a line of credit, bin Saud had issued Ydlibi a letter of authority to apply for oil concessions in his territory in areas not covered by the Al Hasa concession. Ydlibi had formed the Arabian Development Syndicate. Frank Holmes's colleague in E&GSynd, Edmund Janson, appeared on the registration papers as representing a substantial financial interest in the new syndicate. In January, before he returned to Kuwait, Holmes had joined Janson and Ydlibi for discussions in E&GSynd's London office. With a draft concession for the Saudi-Kuwait Neutral Zone, Ydlibi had gone back to Jedda. At the end of March, he had returned to London and was now believed to be heading for Kuwait.

The British minister in Jedda, Sir Andrew Ryan, reported the Saudis were keen on Ydlibi's offer of twenty thousand sterling on signature, ten thousand annual rent, and a loan of fifty thousand. Ryan said that the Saudis "seem to think Shaikh Ahmad has already given the Kuwait half of this concession to Frank

Holmes." Standard Oil of California was claiming to have "rights of first refusal" on bin Saud's side, although the Saudi Government was disputing this. However, the Americans said they were prepared to talk to Ydlibi. Standard Oil of California, and possibly E&GSynd through its somewhat mysterious association with Ydlibi, appeared to be well in front.

Andrew Ryan was advised to immediately instruct the Saudi Arabian government that they must approach the British government because the Neutral Zone "is a condominium" and could only be granted jointly. Ryan should say: "This is, of course, in accordance, with the Shaikh of Kuwait's obligations to HMG not to deal with any foreign government except through us." He should tell the Saudi Government: "It has recently been decided in another connection to insist more stringently than in the past on this obligation as regards dealings between the Shaikh of Kuwait and bin Saud." Although "there will obviously be great difficulty with bin Saud over this," Ryan was told he should stress that the British government would insist that "protection of the oil company's operations in the Saudi-Kuwait Neutral Zone, and generally as regards administrative action, must be carried out by His Majesty's government on behalf of the two rulers."

The India Office wanted the Anglo-Persian Oil Company to move quickly in order to block the Americans and E&GSynd in the Neutral Zone. The Foreign Office contended they could tell the Americans they would be supporting Anglo-Persian/Iraq Petroleum on the grounds that "it is legitimate for us to support a company with British participation in a territory such as the Saudi-Kuwait Neutral Zone in which we have a special interest owing to our special relations with the Shaikh of Kuwait."

At a hasty meeting called to plot how the Anglo-Persian Oil Company could immediately secure the Kuwait half of the Zone, William Fraser explained: "It is difficult for our negotiator, Archibald Chisholm, to approach Shaikh Ahmad about the Neutral Zone concession until Major Holmes has left Kuwait. We know Holmes wants to enter into negotiations on his own account for the Kuwait half of the Zone." The India Office advised that the Kuwait Oil Company should finalize its discussion with Ahmad as "soon as possible" so that Anglo-Persian could jump in for the Neutral Zone. Fraser pointed out that progress was slow because "the Shaikh of Kuwait has not been given a copy of the 'Political Agreement.' Because of this he is becoming suspicious. That is what is holding up the negotiations."

The India Office had another matter they wanted cleared up. They said they had heard the Kuwait Oil Company was suggesting, as a negotiating "sweetener," that the shaikh of Kuwait might be invited to England as a guest of the company. This would be severely frowned upon by the India Office, they told Fraser, and "would raise political questions." Fraser mumbled a reply that it had not really been decided although it had been thought the invitation might "clinch" the negotiations. He promised "nothing would be done without prior consultation" with the India Office.[4]

POLITICAL AGREEMENT REVEALED TO AHMAD

On April 12 the Political Agreement was given to the shaikh of Kuwait, together with the two clauses that would now have to be inserted in the concession agreement.

The shaikh was also reminded that he could not "in any way" commit himself to a concession in the Saudi-Kuwait Neutral Zone without prior approval of the High Government. Moreover, he should immediately inform Dickson if he was "approached, from any quarter whatsoever" for a concession in the Zone.

"The shaikh was very offended," Dickson reported in what must have been the understatement of the year. Ahmad railed at the suspect ethics of a company that would send its own "supposed trusted agents" to negotiate a commercial contract without being told there would be "other dimensions" to any agreement they might conclude. He said that he, Holmes, and Chisholm had negotiated in good faith, going over each clause of the draft "time and time again" over forty-eight days and had come very close to a successful conclusion.

Shaikh Ahmad made it clear he understood very well that insistence on the two clauses now being inserted was intended to make the commercial agreement permanently subordinate to the Political Agreement that the Kuwait Oil Company had made with His Majesty's government, without his knowledge. He said it was also obvious that the content of the agreement had been deliberately withheld from him until the very last moment.

He knew, he said, that he was in a precarious political position. He was following bin Saud's war in the Yemen and also his threats and forced collection of a substantial financial "tribute" from the shaikh of Qatar, whose territory bin Saud also coveted. Ahmad said he was not, therefore, in any position to complain about the High Government "tying him up." Nor could he have "anything whatsoever to say" about the imperial directives spelled out in the Political Agreement.

The concession would be "published to the world," Shaikh Ahmad said, and the two clauses mentioning the British government by name and clearly giving "ultimate control of the concession" to the British would crucify him in the eyes of Arab rulers. Ahmad said that his fellow rulers would see, correctly, his acceding to the two clauses as "signing away my independence." He said he would become a joke, a laughingstock to bin Saud and to all the Arab shaikhs. He said they would conclude that he was not even "allowed" to make a simple commercial agreement with a foreign business firm "without the British stepping in and forcing me to insert a 'by our leave' clause."

Dickson reported that Ahmad told him he had "received another approach for his concession." In reply to Dickson's agitated questioning, Ahmad told him he "understood perfectly well" that he could not deal with Standard Oil of California because "only an all British company or at least a 50-50 British company will be allowed to compete" for his concession.

In his report, Dickson took a few swipes at the conduct of the negotiations from which he was excluded. He said that in this meeting with Ahmad, which took place at Ahmad's desert hunting camp, "I was not assisted by an interpreter nor was anyone else present." He claimed Ahmad said that both negotiators "misunderstood" him.

Dickson tried to show how much better it would be if he were in charge. "The shaikh appeared to me by no means as difficult or unreasonable as the negotiators of the Kuwait Oil Company seem to have led their London principals to understand," he intoned. "This is probably due to the fact that neither of the negotiators know Arabic and both have to rely on interpreters."

Curiously, Dickson admitted to the resident: "I asked Shaikh Ahmad if he could not possibly, as a personal favour to me and because he knew I wished him well, agree to the insertion of the two clauses, or at least one of them. Rather unfairly perhaps, I stressed the fact that by doing so he would be winning the High Government's entire approbation. The shaikh in reply, and with some emotion, said he could not and would prefer to have no oil concession at all than act against his own and his family's interests."[5]

HOLMES AND AHMAD

There were signs that Holmes and Ahmad were working together. For some time, nobody seemed to notice. The signals were Holmes's consistent higher estimate than Chisholm's of what the shaikh would accept on a given detail. His predictions that Ahmad would be immovable on a certain point, such as the royalty, were invariably correct. And the shaikh appeared to have developed an uncanny knack for throwing in an additional demand such as an annual petrol provision at no cost each time he seemed to be giving in on a detail—a negotiating tactic that had long been Holmes's own.

Several times Chisholm remarked how correct Holmes was in his forecasts of Ahmad's behavior, for example, that "he would never lower his royalty figure from an amount just above that paid in Iraq and Al Hasa." Chisholm put this down to Holmes having a "sixth sense" arising from his eleven years of close acquaintance with Kuwait affairs and his long personal friendship with Shaikh Ahmad.

Once they knew the content of the Political Agreement, there was little either Ahmad or Holmes could do except hold out for the best terms possible for the actual concession. With some justification, the shaikh no longer trusted the Kuwait Oil Company. He demanded the right to nominate a director to the board whose sole objective would be to protect Ahmad's interests and "obtain information" on his behalf.

Fearing this appointment, or that of chief local representative, might be given to Holmes, Dickson went into a flap. In a pointed although veiled reference to Holmes, he claimed Ahmad had told him that a "person who was ignorant of the Arabic language, or Arab customs, should not be appointed as the Chief Local Representative." Perhaps Dickson truly believed Holmes did not speak Arabic. More likely is that his own long-standing claim of an unrivaled ability to "handle Arabs" was based on a self-proclaimed knowledge of the language and customs —and Dickson was getting very close to mandatory retirement—Dickson was angling for one of these jobs himself.

In London, the Anglo-Persian men attached to the Kuwait Oil Company reacted to the shaikh's demand as if bitten by a snake. From their lofty height, the directors informed Ahmad:

> We cannot accept that your representative could have access to the company records, including the agenda of board meetings. Records will naturally contain much information regarding materials, construction and other matters which would be of no interest to Your Excellency and information concerning personnel, sales of oil, and other things which in their nature are private and confidential. Besides, if your representative had access, the records of the company might become confused or lost, the conduct of the company's business at its office in London might be disarranged.

Patronizingly, they explained that the six-member board was designed as three appointed by Anglo-Persian and three by Gulf Oil. They said:

> All business transacted by the Board must receive the approval of the directors representing each of the two interests. Your Excellency will understand it is impossible for us to accept a member (appointed by you) who would have a deciding vote in the event that the directors of one side were in favour of a certain resolution while those on the other side were opposed. In such a situation, Your Excellency would, in effect, through a director appointed by you be determining the policies and managing the affairs of this company.

That sounded pretty good to Ahmad. He would not relent. Grudgingly, the directors compromised, saying they would "consider" meeting the cost for a London-based representative of the shaikh of Kuwait, without board powers, possibly entitled to obtain some type of information from the company, on the shaikh's behalf, "which the directors believe will be adequate for your purposes."[6]

SHAIKH AHMAD BALKS

At the beginning of May, Shaikh Ahmad dug in his toes. He dropped a reminder that he did have other offers, including one from a "100% British company." He stated unequivocally that he expected his current financial and other requests to be met, including that of a personal representative. In addition, he would refuse any reference to the Political Agreement in the concession.

He put it in writing. "I have informed you of the terms and conditions which I require in the concession and the amount of payments and royalties which are final and I cannot reduce or alter any of them." He said he now wanted an immediate answer "as I see no good in delaying."

Holmes backed him totally. He advised Kuwait Oil in London that unless they could agree to all of Shaikh Ahmad's financial and other demands as incorporated in the latest draft, either the company should put forward a definitive and final counteroffer or the negotiators should return to London "for consultation." A break in the negotiations, Holmes said, would allow Ahmad time to work through the clauses of the agreement with the resident and the political agent in Kuwait.

Gradually it dawned on Chisholm, and also on the men at the Kuwait Oil Company's office in London, that Holmes may be "less than wholehearted" in trying to reduce the shaikh's royalty and financial and other demands. Slowly, they came to the realization that Ahmad's demands "were remarkably identical" with the attitude and advice the Kuwait Oil Company was receiving from Holmes.

The India Office quickly sent out a directive saying the policy on the Saudi-Kuwait Neutral Zone would now be one of playing for time. The directive said: "The best course will be, once the Kuwait Oil Company negotiations are over, Anglo-Persian/Iraq Petroleum Company will endeavour to secure the option from the Shaikh of Kuwait in respect of his interests in the Neutral Zone. We will now play for a stalemate. If all else fails, the Neutral Zone might have to be divided between the rulers."

Completely, perhaps congenitally, incapable of imagining that any Englishman (or in Holmes's case, a colonial) could actually be working for the benefit of the shaikh, the India Office concluded that Holmes could only be aiming for some benefit to himself. The India Office, the political officers, and the Kuwait Oil Company's directors openly debated their assumption that Holmes must be angling for, say, the position of personal representative that Ahmad seemed so determined to get, or maybe, as Dickson feared, the job of chief local representative.

Archibald Chisholm was thirty-two years old and on his first "big" assignment. With the arrogance of a young man, together with the insensitivity of the born aristocrat, he gave his analysis of the situation. He said Holmes was close to sixty years old and securing the Kuwait concession had "long been the final goal" of Holmes's career. He pointed out that Holmes was not a wealthy man. From Chisholm's viewpoint this automatically meant that Holmes must be "counting on his share of the rewards" for a successful negotiation and the shaikh's "personal representative position in the concessionaire company to be the bulwarks of his old age." Chisholm assumed that the formation of the Kuwait Oil Company, and what he called the "forced" release of E&GSynd, had "disappointed" Holmes's hopes and therefore he must now be "counting on" getting the job of the shaikh's personal representative.

With supreme disdain Chisholm explained that being at the beginning of his own career meant he had a stake in "persuading" the shaikh of Kuwait to accept the lowest possible terms. In contrast, he said, Holmes would gain no direct benefit from the negotiations "apart from any favours to come from his old friend Shaikh Ahmad."

Chisholm said his theory had been confirmed when he commented craftily to Holmes that the salary proposed for Ahmad's representative was "too high." With an assumed air of innocence he had asked Holmes, "Have you any idea for whom the shaikh is creating this lucrative position?" He reported that the "implication" of this question was "particularly resented" by Holmes.

It never seems to have occurred to Archibald Chisholm that it may have been the shaikh of Kuwait's experience of the British, of the Anglo-Persian Oil Company, and now of the Kuwait Oil Company, that had led him to conclude he would certainly need someone to guard his interests and obtain information on his behalf.

Soon after sending off his analysis of Holmes's psychology and position, Chisholm was mortified when he heard there were rumors in Bahrain that the Kuwait negotiations were in trouble. The rumors speculated that the problems were caused "because the Shaikh of Kuwait's friend, *Abu Al Naft* (Father of Oil, Holmes), encouraged Shaikh Ahmad to demand high terms. This makes it very difficult for Anglo-Persian's man, *Al Tawil* (The Tall One, Chisholm)."[7]

More than coincidental with Holmes's advice to London that he and Chisholm should return, Shaikh Ahmad declared he would not discuss the concession for several months. A week later he summoned both men and asked them why they were still in Kuwait. Holmes said he was leaving in forty-eight hours and advised Chisholm to do the same.

INDIA OFFICE REACTS

Dickson breathlessly wired the news followed by another flurry of "secret" cables and letters. The resident cabled the India Office: "Shaikh has informed Kuwait Oil Company that if they cannot agree to certain final terms of his they should return to England until the end of September. During their absence he will not discuss oil with any other competitors."

Ominously, the resident added: "Dickson sees the hand of Frank Holmes in this, and so do I. My own opinion is that difficulties which both Kuwait Oil and we have had with the shaikh are largely due to Holmes. He has probably been working with the shaikh on behalf of himself (to become for instance the shaikh's director in London or Chief Local Representative in Kuwait) or on behalf of the Eastern & General Syndicate for the Neutral Zone. This is entirely the fault of the Kuwait Oil Company who appointed Holmes against the wishes of His Majesty's government."

The India Office called in the Kuwait Oil Company, including Guy Stevens, the American who was Gulf Oil's senior director in the company. The Anglo-Persian directors in Kuwait Oil warned the India Office not to give "any information to Mr. Stevens or any other American interests in regard to the Neutral Zone."

They claimed that "while American companies might compete against one another any information given in this area to American nationals appears to be passed on." The Anglo-Persian men were "anxious" that the India Office "should avoid any reference to, or discussion of, the Saudi-Kuwait Neutral Zone in Guy Steven's presence."

Louis Kershaw himself presided over the meeting and assured the participants that the problems had not arisen because of the Political Agreement. He said this matter had now been "satisfactorily settled." So what, he wanted to know from the Kuwait Oil Company, was the problem?

William Fraser said that while most financial points were more or less agreed upon, the shaikh of Kuwait now wanted the basis of the royalty not on oil exported but on that "won and saved," which would mean an increase of 15 percent to 20 percent in the value of royalty. He did not point out that "won and saved" was the formula used in Persia, Iraq, Bahrain, and Al Hasa. He did say that the Kuwait Oil Company would not agree to this. Worse still, he told the India Office, was Shaikh Ahmad's insistence on a personal representative in London, at a salary of eighteen hundred sterling per annum, for a period of twelve years on the expiry of which he would be replaced by a director personally representing the shaikh. Fraser declared that "this demand of the Shaikh of Kuwait is most objectionable."

Kershaw agreed that this was an unheard-of request. He asked Fraser if he knew what could possibly make this point so important to Ahmad? What intention could he possibly have in insisting on it? Fraser replied that he had absolutely no idea why Ahmad should want a personal representative. But he did have "certain suspicions." Kershaw then read out the resident's latest cable, with its accusations against Frank Holmes. "Is it possible the Resident has hit the right nail on the head?" he asked.

The Gulf Oil Corporation's Guy Stevens, oblivious of the fact that to the men from Anglo-Persian he was as suspect as was Frank Holmes, remarked that although he "disliked having to say it" he had to "frankly admit" he had found it "difficult not to entertain suspicions of a somewhat similar kind." He said he had asked the two negotiators who the shaikh had in mind for "this lucrative post" and whether it would not be best to ensure that the position-holder was a Kuwaiti. "The negotiators first sent an obscure reply," Stevens told Kershaw, "and on being pressed replied that the shaikh had now issued his 'ultimatum.'"

Stevens claimed that Holmes "has apparently established himself in the shaikh's good graces. He has consistently asserted that Shaikh Ahmad wants the concession negotiated by him and the company should let him get on with it."

Stevens commented that "all this looks extremely shady." He seemed genuinely puzzled as he pointed out that in Kuwait Archibald Chisholm had agreed with every report and had cosigned every document.

Stevens told the India Office that it was "only a few days ago" that William Fraser had told him the British government had any objections to Holmes. Fraser jumped to his own defense, declaring it was Stevens who had wanted Frank Holmes as Gulf Oil's negotiator. He charged it was Stevens who had had confidence in Holmes and believed he had influence with the shaikh of Kuwait.

Fraser vowed that, personally, he was inclined to doubt Holmes, particularly in view of his "previous activities" in Bahrain of which, he said archly, the India Office was all too well aware.

Louis Kershaw was happy to discount the fact that all the reports from Kuwait had been signed by both Holmes and Chisholm. It was possible, he said, "that Major Holmes might have had previous consultation with the shaikh before the shaikh's interviews with Chisholm and Holmes together." He also had no trouble in jumping to the conclusion that Holmes must be about to personally benefit in some way. Suspicion being as good as fact, the India Office's minutes recorded: "It seems quite wrong that one of Kuwait Oil's negotiators should stand to receive some financial consideration from the shaikh out of the negotiations."

Without letting Gulf Oil's Guy Stevens know that the Anglo-Persian Oil Company planned to move in on the shaikh of Kuwait for the Neutral Zone, Kershaw said it was unacceptable that the negotiations be suspended until September. The right response to Shaikh Ahmad's instruction that the negotiators should leave for several months, he decreed, would be "to withdraw Major Holmes from Kuwait, leaving Mr. Chisholm to hold the fort." William Fraser enthusiastically agreed, adding: "If the company's suspicions of Holmes prove to be well founded, he will not be allowed to return to Kuwait." Fraser said they would cover themselves by telling Holmes he was "being recalled for a brief period of consultation only."

The resident wrote a personal letter to the India Office declaring himself satisfied with this plan: "I am very glad to get your cable saying that Holmes has been recalled from Kuwait." He was, he said, "sure Holmes had created in Kuwait the various difficulties, both for ourselves and the company, which is entirely the latter's own fault. Apart from other warnings which they received on the subject, I personally told Sir John Cadman when he asked my opinion on the subject that if Holmes was sent out to Kuwait it would certainly be a mistake." He said he and Dickson would "handle the matter of the two clauses" with the shaikh of Kuwait.

But all the India Office machinations came to nothing when Shaikh Ahmad threatened to open negotiations "with other parties" unless Archibald Chisholm followed Frank Holmes out of Kuwait, immediately. This caused the India Office and the resident to suspect that Holmes had "extracted a secret promise" from the shaikh of Kuwait in regard to the Neutral Zone. If so, they decided, they would simply "refuse to recognise it."[8]

DICKSON'S CONVERSION

Harold Dickson was on home leave in London at the same time that Holmes and Chisholm arrived back from Kuwait. He enjoyed the hospitality of the Kuwait Oil Company and was the star performer at a meeting at the India Office, giving the benefit of his experience and self-proclaimed unique insight into "Arabs."

Something happened to Dickson during these weeks. He was feted at a number of lunches, dinners, and cocktail parties with the executives of the Kuwait Oil Company. It may have been the result of observing the plush lives enjoyed by these men, few of whom, if any, had ever visited the Arab world. The experience may have made Dickson acutely aware that he had more in common with Frank Holmes than he had with the well-fed and overpaid executives at the London offices of the Kuwait Oil Company.

He was invited to the India Office. This was a mirror image of the Kuwait Oil Company. Few, if any, of these men had ever visited the Arab world either. And they also seemed to exist in a comfortable round of lunches, cocktails, prestige, and power. Perhaps it hit Dickson that there was an enormous gap between the views and pronouncements of these officials and the reality of the situation in the Arab shaikhdoms of the Persian Gulf, with which he, and Holmes, had been associated for the last twenty years.

Despite dropping heavy hints about the need to ensure the person selected for the job of chief local representative for the Kuwait Oil Company was well versed in "the Arabic language and Arab customs"—coincidentally Dickson's own prime qualifications—and reminding everybody that he was due to retire very soon, Dickson was not offered anything by the Kuwait Oil Company, the Anglo-Persian Oil Company, or the British officials.

Invited to address a combined gathering of Kuwait Oil Company executives and officials from the India Office, the Foreign Office, and the Petroleum Department, Harold Dickson performed a complete about-face on his opinion and assessment of Maj. Frank Holmes.

He began with the subject he loved best, and on which he believed himself to be the outstanding world expert—"Arab mentality in general"—before moving on to the "specific psychology" of the shaikh of Kuwait. All Arabs were motivated by *Sharaf*, meaning pride or honor, and this affected their every interaction, he explained gravely to the assembly. He said all Arabs believed they had "rights," a strange and distinctly foreign view that nevertheless had to be stepped around carefully.

Dickson said the shaikh of Kuwait was a perfect illustration of the Arab male. He was a man of "sensitive pride" and was "suspicious and shrewd." Apparently forgetting his 1931 description of the shaikh of Kuwait as "a simple Bedouin" who was "easily swayed and not possessed of much strength of character," Dickson now declared him to be a capable negotiator "in a different category, as far as ability and business acumen goes, from the other shaikhs of the Persian Gulf." Shaikh Ahmad, said Dickson, could "see through any bluff."

He offered some practical advice. While he agreed the company should not give in on matters such as increased royalty, he said, "the further the company goes to meet the shaikh the better." He advised the oilmen to be "straightforward" and frankly explain their objections to the shaikh's demands. They should not present him with an ultimatum or "rush him too much." He reminded the gathering that the shaikh of Kuwait needed "careful handling." Moreover, Dickson said, Ahmad had now realized "that his horse is apparently so valuable" and therefore was in less of a hurry to sell it than he might have been before.

Dickson suddenly announced that the one man who really understood Shaikh Ahmad of Kuwait was Frank Holmes. Many in his audience were taken aback as Dickson continued, saying that "Major Holmes, whatever his faults, has great influence with the Shaikh of Kuwait and is very skilful in his handling of him." From this point of view, he said, "it will probably be difficult to find a better negotiator."

"The possibility that Frank Holmes will put the interests of Shaikh Ahmad before those of the Kuwait Oil Company, and that his own interests will not be overlooked, must be borne in mind," Dickson told the now thoroughly dazed gathering. "On the whole," he advised, "it will be better if Frank Holmes returns to continue the negotiations in Kuwait in September" together with Archibald Chisholm if required. Dickson said he was sure the two men could "anticipate a friendly reception from Shaikh Ahmad on their return." He stated that the suspension of the negotiations had been caused by Ahmad's "genuine tiredness and irritation" and many outside pressing problems.

When they recovered from the shock, it was this ringing endorsement from such an unexpected source that finally convinced the Kuwait Oil Company and the India Office that Holmes was their only chance. Only Frank Holmes, it seemed, had the skill and ability, and the personal influence, to conclude the negotiations for the Kuwait oil concession.[9]

SAUDI-KUWAIT NEUTRAL ZONE

In August the Saudi Arabians dropped by the Foreign Office to ask if any progress had been made in regard to the Kuwait side of the Neutral Zone. Bin Saud's minister, Yousuf Yassin, visiting London, put it quite openly. "Now that we have just won a war," he said, "we must look about us to see what money we can lay our hands on."

In May 1934 the Treaty of Taif had been signed between bin Saud and the defeated Imam Yahya, giving the Saudis the Assir, Najran, and Jaizan, the Idrissi's old historic base that the Imam Yahya had himself captured. Two years later, in the Grand Mosque in Mecca, two Yemenis would make an attempt on Abdul Aziz bin Saud's life; he would be saved when his eldest son, Saud, threw himself between the would-be assassins and his father. Despite Faisal's proven military and strategic abilities, Saud would be rewarded by being named king on the death of Abdul Aziz bin Saud in April 1950.

The British continued to obfuscate over the Saudi-Kuwait Neutral Zone. They skillfully manipulated the artifice that they were entitled to take over all security and administration in the Zone on behalf of both rulers. They could be pretty sure that Abdul Aziz bin Saud, having just "won a war," would never agree to that.

Shortly after the Kuwait concession was finalized, the Anglo-Persian/Iraq Petroleum negotiator would arrive to open discussions with Shaikh Ahmad for the Kuwait half of the Neutral Zone. The negotiator was Stephen Longrigg who had been so misled by Philby over the Al Hasa concession. Ahmad didn't want to know about it. Dickson, working out his last months as political agent, would tell Longrigg to let it go until the "position is clarified on the Saudi side." As the British would not relent on their demand to be the "protectors" of the whole area, nothing happened.

The two thousand square miles of the Saudi-Kuwait Neutral Zone concession would not be tapped until 1947, coincidental with the Independence of India and the subsequent withdrawal of India Office control over the Arab shaikhdoms of the Persian Gulf. In an uneasy combination, the Kuwait side would be taken by a consortium known as "Aminoil" (American Independent Oil Company) while J. Paul Getty's Pacific Western would obtain the Saudi side. They would strike oil in 1953, discovering a field that would later be described as "somewhere between colossal and history making."[10]

NOTES

1. IOL/R/15/1/645, vol. IX, November 19, 1934, India Office (Laithwaite) Kuwait Oil, "It will be remembered that at Lord Lloyd's speech at the Central Asian Club dinner he said . . . British government should have influenced Anglo-Persian more strongly in this direction and should have refused to allow American interests to get a foot in the Gulf"; IOL/L/PS/12/10R.1007.8, vol. 3808, January 24, 1934, Admiralty (Barnes) to secretary of state for India (Kershaw), "it will mean that all the oil which has hitherto been found on the Arabian shore of the Persian Gulf has passed to partly foreign (or at best international interests)"; and January 27, 1934, general manager, Anglo-Persian Abadan (Elkington) to Petroleum Department (Hearn), "full protection . . . only guaranteed if the Anglo-Persian Oil Company is exploiting it under a concession"; and February 1, 1934, resident (Fowle) to secretary of state for India (Kershaw), "HMG cannot agree to the shaikh giving a concession to the group unless operations are all British. The interest of the American section of the group is, or should be, purely financial."

2. IOL/L/PS/12/10R.1007.8, vol. 3808, February 13, 1934, "Secret," secretary of state for India (Kershaw) to resident (Fowle), "It seems desirable . . . at some stage prior to conclusion of agreement between the shaikh and KOC that the shaikh should be informed of the 'Political Agreement' between the company and HMG"; and March 5, 1934, India Office—confidential, Political Agreement between His Majesty's government in the United Kingdom and the Kuwait Oil Company; note that a virtually identical Political Agreement would be imposed on Qatar.

3. Archibald H. T. Chisholm, *The First Kuwait Oil Concession Agreement: A Record of the Negotiations, 1911–1934* (London: F. Cass, 1975), p. 39, "the only other guest was Muhammad Yateem whom Shaikh Ahmad knew was extremely disliked by Williamson" and p. 110, "antipathy felt by Williamson was reciprocated by Holmes and Yateem and was connected with a bitter business antagonism that dated back to 1922"; IOL/L/PS/12.10R.1007.8, vol. 3808, April 10, 1934, "Secret," Admiralty to India Office, "we do not want the terms of this agreement to be made more public than can be helped." Chisholm, p. 48, "the shaikh would obviously wish to study the Political Agreement before agreeing the proposed additional clauses."

4. ILO/R/15/1/641/86.IV, February 2, 1934, India Office (Laithwaite) confidential note of interview with Mr. Janson and Major Holmes, "Janson said Ydlibi had obtained a letter from bin Saud in substitution for repayment of an obligation of 10,000 sterling and "Holmes said . . . at the time of the negotiations last year between Standard Oil of California and bin Saud . . . the Neutral Zone had been excluded"; PRO/FO/371/17919/170681, April 3, 1934, British minister in Jedda (Ryan) to Foreign Office, "they seem to think the shaikh has already given Kuwait 'to Holmes'"; ILO/R/15/1/641/86.IV, April 9, 1934, Foreign Office (Warner) memo of meeting at the India Office with Anglo-Persian (Fraser and Lefroy), "British government would insist that protection of the oil company's operations and generally as regards administrative action must be carried out by His Majesty's government on behalf of the two rulers"; and April 4, 1934, Foreign Office (Warner) record of a meeting at the India Office with representatives of Anglo-Persian/Iraq petroleum company cc India Office, Admiralty, Petroleum Department, Department of Trade, "legitimate for us to support a company with British participation"; and April 9, 1934, Oil in Kuwait Neutral Zone, record of discussion at India Office with representative of Anglo-Persian Oil Company cc Foreign Office, Admiralty, Petroleum Department, "difficult for Chisholm to approach Shaikh Ahmad about the Neutral Zone concession until Holmes has left Kuwait" and "Ahmad is becoming suspicious."

5. IOL/L/PS/12.10R.928.29, vol. 3811, April 24, 1934, "Confidential," political agent Kuwait (Dickson) to political resident (Fowle), copy to India Office, "become a joke, a laughing stock, to bin Saud and the other Gulf Shaikhs" and "prefer to have no oil concession at all than act against his own and his family's interests"; PRO/FO/371/17919/170681, April 28, 1934, "Confidential," political agent Kuwait (Dickson) to shaikh of Kuwait, "if you are approached, from any quarter whatsoever for a concession in the Neutral Zone." For the shaikh of Qatar's enforced "tribute" paid to bin Saud see, for example, IOL/R/15/1/641/86.IV, February 2, 1934, India Office memo (Laithwaite), "Shaikh of Qatar is well off, but has to pay substantial sums (estimated at 10,000 sterling a year) to bin Saud. Having paid this rent to bin Saud for some years, he has been called to Riyadh by bin Saud and warned to be careful in disposing of his oil concession. The Shaikh of Qatar stands in great fear of bin Saud and will certainly be swayed by an

instruction or request he receives from him. The Shaikh of Qatar's southern boundary is rather uncertain. bin Saud has given him clearly to understand that there must be no activities on behalf of Qatar such as investigations by Anglo-Persian Oil Company geologists south of a certain line" any such boundary would of course be inconsistent with the line indicated to bin Saud by Sir Percy Cox in 1922.

6. Chisholm, *First Kuwait Oil Concession Agreement*, p. 207, reprinting September 30, 1934, Kuwait Oil Company Argument to be Presented to Shaikh Ahmad, "and other matters which would be of no interest to Your Excellency and information concerning personnel, sales of oil, and other things which in their nature are private and confidential" and "Your Excellency will understand it is impossible for us to accept a member of the Board (appointed by you)" and "consider meeting the cost for a London based representative of the Shaikh of Kuwait—without Board powers. . . ."

7. PRO/FO/371/17919/1706, May 8, 1934, Oil in Kuwait Neutral Zone, record of discussion at India Office with representative of Anglo-Persian Oil Company cc Foreign Office, Admiralty, Petroleum Department, "play for a stalemate. If all else fails, the Neutral Zone might have to be divided between the rulers"; Chisholm, *First Kuwait Oil Concession Agreement*, pp. 59–60, "Less than wholehearted" and "remarkably identical" and "Holmes must be counting on his share of the rewards . . . to be the bulwarks of his old age" and "the shaikh's friend Abu Al Naft encourages him demand high terms making it very difficult for *Al Tawil*."

8. IOL/L/PS/12.10R.928.29, vol. 3811, "Secret," June 7, 1934, political resident (Fowle) to India Office, "entirely the fault of the Kuwait Oil Company who appointed Holmes against the wishes of HMG"; PRO/FO/371/17919/170681, India Office (Laithwaite), note of conversation with Mr. Lefroy of the Anglo-Persian Oil Company, "avoid any reference to, or discussion of, the Neutral Zone in Guy Steven's presence"; IOL/L/PS/12.10R.928.29, vol. 3811, June 8, 1934, and June 12, 1934, discussion at the India Office with representatives of the Kuwait Oil Company, "withdraw Major Holmes from Kuwait, leaving Mr. Chisholm to hold the fort" and "tell Holmes that he was being recalled for a brief period of consultation only"; and June 13, 1934, "Extract of Private and Personal Letter" from resident (Fowle) to India Office (Laithwaite), "I personally told Cadman . . . if Holmes was sent to Kuwait it would certainly be a mistake"; PRO/FO/371/17919.170681, June 25, 1934, India Office memorandum, Kuwait Neutral Zone Oil Concession, "Holmes may have extracted secret promise from sheikh . . ." and June 27, 1934, India Office to resident (Fowle), "We can refuse to recognise any secret undertaking the shaikh may have given to Holmes" Yossef Bilovich, "Great Power and Corporate Rivalry in Kuwait, 1912–1934: A Study in International Politics" (PhD diss., University of London, 1982), p. 357, citing June 15, 1934, India Office meeting with KOC.

9. IOL/L/PS/12.10R.928.29, vol. 3811, July 6, 1934, memorandum of an informal meeting held on Kuwait Oil, present were the political agent Kuwait, Colonel Dickson, Foreign Office (Walton), India Office (Laithwaite), Petroleum Department (Hearn), Kuwait Oil Company (Fraser, Stevens, Lefroy), "Holmes, whatever his faults, has great influence with the Shaikh of Kuwait and is very skilful in his handling of him" and "it will be better if Holmes . . . returns to continue the negotiations in Kuwait."

10. PRO/FO/371/17919.170681, August 7, 1934, Foreign Office (Calvert) cc British minister Jedda, India Office, Admiralty, Petroleum Department, Department of Trade, "Yousuf Yassin . . . asked me whether I had anything fresh to tell him with regard to the

question of the Kuwait Neutral Zone. I replied that I had no instructions on the subject whereupon he requested me to write to remind HMG"; David Holden and Richard Johns, *The House of Saud* (London: Sidgwick and Jackson, 1982), pp. 97–99, "two Yemenis would make an attempt on bin Saud's life in the Grand Mosque in Mecca"; Stephen Helmsley Longrigg, *Oil in The Middle East: Its Discovery and Development* (London: Oxford University Press, 1954), p. 112, "until the position is clarified on the Saudi side"; Daniel Yergin, *The Prize: Epic Quest for Oil, Money, and Power* (New York: Simon and Schuster, 1991), pp. 441–42, "*Fortune* magazine described this as 'somewhere between colossal and history making.'"

CHAPTER 23

CLOSURE

A t the end of September 1934, Holmes and Chisholm traveled together back to Kuwait. Lugging a huge radio installation as a gift for Shaikh Ahmad, the pair spent two weeks in Alexandria and Cairo before catching a flight to Baghdad.

During the three-day stopover at Basra, Frank Holmes and his friend from the British consulate met for a long lunch. The English judge of the Basra High Court joined them. The conversation turned to the Kuwait concession. Out of the blue, the judge interrupted the conversation. Fixing his lunch companion with a stern look, he leaned across the table and declared: "Holmes, do you really think for one moment that you will be permitted to secure that oil concession for a joint Anglo-*American* group?" While Holmes digested the import of this, the judge confided that he personally knew "a new company" had "practically agreed" upon the concession with Kuwait. From his fellow diners Holmes learned that while the negotiators and Dickson had been in London, someone had been in Kuwait schmoozing the shaikh. Holmes would soon discover that the Anglo-Persian Oil Company had set up an intricate double-cross designed to knock out its American partners, the Gulf Oil Corporation.

THE "TRADERS" PLOT

The "new company" turned out to be the mysterious "100% British company," about which Shaikh Ahmad had dropped such heavy hints. It was called the Trade Arts Company Limited, Traders for short. On investigation, its participants were found to include the elderly president and former chairman of Anglo-Persian, Charles Greenway, and a number of men who, like Greenway, had strong connections to the longtime Anglo-Persian-associated Shaw, Wallace & Company of India. Although he was by now implacably hostile to the Anglo-Persian Oil Com-

475

pany and never missed a chance to berate it in Parliament, Lord John Lloyd had been persuaded to adopt the role of mentor to Traders in its effort to obtain the Kuwait concession. Lord Lloyd appeared to be unaware of the connection with Anglo-Persian. He, too, believed Traders was merely a "100% British company."

Holmes asked his Kuwaiti friends and found they were all firmly convinced that Traders was a front for the Anglo-Persian Oil Company. The negotiator for Traders was J. Gabriel, Shaikh Ahmad's Armenian legal adviser; the same Gabriel who had earlier inspired Ahmad to split the Kuwait concession. Gabriel's partner in their Basra law firm, Mirza Muhammad, was legal adviser to the Anglo-Persian Oil Company. Holmes concluded Gabriel was acting under the instructions of Mirza Muhammad.

Holmes learned that although the terms and conditions were much the same as Kuwait Oil Company's, Traders' offer contained higher initial payments and higher royalties. Most attractive to the shaikh of Kuwait was that Traders made no mention of a political agreement.

If there was any way Ahmad could have been spared the ignominy of a political agreement, and get better terms for his concession, Holmes would have been all for it. Even though he did not know of the restrictive secret agreement signed between Anglo-Persian and the Gulf Oil Corporation, Holmes well knew that left to its own devices, the Anglo-Persian Oil Company's aim was not to develop but to retard Kuwait's oil development. Their goal would be to block the identification and development of oil in Kuwait, not facilitate it.

Holmes had had suspicions that all was not right with the supposed "50-50" partnership between Gulf Oil and Anglo-Persian. In May he had received a cable from London instructing him to take all the confidential Arabic translations of the negotiations to Anglo-Persian's lawyer in Basra, Mirza Muhammad. Holmes ignored it, and the fact that he heard nothing further convinced him that the directive had not emanated from both sides of the joint venture.

Now Holmes further suspected that the India Office was colluding with Anglo-Persian in the Traders double-cross. He sent a package to his wife, Dorothy, and asked her to discreetly hand-deliver it to Guy Stevens at Kuwait Oil's London office. "This letter," Holmes wrote Stevens, "is 100% confidential. If the British government become aware that I have given the game away, I will be in a very awkward position."

"Your associates (Anglo-Persian) are certainly beyond the pale with their perfidy and deceit," Holmes commented. From the depths of his personal experience he ruefully advised the American Guy Stevens: "The English, when engaged in a questionable piece of work, can always manage to salve their consciences by deluding themselves that what they are doing is for the benefit of their country, and therefore a virtue."

He laid out the circumstantial evidence of the plot to deprive the Gulf Oil Corporation of a stake in Kuwait's oil. He told Stevens that Gabriel's partner,

Mirza Muhammad, had tried the extraordinary move of attempting to recruit Holmes's own colleague, Muhammad Yateem. Yateem was certain Anglo-Persian was behind this approach to him. Gabriel himself had tried to "enlist the goodwill and active support" of Mullah Saleh, Shaikh Ahmad's senior adviser and Yateem's best friend. Mullah Saleh was also convinced Traders was actually Anglo-Persian. Saleh believed Shaikh Ahmad did not suspect anything. Saleh told Holmes that Ahmad appeared to think Traders was merely some wealthy group in London.

When Shaikh Ahmad dropped by his house for dinner and a game of cards, Holmes asked him about "Gabriel's company." Ahmad confirmed that he had indeed been approached. When Holmes suggested a check into the background of this company might be advisable, Ahmad said he had "avoided finding out" who was behind Traders because he wanted to keep an "open mind" until the Kuwait Oil Company either accepted or rejected his current demands. Although not wanting to believe he had been tricked, Ahmad assured Holmes he "had no intention of permitting the English to stab the Americans in the back."

Holmes urged Guy Stevens to take action in London. He told Stevens, "if we are to prevent being double crossed" the three Gulf Oil Corporation directors must "force" the three Anglo-Persian directors to accept the shaikh's current terms. "If you do not accept my advice," Holmes warned, "you will play into the hands of those who are not our friends, no matter how long they profess to the contrary." Nor the friends of Kuwait, he may well have added. Holmes said he had not "given the slightest hint to Chisholm that I know of the perfidy of his crowd."[1]

DICKSON DISCOVERS THE PLOT

Dickson had, independently, reached the same conclusion as Frank Holmes that the Anglo-Persian Oil Company was attempting to double-cross their American partners.

On his return to Kuwait, he had learned of the shenanigans and had pursued his own investigations. He forwarded his analysis to the resident, saying: "Anglo-Persian smarted under the fact that they were unable themselves to get an oil concession out of the Shaikh of Kuwait because of the Americans. For appearances sake they leagued themselves with these selfsame Americans." His analysis continued: "While outwardly agreeing to work with them under the name Kuwait Oil Company, Anglo-Persian secretly set themselves the task of hindering progress from the start with the deliberate intention of preventing the shaikh from giving the concession to the Kuwait Oil Company."

Dickson concluded: "At the psychological moment and when the negotiations with the Kuwait Oil Company reached deadlock, Anglo-Persian already had a new and secret subsidiary company to throw into the breach and so walk off with the prize. As far as the world is concerned, this company, Traders, could in no way be connected with the Anglo-Persian Oil Company. It would be all-British. It would be registered in London and Iraq."

Dickson went further. He accused the former spy and covert operator, Hajji Abdulla Williamson, of being actively involved.

Dickson also charged that Archibald Chisholm had, at the very least, "known about" the plot since May. Dickson said he had personal knowledge that Chisholm had had many "secret" meetings with Anglo-Persian in Basra.

Dickson explained to the resident (Fowle) that being put in this position had a bad effect on Archibald Chisholm. When he returned to Kuwait soon after Holmes and Chisholm, Dickson said, he observed that "from being the very open and delightful friend of the family that he used to be, Chisholm suddenly adopted a close and decidedly aloof attitude. I now put this down to the fact that he had received his 'orders' in London and not only disliked playing a dual role, but definitely felt he was playing a 'dirty' game and was ashamed of himself. He certainly was continually at loggerheads with Holmes and whatever progress Holmes seemed to make in the negotiations, he, Chisholm, appeared to want to spoil."

After the concession was signed, Dickson would write the resident: "The American side of the Kuwait Oil Company had a very narrow escape, they were very nearly victims of a barefaced attempt to 'double cross' them, on the part of their allies and friends, the British side, the Anglo-Persian Oil Company." Shaikh Ahmad would later claim that he "saw the hand of Anglo-Persian in Traders" from an early stage but had decided "to go all out and play with them." Dickson told the resident, "I take a little credit myself for having 'saved the shaikh's bacon.'"

Dickson's complete reversal from his once ardently pro-Anglo-Persian position that began with his India Office address was now final. He confided to Fowle: "Of course, the Anglo-Persian Oil Company have covered their tracks well. One will never, I suppose, be able to prove anything against them. Which in any case would be undesirable."[2]

KUWAIT CONCESSION SIGNED

During the weeks that followed Holmes's and Dickson's discovery of the plot, the services of Hajji Abdulla Williamson were dispensed with and Chisholm was gently eased out. Holmes and Shaikh Ahmad had frequent meetings, to which Chisholm was not invited. There was not a peep from Dickson. By November 7, Holmes and Ahmad had drawn up two alternatives from which the Kuwait Oil Company should choose.

In London, Guy Stevens was noticing that his fellow directors from Anglo-Persian seemed to be in the habit of delaying matters by putting off decisions on even minor points. Alerted by Holmes's warning, discreetly delivered by Dorothy, he made a personal visit to the India Office.

There he inquired whether they could confirm "rumours" that another company had applied for permission to negotiate with Shaikh Ahmad in competition with the Kuwait Oil Company. After beating around the bush for some time, the India Office reluctantly confirmed that this was so. Stevens realized that Holmes's suspicion was well-founded.

He returned to the office and immediately tackled the members of the board, demanding that they select one of the alternatives put forward by Holmes and Ahmad. On December 13, 1934, the board advised the negotiators in Kuwait, in writing, that all of Shaikh Ahmad's proposals were accepted and they were authorized to sign the concession.

With Harold Dickson as the witness, the Kuwait oil concession, in favor of the Kuwait Oil Company, was formally signed on December 22, 1934, on terms that Holmes had forecast Shaikh Ahmad would ask for back in March and had advised Kuwait Oil Company to accept.

Through the device of being transformed into "an exchange of letters" between the shaikh of Kuwait and His Majesty's government, and "endorsed" by the Kuwait Oil Company, the Political Agreement would remain hidden until Kuwait declared independence in 1961.[3]

The final terms of the Kuwait concession were $173,000 on signature with annual rent of $35,000 before commercial production and $18,000 after. Royalty was to be paid according to the "won and saved" formula.

Holmes did get Shaikh Ahmad of Kuwait better terms than Abdul Aziz bin Saud achieved from Standard Oil of California for the resale of the much larger area of the Al Hasa concession. Bin Saud's $150,000 on signature wasn't even a payment, it was a loan to be repaid over five years. And Holmes had obtained $10,000 more for Ahmad's first year's rent than the $25,000 Standard Oil of California paid on Al Hasa in Saudi Arabia.

Yet, the final settlement was not as attractive as the terms Holmes had originally offered on behalf of E&GSynd. In 1923 Holmes had offered Kuwait 20 percent participation in the development of its own oil. This had been eliminated when Holmes was negotiating on behalf of Gulf Oil and was not reinstated in the Kuwait Oil Company terms. Holmes's agreement had also provided for the shaikh of Kuwait to appoint his own director to the development company, with full board powers.

When he published his record of the Kuwait negotiations in 1975, Archibald Chisholm defended the terms of the Kuwait Oil Company's purchase of the concession.

Chisholm claimed: "This was not a case of the British protecting power preventing the Shaikh of Kuwait reaching agreement with either Frank Holmes and his E&GSynd or with the Gulf Oil Corporation, and then encouraging its own Anglo-Persian Oil Company to intervene, and subsequently pressuring the shaikh into awarding it cheaply to an Anglo-American combine." Chisholm also strongly denied that Traders was a front for the Anglo-Persian Oil Company.

In spite of all the evidence to the contrary, Chisholm wrote:

> During the years following Holmes's appearance in Kuwait in 1923 to compete with the Anglo-Persian Oil Company, first on behalf of E&GSynd and then from 1927 for Gulf Oil, the British government although understandably reluctant in those days to facilitate American commercial penetration of the Gulf area, went to great lengths to hold the ring fairly between the two competitors and to get the best terms possible for Kuwait. And even in 1934, on the last minute appearance of Traders as a new competitor, the British government was ready to sanction its intervention as being in Shaikh Ahmad's interest in the auction of his concession, if the shaikh himself had not rejected it.[4]

After the concession was signed, Chisholm returned to Anglo-Persian in Abadan for twelve months. He resigned in 1936 and was editor of the *Financial Times* from 1937 to 1940. From 1940 to 1945 he served with the British army. In 1946, "at their invitation," he returned as public affairs manager to what was now called the Anglo-Iranian Oil Company. From 1962 to 1972 he was adviser to British Petroleum, the newest name of the entity that had originally been the Anglo-Persian Oil Company.

HOLMES GETS HIS JOB

Ahmad never wavered in his demand for a personal representative who would guard his interests with the Kuwait Oil Company. The final wording of the agreement gave him "the right to appoint a representative in London to represent the shaikh in all matters relating to this Agreement with the company in its London office." What the representative would be allowed to do was spelled out in niggardly tones. "The representative shall have full access to the *production* records of the company [*sic*], including the agenda of the board meetings and shall be entitled to attend those board meetings at which the shaikh's interests are discussed. Salary of $821 a month shall be paid by the company and not by the shaikh. But the travelling and general expenses of the representative shall be defrayed from the salary."

In a letter dated January 5, 1935, and marked "Personal," Shaikh Ahmad appointed Frank Holmes. Ahmad wrote: "As I have great confidence in your friendship, and depend upon your honesty, I have decided to appoint you my Representative in London and will be pleased if you kindly inform me of your acceptance at your earliest convenience."

Ahmad added: "I also shall be much obliged if you ask your company to commence geological exploration at the earliest as possible, as the weather of these days is comfortable and cold. Please inform your company my wishes and oblige." Working in this capacity and frequently traveling to Kuwait and the other Arab states until his death in 1947, Frank Holmes would continue to be associated with Arabia's oil.[5]

AHMAD GETS HIS LONDON TRIP

Shaikh Ahmad had almost made it to London twice, once when he was invited by Edmund Janson and again during the negotiations when the Kuwait Oil Company offered to host his visit. When the India Office objected, the company had withdrawn this offer. Included in the concession under "other payments" was an amount of $18,000 "in lieu of expenses for visit to London."

After the concession was signed, the India Office thought their award of an Indian empire Knighthood (KCIE) would satisfy "Sir" Ahmad. But Ahmad was not going to be stymied again. He wanted to go to London. And he wanted to go with all the pomp and ceremony he believed to be his due.

The India Office was still decidedly cool to the idea. Perhaps they worried about what Shaikh Ahmad might learn there. Perhaps they worried about what the British press might learn about the administration of Kuwait under the political officers of the Government of India.

In her memoirs, Harold Dickson's wife, Violet, recalled that "for reasons that have never been clear to me, the British government did not view the idea favourably. Harold was instructed to do all he could to dissuade the shaikh from going." Violet thought it odd that "the authorities even suggested it would be dangerous for the shaikh to travel through Iraq since he had political differences with the government there."

"But," Violet observed shrewdly, "Shaikh Ahmad had made up his mind. As for Iraq he knew he would be treated royally there according to the Arab laws of hospitality. In due course he made the journey and all went well. It was the first time a Shaikh of Kuwait travelled to Europe. Harold was more than a little disappointed that he was not allowed to accompany the ruler and help escort him around London."

In England, although appreciating what he described as "the glorious program of receptions in my honour," Ahmad elected to spend the majority of his time with Frank Holmes. He cut quite a swathe in Essex when he visited with Frank and Dorothy at their farm.

Holmes had purchased MillHill in 1922 because, he said, he needed to provide a home for the pedigree Arab horses given him by bin Saud and for the saluki hunting dogs of the Arabian Gulf that he and Dorothy eventually bred. Dorothy ran the farm while Frank was in the Persian Gulf, also maintaining a pedigree herd of some thirty Friesian milking cows and twenty breeding stock. Despite the fabulous wealth that would be generated by the oil fields he discovered, Frank Holmes's only asset would be MillHill. The value of the entire estate left to Dorothy on Frank's death would be thirty thousand sterling.

Writing to Sir William Fraser on his return from England, Shaikh Ahmad said: "Yes, I am very happy indeed to be back home again. But I can assure you that I will never forget the happy days when my friends in London entertained me and organised such a glorious program of receptions in my honour." Considering that all the profits anticipated by the Kuwait Oil Company would come from the oil of the country he ruled, Ahmad added pointedly, almost sarcastically: "I am further glad to hear of your placing my photograph in your office."[6]

DICKSON GETS HIS JOB

Ahmad offered the job of chief local representative to Holmes's colleague, Muhammad Yateem. Perhaps remembering the trials the political officers of the Government of India had heaped on Holmes's head when he held that job for Bahrain, Yateem refused it. He returned to work with the Yateem family business, headquartered in Bahrain and successfully operating right across the Persian Gulf.

Muhammad Yateem may have been a wise man, because no sooner had the shaikh given the job to Abdulla, the son of Mullah Saleh, than the men at the Anglo-Persian Oil Company and at the India Office—ignoring the fact that in Bahrain they had replaced Holmes with an American—began to agitate against Abdulla's appointment, using the argument that the position should be held by a British subject.

At the end of 1935, when Harold Dickson retired at the Government of India–mandated age of fifty-five, he took over from Abdulla Saleh. Surprisingly, it was Frank Holmes who clinched the job for Dickson. Holmes had actually recommended Dickson in October 1934, the month both men realized the extent of what Holmes called the "perfidy" of Anglo-Persian's double-cross of Gulf Oil through the Traders attempt. At that time Holmes had written to Guy Stevens:

> I have recently felt that, if the Kuwait Oil Company could secure the good will and later the active support of Colonel Dickson, it would be a good thing. I had in mind the post of Chief Local Representative.
>
> The shaikh has the idea that the question of Chief Local Representative has loomed large in our discussion, the more so because the Anglo side of the Kuwait Oil Company wants to have as much in their hands as possible. The shaikh does not think the India Office is pushing the status and mode of appointment of the Chief Local Representative, but he is sure the Anglo-Persian side is.
>
> I cannot say there has been, in the past, complete confidence between Colonel Dickson and myself. No doubt this is due to his having to look upon me, from an official viewpoint, as an interloper and one who was not working along the lines and policy of HM Government. Therefore many of my actions, although innocent, were officially suspect. This is now all altered and we know where we stand.

The standing of Dickson with the shaikh is high and secure. It has grown perceptibly during the last twelve months. The shaikh has entire confidence and places implicit trust in Dickson. He is now rated in Kuwait as the bosom friend of the shaikh. The shaikh knows that Dickson is now his friend and can now visualise the outlook and aspirations of the shaikh, and a more important point, of the Kuwait people.

The Anglo side of the Kuwait Oil Company is suspect and the reason the shaikh is difficult about the Chief Local Representative is his fear of the treatment he is likely to receive at their hands.

Dickson told me that when he was in London he mentioned to the Gulf Oil Corporation's Major Davis that he would be retiring from the service in about 18 months time. He told Davis he would be available for a post with the Kuwait Oil Company. Dickson says he spoke to Davis because you were busy with bin Saud's London Minister, Hafiz Al Wahba. But he understands that you are the senior man.

There is no doubt in my mind that, if you could manage to arrange for Dickson to take up the Chief Local Representative job as a full time job, say from the beginning of 1936, and in the meantime let it be arranged to be carried on, temporarily, by someone suitable, the American interests will be safely guarded by Dickson and the shaikh. The shaikh would like the arrangement. He does not care for the idea of having the local manager in the dual capacity of Chief Local Representative and Manager. I have told Dickson I would write this letter to you. He wishes to write you direct. I told him I would let you know and that I was sure you would be glad to hear from him.[7]

Although repeating Dickson's own inflated version of how his appointment came about, Violet Dickson acknowledged in her memoirs Frank Holmes's role in getting the job for her husband. "With the search for oil getting underway, Shaikh Ahmad was anxious about the effects on the traditional Arab way of life," Violet wrote. "Ahmad told Harold he wanted to be sure that he had in Kuwait one good English friend whom he could trust and who would be his link with the Kuwait Oil Company. He asked the oil company to employ Harold as one who would be *persona grata* in the position of an intermediary." Violet further explained: "Frank Holmes was also in favour of Harold's appointment and his opinion carried great weight with the Kuwait Oil Company."

Although she probably knew more about the life of the people of Kuwait than did her husband, Violet faithfully parroted Dickson's own long-standing claim to fame. She continued: "Shaikh Ahmad said it would be part of Harold's job to explain to oil company personnel the manners and customs of the Arabs."

Harold Dickson died in Kuwait on June 14, 1959, aged seventy-eight. Violet stayed on. Shaikh Ahmad and his successors made sure she wanted for nothing, awarding her a state pension and meeting her expenses until her death in 1989.

NO RUSH TO KUWAIT'S OIL

When Holmes returned to London after the conclusion of the concession negotiations, he told the officials at the India Office that, despite the "unfavourable opinions that have been expressed in the past," he definitely considered there was a "good prospect of valuable oil in Kuwait."

Holmes held to this opinion as eighteen months passed before the Kuwait Oil Company proceeded to drill—in the least likely place. That the Kuwait Oil Company began drilling near Bahra, without results, could be seen as willful delay. Certainly it is otherwise difficult to comprehend just how it would take another eleven years, until 1946, before the first barrel of oil was exported from Kuwait.

It was Anglo-Persian's geologists who conducted this, their eighth survey of Kuwait, prospecting the fields and selecting the first site. The geologists included Peter T. Cox, who had delivered the seventh negative report on Kuwait in 1932, after four months including drilling. Ralph Rhoades, who confirmed Holmes's positive opinion of Bahrain in 1928, made an initial short visit as Gulf Oil's geologist. Anglo-Persian's geologists would spend all of 1936–37 looking very busy as they minutely reexplored the whole of Kuwait's territory by magnetometer, gravimeter, and seismograph.

In August 1938 Holmes was visiting New York and had a lunch date with Thomas E. Ward, the entrepreneur and broker of the Bahrain concession to the Americans. Ward recorded that "Holmes was confident about finding oil in Kuwait and said that he had not been in favour of drilling at Bahra, where the first well was abandoned at 8000 feet." Ward said Holmes "had wanted the first well to be drilled at Burgan." From his very first visit, Frank Holmes had advocated Burgan as the site where Kuwait's oil would be found.

Holmes told Ward he was "positive that the well now being drilled at Burgan will strike oil." Two months later the Burgan well struck oil. Jubilantly, Ward recorded, "Holmes was right!" It was Burgan that Ahmad's grandfather, Shaikh Mubarak, had wanted Winston Churchill's Slade Commission to see in 1913. The Kuwait Oil Company had ignored both Holmes's proven instincts and those of Shaikh Mubarak.[8]

Kuwait's Burgan field, thirty miles square and fifteen miles long, would contain one-third of the world's then-known oil supplies. The Burgan field would hold one and a half times the total oil resources possessed by the United States, until then the world's greatest producer.

FRANK HOLMES—*ABU AL NAFT*

Abu Al Naft had earned his title Father of Oil. He had proved his 1918 prediction written to his wife. There was indeed "an immense oilfield" running from Kuwait right down the coast of the Arabian mainland. After Frank Holmes's unwavering conviction that the Arab shaikhdoms of the Persian Gulf were rich in oil had been spectacularly proven correct, and severely embarrassed the experts of the day, he was asked how he had known.

Archibald Chisholm described a 1945 London meeting of oil company representatives where *Abu Al Naft* was asked how he had been so certain that the world's leading petroleum geologists were wrong in their unanimous opinion there was "no oil in Arabia." Tapping a finger to the side of his nose, Frank Holmes replied enigmatically, "This was my geologist."

According to Dorothy, while Holmes had an abiding love for the people, the deserts, and the oasis of the Arabian Peninsula where he had lived and worked since 1914, it was his New Zealand birthplace that remained his "first and last love." Dorothy said they were planning a visit the year Frank died and "he was most disappointed when he realised this could not be done."

Perhaps he would have had some reservation at being described by Chisholm as "an outstanding *British* personality" in an obituary in the *Times* of London. Eye-catching at Frank Holmes's funeral was a gigantic floral tribute sent by Ahmad. Shaikh Ahmad of Kuwait would outlive Holmes by three years. Following a short illness, he died of a heart attack in January 1950.

The obituary that appeared in the *Times* of February 5, 1947, read:

> Major Frank Holmes, who died last week, was uniquely responsible for discovering the vast petroleum resources of Arabia, whose development is of such current interest today. Though avoiding publicity in this country, he was an outstanding British personality in the Middle East especially among the Arab Shaikhs of the western shore of the Persian Gulf who appreciated both the formidable personality of this rugged New Zealander, and the great riches which the initiative of "the Father of Oil" (*Abu al Naft*), as they called him, brought to their coffers.
>
> A mining engineer by profession, Holmes staked his own opinion against that of experts in petroleum geology (who had pronounced Arabia "oil-dry") when he obtained a concession for Bahrain Island in 1923. For over five years thereafter he sought in vain to find a British or American oil concern to back his fancy and exploit the concession. Eventually he succeeded and the discovery of a major oil field in Bahrain in 1932 was rapidly followed by still greater discoveries in Saudi Arabia and Kuwait (where Holmes himself was concerned in obtaining a concession from his old friend Shaikh Ahmad al Sabah in 1934). Of powerful physique, blunt speech, and great strength of character, Frank Holmes had also those qualities of generosity, friendliness and frankness which Arabia most admires.[9]

NOTES

1. Archibald H. T. Chisholm, *The First Kuwait Oil Concession Agreement: A Record of the Negotiations, 1911–1934* (London: F. Cass, 1975), pp. 65–67, 207–10, citing Holmes to Guy Stevens, Gulf Oil/KOC, October 15 memorandum and October 16 letter and cable. Note that Chisholm devotes substantial space throughout his text to a detailed refutation of the claim of Holmes, and Dickson, that Traders was a front company for Anglo-Persian. Even if one accepts Chisholm's argument, and his case is not proven, he does not dispute that, if Traders got the concession, Anglo-Persian was to be the operating company.

2. Ibid., p. 220, November 17, 1934, and pp. 213–15, December 28, 1934, Dickson to political resident (Fowle), "While outwardly agreeing to work with them under the name Kuwait Oil Company, Anglo Persian secretly set themselves the task of hindering progress from the start with the deliberate intention of preventing the shaikh from giving the concession to Kuwait Oil" and "Chisholm had received his 'orders' from London" and "American side of the Kuwait Oil Company had a very narrow escape"; note that Yossef Bilovich also concludes in "Great Power and Corporate Rivalry in Kuwait, 1912–1934: A Study in International Politics" (PhD diss., University of London, 1982), pp. 380–82, "there is evidence of underhand dealing by the Anglo Persian Oil Company."

3. Chisholm, *First Kuwait Oil Concession Agreement*, pp. 74–76, for Guy Stevens's reaction, visit to the India Office, and subsequent pressure on the Anglo-Persian directors in the Kuwait Oil Company; p. 58, "Political Agreement was transformed into 'an exchange of letters' between the Shaikh of Kuwait and HMG, and 'endorsed' by the Kuwait Oil Company."

4. Chevron Archives, box 0824611, December 23, 1934, precis of Kuwait concession gives final terms agreed; Chevron Archives, box 0426154, Principal Documents in Regard to Kuwait Organization (1951), pt. 11, Background of the Kuwait Oil Concession, Its 50-50 Ownership by the D'Arcy Exploration Company (Anglo-Iranian Oil Company Ltd.) and Gulf Exploration Company (Gulf Oil Corporation) through the Kuwait Oil Company, and the Restrictions of the Agreement of December 14, 1933, March 24, 1934, Holmes and Chisholm to KOC London, "it was arranged for Chisholm to broach this subject . . . as you are aware Holmes in his previous negotiations had agreed to this point (director) with the shaikh"; Chisholm, *First Kuwait Oil Concession Agreement*, p. 81, "not a case of British protecting power first preventing the Shaikh reaching agreement . . ."

5. IOL/L/PS/12/102/928/29, vol. 3811, December 12, 1934, India Office files "Article 6(c) right to appoint . . . a Representative in London to represent the Shaikh in all matters relating to this Agreement with the Company in its London office . . ."; Chisholm, *First Kuwait Oil Concession Agreement*, p. 250, reprints January 5, 1935, Ahmad al Jabir al Sabah, Kuwait, to Frank Holmes in London, "As I have great confidence in your friendship, and depend upon your honesty, I have decided to appoint you my Representative in London."

6. Violet Dickson, *Forty Years in Kuwait* (London: Allen and Unwin, 1971), p. 130, "Harold was instructed to do all he could to dissuade the shaikh from going" and "even suggested it would be dangerous for him to travel through Iraq"; personal interview with Archibald Chisholm, London, September 1987, for detail Holmes's later life, his estate, and Shaikh Ahmad's 1935, see Scholefield Collection, Dorothy Holmes to Scholefield,

October 27, 1959, for detail 1922 purchase of Millhill and "pedigree Friesian herd"; Chisholm, *First Kuwait Oil Concession Agreement*, p. 249, reprints October 25, 1935, Shaikh Ahmad Al Sabah to Sir William Fraser, "glad to hear of your placing my photograph in your office."

7. Dickson Papers: "Confidential," October 24, 1934, Holmes in Kuwait to Guy Stevens, KOC, London, "cannot say there has been, in the past, complete confidence between Colonel Dickson and myself" and "if you could manage to arrange for Dickson to take up the Chief Local Representative job as a full time job" (Guy Stevens died in 1945). Dickson, *Forty Years in Kuwait*, p. 131, "Ahmad told Harold he wanted to be sure that he had in Kuwait one good English friend whom he could trust."

8. Chisholm, *First Kuwait Oil Concession Agreement*, p. 250, March 31, 1935, Anglo-Persian general manager Abadan (Elkington) to Anglo-Persian London, "geologists arrived in Kuwait on March 17, 1935, and included Cox, Shaw and Crowl. Ralph Rhoades was Gulf Oil's geologist for this survey"; Ralph Hewins, *Mr. Five Per Cent: The Story of Calouste Gulbenkian* (New York: Rinehart, 1958), p. 224, "magnetometer, gravimeter and seismograph"; IOL/R/15/1/645, vol. IX, February 11, 1935, "Confidential," India Office (Laithwaite) note of conversation with Maj. Frank Holmes "he definitely considered there was a good prospect of valuable oil in Kuwait"; and Thomas E. Ward, *Negotiations for Oil Concessions in Bahrain, El Hasa (Saudi Arabia), the Neutral Zone, Qatar and Kuwait* (New York: privately published, 1965), p. 244, "Holmes had wanted the first well to be drilled at Burgan and was positive that the well, then being drilled, would strike oil. . . . He was right!" Note that Chisholm, p. 80, claims that the Bahra site was selected from KOC surveys "and evidence derived from Holmes' 1927 water wells." Apart from his confusion about the "traces of oil" in the Bahrain wells, not those of Kuwait, Chisholm's implication is not correct. From the very beginning, Holmes had advocated Burgan as Kuwait's prime site. See also Ward, p. 231, citing KOC annual report of December 1951, "all Kuwait production has thus far come from the Burgan field"; Chisholm, p. 161, "this was my geologist."

9. Chisholm, *First Kuwait Oil Concession Agreement*, pp. 94–95, reprints this obituary.

EPILOGUE

The policy set in place at Churchill's 1921 Cairo Conference still prevailed in the Persian Gulf in September 1939 when Britain declared war on Germany; administration was in the hands of the Government of India through her political officers and policy was with the India Office, taken over from the Colonial Office six years earlier. The Arab shaikhdoms played no part in determining their own role in this conflict. They were unable to protest when Britain's fear that the Germans might strike out from North Africa for the Arabian oil fields resulted in the shutting down of the oil wells in Bahrain and Kuwait.

But the Persian Gulf of World War II was not the Persian Gulf of World War I. The region's war effort was not under the management of the Government of India that had in World War I so repressed the people of the area. Arab nationalism had bubbled to the surface. This time, Iraq, Kuwait, and Saudi Arabia made cautious advances to the Germans and Italians. Young men in Iraq and Kuwait, concluding there could be little difference between "bossy Englishmen and bossy Germans," openly gave vent to anti-British sentiments.

When the Italians bombed the Bahrain oil installations, with little success, a group of Kuwaitis were inspired to seize the Kuwait arsenal that they held for several days, meeting almost no opposition from Kuwait's officials. In April 1941 an insurrection broke out in Iraq that deposed the pro-British premier. Britain responded by bringing in the Arab Legion from Jordan—and flying in reinforcements from India. Suspicion of neutral Iran's loyalties resulted in the August 1941 Anglo-Soviet invasion of Iran.[1]

In comparison, Bahrain appeared an island of pro-British loyalty. Shaikh Hamad, whom the British had installed in 1923 when they so unceremoniously deposed his father, was, said Charles Belgrave, "more pro-British than any of the other shaikhs." Belgrave, adviser in Bahrain since 1926, reported the ruler of Bahrain "declared himself and his people wholeheartedly on the side of the Allies; he made a generous gift to war funds." Such loyalty marked out Bahrain as the center of Britain's Persian Gulf Administration after World War II, and the home of the Government of India's ideology during the war.[2]

AMERICAN CONSUL REJECTED

The Second World War did nothing to open up the thinking of the Government of India in relation to the Arab shaikhdoms. Belgrave observed that several of "the 'Politicals' who served in Bahrain during the war were badly chosen and difficult to deal with . . . so often the attitude was as though the Resident was the headmaster of a school, the Political Agent the form master, the shaikh the head boy." Belgrave understood well the Indian dimension of the British in the Persian Gulf. "Some of the Political Officers who had served in the states of Indian Princes expected the same formality as an Indian Court. I was in favour of a certain amount of state and ceremony on appropriate occasions, provided it also applied to the shaikh and not only to British officials."[3]

Maintaining hegemony and keeping out "foreigners" was still the driving force. In 1943 the Americans proposed installing an American consul at Bahrain to service Americans working in the oil companies at Bahrain, Kuwait, and Saudi Arabia. In an initial attempt to dissuade the Americans, the Foreign Office adopted a reasonable tone, explaining:

> The circumstances in Bahrain are exceptional. Like a number of other Arab Shaikhdoms in the Persian Gulf, Bahrain is a British-protected State in special Treaty relations with His Majesty's Government. In these Arab States the local Arab ruler exercises jurisdiction over his own subjects, and in some cases over other Arabs in his territory, but all other jurisdiction is exercised by His Majesty's Government through the British Political Resident and the British Political Agent. The local Arab Ruler, moreover, does not enter into relations with any foreign Power.
>
> We have hitherto not permitted any foreign Consuls or other Government's Agents to reside in these Arab States. For many reasons we should not wish to now alter the attitude which we have consistently maintained on this point. If a foreign Consul were appointed to reside in Bahrain, it would of course be necessary to make it clear that he should deal exclusively with the British authorities and not with the Arab ruler. But there would always be a tendency on the part of the Arab ruler, if a foreign Consul were to reside in his territory, to have dealings with him.
>
> His Majesty's Government would greatly prefer that there should be no change in the existing arrangement whereby no representative of other Governments reside in Bahrain. This system, we are convinced, has in the past greatly contributed to the maintenance of peace and good order. It has enabled the British representatives in this area to exercise their influence to the fullest extent. The growth of American interests in Bahrain entitles us to claim that the system has not adversely affected the United States.[4]

The idea of an American consul in Bahrain alarmed the India Office. The resident, political agents, and the Government of India were instructed to urgently forward "any considerations, or alternative views, that could be used to justify a refusal to meet the wishes of the Americans in this matter." The acting political agent in Bahrain, Capt. Michael Dixon, obligingly supplied a number of arguments while warning that, if the shaikh of Bahrain came to know of the proposal, "he would probably regard it with approval as emphasising the importance of Bahrain."

He was personally aghast as he confided to the resident, Charles Prior: "The suggestion that an American Consul should be appointed to act *jointly* with the Political Agent in hearing American cases is objectionable, from both political and legal points of view." Nobody could ever contemplate this, he said, because "the Political Agent's loss of prestige would be very considerable. In no time he could be sharing the bench with a *Persian* colleague." He couldn't see what all the fuss was about anyway. The Americans should be quite satisfied, he said, after all "an American Consul can attend judicial proceedings here, as he can elsewhere in the Empire, and can also act as a juryman or assessor."

Prior—who had maintained during the war the Government of India's habit of deporting "suspicious" persons to Bombay, including an Italian Catholic priest—was himself already in full swing. He strongly urged the India Office to "resist the US government in every way possible." He charged that the Americans had "a policy of penetration" and to allow them a consul in Bahrain "would undermine our whole position in the Persian Gulf." Besides, he sniffed, the Bahrain Oil Company was "British, not American" and therefore "bound to employ the *minimum* number of USA subjects."

The Foreign Office had claimed to the Americans that Bahrain was "a British-protected State." This might have fooled the Americans, but the British knew the difference.

Almost hysterically, Prior now declared that an American consul would result in "a ferment which would almost certainly compel us to declare a Protectorate over Bahrain." He said this would be "an act which would have very serious political repercussions in every other state in the Persian Gulf."

From New Delhi, the Government of India replied angrily that "the agreement between the oil company and HMG makes this a *British* company and its interests in Bahrain are a *British* charge." They warned shrilly: "The US government have other objectives in view beyond those which they have so far disclosed, including possible visions of oil and air hegemony along the entire length of the Arab coast."[5]

The proposal was resoundingly rebuffed. The British suggested the Americans should install their consul in Dhahran in Saudi Arabia and, in 1944, they did just that. But even Dhahran seemed too close for comfort.

Maj. Tom Hickinbotham, political agent in Bahrain, appeared unaware his acting replacement had noted that the shaikh of Bahrain would be flattered by American interest in his country. He now claimed that the prospect of visits by an American consul "considerably alarmed" the shaikh of Bahrain. He vowed the shaikh of Bahrain had confided in him that "he relies on us, in whose hands the conduct of his Foreign Affairs rests, to protect him from such penetration . . . although the United States is a great and powerful ally he does not desire that our influence in his country should be shared by anyone. . . . In spite of my assurances, he is still deeply suspicious of even private visits by diplomatic representatives of our friends, the Americans."

The official letter introducing the new American consul in Dhahran confirmed the political officers' worst fears. The letter stated that his duties would include "a visit to the Bahrain Islands, at least once a week, to perform such services as may be required." The resident was soon reporting, "Top Secret," to the India Office in London, urging that the activities of the American consul "be confined to matters strictly within his purview." If not, warned Prior, "our whole position along this coast will be rapidly undermined, more especially as he has expressed the desire to tour to Muscat, Qatar and the Trucial Coast."

With supreme disdain Prior reported the new American consul was an "inexperienced and uncouth" person—characteristics apparently illustrated by his having called on the resident "in a shirt and trousers."

He assured the India Office that "we are extremely fortunate in that this individual has no knowledge of the Arab world." Prior didn't think much of the State Department either. He added that as long as the United States "are content to employ officers with these qualifications, the danger to our interests is minimised, although an intelligent man could secure very full details of local politics from the members of the American Mission who are extremely well informed."

Nevertheless, he warned the India Office that once the Americans were "able to post officers with experience of the Middle East and a fluent knowledge of Arabic, the position will become entirely different."[6]

GULF RAJ REGROUPS ON BAHRAIN

In late 1946, preparatory to Indian independence, the establishment of the political resident in the Persian Gulf was taken from Bushire in Iran and reset in Bahrain. As it was "inappropriate to hand over responsibility for dealing with the Gulf Arabs to Indians or Pakistanis," the British government announced that it would be taking over control of the affairs of the Residency, from London. In April 1947 the Gulf rulers were informed that, owing to the constitutional changes affecting India and Pakistan, Britain would now deal with them directly.

Even after its merger into the Commonwealth Relations Office, the India Office retained responsibility until late 1948 when the Foreign Office was appointed the responsible ministry. A renewed claim to sovereignty by Iran, prompted by the transfer of the Residency, ensured the dependency of Bahrain on the "goodwill" of the British government.

Bahrain was now the repository of what remained of the Gulf Raj. The previous hierarchy was maintained with the political resident and the officers of the Indian Political Service continuing to serve in their posts until they were gradually replaced by appointees of the Foreign Office. The last member of the old Indian Political Service remained at Bahrain until his retirement in 1958. The force on which the Resident had relied for so many years, the Indian marines, became the Indian navy. Nevertheless, the resident was far from bereft.

There were usually three British frigates in the Gulf, under the command of the "Commodore Arabian Sea and Persian Gulf," and although "force was less often used," a ship of the British Royal Navy could be called on "to stand by for the protection of lives and property" or for "mediation" at the resident's discretion.

Under the new arrangement at Bahrain, the resident and the commodore lived side by side so as to achieve "the closest cooperation." Conveniently, the Royal Air Force headquarters was just a stone's throw away. The clause in the Bahrain oil concession, and in Kuwait's secret Political Agreement, decreeing that the oil companies' contact with the rulers must pass solely through the British political agent remained in force.[7]

RENEGOTIATION WITH THE OIL COMPANIES

The Persian Gulf chafed under the arrangements with the oil companies. From 1947 to 1952 (as detailed below) the American-inspired Marshall Plan depended on the supply of Persian Gulf oil for the rehabilitation of war-torn Europe. Between 1945 and 1950 Anglo-Iranian made a clear profit, after taxes, of two hundred fifty million sterling. The Iranian government received a mere ninety million in royalties. The British government received more in taxes than Iran did in royalty. Additionally, the British government made further substantial earnings through the dividends on its 51 percent shareholding, and the company sold large amounts of Iran's oil to the British navy, at a heavily discounted price.[8]

Like the Persians with the Anglo-Iranian Oil Company, the Saudi Arabians realized in 1949 that it was the American oil company that was reaping the riches from their oil. In 1949, for example, Aramco's clear profits were three times the amount the Saudi Arabian government received. Moreover, the US government was making more from the concession than was Saudi Arabia. Taxes paid by Aramco to the US government for that year were $43 million, $4 million more than the Saudis received in royalties. Using the same tactic as had the Turkish Petroleum Company in Iraq, Aramco paid no taxes in Saudi Arabia on the grounds that it was nonprofit-making because only its four American corporate owners took up its oil. A March 1949 article in *Life* magazine recorded the detail that Saudi Arabia received some $110,000 a day from its oil—while Aramco's take was one million dollars each day, possibly more, because "Aramco is secretive about its profits."

Devaluation of the Indian rupee in 1949, following India's independence, reduced Kuwait's royalty on its oil to about nine cents per barrel.[9]

The Americans blamed the oil countries' unhappiness on the United Nations. A Standard Oil of California executive explained to the company's historian that "in the beginning, doing business was about the same in each Middle East country, but after they got rich, they got difficult. Once the money had started to come in, agreements were ignored in favour of heavier and heavier demands for more money." Trouble with oil-rich countries began, he said, "about a year after the UN was formed in 1945 . . . delegates from various countries tend to gravitate towards others with common interests and this leads to a pooling of information as to how much each is receiving in the way of oil royalties and other income."[10]

The Persian Gulf was aware that Mexico had nationalized its oil industry and Venezuela had improved its receipts from the oil companies. As the Standard Oil of California executive observed, "All the Middle East countries wanted the same increased share. They all got it, too, except Iran, where the British held out—and lost."

Both the Saudis and the Kuwaitis successfully pushed for new arrangements based on the Venezuela 50-50 formula. Under this method the oil company's earnings became subject to a local income tax adjusted so that the combination of royalty payments and the income tax amounted to one-half the oil company's production profits, *before* deduction of foreign taxes.

For the year 1951, Kuwait's receipts leapt from $30 million to $140 million, the royalty going up from nine to fifty-two cents per barrel; in return the Kuwait concession was extended an additional seventeen years, making its term ninety-two years from 1934. For the same year, Saudi Arabia's collection jumped from $39 million to $110 million. Iraq achieved a similar improvement by 1952, raising its oil income from 13.5 million sterling to thirty-three million in the next twelve months.[11]

Bahrain had some success. After devaluation of the Indian rupee, renegotiation of the royalty to the Bahrain government raised this to twenty-nine cents per barrel from January 1950. Negotiations for the 50-50 agreement with the Bahrain Petroleum Company were conducted by the newly knighted Sir Charles Belgrave.

"It was difficult," he reported, "explaining to the shaikh the ramifications of American tax laws that affected the discussions."

The shaikh of Bahrain may well have been puzzled by Belgrave's insistence that American taxes be taken into account. Bapco was registered in Canada, which did not levy corporate tax, and its marketing subsidiary, Caltex, was registered in the Bahamas where US taxes were not collected. Bapco did not pay American tax. The new arrangement reached in Bahrain raised the oil income from the 1951 $4 million to $6.5 million two years later. As the oil produced by Bapco was sold to Caltex, its own subsidiary, at an artificially low price, Bahrain's take on its oil bore no relationship to market prices.

In early 1953, at the height of anti-imperialist fever in Iran, which found its target in the Anglo-Iranian Oil Company, the ruler of Bahrain asked the oil company to be more revealing of its activities and to explain to the citizens of Bahrain "the company's contribution to the country." The Bahrainis, with a documented four-thousand-year history, were somewhat surprised when Bapco responded by widely distributing a propaganda leaflet claiming the company's contribution as "the story of Bahrain."[12]

Iran chose the Mexican option. Three days after becoming prime minister of Iran, Dr. Muhammad Musadiq passed a law nationalizing the oil industry on May 1, 1951, while declaring the Iranian people were opening "a hidden treasure upon which lies a dragon." By "dragon" Musadiq meant the 51 percent British government–owned Anglo-Iranian Oil Company.

Once again Kuwait was used by Britain in argument with Iran. Sir Anthony Eden, British foreign secretary at the time, recorded in his memoirs that he told the Iranian ambassador in London: "I hoped the Iranians would have noticed our agreement with the Shaikh of Kuwait. It seemed a great misfortune that Iran was not enjoying a like increase in revenues." Eden does not say whether or not he mentioned to the ambassador the "obdurate objections" of the Anglo-Iranian Oil Company to any increase in payments to Kuwait—or the still secret Political Agreement between the Kuwait Oil Company and the British government. In Eden's view "British authority throughout the Middle East had been violently shaken" causing them to "move land forces and a cruiser to the vicinity of Abadan where the fate of the largest oil refinery in the world was at stake." Only the urgent entreaties of the United States, concerned about a possible Soviet response, prevented their immediate use.[13]

The nationalization of the Iranian oil industry focused the Persian Gulf on the realization that the reign of the Gulf Raj, enforced and policed by the Government of India, was all but over. Had the Indian army and the Indian marines still been available to back up British rule, as happened in the past, the outcome in Iran might have been very different. There were many who regretted the passing of the old regime.[14]

BETTER IN THE OLD DAYS

In 1956 rioting and rebellion broke out in Bahrain where the remains of the Gulf Raj now resided. The political uprising was aimed equally at the British political hold on the island's affairs, represented by the political resident and the British adviser, and the commercial hold of the Bahrain Petroleum Company. Belgrave, adviser from 1926 to 1957, noted in his memoirs, "When the Gulf was controlled by the India Office, and in the Residency and Political Agencies there were only a handful of British officials of the Indian Political Department and a staff of excellent hardworking Indians, the affairs of the Gulf were better managed than they are now."

Unconsciously, Belgrave echoed the Orientalist views about the "Eastern Mind" expressed in 1924 by Anglo-Persian's general manager as he dispatched agent provocateur Hajji Abdulla Williamson to Kuwait to stir up trouble for Frank Holmes. Charles Belgrave declared that affairs used to be better managed because "the British officials who were in the Gulf in those days, and in some cases their fathers before them, had spent all their working years in Eastern countries, in India, Persia and the Gulf. They knew the people and understood how to deal with them."

As adviser, Belgrave had manipulated himself into the position where he had control of both the judiciary and the police. In the 1956 rioting he deported politically troublesome Bahrainis, by British naval frigate, to imprisonment on St. Helena. Napoleon's place of banishment was chosen because the traditional destination of the Arab shaikhdoms' deportees, Bombay, was no longer an option. The 1956 deportations to St. Helena were questioned in the British parliament, which perhaps explains Belgrave's conviction that the authority exercised by the Government of India had been a good thing because "the opinions of the men on the spot carried weight and their decisions were not constantly countermanded from London."

Belgrave praised the reign of the Government of India, and its political officers, on the grounds that "the British, who controlled the Gulf, were respected . . . because they were represented in the Gulf by men who understood the Arabs." In his opinion, he said, forcibly connecting the Gulf Arabs to India had successfully ensured "they were not interested in the affairs of the Arab world." Belgrave seemed to display the same Government of India xenophobia about "foreigners" as he concluded the rot had set in when "education, travel and most of all the propaganda power of the radio exposed the Gulf Arabs to outside influences."[15]

ARAB INDEPENDENCE

Following the July 1958 revolution in Iraq that ended the British-imposed monarchy there, Kuwait was the first of the Arab shaikhdoms to declare independence from Britain, in June 1961 abrogating all previous treaties including the infamous Political Agreement. The Yemen revolution broke out in September 1962, infected Aden, and continued in armed anarchy. In 1967, the British cabinet, under a Labor government, issued its decision to quit Aden by the end of that year. The Gulf rulers were informed that, by March 1971, Britain would also withdraw from the region of the Persian Gulf. Within a month of this announcement, the shaikhdoms of the Arabian Peninsula held their first meeting aimed at achieving an independent federation.

Their aspirations were soon nipped in the bud when Britain's opposition Conservative Party announced that, if they were returned to power, they would reverse the decision to withdraw from the Persian Gulf. In the election of May 1970, the British government did indeed change, causing consternation among the Gulf rulers. Kuwait announced its "insistence" on British withdrawal. After touring the Gulf, the foreign minister of Iran declared there was concerted opposition "to a British military presence in the Persian Gulf after 1971," adding, "all the states maintain the view that Persian Gulf affairs must be handled by the countries of the region, without outside interference."[16]

The new Conservative Government appointed Sir William Luce, ex-governor of Aden and ex-resident in the Gulf, to report. In July 1967 Luce had commented in newspaper articles that Britain's "continuing interest" in the Persian Gulf included the "security of oil supplies for the industrial world."

Luce had predicted the British withdrawal would result in "the whole area rapidly becoming a jungle of smash and grab." From September 1970, however, his visit to the area led him to a different conclusion based on the realities now apparent in the region. Britain should withdraw on schedule, he recommended— and afterward station a naval force in the area.[17]

In August 1971 Bahrain ended all political and treaty relations with Britain and declared independence. Qatar followed in September and in December 1971 the old Trucial States adopted the name United Arab Emirates and were proclaimed an independent federation. The final judgment on how the shaikhdoms viewed their history of being ruled by the Government of India, and her political officers, can be gleaned from the fact that not one of the new independent states of the Persian Gulf applied to join the Commonwealth. They did, however, join the Arab League.[18]

A "RIGHT" TO ARABIA'S OIL

The advantage obtained through erasing Frank Holmes's achievements from the record of discovery and development of Arabia's oil comes into context when it is seen that both the British and the Americans, rather than admitting a purely commercial and/or military interest in control of Arabia's oil, professed to be driven by altruism.

Historically, the British based their claim to a moral right to Arabia's oil on assertions they were motivated by superior goals in that they were acting for the betterment of all humankind. The Americans based their claim on declarations that they had earned it; the Arabs owed it to them on the basis of the blood, sweat, and tears they expended to find it.

The splendid vision of lofty British goals could not be spoiled by any mention of the gunboat diplomacy, repression, intimidation, and sheer political bullying that had persuaded both Persia and Iraq to give up their oil and had prevented Bahrain, Kuwait, and Saudi Arabia benefiting from theirs. Nor would Britain ever want to be reminded that she had missed out on the great oil fields of Arabia because the Government of India had been allowed a free hand in the Arab shaikhdoms; a free hand that was used to persecute what they saw as a colonial interloper, geologist extraordinaire Frank Holmes.

The Americans would not wish to mar their created mythology in which virtuous, hardworking American pioneers self-sacrificingly toiled in the heat and sand of the deserts, on behalf of the Arabs, until they discovered oil. As a Pulitzer Prize–winning American journalist gushed in 1971, "the men of Aramco transformed Arabia" and what they did for the Saudi Arabians should "go on record as one of the outstanding jobs . . . in the history of the world."

Frank Holmes could have no place in folklore such as this. The fact that it was Holmes who identified and mapped the fields, and held the original concessions that the Americans later purchased—and where they found oil—was highly inconvenient. The record of Holmes's achievements had to go in order to make way for the American claim to have earned a right to Arabia's oil.[19]

The original concessions, fondly known as the Colonial Office model, gave companies the right to explore for, own, and produce oil in a given territory. The oil did not belong to the country in which it was found, but belonged by "right" to those who extracted it. The principle reverberates in the American claim to have earned a "right" to Arabia's oil by dint of having brought it out of the ground.

This concept would not be overturned until the Organization of Petroleum Exporting Countries (OPEC), formed in 1960, succeeded in gaining international acceptance of the idea of national ownership of natural resources. The authors of an influential 1978 publication *The Pressures of Oil: A Strategy for Economic Revival* observed the argument that surrounded passage of the relevant resolutions through the United Nations. They noted that many industrialized nations supported the view that "such sovereignty was by no means absolute, in order, of course, to allow for the possibility of foreign companies' rights over resources they discovered."[20]

There was little difference between the British and the Americans in their initial approach to Arabia's oil. Neither offered participation in the hugely exorbitant profits being made. Both manipulated production and supply in order to maintain the price of oil, though without extending any financial benefit to the countries in which the oil originated.

This united view of the West's right to Arabia's oil is apparent in the tenets of the 1947–52 European Recovery Plan. Known after its originator as the Marshall Plan, this scheme aimed at rehabilitating a Europe devastated by World War II—a war in which the Arabs played no part—through application of American aid, and "a lavish and continuous supply of Middle Eastern oil."

In implementing the Marshall Plan, the price of Arab oil was pushed below that of American oil, resulting in "a great surge of cheap . . . Middle East oil." As a US government report of the time observed, "without the petroleum, the Marshall Plan could not have functioned." Here indeed was Arabia's oil being utilized for the betterment of civilization, but only that half of civilization that existed in the (non-Communist) industrialized world.[21]

Alignment of thinking, and action, was again clear when investigations in 1951 reported that the international oil companies combined to set prices. The price-fixing and exploitative actions revealed involved five American companies—Standard Oil of California, Texas Oil, Standard Oil of New Jersey, Socony Vacuum, and Gulf Oil—together with British Petroleum and the Dutch Shell groups. The tactics of juggling quotas and production restrictions, first mooted by Sir John Cadman in 1933 and perfected by the international petroleum cartel that became known as the Seven Sisters, was the model on which OPEC would later draw.[22]

Reaction was aggressively hostile, however, when it was mainly the Arabs implementing Cadman's cartel principle, in the form of OPEC, rather than the Anglo-American Seven Sisters. *The Pressures of Oil* revealed the united, and concerted, attempts to "break up" OPEC conducted by the members of the Organization for Economic Cooperation and Development (OECD). "In the view of the United States and its allies," the authors stated, "OPEC has neither the right to exist nor even a right to insist on charging a 'fair' price for its oil."[23]

The legacy of the early British and American claims is still apparent today. The industrialized world appears to believe it has a right to plenty of Arabia's oil, preferably at a price set by the buyer not the seller (and a right to make a profit from it as every government does with the imposition of heavy domestic petroleum taxes).

Echoes of the original American justification of the "right" to cheap Arab oil can still be heard when the Arabs are depicted as ungrateful—"after all we've done for them"—and the original British justification of their "right" to cheap oil is heard when the Arabs are accused of pushing up the price of oil, and so threatening the good of humankind and the very fabric of civilization. Both justifications are as spurious today as they were then.

THE BRITISH CLAIM EXAMINED

The best known of the Government of India's political residents, Sir Percy Cox, once declared that "the Arab potentates" in the Persian Gulf were totally "the creation of Great Britain." And the last of the Government of India's political residents, Sir Rupert Hay, who retired in 1958, spoke of "the partnership between the Arabs of the Gulf and the British, that has been of most benefit to the Arabs, who have not only preserved their independence but acquired wealth through the development of their oil resources . . . which only the Pax Britannica has made possible."

The first chairman of the Anglo-Persian Oil Company, Sir Charles Greenway, in a 1916 tribute to Sir Boverton Redwood, Britain's éminence grise of petroleum, said he deserved the gratitude of the nation because of his role in the acquisition, exploration, and exploitation of the oil fields of Persia. Redwood was a true patriot, the chairman said, because he had worked "to secure for the British nation that share in the oil industry to which we are entitled—by virtue of our enormous consumption."

The British were not shy about what they thought they could do with oil. In an article curiously titled "The Romance of Persia's Oil," a British expert pointed out in 1920 that because of its indispensable value in warfare "the nation which controls the largest oil supplies could in the end dictate terms to the rest of the world."

Anglo-Persian's second chairman, Sir John Cadman, addressing the American Petroleum Institute in 1921, referred to "those high qualities that are preserved by our two races"; he meant the British and the Americans. Speaking of the need for "cheap power produced from cheap petroleum," he said cheap oil must be available "in the interests of civilisation and for the good of mankind."[24]

The early British view of Arabia's oil was clear. First, the countries of the Persian Gulf where the oil lay were "the creation of Great Britain." Consequently, the creator had a prior right, as every parent had. Second, Britain was entitled to the oil because she used a lot of it. The third justification was that the oil should be given without restriction and cheaply to Britain because she would use it for the good of humankind and to preserve civilization as we know it.

The philanthropic slant on Britain's actions in the matter of oil was still being propagated in 1951 when Iran nationalized its oil industry. When Anthony Eden drew up four "minimum" requirements for resolving this dispute, and forwarded them to the American mediators, he claimed to be acting to protect "all countries who have similar interests in foreign countries" as well as the Iranian people. Eden had experience in managing Britain's oil relationships. He was undersecretary at the Foreign Office in 1933 when the Persian government denounced the Anglo-Persian Oil Company's concession. In his memoirs, Eden claimed it was he who advised an immediate appeal to the League of Nations and later took part in "the negotiations that followed culminating in a new agreement between the company and the Government of Persia . . . Sir John Cadman represented the company."[25]

The British claim to a moral right to Arabia's oil—on the basis of a superior intention to use it to make the world a better place—was very much to the fore following the oil embargo imposed by the Arabs in response to the West's support of Israel in the 1973 war.

British writer Leonard Mosley, in the viciously anti-Arab 1974 bestseller *Power Play: Oil in the Middle East*, advised of "the need" to place the Arab's oil resources under "some sort of international control." In return, Mosley said, the Arabs should be given "raised standards of living and freedom for *all* [*sic*] their people, which will turn them away from war, militancy and blackmail." An influential work, on many university reading lists for a decade or more, was J. B. Kelly's 1980 *Arabia, the Gulf and the West: A Critical View of the Arabs and Their Oil Policy.* Kelly advocated everything from neo-imperialism to Western military occupation of the Arabian oil fields, justifiable, he said, under something he called "the doctrine of necessity in international law."

The Persian Gulf and its oil fields "are one of the great strategic prizes in the world," Kelly declared, and could not be left in charge of the Arabs because "no Middle Eastern state, least of all any of those bordering on the Persian Gulf, is the peer of any of the major powers of Europe." Describing Arab oil policy as "the tactics of larceny and intimidation practised by the Middle Eastern oil producing states since 1970," Kelly invoked the "moral right" to Arabia's oil. Even though "desperate measures may be required to retain control over the Gulf's oil for the West," he advised, this was preferable to "wasting effort" endeavouring to "cajole" the "regimes of the Persian Gulf into acting with a sense of responsibility to the world at large."[26]

THE AMERICAN CLAIM EXAMINED

The seeds of the mythology and folklore claiming an American right to Arabia's oil on the basis of being the area's "oil pioneers" were sown in appearances before some twenty congressional investigations into the petroleum industry. When the Americans moved into Arabia's oil they were ideologically unable to expound a purely imperial rationale of their right to Arabia's oil—in fact they repeatedly propagated the opposite. They claimed the Arabs welcomed them with open arms because of their republican status, because they were not imperialists.[27]

It is true that Abdul Aziz bin Saud never forgave the Anglo-Persian Oil Company for the personal slights of 1922–23. The 1933 concession signed with Standard Oil of California did contain a clause prohibiting the transfer of rights and obligations, without the written consent of the Saudi Arabian government. And certainly in originally granting this same concession to Frank Holmes in 1923, bin Saud had required Holmes to sign a document specifically excluding Anglo-Persian from his territory. In their mythology, however, the Americans extended bin Saud's specific objections to the Anglo-Persian Oil Company into a mistrust of the British in general.[28]

The American claim that bin Saud preferred Americans because they were republicans was prominent in 1939 after Standard Oil of California obtained an expansion of the original Al Hasa concession. A *New York Times* article quoted company executives declaring that the king of Saudi Arabia had personally told them no European or Asian power could be trusted because they all had "political motives."

According to Standard Oil of California, Abdul Aziz bin Saud said he "preferred to give all rights in the entire kingdom to us—for much less than he could have had from others—because . . . the United States has no political designs on his country." Immediately after the king had signed this new concession, the United States accredited the first US minister to Saudi Arabia.[29]

That Standard Oil of California and the Texas Oil Company, early partners in the American Arabian Oil Company (Aramco), did not actually believe their own propaganda claiming bin Saud preferred them above all others becomes apparent in the events surrounding the involvement of the US government in advancing finances to bin Saud during World War II.

In 1940, when the war curtailed production at Aramco and prevented travelers attending the annual pilgrimage at Mecca, bin Saud requested financial assistance from the American oil company and, to the company's chagrin, from the British government. The company did make some advances against future royalties, secured by a two-year extension of the concession period. But they were aware the British government was granting subsidies to the Arab shaikhs, and to bin Saud, that were not in the form of loans and did not have to be returned. The Americans at Aramco were concerned that bin Saud might repay his British benefactors by bestowing on them the only thing he had of value, an oil concession in his territory. They worried that while bin Saud might not cancel their concession outright, he might reduce it and transfer part of it to the British.

In January 1941 Aramco set out to counter British influence by promising bin Saud a loan of $6 million. Two months later, in March 1941, when the US government promulgated the Lend Lease Act, a program of military and economic aid to nations warring against the Axis powers, Aramco saw a means by which it could cast its own US government as bin Saud's savior, and be relieved of making the advance from its own funds.

In appealing to the US government to make this transaction under Lend Lease, the company initially played down its fear that the concession might be canceled if bin Saud did not get the loan, or that the British might get part, or all, of the concession. Aramco professed to be humanistically concerned for the region as it warned dramatically, "We believe that unless this is done, and soon, this independent kingdom—and perhaps with it the entire Arab world—will be thrown into chaos."[30]

Response to Aramco's request was lukewarm. It was suggested that "some of it" might be done under Lend Lease by shipping food aid directly to Saudi Arabia; this suggestion was accompanied by the comment "although just how we could call that outfit a democracy I don't know."

In July 1941, when a $425-million loan was approved to the British government, President Franklin Roosevelt requested the administrator of the Federal Loan Agency to "tell the British I hope they can take care of the King of Saudi Arabia. This is a little far afield for us!"[31]

Aramco was relieved they had avoided financial responsibility for bin Saud. The British, drawing on the American loan, would take over. But they had not succeeded in their parallel aim of lessening British influence and increasing American power in the area. And, despite their claims since gaining the concession in 1933 that bin Saud would deal only with Americans, they were anxious the British might gain ground. Appearing before the Senate Committee Investigating the National Defense Program (Part 41 Petroleum Arrangements with Saudi Arabia), Aramco executives claimed the British were "increasing their influence with Saudi Arabia at American expense."[32]

In a rare admission, in light of past (and future) propaganda, the chairman of Aramco told the committee, "We have been afraid all the time over here of the encroachment of the British into the oil picture of Saudi Arabia." In December 1942 the president of Standard Oil of California and the chairman of the Texas Oil Company, parents of Aramco, told the US Secretary of the Interior, now acting as Petroleum Administrator for War, they feared the American concession would be canceled and given over to British interests.

Their unrelenting three-year lobbying effort for Saudi Arabia to be granted status under Lend Lease finally succeeded. They were instrumental in the opening of America's first legation, in Jedda, in May 1942. And on February 18, 1943, President Roosevelt declared "to enable you to arrange Lend Lease aid to the Government of Saudi Arabia, I hereby find that the defence of Saudi Arabia is vital to the defence of the United States."

Standard Oil of California and the Texas Oil Company had achieved their goal; they had committed the United States to the protection of the American concession in Saudi Arabia and removed the possibility of British encroachment.[33]

There is a postscript to this episode. The two oil company CEOs had deeply impressed the US Secretary of the Interior with their warning of the danger to the American national welfare if Saudi Arabian oil was lost to the British. The secretary was convinced America was running out of oil.

The argument of the two CEOs had so swayed the Secretary of the Interior that he concluded that if Saudi oil was so plentiful, and so important for the American national welfare, then the only guaranteed protection must be for the US government itself to take over the Saudi Arabian concession.

He recommended to the US president that the Petroleum Reserves Corporation be organized to acquire and participate in the development of foreign oil reserves. The first order of business, the Secretary of the Interior said, should be "the acquisition of a participating and managerial interest in the crude oil concession now held in Saudi Arabia by an American company." He added that this move "will also serve to counteract certain known activities of a foreign power which presently are jeopardising American interests in Arabian oil reserves."

The proposal was put to Standard Oil of California and Texas Oil. The CEOs of both companies were shocked to realize just how persuasive their warnings and arguments had been. They could see that the enormous wealth-producing concession they had fought so hard to protect from the British was about to slip from their grasp, hijacked by their own government. Negotiations continued from August to October 1943 with the government dropping its demand from 100 percent ownership to 70 percent, to 51 percent, and the company holding out for 33 percent. Negotiations were broken off after the tide of war turned to the Allies' favor in North Africa, making the Middle East look secure again.[34]

After World War II Standard Oil New Jersey and Socony (the old Standard Oil New York) joined Standard Oil of California and the Texas Oil Company as equal partners in the Saudi Arabian concession. This took place against a background of renewed claims that Abdul Aziz bin Saud would deal only with Americans.

The US Secretary to the Navy confidently declared that Abdul Aziz bin Saud did not "care which American company or companies developed the Arabian reserves" so long as they were "American." And after Aramco executives put the merger proposal to bin Saud they reported to the American public that the king of Saudi Arabia was "interested in only one point and on that he was insistent; he wanted to be certain neither Jersey nor Socony were 'British controlled.'"[35]

Seemingly forgetting their panic of 1940–43 that bin Saud might redistribute the American concession to the British, the oil company cast itself as supercitizens, implying that it was through its efforts that the US was respected and admired on this foreign shore.

This story was most often spun about the countries of the Middle East where the oil companies claimed to have won over the Arab rulers from a long association with the British. And, as has been seen, the State Department was not averse to claiming credit for itself for the American presence in Arabia. That even American officials might not have actually believed the propaganda about Saudi Arabia preferring them "above all others" was again shown by the comment of the Assistant Secretary of State appearing before a Senate hearing in 1974. The 1951 "50-50" arrangement he had helped negotiate in Saudi Arabia was necessary, he said, because "the threat was the loss of the concession."[36]

Those who worked in the Arabian Peninsula, and did believe the mythology, were sometimes pulled up short. Saudi Arabia took a 60 percent share in Aramco in 1974, obtaining the remaining 40 percent after a further eighteen months negotiating. In 1980 Saudi Arabia paid compensation, based on net book value of all their holdings within the kingdom, to the four American ex-owners of Aramco. After almost half a century, Saudi Arabia reversed the situation by informing Aramco it could continue to be the operator and service provider—"for which it would receive 21 cents per barrel."[37]

The Kuwait government bought 60 percent of the Kuwait Oil Company in 1974, moving the next year to obtain the final 40 percent. The joint founders, Gulf Oil Corporation and British Petroleum (the renamed Anglo-Iranian) demanded $2 billion in compensation. Kuwait gave them $50 million. Even so, the two companies assumed they would still have preferential access to Kuwait's oil. Gulf Oil sent one of its best men. He was shocked to learn not only would there be no preferential access, but that Kuwait now intended to sell directly into Gulf Oil's own markets, including Japan and Korea.

Apparently believing the mythology, the American from Gulf Oil expected he could soften Kuwait's position by running through "the history . . . of all that Gulf Oil had done for Kuwait." After listening to this, the Kuwait official became very angry. Announcing "you never did us any favours," he walked out of the room.

The American could not grasp that it was his illusory picture of the discovery and development of oil in Arabia that had generated the offense. After reflecting on the episode, this Gulf Oil executive said, he thought maybe the misunderstanding could have been caused by "the conceit of the Americans that we were loved—because we had done so much for these people."[38]

NOTES

1. Benjamin Shwadran, *The Middle East: Oil and the Great Powers* (London: Atlantic, 1956), p. 388, in July 1942 Kuwait operations were suspended by the British military as a war measure and all the wells were shut in; Angela Clarke, *Bahrain Oil and Development, 1929–1989* (London: Immel Publishing, 1991), pp. 174–76, for Britain's "oil denial" policy that included Kuwait, Bahrain, and Saudi Arabia and pp. 165–69, for Italian air raid; Robert Lacey, *The Kingdom* (London: Hutchinson, 1981), pp. 256–58, for bin Saud's flirtation with Germany; Ralph Hewins, *Mr. Five Per Cent: the Story of Calouste Gulbenkian* (New York: Rinehart, 1958), pp. 225–27, "bossy Englishmen and bossy Germans," "seized the Kuwait arsenal," and "reinforcements from India"; Shwadran, p. 60, "combined Russian-British force."

2. Charles Belgrave, *Personal Column* (Beirut: Librarie du Liban, 1960), pp. 120–24, "more pro-British"; Clarke, *Bahrain Oil and Development*, p. 167, goes further stating, "Britain declared war on Germany. Next day, the Ruler of Bahrain declared war on the Axis Powers"; note that Shaikh Hamad died in February 1942, was succeeded by his son, Sulman, who died in 1961. Sulman was succeeded by his son, Isa (1933–1999).

3. Belgrave, *Personal Column*, p. 122, "difficult to deal with" p. 125, "of an Indian Court."

4. IOL/R/15/2/854, vol. 36/2, March 18, 1943, Foreign Office to United States Embassy London, "greatly prefer no change."

5. IOL/R/15/2/854, vol. 36/2, April 21, 1943, India Office London to resident, political agent Bahrain, Government of India, New Delhi, "to have your views . . . or any alternative suggestions re proposed USA Consulate . . . to justify a refusal"; and also April 21, 1943, acting political agent Bahrain (Captain Michael Dixon) to resident, "loss of prestige . . . in no time he could be sharing the bench with a Persian colleague"; and April 27, 1943, resident (Prior) to India Office London cc Government of India New Delhi, "resist in every way possible"; and May 6, 1943, Government of India New Delhi to India Office London cc resident, "other objectives"; See Clarke, *Bahrain Oil and Development*, p. 169, for deportation of Italian Catholic priest.

6. IOL/R/15/2/854, vol. 36/2, August 30, 1944, secret, political agent Bahrain (Maj. Tom Hickinbotham) to resident, "considerably alarmed"; IOL/R/15/2/854, vol. 36/2, September 3, 1944, American Consulate Dhahran, vice consul (in charge) Parker T. Hart, to His Britannic Majesty's political agent Bahrain, "at least once a week"; November 22, 1944, resident (Prior), top secret, to India Office London, "inexperienced and uncouth."

7. Sir Rupert Hay, *The Persian Gulf States* (Washington, DC: Middle East Institute, 1959), p. 18, "inappropriate" and p. 26, "force less often used."

8. Daniel Yergin, *The Prize: Epic Quest for Oil, Money, and Power* (New York: Simon & Schuster, 1991), pp. 451–52, "Anglo-Iranian profit 250 million."

9. Ibid., p. 445, "Aramco's profits were three times"; Anonymous, "Aramco: An Arabian-American Partnership Develops Desert Oil and Places US Influence and Power in the Middle East," *Life*, March 28, 1949, pp. 62–64, 66–78, "Aramco is secretive about its profits"; Shwadran, *The Middle East*, p. 390, "nine cents per barrel."

10. Chevron Archives, box 120797, April 17, 1958, in-house interview with J. H. MacGaregill, headed Middle East Problems, "leads to a pooling of information"; Clarke, *Bahrain Oil and Development*, p. 191, "June 1945 delegates from 50 countries signed in San Francisco the World Security Charter establishing an international peacekeeping body and forum to be called the United Nations Organisation, Egypt, Iraq and Saudi Arabia were immediate members."

11. Chevron Archives, box 120797, April 17, 1958, in-house interview with J. H. MacGaregill, "they all got it too"; Shwadran, *The Middle East*, p. 391, table, Direct Oil Payments to Kuwait government; Yergin, *The Prize*, pp. 446–47, Saudi Arabia/Aramco December 1950 agreement, "the heart of which was the Venezuelan 50-50 principle"; Shwadran, p. 260, Iraq's new agreement was signed in February 1952, p. 271, in May 1950 Iraq had created the Development Board, assigning 70 percent of oil revenues.

12. Shwadran, *The Middle East*, p. 376, "29 cents per barrel"; Belgrave, *Personal Column*, p. 179, "ramifications of American tax laws." Note in Shwadran, p. 382, that Bapco was not liable for any tax, Canada levied tax only on dividends and Bapco never declared dividends; Shwadran, p. 376, table, Royalty Payments to Bahrain; Clarke, *Bahrain Oil and Development*, p. 209, "the story of Bahrain." Much is written on Bahrain's history, see, for example, Geoffrey Bibby, *Looking for Dilmun* (London: Collins, 1970); Belgrave, *Personal Column*, p. 213, after helping to put down the 1956 Bahrain rioting, much of it targeting the Bahrain Petroleum Company, Belgrave sailed with his wife in June 1957, "on the *Queen Elizabeth* as guests of the Standard Oil Company of California" for a month touring the United States, on p. 217 Belgrave notes his son, James, was employed in the Public Relations Department of the Bahrain Petroleum Company from 1955 to 1957; see Shwadran, pp. 382–83, for a discussion of the "exorbitant profits" gleaned by Bapco; "the Bahrainis received only a very small portion of the income from oil, compared with the company's profits."

13. Sir Anthony Eden, *The Memoirs of Sir Anthony Eden: Full Circle* (London: Cassell, 1960), p. 194, "dragon," p. 204, "would have noticed our agreement"; Yergin, *The Prize*, p. 448, "obdurate objections"; Eden, pp. 194–98, "moved land forces and a cruiser." For the Iranian nationalization of oil see also Shwadran, *The Middle East*, pp. 103–93, and Yergin, pp. 450–78.

14. Iran paid a heavy toll for nationalizing its oil industry. The nationalization crisis of 1951–53 resulted in the Anglo-American engineered-overthrow of the elected reformist

government of Musadiq, and the drift of Iran under the shah into the American orbit, with consequences for the entire region. See Amin Saikal, *The Rise and Fall of the Shah, 1941–1979* (London: Angus and Robertson, 1980).

15. Belgrave, *Personal Column*, p. 223, "the affairs of the Gulf were better managed," p. 233. Questions were raised in the British House of Commons about the legality of the procedure by which three politically activist Arabs, subjects of the shaikh of Bahrain, were removed to and imprisoned on St. Helena, a British colony; p. 236, "British were respected," p. 237, "outside influences." See Muhammad G. Rumaihi, *Beyond Oil: Unity and Development in the Gulf* (London: Al Saqi Books, 1986), in which the author argues that artificial attachment to India, with its accompanying imposed isolation, "distorted the Arab character of the Gulf."

16. The new Republic of Iraq made a claim on Kuwait, stating it had traditionally been within the *vilayet* of Basra, and if the British were going to give it up, then Iraq wanted it back; William Harold Ingrams, *The Yemen, Imams, Rulers and Revolutions* (London: John Murray, 1963), p. 129, "the Yemen revolution started during the evening of September 26, 1962"; J. B. Kelly, *Arabia, the Gulf and the West* (London: Weidenfeld and Nicolson, 1980), p. 47, "immediately quit Aden"; Kelly, pp. 57–60, 78–81, for Conservative Party announcements, actions, and policies, p. 80, "Kuwait insistence" and "without outside interference"; Rumaihi, *Beyond Oil*, pp. 57–65, the first meeting of the Gulf states to discuss federation was held in Dubai, February 25–27, 1968, just one month after Britain's announcement of withdrawal.

17. Kelly, *Arabia, the Gulf and the West*, p. 81, "smash and grab," p. 82, "withdraw on schedule"; Rumaihi, *Beyond Oil*, p. 59, the shah of Iran in July 1968 declared Bahrain's stated intention to join the union was a provocative act to which the Iranian government would respond; the shaikh of Bahrain lost his nerve saying, according to Kelly, p. 92, "Britain is weak now where she was once so strong. You know we and everybody in the Gulf would have welcomed her staying"; Rumaihi, p. 59, a UN commission of inquiry recommended Bahrain be recognized as an Arab country whose people desire independence, the Security Council resolution was dated May 11, 1970; Rumaihi comments, "the recognition of a more prominent Iranian role in the Gulf was formalised in exchange for Iran's renunciation of its claim to Bahrain."

18. Middle East Economic Digest (MEED) 1980's "Practical Guide" series on Kuwait, Bahrain, Qatar, Saudi Arabia, Oman, and the UAE for independence details; Rumaihi, *Beyond Oil*, p. 62, UAE constituent emirates were Abu Dhabi, Dubai, Sharjah, Umm Al Qaiwan, Ajman, and Fujairah; Ras Al Khaimah became the seventh member in 1972; Lawrence James, *The Rise and Fall of the British Empire* (London: Little, Brown, 1994), pp. 559–87, covers Suez, Aden, Iran, and the withdrawal from the Gulf, p. 586, Lawrence comments, "unlike Africa, or India, Arabia and the Gulf had never felt . . . Britain's 'civilising' mission." Note that in Oman in July 1970, the British engineered a coup in which Sultan Qaboos succeeded his father, Sultan Said bin Taimour; on his succession Qaboos did not abrogate the many treaties and alliances with Britain; See Rumaihi, *Beyond Oil*, p. 53, quoting Sir Geoffrey Arthur, the last political resident in the Gulf, stating in a 1973 address given at Durham University, "when Britain attempted to look up all the treaties and alliances binding it to the Gulf Amirates, it proved impossible to produce a comprehensive list. Therefore, when Britain undertook to agree to independence, it made do with the stipulation that all prior treaties were to be annulled, without listing them."

19. Wallace Stegner, *Discovery! The Search for Arabian Oil*, abridged for *Aramco World* (Beirut, Lebanon: Middle East Export Press, 1971), p. xii, "in the history of the world."

20. OPEC was formed in 1960 with twelve members: Algeria, Gabon, Indonesia, Iran, Iraq, Kuwait, Libya, Nigeria, Qatar, Saudi Arabia, United Arab Emirates, and Venezuela (Ecuador joined in 1973 and left in 1992); Peter R. Odell and Luis Vallenilla, *The Pressures of Oil: A Strategy for Economic Revival* (New York: Harper and Row, 1978), pp. 63–64, "to allow for"; United Nations resolutions concerning permanent sovereignty over natural resources are mainly contained in General Assembly resolutions 3201 (S-VI) and 3202 (S-VI) of May 1, 1974, containing the Declaration and the Programme of Action on the Establishment of a New International Economic Order and 3281 (XXIX) of December 12, 1974, containing the Charter of Economic Rights and Duties of States. The resolutions agreed that countries possessed undisputed rights to resources within their own borders.

21. The plan was conceived by US secretary of state George Marshall and announced in June 1947; Elizabeth Monroe, *Britain's Moment in the Middle East, 1914–1971* (London: Chatto and Windus, 1981), pp. 95, 113, "the Marshall Plan depended on a lavish and continuous supply of Middle Eastern oil"; Yergin, *The Prize*, pp. 423–26, "Middle Eastern oil was being pushed down to price levels below what had until then been the benchmark US Gulf Coast price" and "great surge of cheap production from the Middle East," on p. 425 Yergin comments, "in 1946, 77% of Europe's oil supply came from the Western Hemisphere . . . by 1951 it was expected 80% of supply would come from the Middle East." Note that Europe was the major market for American goods; without a prosperous Europe, the United States faced the prospect of severe economic depression. Note also that sixteen European countries founded the Organization for European Economic Cooperation (OEEC) in April 1948 to administer the Marshall Plan (the OEEC was superseded by the OECD in 1960). However, the largest amount of the $13 billion of American aid went to Britain, France, Italy, and West Germany, in that order. After 1949 as Cold War tension rose, the funds increasingly went into military expenditure rather than industrial rebuilding.

22. Anthony Sampson, *The Seven Sisters: The Great Oil Companies and the World They Made* (Hodder and Stoughton, 1975), p. 137–40, the US Federal Trade Commission investigated the foreign agreements of the oil companies, the report released in August 1951 charged the seven companies controlled all the principal oil-producing areas outside the United States; all foreign refineries, patents, and refining technology; that they divided the world markets between them, and shared pipelines and tankers throughout the world—and that they maintained artificially high prices for oil. The report said the companies were "engaged in a criminal conspiracy for the purpose of predatory exploitation."

23. When OPEC was formed in 1960, seven of the twelve founding members were Arab countries, plus non-Arab Iran. The nations that signed the Convention of the OECD in December 1960 were Austria, Belgium, Canada, Denmark, France, Great Britain, Greece, Iceland, Ireland, Italy, Luxembourg, the Netherlands, Norway, Portugal, Spain, Sweden, Switzerland, Turkey, the United States, and West Germany. Japan, Finland, Australia, and New Zealand joined later; Odell and Vallenilla, *The Pressures of Oil*, pp. 62–65, "neither the 'right' to exist . . . nor . . . charge a 'fair' price."

24. PRO/CO/730/26, December 20, 1922, secret, Cox to duke of Devonshire, Colonial Office, "creation of Great Britain"; Hay, *Persian Gulf States*, p. 18, "most benefit to the Arabs"; Owen Papers, box 1a, April 10, 1916, typed manuscript, "Tribute to Sir

Boverton Redwood" by Charles Greenway, and John K. Barnes, "The Romance of Persian Oil," *The World's Work* (n.p.: June 1920): 143–52, "dictate terms to the rest of the world" and "As John Bull Views It," an address by Sir John Cadman, American Petroleum Institute, December 8, 1921, "interests of civilization."

25. Eden, *Memoirs of Sir Anthony Eden*, p. 199, four "minimum" requirements, pp. 191–92, "took part in."

26. Leonard Mosley, *Power Play: Oil in the Middle East* (New York: Random House, 1973), p. 427, "international control." Mosley's work is studded with vindictive allegations about the personal behaviour and character of the Gulf Arabs, particularly the shaikhs, for which he provides neither source nor substantiation. There are many factual errors; Kelly, *Arabia, the Gulf and the West*, p. 499, "doctrine of necessity in international law," p. 500, "the major powers of Europe," pp. 502–504, "larceny and intimidation" and "to the world at large."

27. Yergin, *The Prize*, p. 409, "twenty Congressional hearings"; curiously, the *Life* magazine article of March 28, 1949, pp. 62–78, refers throughout to "American colonisers" and says, "like most other colonists Aramco's Americans spend their private lives apart from the natives" while also claiming "Aramco's bosses have tried to avoid the odium of old style colonialism. They want Arabs to like Americans . . . Aramco minimum pay to its Arabs is 90 cents a day, skilled pay $270 a month, this means unprecedented fortune . . . skilled Arab employees do not often attain top rank . . . Aramco has even succeeded in tempering harsh Saudi Arabian justice with mercy . . . a doctor from Aramco now attends to paint the culprit's forearm with iodine before the hand is chopped then bandages the stump . . . Aramco . . . will soon have to meet one of the chronic social questions of any venture in internationalism; whether or not to mix Arab and American students when it starts a high school, Aramco, which wants to show enlightenment, has not yet decided."

28. Abu Dhabi Documentation Center, Copy of Records of the US State Department Relating to Internal Affairs of Saudi Arabia 1930–44, Film-T1179/2, August 6, 1930, from US consul Alexander Sloan, Baghdad, to Department of State, "King bin Saud will not permit Anglo-Persian to obtain any concession within the boundaries of Nejd. Furthermore (Bill Taylor, Standard Oil of California geologist visiting Bahrain) was informed the King had stated that if he gave an oil concession to any company and later found that company was affiliated in any way with APOC or had sold out to any company affiliated with APOC he would immediately cancel that concession." Shwadran, *The Middle East*, p. 293, "American concession . . . contained a clause."

29. Joseph M. Levy, "US Company Wins Arabian Oil Grant," *New York Times*, August 8, 1939, p. 1, "this Arab potentate mistrusts all the European and Far East Powers"; Shwadran, *The Middle East*, p. 296, the 1939 agreement, known as the "supplemental agreement" covered the neutral zones between Saudi Arabia and Kuwait and between Saudi Arabia and Iraq, for which the company made a down payment of $1.5 million with annual rent at $750,000. Compared to the terms of the company's 1933 purchase, this can hardly be construed as "much less" than bin Saud may have been able to get from "others"; *Fortune*, May 1947, claimed, "The best ambassadors the US has in the Middle East can be the American oil companies. . . . The west is still on trial in the Middle East among peoples under constant pressure to choose between vying ways of life. Will it be poverty, chaos, and then the heel of total authority? The Communists can offer words and propaganda. The oilman can offer a living example of what freedom and enterprise

can do"; the American minister in Cairo was instructed to add Saudi Arabia to his portfolio in June 1939.

30. The Lend Lease deal with Saudi Arabia is covered in Shwadran, *The Middle East*, pp. 301–307, and Yergin, *The Prize*, pp. 393–99. Note that the Lend Lease Act of 1941 empowered President Franklin Roosevelt on behalf "of any country whose defence the President deems vital to the defence of the United States, to sell, transfer title to, exchange, lease, lend, or otherwise dispose of, to any such government any defence article not expressly prohibited." This included cash amounts. When the war ended in August 1945, Lend Lease appropriations totaled about $48 billion, repayment to the United States by all countries was virtually complete by 1960; Shwadran, p. 304, "thrown into chaos."

31. Shwadran, *The Middle East*, p. 305, "call that outfit a democracy," p. 306, "take care of the King."

32. Ibid., p. 307, "at American expense."

33. Ibid., p. 307, "encroachment of the British" and p. 308, Standard Oil of California and Texas Oil Company December 2, 1942, meeting with Harold Ickes, Secretary of the Interior, "feared that the American oil concession would be cancelled and given over to British interests," p. 309, "defence of Saudi Arabia is vital to the defence of the USA."

34. Yergin, *The Prize*, p. 395, refers to an article authored by Ickes in December 1943 titled, "We're Running Out of Oil!" Shwadran, *The Middle East*, pp. 310–15, "Rommel was chased out of North Africa and the concession was secure"; see "Politics Has a Part in International Oil," *Life*, March 28, 1949, p. 78, reports that in February 1949 James A. Moffett (in 1936 chairman of the Bahrain Petroleum Company and Caltex), a longtime friend of President Franklin Roosevelt, successfully sued Aramco in the federal court for services involving using his influence with Roosevelt on Aramco's behalf in the Lend Lease deal. The jury awarded Moffett $1.5 million. Moffett's offer, he claimed, was obtained from Aramco at Roosevelt's request, had included petroleum products to the US Navy at "attractive prices" in return for the US government advancing $6 million annually for five years to the king of Saudi Arabia. A US Senate committee, investigating Aramco oil sales to the US Navy, found Aramco had overcharged to the tune of $25 million. What the author terms "the Moffett Affair" and the extension of Lend Lease to Saudi Arabia is covered well in Aaron David Miller, *Search for Security: Saudi Arabian Oil and American Foreign Policy, 1939–1949* (Chapel Hill: University of North Carolina Press, 1980).

35. Yergin, *The Prize*, p. 412, "so long as they were American," and p. 415, "on that he was insistent."

36. Ibid., p. 449, "the loss of the concession."

37. Ibid., pp. 651–52, "21 cents per barrel"; Anthony Cave Brown, *Oil, God, and Gold: The Story of Aramco and the Saudi Kings* (Boston: Houghton Mifflin 1999), p. 366, describes Aramco after the Saudi Arabian takeover extraordinarily as "under Wahhabi management."

38. This episode is recounted in Yergin, *The Prize*, pp. 647–48, citing an April 2, 1975, personal interview with the Gulf Oil executive concerned; p. 653, Yergin comments, "in the 1950s and 1960s cheap and easy oil had fuelled economic growth and thus, indirectly, promoted social peace. Now it seemed expensive and insecure oil was going to constrain, stunt, or even eradicate economic growth. Who knew what the social and political consequences would be? . . . After 1973, the very substance of power in international politics seemed to have been transmuted."

GLOSSARY

aba. Also called **bisht**, formal, lightweight cloak worn by men, usually black or dark brown, frequently embroidered with gold thread along the edges (women wear *abaya*).

Abu Al Naft. Father of Oil (honorific awarded by the Gulf Arabs to Frank Holmes).

amir. Leader of a defined area containing a number of tribes (i.e., amirate). Note that the British frequently used the English *e* to express the Arabic *a/alif*, as in *Bahrein*, *Koweit*, and, most notably, *Emir*. This spelling was retained, for example, in the name "United Arab Emirates."

Arabian Gulf *versus* **Persian Gulf.** Both titles are now used interchangeably. Sayed Amin discusses this issue in legal and historical detail in *Political and Strategic Issues in the Persian-Arabian Gulf* (Glasgow, Scotland: Royston, 1984, pp. 81–86) and notes that "Arabian Gulf" is the common usage in the Gulf states. He points out that by 1968 all the Arab states had passed laws and issued decrees making the use of "Arabian Gulf" compulsory in all official communications. Peter Mansfield in *The Arabs* (London: Penquin Books, 1982, p. 371) observes that "the Arabs of the area reject the term Persian Gulf and insist on 'Arab' or 'Arabian Gulf.' Persian-Arabian Gulf would be a suitable compromise but it is cumbersome."

balyoz. Descriptive term used by the Turks for the representative of the Venetian Republic to the court of the Ottoman Sultan in Constantinople.

bin. Son of (female equivalent is *bint*). Note that the term **ibn** was the Turkish equivalent used by the Ottoman Empire.

diwan. Commonly used to describe an administrative office, as in "the Amiri Diwan"; literally it is the reception room of a palace or large private house, also called *majlis*.

dyarky. Or "double rule"; Britain's term for the system implemented in Iraq from 1920 to 1928 in which all Iraq institutions including government ministries and the army had double staffs. At the top, an Iraqi national would hold the title and ostensibly also hold executive authority. His British colleague was termed an adviser. In reality, the adviser exercised all power and decision making.

ighal. Headband worn by men over the large cotton headscarf (*ghutra*); nowadays this is usually a heavy black silk loop, with or without cords extending down the back; originally the *ighal* was a double loop of rope that could be used to hobble camels; can also be made with gold strands to denote high rank.

Ikhwan. Literally "Brotherhood"; can be a group of brothers come together to achieve a religious or political goal. In early Saudi Arabia, the Ikhwan were men from the Bedouin tribes whom Abdul Aziz bin Saud galvanized into missionary-warriors who fought with him in the campaigns beginning with Hasa in 1913 to the 1924 capture of Mecca and the subsequent surrender of Medina and Jedda. Bin Saud had also sought to bring the Bedouin tribes into settlements, the first was established in 1913; there were sixty within a decade. After the capture of Mecca and Jedda, the Ikhwan rebelled against bin Saud. Successive waves of Ikhwan insurrection were only quelled, in 1930, with active British military assistance.

imam. A learned religious leader.

Majlis al Urf. Roughly "commercial court," a traditional meeting to resolve commercial disputes.

Mesopotamia. Modern Iraq occupies most of the area of ancient Mesopotamia, the plain between the Tigris and Euphrates rivers. Before the seventh century, Mesopotamia was the center of the Babylonian and Assyrian civilizations. The Turks initially conquered Mesopotamia in 1534 but did not secure true control until the seventeenth century when Mesopotamia was brought under the direct administration of the Ottoman Empire. Mesopotamia had three provinces, Mosul, Basra, and Baghdad. The name Mesopotamia continued to be used by the British until, following the uprising of 1920, plans were drawn up to institute a provisional government for what would be called the State of Iraq. The area had been known as Al Iraq in antiquity, and the Arabs had always used both Al Iraq and Mesopotamia interchangeably. In modern history others knew only Mesopotamia; the Ottomans used Mesopotamia. The revived name came into official usage in August 1921 when Faisal was proclaimed king of Iraq.

qada. (Turkish, in Arabic, *qarya*) village, designation used by the Ottoman Empire.

qaimaqam. (Turkish) governor of a subdistrict. The governor of a province was called *mutassarif.*

shaikh. Was traditionally an awarded title for a tribal or religious leader (literally an honorific meaning "old man"), has come into use as an inherited title designating close members of the ruler's family (female equivalent is shaikha).

vilayet. (Turkish, in Arabic, *wilaya*) province, designation used by the Ottoman Empire.

Wahhabi. Originally a reformist movement led by Muhammad bin Abd Al Wahhab (died 1787) dedicated to rooting out what he saw as the decadence of popular Islam and achieving a return to the purity of Islam as it existed in the time of the Prophet and the four caliphs who followed. In 1806 Wahhabi fighters led by Amir Muhammad Al Saud (ancestor of Abdul Aziz bin Saud) occupied Mecca but were ejected by forces sent by Ottoman sultan Mahmoud II. When the twenty-one-year-old Abdul Aziz bin Saud recaptured Riyadh in 1902, he set in motion a revival of the Wahhabi movement that gave birth to a warrior brotherhood with which he created, by conquest, the kingdom of Saudi Arabia. (Bin Saud's fighting followers were known as Wahhabi Warriors or the Ikhwan.)

WHO'S WHO

Ahmad, Hashem bin, of Kuwait, personal secretary and scribe of the *Diwan* to Abdul Aziz bin Saud.

Albuquerque, Afonso de (1453–1515), Portuguese navigator, statesman, and founder of the Portuguese empire in the Orient. He was born in Alhandra, near Lisbon, and spent his youth at the court of King Alfonso V of Portugal. He took part in the expedition against the Turks that culminated in a Christian victory at Otranto, Italy, in 1481. In 1503 he made his first trip to the East, traveling with a Portuguese fleet around the Cape of Good Hope to India. He was appointed viceroy of all Portuguese possessions in Asia. (His predecessor, Francisco de Almeida, refused to give up his office and imprisoned Albuquerque from 1508 to 1509.) As viceroy, Albuquerque captured the Indian district of Goa in 1510, and then conquered Malabar, Sri Lanka, the Sunda Islands, the peninsula of Melaka, and the island of Hormuz at the entrance to the Persian Gulf. He was the victim of intrigue at the Portuguese court, and in late 1515 King Emanuel appointed one of Albuquerque's enemies as his successor. A few days after receiving notice that he had been superseded, he died at sea off the Malabar Coast near Goa.

Al Gosaibi, Abdul Aziz (d. 1950), Abdul Rahman (d. 1976), Abdulla, Hassan, and Sa'ad (known as "the five brothers"). The family originated in Hofuf, where they were involved in agriculture and traded dates to India through Bahrain, diversifying into trade in pearls. As young men Abdul Aziz, Abdul Rahman, and Abdulla spent time in Bombay, where the family had an office, learning English and the merchant trade; Abdul Rahman was frequently in Paris dealing with the family's pearl trade. The branch of Abdul Aziz Al Gosaibi represented the interests of Abdul Aziz bin Saud and his territories in Bahrain from 1908 until the establishment of the Saudi Arabian Foreign Service. In Bahrain they arranged accommodation and transport for all visitors to and from bin Saud and provided

515

a channel of communication; the British paid their subsidy to bin Saud through the Gosaibis. For some years, the Gosaibis in Bahrain had close to a monopoly on the carrying trade between Bahrain and the mainland, and supplied bin Saud with foodstuffs from India and general provisions.

Al Hashem, King Faisal bin Husain (1885–1933), son of Sharif Husain of Mecca, 1916–18 commanded the Hijaz army in operations with T. E. Lawrence. In 1919 represented the Hijaz in the post–WWI Paris Peace Conference. March 8–July 25, 1920, head of the Provisional Arab Administration in Syria before the French expelled him in order to entrench their mandate over Syria. August 23, 1921, proclaimed king of Iraq.

Al Hashem, Sharif Husain bin Ali of Mecca (1856–1931), father of Faisal of Iraq and Abdulla of Transjordan. In 1908 became amir of Mecca. In 1914–16 Britain's main ally in the Middle East theater of operations during WWI. In 1916 proclaimed king of the Arab countries by his own followers; the French objected to what they saw as Britain's promotion of Hashemite leadership; in January 1917 a compromise was reached in which both Britain and France recognized Husain as king of the Hijaz. October 3, 1924, forced to abdicate on Abdul Aziz bin Saud's military conquest of Mecca and went into exile in Cyprus. Died in 1931 while visiting his son in Amman.

Al Idrissi, Muhammad bin Ali (1876–1923), Saiyed (religious leader and Sufi) and ruler of Assir, Tihama, Hodeidah. Allied to the British during WWI he could muster a fighting force of thirty thousand men. Signed two treaties with the British, the first in 1915 guaranteed protection of his coastal towns from foreign attack, provided a subsidy and arms. The second in 1917 related to Farasan Islands, which he had wrested from the Turks. On his death in 1923 Muhammad's brother Hassan, then in his forties, refused the succession, preferring a Sufi spiritual life. The rule went to Ali, born in 1905. Within two years Ali had ceded Hodeida and other towns to the Imam Yahya and lost the support of the tribes in Assir. Ali was challenged by his kinsman, Mustafa, and in 1926 was deposed. His uncle, Hassan, took up the rule but placed Assir under the "protection" of Abdul Aziz bin Saud, shortly after formally annexed to Saudi Arabia.

Al Khalifa, Shaikh Hamad bin Isa (d. 1942), although holding the title of deputy ruler until the death of his father in 1935, Shaikh Hamad was ruler of Bahrain from 1923 when the Government of India forced his father to abdicate. Hamad and his brother Abdulla negotiated the Bahrain oil concession.

Al Khalifa, Shaikh Isa bin Ali (1869–1935) signed two treaties with Britain, in 1880 and 1892, not to enter into relations with any "foreign" government, and not to cede, sell, or lease any part of his territories without British consent. In 1914 Shaikh Isa signed the exclusion "convention" agreeing not to grant an oil concession except to an individual, or company, approved by the British government. The Government of India forced Shaikh Isa to hand authority to his son, Hamad, in 1923.

Al Sabah, Shaikh Ahmad bin Jabir (1885–1950), succeeded his uncle Shaikh Salim following the latter's death in February 1923 (official date of accession is March 29, 1923). Shaikh Ahmad negotiated the Kuwait oil concession.

Al Sabah, Shaikh Mubarak bin Sabah, "Mubarak the Great" (1896–1915). Kuwait signed in 1899 a treaty with Britain agreeing not to conclude treaties with any other powers, not to admit any "foreign" agents and not to cede, sell, or lease any part of Kuwait's territory without British consent. Britain promised to protect both the Al Sabah ruler and his heirs. In 1913 Mubarak signed with Percy Cox the exclusion "convention" agreeing not to grant an oil concession except to an individual, or company, approved by the British government.

Al Saud, Abdul Aziz bin Abdul Rahman (1880–1953), after a childhood spent in exile in Kuwait, in 1902 bin Saud regained the Saudi patrimony of Nejd. In 1924–26 he took over the Hijaz, with Mecca, driving out the Hashemite rulers. In January 1926 he was proclaimed king of the Hijaz and sultan of Nejd and its dependencies (soon after he annexed the Assir from Yemen), September 18, 1932, proclaimed king of Saudi Arabia. From 1926 bin Saud dealt with waves of internal rebellion until the British and the Government of India assisted him militarily in 1929–30. Bin Saud signed with Percy Cox the Treaty of Darin in 1915, including the usual clauses of not entering into any correspondence, agreement, or treaty with any "foreign" power and agreement not to cede, sell, lease, mortgage, or otherwise dispose of any part of his territory without British consent. Bin Saud granted the Al Hasa oil concession to Frank Holmes in 1923 and, after it had lapsed, negotiated it again in 1933 with Standard Oil of California.

Al Thani, Shaikh Abdulla bin Qasim (d. 1949), ruler of Qatar. From 1906–13 was governor of Doha and successful pearl merchant. From 1913–49 succeeded his father as ruler of Qatar. Signed treaty with Britain in 1916 placing Qatar on same footing as the other Gulf shaikhdoms. Abdulla secretly paid an annual subsidy to Abdul Aziz bin Saud to protect his territory from bin Saud's expansionism. In 1922, Abdulla tried to conclude an oil concession with Frank Holmes but was prevented from doing so by the Government of India's political resident. In 1926, under pressure from the political agent, he signed a prospecting license with D'Arcy Exploration Company (a subsidiary of the Anglo-Persian Oil Company). In May 1935, in return for a formal document from the British government guaranteeing recognition of his son and heir and protection from "outside" attack against his territory, he signed a concession with Petroleum Development (Qatar) Limited, the subsidiary of the Anglo-Persian Oil Company.

Belgrave, Sir Charles (1895–1970). Educated at Oxford. With the British army in Sudan, Palestine, Egypt. Joined the Frontier Districts Administration (Anglo Egyptian Civil Service). Joined the Colonial Service in East Africa. Married the daughter of Lord Bristol. In 1926 hired by the India Office and served as financial adviser to Shaikh Hamad, ruler of Bahrain, and on Hamad's death, to his son Shaikh Salman, until Belgrave's retirement in 1957. Although the Government of India recommended the appointment of a personal adviser to the shaikh because of the impropriety of the political agent running all the affairs of the island, Belgrave actually reduced Bahraini input. Belgrave, with the assistance of a handful of British nationals, ran the Administration, the police, and the Judiciary, as well as controlling all commercial and financial interests. Wrote *Personal Column*, published 1960.

Bell, Gertrude Margaret Lowthian (1868–1926) was educated at Oxford, the first woman to obtain a first-class degree in modern history. She became a competent archaeologist visiting Persia and the Middle East and published three books of her travels. In 1914 she joined the Arab Bureau in Cairo for the task of researching the Bedouin tribes and shaikhs of Northern Arabia. In 1916 she joined the staff of military intelligence and was promoted to Oriental Secretary, a post she retained when she moved to Baghdad after its capture in 1917 as adviser to Sir Percy Cox. She was responsible for instituting the respected Iraq Antiquities Department and the Baghdad Museum.

British Oil Development Ltd. (BOD) registered in London 1928 with mainly British and Italian shareholding. Reorganized with increased capital after obtaining in May 1932 from the Iraq government an oil concession west of the Tigris. Mosul Oilfields Ltd. with British, Italian, German, Dutch, French, and Iraq shareholding was registered in December 1932 to acquire the shares of BOD; the Iraq Petroleum Company moved to acquire the shares of BOD in 1936, dissolving both BOD and Mosul Oilfields in 1941.

Burma Oil (originally Burmah Oil), the first British oil company. Founded by a group of Scottish traders and investors after the 1885 annexation of Burma to India. Burma Oil began by appropriating the oil-gathering activities of Burmese village people, an appropriation on which it built a commercial industry and a refinery in Rangoon. The growth of Burma Oil was aided by the monopoly, given in 1889 by the Government of India, over the sale and extraction of oil products and protection from all foreign competitors throughout the Indian empire.

Chairman Lord Strathcona financed Canadian Pacific Railways. Before D'Arcy was producing in Persia he called for financial assistance. The British Admiralty feared the concession might fall into the hands of American or Dutch oil trusts and so persuaded Lord Strathcona to have Burma Oil cooperate with D'Arcy. In May 1905 Concessions Syndicate Ltd. was formed for operations in Persia. After the 1908 oil strike at Masjed Soleyman, on April 14, 1908, Concessions Syndicate Ltd. was relegated to a subsidiary and the Anglo-Persian Oil Company formed with Lord Strathcona as chairman. Burma Oil was the majority shareholder in Anglo-Persian until the British government's own 1914 purchase of 51 percent of the stock. Burma continued to hold 23 percent of Anglo-Persian/British Petroleum until 1974 when their shareholding was taken over by the Bank of England.

Cadman, Sir John, first Baron Cadman of Silverdale (1877–1941). Educated in science at Durham University. In 1902 Inspector of Mines (coal) East Scotland and Staffordshire. From 1904–1907 set up the Mines and Petroleum Department for the Government of Trinidad. In 1908 with the Royal Commission on Mines; 1910 professor of mines at Birmingham University where he established the Department of Petroleum Technology. In 1913 appointed to the Admiralty Commission (with Admiral Slade), sent by Churchill to report on the operations of the Anglo-Persian Oil Company; 1914 director Petroleum Executive, member Inter-Allied Petroleum Committee. In 1921 joined Anglo-Persian Oil Company; 1923 director and 1925 deputy chairman, then chairman Anglo-Persian Oil Company. From 1927–41 chairman of both Anglo-Iranian Oil Company and the Iraq Petroleum Company. At the time of his death he was honorary principal adviser on oil to the British government.

Chisholm, Archibald H. T. (1902–88). Educated at Oxford, descended from Lord Byron, with family connections to the family of Anglo-Persian's John Cadman. Pursued journalism with the *Wall Street Journal* (his father was editor of the *Times of London* and chairman of the Athenaeum Club, Arnold Wilson's London haunt). From 1927–32 hired by Cadman and sent to Persia for the Anglo-Persian Oil Company; 1932–34 Anglo-Persian's negotiator for Kuwait oil concession; 1935–36 with Anglo-Persian at Abadan. From 1937–40 editor of the *Financial Times* London; 1940–45 with the British army. From 1945–62 manager public relations Anglo-Iranian Oil Company/British Petroleum. From 1962–72 adviser British Petroleum. From 1973–75 researched and published *The First Kuwait Oil Concession Agreement: A Record of the Negotiations, 1911–1934*.

Churchill, Sir Winston Leonard Spencer (1874–1965). Entered Parliament in 1900. Junior officer with British Forces abroad and war correspondent Boer War.

From 1905–1908 undersecretary for the Colonies; 1908–10 president of the Board of Trade; 1910–11 Home Secretary; 1911–15 and 1939–40 First Lord of the Admiralty; 1915 chancellor of the Duchy of Lancaster; 1917 Minister of Munitions; 1918–21 Minister of War; 1919–21 Minister of Air. From April 1921 to October 1922 secretary of state for the Colonies; 1924–29 Chancellor of the Exchequer; 1940–45 Prime Minister and Minister of Defense; 1951–55 Prime Minister.

Cox, Maj. Gen. Sir Percy Zachariah (1864–37). Early career included six years in India with the Indian army, four years as political officer in the Indian state of Baroda, six years in Somalia. From 1904–13 political resident Persian Gulf; 1914 secretary Foreign Department Government of India; 1914–18 chief political officer Indian Expeditionary Force "D." March 1918–20 acting minister Tehran. From 1920–23 as Sir Percy Cox, high commissioner Mesopotamia/Iraq (Cox was also titular Political Resident until October 1920 although absent in Baghdad. On his 1918 appointment to Tehran, Arnold Wilson took over as absentee Resident in Baghdad.)

Crane, Charles R., American millionaire, heir to the Crane Bathroom Equipment Company of Chicago and colleague of Henry Ford, Thomas Edison, and George Westinghouse. Was US minister to China 1921. Crane was appointed by US president Woodrow Wilson to the 1919 King-Crane Commission sent to ascertain local reaction to the proposed post–WWI mandates in Palestine, Iraq, and Syria, and particularly to the Balfour Declaration granting a Jewish homeland in Palestine. The report was never tabled at the Versailles Peace Conference. From his travels for the King-Crane Commission, and from time spent in Egypt as a young man, Crane developed a fascination for the Arab world and an empathy for the people. He believed the two important factors that could assist the Arabs out of poverty were education and agriculture. At his own expense he educated a number of Syrian and Lebanese young men in America. He was a patron of the American Mission hospitals and schools in the Arabian Peninsula and was involved in the setting up of American higher education facilities in Lebanon. He initiated a date farm on his ranch in California devoted to researching and improving Arabian strains and propagation techniques and did similar research on the breeding of Arabian horses. He paid for a three-year fact-finding and development mission to Yemen, which he visited twice, and similar to Jedda, which he also visited. The two programs were led by the American engineer Karl Twitchell.

Craufurd, Commander C. E. V. Royal Navy (Ret.). In WWI commanded HMS *Minto* patrolling the Red Sea. Employed personally by Sir Edmund Davis, Eastern & General Syndicate's chairman and chief investor, accompanied Frank Holmes to Aden in 1921. He became obsessed with locating the site of King

Solomon's Mines, eventually becoming convinced he had traced them to the port of Mukalla, not far from Aden. Delivered lectures and speeches and wrote articles on this subject in London until the mid-1930s. Infected American engineer Karl Twitchell with a similar obsession when he came to Yemen in 1927 on a mission for American philanthropist Charles Crane; Twitchell became convinced King Solomon's Mines were in Saudi Arabia.

Curzon, George Nathaniel, Lord Curzon of Kedleston, First Marquess (1859–1925); statesman and administrator. From 1898–1905 viceroy and governor-general of India; 1915–16 Lord Privy Seal; 1916–18 member of Lloyd George War Cabinet; 1916–24 leader of the House of Lords; 1919–24, secretary of state for Foreign Affairs.

Daly, Sir Maj. Clive Kirkpatrick. From 1916–20 served in the Indian Political Service as lieutenant colonel, was political governor Diwania (Iraq). In 1920 demoted to major following the outbreak of the Iraqi Revolt in his area. From January 1921 to September 1926 political agent Bahrain, wounded in a local uprising against his oppressive policies and aimed, in part, at the Persians he had gathered into a police force, answerable only to him, to control the Bahrainis. The Government of India knighted Daly on his return to London.

Damlouji, Dr. Abdullah, Iraqi physician, practicing in Baghdad. Medical consultant to Sir Percy Cox's administration in Baghdad. Personal friend and frequently physician to Abdul Aziz bin Saud. Saudi Arabia's first minister of Foreign Affairs.

D'Arcy, William Knox (1849–1917). Solicitor, born in Devon, UK. Held principal interest in Mount Morgan Gold Mine, Queensland. May 28, 1901, obtained oil concession from the Persian government. May 21, 1903, formed First Exploration Company. May 5, 1905, formed, with Burma Oil Company, Concessions Syndicate Ltd. From 1909–17 director Anglo-Persian Oil Company (now incorporating D'Arcy Exploration and Concessions Syndicate Ltd.).

Dickson, Harold Richard Patrick (born Beirut 1881, died Kuwait 1959). In 1908 with the Indian army; 1912–13 tutor and guardian to eldest son of maharaja of Bikaner; 1914–15 with 29th Lancers to Mesopotamia. From 1915–19 selected by Arnold Wilson as political officer Nasiriyah (Iraq) press-ganging Arab laborers for railway construction. From November 1919 to November 1920 as major, political agent Bahrain. In 1921 political adviser at Hillah (Iraq), including Karbala and Najaf until his removal in July 1922. From August 1922 to December 1922 in Bahrain, no official post, but loose title of personal representative of Percy Cox, expected appointment as agent to bin Saud. From December 1922 to

August 1923 pursued unfair dismissal case in London. From 1924–28 reinstated to Indian army in Quetta then guardian and tutor to second son of maharaja of Bikaner. May 1928–April 1929 secretary to the political resident Bushire. From May 1929 to February 1936 (and temporary May 1941 to August 1941) as lieutenant colonel, political agent Kuwait. From 1936–59 chief local representative, Kuwait Oil Company. (His father, John, [1847–1908] born in Tripoli, Libya, was British consul Beirut 1879–82, consul general Damascus, Homs, and Jerusalem 1882–89. Grandfather was Dr. Edward Dickson [1815–1900], born in Tripoli, Libya, employed by the amirs of Western Tripoli [Al Babat], later joined the Ottoman Military Service as a doctor, died in Istanbul.)

East India Company, with rights, privileges, and trade monopolies granted by Elizabeth I in 1600, gained control in India through conquest and the ceding to it of territories and concessions by local rulers. Its main commercial outlets, known as factories, were in Bombay, Calcutta, and Madras. Charles II (died 1685) further granted the company sovereign rights, allowing it to acquire and administer territory, issue currency, negotiate treaties, and wage war in its own name. The 1773 Regulating Act went some way toward giving the British government some rights to oversee and regulate the affairs of India but was foiled to all intents and purposes in 1774 when the East India Company appointed its own governor of Bengal, Warren Hastings, as first governor general of India. The British Parliament's India Act of 1784 passed executive control of India affairs to a Board of Control whose president was a member of cabinet and answerable to Parliament (Viscount Castlereagh). Following the "Indian Mutiny" against company rule, legislation of 1858 established the British Crown as the government of India with two sources of executive power, the secretary of state for India, who answered to Parliament, and the viceroy, who oversaw everyday administration and lawmaking. Under the powerful viceroy was a hierarchy of presidency and provincial governors, commissioners, judges, magistrates, and so on. The new Government of India re-created most British government departments, including its own Department of Foreign Affairs. The East India Company was wound up in 1874, and Queen Victoria took the title empress of India in 1877.

Greenway, Sir Charles, Baron Greenway of Stanbridge Earls (1857–1934). In 1893 joined Shaw, Wallace & Co. India. In 1897 senior partner; 1910 senior partner R. G. Shaw & Co., managing director Lloyd Scott & Co. From 1910–19 managing director then 1914–27 chairman and 1927–34 president Anglo-Persian Oil Company.

Glubb, Sir John Bagot, known as "Glubb Pasha" (1897–1986), British soldier, born in Lancashire. He trained at the Royal Miltary Academy, Woolwich, served in World War I. In 1920 became the first organizer of the native police force in the new state of Iraq. In 1930 transferred to British-mandated Transjordan, organizing the Arab Legion's Desert Patrol, and became Legion Commandant 1939. He had immense prestige among the Bedouin, but was dismissed from his post in 1956 following Arab criticism. Knighted in 1956, he then became a writer and lecturer.

Gulbenkian, Calouste (1868–1955), Armenian. From 1887–1891 graduated in mining engineering King's College London, worked in the family's oil business in Baku, wrote scholarly articles and a book on Russian oil, authored a report for the Turkish sultan on oil possibilities in Mesopotamia. In 1893 moved to Constantinople. In 1896 moved to Egypt then London where he traded in Baku oil and acted as an oil industry broker. In 1907 returned to Constantinople representing Royal Dutch Shell. In 1912 formed the Turkish Petroleum Company with the Deutsche Bank and Royal Dutch Shell holding 25 percent each and 50 percent held by the Turkish National Bank. Gulbenkian held 30 percent of the Turkish National Bank, which equated to a 15 percent holding in Turkish Petroleum. In 1914 Anglo-Persian bought into Turkish Petroleum with the Deutsche Bank and Shell each retaining their 25 percent and Gulbenkian holding 5 percent. Gulbenkian succeeded in retaining this 5 percent through the various incarnations of Turkish Petroleum and the subsequent Iraq Petroleum Company. He died, aged eighty-five, in Lisbon.

Hardinge, Charles, First Baron of Penshurst (1858–1944). In 1896 secretary Persian Legation. From 1906–10 permanent undersecretary of state for Foreign Affairs; 1910–16 viceroy of India; 1916–20 again permanent undersecretary of state for Foreign Affairs; 1920–23 ambassador in Paris.

Heim, Dr. Arnold Albert (1882–1965), Swiss. Geologist and docent at the University of Zurich. Commissioned in 1924 by Frank Holmes and the Eastern & General Syndicate Ltd to survey for oil and water in Bahrain, Kuwait, Al Hasa, and the Kuwait-Saudi Arabia Neutral Zone. In the late 1920s Heim published a number of magazine articles and mounted lantern slide lectures based on his 1924 visit to Arabia.

Hirtzel, Sir Arthur (1870–1937). In 1894 joined India Office. From 1917–21 assistant undersecretary of state for India; 1921–24 deputy undersecretary of state for India.

Hoover, Herbert Clark (1874–1964), thirty-first president of the United States (1929–33). In 1896–97 worked for a leading engineer in San Francisco. On his recommendation the twenty-two-year-old Hoover got a job with a London mining firm, Bewick, Moreing and Company, to introduce California methods to the company's gold mines in western Australia; he recommended the purchase of Sons of Gwalia gold mine in Kalgoorlie. He turned from technical work to administration, bargaining with labor and negotiating with the Australian government. His company transferred Hoover to China where he became chief engineer in the Chinese government's Imperial Bureau of Mines. In 1900 he held a one-fifth interest in Bewick, Moreing and Company, which then possessed gold, silver, tin, copper, coal, and lead mines in Australia, New Zealand, South Africa, Canada, and Nevada. He also owned a turquoise mine in Egypt. Hoover became a world-renowned consulting engineer, accepting commissions to revive unproductive mines and was managing director or chief consulting engineer in a score of mining companies—and he was a wealthy man. In 1917 he was US food administrator and chairman of the American Relief Administration to assist in the economic restoration of Europe. President Harding appointed him secretary of commerce in 1921. In seven years as head of the Department of Commerce, Hoover extended its control over mines and patents and supported the expansion of American business overseas. For the Republicans, Hoover won the 1928 presidential election in a landslide. He was in office less than eight months when the Wall Street crash occurred. In 1932, he lost to Franklin D. Roosevelt. Herbert Hoover died in 1964, in New York City.

Jacks, T. L. (1884–1966). From 1909–13 oil assistant Strick Scott & Co Abadan. From 1917–20 assistant manager then 1921–22 joint general manager with Arnold Wilson Anglo-Persian Oil Company Abadan; 1926–35 resident director Anglo-Persian Oil Company Tehran.

Khazaal, Shaikh of Muhammerah (1860–1936). The Anglo-Persian Abadan refinery was built in his territory. In return for allowing the refinery, the 1909 undertaking, reconfirmed in 1914, given by Sir Percy Cox and the British to Khazaal promised to protect him and his heirs against encroachment on his jurisdiction and recognized rights from the Persian government. Until 1924 Khazaal was the de facto ruler of the left bank of the Shatt al Arab (Arabistan later Khuzistan). In April 1925, he was removed to house arrest in Tehran by the forces of Riza Khan who then established central government control over Khazaal's area. During a debate in the British parliament it was claimed that no undertaking as to the personal security of Khazaal had ever been given and, as his nationality was Persian, Britain's only interest in him was one based on long-standing friendship. He was never permitted to return to Muhammerah and never regained any of his former rights. He died in Tehran.

Lawrence, Col. T(homas) (Edward), "Lawrence of Arabia." He was the illegitimate son of an Anglo-Irish baron. Educated at Oxford where he was mentored by D. G. Hogarth, influential author and archaeologist, associated with the Ashmolean Museum. In 1914 Hogarth took Lawrence with him to work with military intelligence in Cairo and transferred him into the Arab Bureau (of Intelligence and Diplomatic Officers) when he took over as head. Lawrence was on the mission to Jedda in October 1916 and then joined the Arab Revolt financed and assisted by the British from Cairo, led by Faisal bin Husain Al Hashem, the son of the Sharif Husain of Mecca. Lawrence romanticized his own role in the Arab Revolt in the best-selling *Seven Pillars of Wisdom.*

Lombardi, M. E. (b. 1878), vice president and director Standard Oil of California. His background was in the company's foreign crude oil production department. He negotiated the 1928 purchase from Gulf Oil Corporation of Holmes/E&GSynd's Bahrain oil concession. In September 1932 Lombardi met Karl Twitchell in New York and asked for his assistance in obtaining from Abdul Aziz bin Saud the Frank Holmes–lapsed 1923 Al Hasa concession. Lombardi's colleague, Francis Loomis, had already put the same request to Harry Philby. Eventually both Philby and Twitchell were hired by Standard Oil of California. Lombardi visited Iraq, Bahrain, Kuwait, and Egypt but did not visit Saudi Arabia. He retired in 1941.

Longhurst, Henry. Wrote *Adventures in Oil: The Story of British Petroleum*, published 1959, for the company's fiftieth anniversary. At the time Longhurst was a sports reporter, specifically golf correspondent, for the *London Sunday Times.*

Longrigg, Brig. Stephen Helmsley. In 1920 worked with Harry Philby at Interior Ministry Iraq. In 1921 as major, was political officer at Kut, Iraq; 1922 replaced Harold Dickson as political officer at Hillah Iraq; 1925 joined Anglo-Persian/Iraq Petroleum Company. Wrote *Oil in the Middle East: Its Discovery and Development*, published 1954.

Loomis, Francis B. Was assistant secretary of state under John Milton Hay (Hay, who had been assistant private secretary to President Lincoln, was secretary of state to William McKinley and Theodore Roosevelt). In 1901, while Hay was ill, Loomis took part in the negotiations for the Panama Canal concession. He joined Standard Oil of California with the title of Washington representative. He dealt with the US State Department in the matter of requesting diplomatic intervention with the British over the Bahrain oil concession. In July 1932 he approached Harry Philby in London, requesting him to use his friendship with Abdul Aziz bin Saud to assist Standard Oil of California to purchase Frank Holmes's lapsed 1923 Al Hasa concession and in September 1932 agreed to the same deal with Karl Twitchell, who had already been approached by Loomis's colleague M. E. Lombardi. Eventually both Philby and Twitchell were hired by Standard Oil of California.

Madgwick, Thomas George. British-born engineer. During WWI was involved in sinking water wells in Salonika and Gallipoli. Lectured at Birmingham University and set up a London-based consulting partnership. In 1925 hired by Holmes/E&GSynd on a four-month contract for the artesian water-drilling project in Bahrain. Introduced American Thomas Ward, whom he knew from Trinidad, to Holmes/E&GSynd. In 1926 through Ward, obtained a short-term contract as petroleum consultant for the Canadian government in Calgary; this eventually became a permanent position in Ottawa. Madgwick's claim to have discovered the oil of Bahrain became more and more elaborate over the years.

Mann, Dr. Alexander, physician. Nearing retirement on the staff of Sir Percy Cox in Mesopotamia, in 1921 he was sent by Cox on a medical visit to Abdul Aziz bin Saud. Subsequently bin Saud appointed him (with Cox's approval) his personal representative and agent in London. In 1922, Mann brought Frank Holmes to Abdul Aziz bin Saud. Holmes's personal aim was to explore for oil and also report on development and investment opportunities such as ports. Meantime, Mann was proposing his own ideas to bin Saud for increasing revenue, such as issuing visas and passports and establishing a postal service. When Mann's activities came to the attention of the Government of India's political officers in the Persian Gulf they successfully lobbied the Colonial Office to have him removed from bin Saud's service. Bin Saud was informed that His Majesty's government would not recognize Dr. Mann in any official capacity, or permit correspondence to be conducted through him. A dismissal letter was then arranged and given to bin Saud for signature. In November 1923, the political officers became aware Mann was in Bombay, and were convinced he intended traveling to Bahrain and Nejd looking for oil concessions. While he was onboard ship, they cabled an alert to every Gulf port describing Mann as a "notorious character." On arrival in Bahrain, he was detained, his passport canceled, and he was told to move on. In July 1924, he arrived in Tehran, representing Sir Norton Griffiths and the Phoenix Oil and Transport Company. Again he aroused the enmity of the political officers and quickly departed.

Morgan, C. Stuart. British born, Morgan was with Wilson's and Cox's administration in Mesopotamia where he worked with Harry Philby at the Ministry of Interior. He moved on to join the Anglo-Persian Oil Company. T. D. Cree, also ex-Mesopotamia and one of Philby's partners in Sharqieh, recommended Morgan to the head of Standard Oil New Jersey (Teagle), which hired him as adviser of Near Eastern work. He became secretary of the Near East Development Corporation, the entity formed to represent the American interests in the Turkish Petroleum Company/Iraq Petroleum Company.

Muhammad, Khan Bahadur Mirza (1885–1974). From 1921–45 was legal adviser in Basra to the Anglo-Persian Oil Company. He Arabized his name to

Muhammad Ahmad when he became an Iraqi subject after 1934. His colleague, J. Gabriel, was the attorney for Shaikh Ahmad of Kuwait. Both men played a somewhat murky role in Traders Ltd., the entity suspected of being created by the Anglo-Persian Oil Company in 1934 to deal the Gulf Oil Corporation out of the Kuwait oil concession.

Musadiq, Dr. Muhammad Khan (b. 1873). Persian. Studied law in Switzerland. In 1920 was governor general of Fars; 1921 Minister of Finance; 1922 governor general Azarbaijan; 1923 Minister of Foreign Affairs; 1925 opposed the change of regime. May 1951 became prime minister of Iran; presided over nationalization of the oil industry.

Philby, Harry St. John Bridger "Jack" (born Ceylon, April 1885, died Beirut, September 1960). From 1904–1907 educated at Cambridge; 1908–13 various jobs with Indian civil service including Criminal Investigation Department (CID) Simla. November 1914 to Mesopotamia in response to Percy Cox's urgent demand for linguists to help with the occupation; Philby's job was to collect revenues, tax the Iraqis, and set up a system of accounting. In 1916 revenue commissioner. May 1917 personal assistant to Percy Cox. When Arnold Wilson moved from Basra to Cox's office in Baghdad, Philby was sent, November 1917 to October 1918, to (unsuccessfully) encourage Abdul Aziz bin Saud to extend military assistance for British campaign against the Turks. March 1920 failed in an application to join the financial mission to Persia. May 1920 failed in an application to join the staff of Sir Herbert Samuel, high commissioner to Palestine. On Cox's appointment as high commissioner Iraq became British adviser to the Iraqi Minister of Interior. Because of his active hostility to King Faisal, in July 1921 Philby was dismissed. October 1921 chief British representative to Amir Abdulla in Transjordan charged with setting up an administration and controlling revenues. Resigned in January 1924, just days before Herbert Samuel was scheduled to dismiss him. From October 1925 to March 1926, backed by small group of London financiers in Jedda scouting possibilities of concessions in Saudi Arabia. October 1926–56 Jedda representative of Sharqieh Ltd., "the Company of Explorers and Merchants in the Near & Middle East," originally with London financing. February–May 1933 assisted Standard Oil Company of California to obtain Al Hasa concession and thereafter retained as local consultant. Philby caused a sensation in London by converting in 1930 to the strict Wahhabi Islam of Saudi Arabia. His son was Harold "Kim" Philby, born in the Punjab in 1912, died in Moscow, 1988. Kim rose to head of anti-Soviet Counterespionage in British Intelligence. Exposed as a Soviet spy in 1963, Kim was one of the most successful double agents in history.

Rajab, Sayid Talib Pasha bin Saiyid, Arab nationalist in Basra and deputy in the Turkish parliament; joined the Committee of Union and Progress (the "Young

Turks" reform movement); son of the Najib Al Ashraf (chief of the nobles); was anti-Turkish ruler of Mesopotamia and became fervently anti-British. In 1913, together with Shaikh Mubarak of Kuwait and Shaikh Khazaal of Muhammerah formed the Reform Committee actively seeking transfer of power from the Turks and independence from the British. Moved to India on the outbreak of WWI, returned to Basra in 1920 and on to Baghdad when his father was appointed president of what Sir Percy Cox called a Provisional Arab Council of State with British Advisers. Talib became Minister of Interior (with Harry Philby as his adviser). Talib soon had a powerful following, and a movement developed campaigning to elect him president of a republican Iraq. In April 1921, after taking afternoon tea with Lady Cox and Gertrude Bell, Talib was secretly arrested by Percy Cox who deported him to Ceylon, leaving the way open for Sharif Husain's son, Faisal, to became king of Iraq.

Rihani, Ameen F. (1876–1940). Born in Freike, Lebanon, one of six children, the eldest son of a Lebanese Maronite raw silk manufacturer. In 1888 Ameen, then aged twelve, and his uncle migrated to America; his father followed a year later. In New York, he worked in the family trading business. In 1895 he joined a touring theatrical group and was stranded in Kansas City. Back in New York he attended night school for a year and in 1897 entered the New York Law School. Twelve months later, following a severe illness, he was sent to Lebanon where he taught English and studied Arabic. In 1911 he published his first book in English, *The Book of Khalid*, with illustrations by fellow Lebanese American Khalil Gibran. In 1916 he married Bertha, an American painter. Between 1924 and 1932 Rihani made three extensive trips through the Arabian Peninsula and Yemen and published three books on his observations and experiences. Classified as travel books, the first works on the region aimed at the general reader, the books are *Ibn Sa'oud of Arabia: His People and Land* (1928), *Around the Coasts of Arabia* (1930), and *Arabian Peak and Desert: Travels in Al Yaman* (1930).

Riza Khan/Riza Shah (1878–1944). Rose to colonel through the ranks of Persian Cossack Brigade, joint leader of coup d'etat. In 1921 Minister of War; 1923 prime minister; April 1926 proclaimed as Riza Shah of the throne of Persia; 1942 abdicated.

Saleh, Mullah, of Kuwait. Personal secretary and secretary of the Council of State to Shaikh Ahmad Al Sabah, ruler of Kuwait. Saleh was also secretary to the two previous rulers of Kuwait.

Slade, Adm. Sir Edmond (1859–1928). From 1907–1909 director Intelligence Division of the Admiralty; 1909–13 commander-in-chief East Indies; 1913 head of the Admiralty Commission appointed by Churchill to report on the operations of the Anglo-Persian Oil Company. In 1914 appointed the British government director of Anglo-Persian Oil Company after the British government bought its 51 percent shareholding. From 1916–28 vice chairman Anglo-Persian Oil Company.

Stevens, Guy (d. 1945). Assistant to the vice president (William Wallace) Gulf Exploration Company New York. February 1934 became senior Gulf Oil director to Kuwait Oil Company, Gulf Oil's joint venture with Anglo-Persian Oil Company.

Thesiger, Wilfred Patrick (1910–2003). Explorer of Arabia, born in Addis Ababa. He studied at Oxford. In 1933 he hunted with the Danakil tribes in Ethiopia and explored the sultanate of Aussa. In 1935 he joined the Sudan Political Service and traveled by camel across the Sahara to the Tibesti Mountains. He was seconded to the Sudan Defense Force at the outbreak of World War II. From 1945 to 1950 he explored the Empty Quarter of South Arabia and the borderlands of Oman with Bedu companions, which he described in *Arabian Sands* (1959). He first traveled in East Africa in 1961 and returned there to live with tribal peoples from 1968 onward. His autobiography, *The Life of My Choice*, was published in 1987.

Thesiger, Frederick John Napier, 1st viscount and 3rd baron of Chelmsford (1868–1933). Colonial administrator, born in London. Governor of Queensland 1905–1909 and of New South Wales, Australia, 1909–13, and viceroy of India 1916–21 where he helped increase the number of Indians taking part in government, but was also noted for his severity against the nationalists. He became First Lord of the Admiralty in 1924.

Twitchell, Karl Saben (b. 1885 in Vermont). From 1908–13 graduated the Kingston School of Mines (Canada), worked as surveyor and manager of gold property in Nevada, sampler at gold mine in Denver, Colorado. From 1914–18 worked for Seeley W. Mudd and his partner Carl Lindberg at a copper mine in Arizona, then at the Skouriatissa Mine (iron pyrite) in Cyprus and with the Cyprus Forest Department. From 1920–25 was with a failed tin mine (the mine had been salted), and copper mining in Portugal. From 1926–27 worked at three concessions in Abyssinia held by Consolidated Gold Fields of South Africa. From 1928–31 sent by American philanthropist Charles Crane to Yemen to report on mineral possibilities. Crane provided training in systems of irrigation and pumps so Twitchell could also advise on agricultural development using seeds provided by US Department of Agriculture. Crane ordered a bridge built as a permanent monument to American friendship. In 1932–33 sent by Charles Crane to Jedda for the purpose of finding water in Saudi Arabia. Reported negatively on water possibilities but was fascinated by ancient gold mine remains; came to believe Saudi Arabia was the site of King Solomon's Mines. Returned to the

United States for the purpose of raising investment capital to develop the ancient mine of *Mahad Dhahab* (Cradle of Gold). In 1933–34 contacted by Standard Oil of California while in New York, appointed negotiator (along with Harry Philby) for that company in their bid to obtain Frank Holmes's lapsed 1923 Al Hasa oil concession. Obtained gold mining concession from Saudi Government, formed the Saudi Arabian Mining Syndicate Ltd., with British and American backing. Began production in 1937, extracted $32 million worth of gold, silver, and copper before closing in 1954. With Edward Jurji wrote *Saudi Arabia, with an Account of the Development of Its Natural Resources*, published in 1947, in which he claimed to have discovered the oil of Al Hasa. In his later years developed a speaking and lecturing circuit as an American "insider" and expert on the Middle East.

Wahba, Shaikh Hafiz (*shaikh* is an awarded honorific title). Egyptian. In Kuwait, was active in education and played a role in local politics. Requested to come to Bahrain by the progressive Shaikh Abdulla Al Khalifa after Abdulla's return from a tour of European and Arab capitals. Wahba became headmaster of the first modern school in Bahrain, set up by Abdulla. In 1921 Wahba was deported by the political agent in Bahrain allegedly for conspiracy against the Order-in-Council. He joined the *Amiri Diwan* of Abdul Aziz bin Saud, charged particularly with instituting a modern education system for bin Saud's territories. Shaikh Hafiz Wahba was Saudi Arabia's first minister (ambassador) to Great Britain in London. Wrote *Arabian Days*, published 1964.

Wallace, William T., vice president in New York of Gulf Exploration Company, Gulf Oil Corporation's subsidiary dealing with its Bahrain and then Kuwait interests. Wallace negotiated the 1927 option on Holmes/E&GSynd Bahrain oil concession.

Ward, Thomas E. British born, president and proprietor of Oilfields Equipment Company New York. Acted as broker in 1927 sale of option on Holmes/E&GSynd's Bahrain oil concession to Gulf Oil Corporation and again in the 1928–29 transfer of this concession from Gulf Oil to Standard Oil of California. Wrote *Negotiations for Oil Concessions in Bahrain, El Hasa (Saudi Arabia), the Neutral Zone, Qatar and Kuwait*, privately published 1965.

Williamson, William Richard, also known as Hajji Abdulla Fadhil (1872–1958). From 1885 to 1890 was merchant seaman and gold miner in California, whaler in the Arctic and Pacific. In 1891 joined Anglo-Indian police force at Aden and converted to Islam; 1893 transferred to Bombay. From 1894–1909 trader between Basra, Kuwait, other Gulf ports and the interior. In 1909 settled in Kuwait as ship owner and pearl merchant. From 1914–19 intelligence officer with the Anglo-Indian army in Mesopotamia. In 1919 assistant collector of customs at Basra. From 1924–37 inspector of Gulf agencies for the Anglo-Persian Oil Company, translator and assistant to Archibald Chisholm during negotiations for Kuwait oil concession. In 1937 retired to Basra where he remained, living as an Iraqi.

Wilson, Sir Arnold Toynbee (1884–1941). In 1908 with the Indian army and was posted to Persia to guard the site of the Anglo-Persian Oil Company from the native Persians. From 1912–14 assistant political officer Basra; 1915–17 deputy chief political officer; 1917–20 acting civil commissioner Mesopotamia/Iraq and acting resident Persian Gulf; presided over the tragic Iraqi Revolt and was removed from Iraq by the British government. In 1921 he was knighted by the Government of India and appointed as Sir Arnold Wilson, general manager Persia, Mesopotamia and the Gulf, Strick, Scott & Co. managing agents of Anglo-Persian. In 1922 became resident director Anglo-Persian Oil Company at Abadan, sharing title of joint general manager with T. L. Jacks. In 1926 transferred to London head office of Anglo-Persian as managing director D'Arcy Exploration Company; 1932 resigned from Anglo-Persian. From 1933–39 member of Parliament. In 1939 joined Royal Air Force; 1940 killed in action.

Yahya, bin Hamad al Din (1868–1948). Imam (religious leader) and hereditary ruler of Yemen, estimated in the 1920s to contain a population of three million, including Sunnis, perhaps one million Shia Zaidis, and Jews. It was believed the Imam could raise, and arm, a force of more than three hundred thousand men. During World War I he was neutral against the Turks with whom he had signed a ten-year treaty in 1911. The Imam shared the ardent Pan-Arab ideology of Sharif Husain of Mecca. Imam Yahya was assassinated in 1948.

Yateem, Muhammad and Ali, of the Bahrain Yateem family, influential merchants with connections throughout the Gulf and in Bombay. From 1922–34 Muhammad was personal assistant to Frank Holmes. (Frank and Dorothy Holmes took Muhammad's nephew, Husain, for education at Brighton.) After the Kuwait oil concession was concluded, Shaikh Ahmad of Kuwait offered the job of chief local representative, Kuwait Oil Company, to Muhammad Yateem. Not anxious to deal again with the Anglo-Persian Oil Company, Yateem refused it. He returned to the successful family business headquartered in Bahrain with branches in every Gulf state.

BIBLIOGRAPHY

INTERVIEWS

Archibald Chisholm, London, September 1987.
Husain Bin Ali Yateem, Bahrain, 1987–1988.
Violet Dickson, Kuwait, June 1987.

LIBRARIES AND ARCHIVE COLLECTIONS

Abu Dhabi Documentation Center, Abu Dhabi.
Bahrain Historical and Archaeological Society, Manama.
Bahrain Historical Documents Center, Manama.
Bahrain Petroleum Company (Bapco) Archives, Manama.
Cadman, Sir John. Papers and personal correspondence. University of Wyoming, American Heritage Center International Archive of Economic Geology, Laramie.
Dickson, Harold, Harry St. John Bridger Philby, and Charles Crane. Personal papers. Oxford University, Middle East Center, St. Antony's College.
Heim, Arnold. Collection and personal correspondence. University of Zurich, ETH Bibliothek.
Hoover, Herbert. Personal correspondence. Hoover Library, Philadelphia.
India Office Library and Oriental and India Office Collections, British Library, London.
Owen, Edgar Wesley. Collection and personal correspondence. University of Wyoming, American Heritage Center (International Archive of Economic Geology). Laramie.
Public Record Office, Kew.
Records of the Gulf Oil Corporation and Standard Oil of California relating to the Bahrain and Kuwait oil concessions and to the Bahrain Petroleum Company (Bapco) and Kuwait Oil Company (KOC).Chevron Oil Archives, San Francisco.
Records of the "Sons of Gwalia" gold mine, Kalgoorlie. J. S. Battye Library of West Australian History, West Australia State Library, Perth.
Scholefield, Guy, ed. *Dictionary of New Zealand Biography*. Papers. Alexander Turnbull Library, Wellington.

Twitchell, Karl Saben. Personal papers. Princeton University Library, Seeley G. Mudd Manuscript Library (Public Policy Papers and University Archives).

Ward, Thomas E. Collection and personal correspondence. University of Wyoming, American Heritage Center (International Archive of Economic Geology). Laramie.

PRIVATE COLLECTIONS

Engberg, Shirley. Correspondence with Holmes family members. Arizona, US, Edward Holmes, Nevada, US, and Dorrie Wilby, California, US.

Holmes, Dorothy. Personal papers, held by Stella Pendlebury, Barnstaple, UK.

Holmes family. Personal papers, held by Philip and Gerald Davidson, Perth, Australia.

Holmes, Frank. Personal papers, including personal diary 1914–1915, held by Peter Mort, Blandford, UK.

Holmes, Percy. Personal papers, held by Michael Stephenson of Cochabamba, Bolivia.

Rihani, Ameen. Personal papers, held by Albert Rihani at the Rihani Museum, Lebanon.

SPECIFIC DOCUMENTS (CHRONOLOGICAL ORDER)

IOL/R/15/1/317, vol. C10. Confidential—Note on interview between His Excellency the Viceroy (Curzon) and the shaikh of Bahrain on the November 27, 1903.

Abu Dhabi Documentation Center, Film M722/7, Records of the US Department of State Relating to Internal Affairs of Asia, 1910–28.

IOL/R/15/5/236, December 29, 1912. Confidential—Correspondence resident, political agent Kuwait, secretary to the Government of India in the Foreign Department Simla; Re: Shaikh Mubarak bin Sabah requesting government assist in furnishing the town of Kuwait with an adequate water supply.

IOL/R/15/5/236, January 14, 1913. Confidential office note from Captain Shakespear. For use of geologist deputed by government to visit Kuwait for investigation.

Australian National Library, Canberra: 1914 House of Commons Tabled Papers, vol. 54 (liv) Cd 7419. Final Report of the Admiralty Commission on the Persian Oil Fields, Explanatory Memorandum; vol. 54 (liv) Cd. 7419, Navy Oil Fuel: Agreement with the Anglo Persian Oil Company Limited. Presented to Parliament.

PRO/FO/371/2721, February–March 1916. Oil Fields in Asiatic Turkey. PRO. Abyssinia and Files of the Board of Trade Abyssinia.

PRO/ BT 66/11, October 1917–December 1918. Complete file, includes background and composition of the Edmund Davis syndicate(s).

PRO/CO/727/3, 1917. Colonial Office memo/correspondence Department of Overseas Trade.

IOL/L/PS/10, vol. 989. Political Intelligence Department, Foreign Office (undated but prepared about 1918). Secret: Memorandum on British Commitments (during the war) to the Gulf Chiefs.

PRO/WO/106/52, 1918. Secret: Correspondence regarding the Mesopotamian Expedition—Its Genesis and Development.

IOL/R/L/PS/11/193, 1921. Report of the Interdepartmental Committee appointed by the prime minister to make recommendations as to the formation of a new department under the Colonial Office to deal with mandated and other territories in the Middle East, January 31, 1921.

IOL/R/15/2/88, August 1923–February 1924. Dawasir Decamp to Nejd, complete file.

IOL/R/15/1/341/C25/19/166, May 1923–July 1923. Disturbance at Bahrain between Nejdis and Persians, complete file.

IOL/R/L/PS/10/1273 (PG [Sub] 18), December 12, 1929. Committee of Imperial Defense, Persian Gulf. Report by a Subcommittee on Political Control, Parts I–II and Appendix; "De Facto" position as regards political arrangements in the Persian Gulf.

Abu Dhabi Documentation Center, Film T1179/2. Reports of the US Department of State Relating to Internal Affairs of Saudi Arabia, 1930–44 and Film T1180/3. Records of the US Department of State Relating to Internal Affairs of Iraq, 1930–44.

IOL/R/15/1/623, vols. 82, 86. Confidential: Record of a meeting held at the Colonial Office on April 26, 1933, to discuss various questions relating to oil in the Persian Gulf.

IOL/R/15/5/242, vol. VII. Final record of a meeting (adjourned from April 26, 1933) held at the Colonial Office on May 3, 1933, to discuss various questions relating to oil in the Persian Gulf.

IOL/L/PS/12/10R/1007/8, vol. 3808, March 5, 1934. India Office: Confidential—Political Agreement between His Majesty's Government in the United Kingdom and the Kuwait Oil Company.

Chevron Archives, San Francisco, Box 0426154. Principal Documents in Regard to Kuwait Organization (1951) Part II, Background of the Kuwait Oil Concession, Its 50-50 Ownership by the D'Arcy Exploration Company (Anglo Iranian Oil Company Ltd.) and Gulf Exploration Company (Gulf Oil Corporation) through the Kuwait Oil Company . . . the restrictions of the Agreement of December 14, 1933.

India Office Library, London, 1876–1940. Administrative Reports of the Persian Gulf Residence, Administrative Reports of the Bahrain Political Agency, Administrative Reports of the Kuwait Political Agency, and Administrative Report on the Persian Gulf Political.

BOOKS AND ARTICLES

Al Baharna, Hussein Muhammad. *The Modern Gulf States: Their International Relations and the Development of Their Political, Legal and Constitutional Positions.* Beirut, 1973.

Al Tajir, Mahdi Abdalla. *Bahrain, 1920–1945: Britain, the Shaikh, and the Administration.* London: Croom Helm, 1987.

Amin, S. H. (Sayed Hassan). *Political and Strategic Issues in the Persian-Arabian Gulf.* Glasgow, Scotland: Royston, 1984.

Anonymous. "Aramco: An Arabian-American Partnership Develops Desert Oil and Places US Influence and Power in the Middle East." *Life*, March 28, 1949, 62–64, 66–78.

Anonymous. "The Great Oil Deals, US Business, Doing Business, Does What Government Cannot Do." *Fortune*, May 1947, 81–84, 143, 175–76, 179, 180, 182.

Antonious, George. *The Arab Awakening.* London: H. Hamilton, 1938.

Armstrong, H. C. *Lord of Arabia*. London: Penguin Books, 1924.

Aziz, Abdul, and Moudi Mansour. *King Abdul-Aziz and the Kuwait Conference, 1923–1924*. London: Echoes, 1993.

Baldry, John. "The Powers and Mineral Concessions in the Idrissi Imamate of Assir, 1910–1929." *Arabian Studies* 11. Cambridge: Middle East Center, University of Cambridge (1975): 76–107.

Barnes, John K. "The Romance of Persian Oil." Filed in Owen Papers, American Heritage Center. *The World's Work*, June 1920.

Beatty, Jerome. "Is John Bull's Face Red!" *American*, January 1939, 32, 33, 110, 111.

Belgrave, Charles. *Personal Column*. Beirut: Librarie du Liban, 1960.

Bell, Lady, ed. *The Letters of Gertrude B.* London: Ernest Benn, 1927.

Bibby, Geoffrey. *Looking for Dilmun*. London: Collins, 1970.

Bidwell, Robin. *The Two Yemens*. Harlow, Essex: Longman, 1983.

Bilovich, Yossef. "Great Power and Corporate Rivalry in Kuwait, 1912–1934: A Study in International Politics." PhD diss., University of London, 1982.

———. "The Quest for Oil in Bahrain, 1923–1930: A Study in British and American Policy." In *The Great Powers in the Middle East, 1919–1939*. Edited by Uriel Dann. New York: Holmes and Meier, 1988.

Blainey, Geoffrey. *The Rise of Broken Hill*. Melbourne: Macmillan of Australia, 1968.

Brown, Anthony Cave. *Oil, God, and Gold: The Story of Aramco and the Saudi Kings*. Boston: Houghton Mifflin, 1999.

Bullard, Reader, Sir. *The Last of the Dragomans: Sir Andrew Ryan, 1876–1949*. London: G. Bles, 1951.

Burner, David. *Herbert Hoover, A Public Life*. New York: Knopf, 1979.

Burton, Richard Francis. *The Gold-Mines of Midian*. London: Kegan Paul, 1878.

———. *The Ruined Midianite Cities*. London: Kegan Paul, 1878.

———. *The Land of Midian*. 2 vols. London: Kegan Paul, 1879.

Busch, Briton Cooper. *Britain and the Persian Gulf, 1894–1914*. Berkeley: University of California Press, 1967.

———. *Britain, India, and the Arabs, 1914–1921*. Berkeley: University of California Press, 1971.

Chisholm, Archibald H. T. *The First Kuwait Oil Concession Agreement: A Record of the Negotiations, 1911–1934*. London: F. Cass, 1975.

Clarke, Angela. *Bahrain Oil and Development, 1929–1989*. For Sixtieth Anniversary of Bahrain Petroleum Company. London: Immel Publishing, 1991.

Collins, Robert O. *An Arabian Diary: Sir Gilbert Falkingham Clayton*. Berkeley: University of California Press, 1969.

Cox, Percy. "Persia." In *Encyclopaedia Britannica* 17, 14th ed., 1929.

Craufurd, C. E. V. "The Dhofar District." *Geographical Journal* 53 (1919): 97–105.

———. "Lost Lands of Ophir." *Journal of the Central Asian Society* 14, pt. 3 (1927): 227–37.

———. *Treasures of Ophir*. London: Skeffingtton and Son, 1929.

———. "Yemen and Assir: El Dorado?" *Journal of the Royal Central Asian Society* 20, pt. 4 (October 1933): 568–77.

Davis, Fred, and Bill Taylor. "We Travel East of Suez." *Among Ourselves*. In-house magazine of Standard Oil of California. December 1930.

Dickson, H. R. P. *The Arab of the Desert*. London: Allen and Unwin, 1949.
———. *Kuwait and Her Neighbours*. London: Allen and Unwin, 1956.
Dickson, Violet. *Forty Years in Kuwait*. London: Allen and Unwin, 1971.
Eden, Sir Anthony. *The Memoirs of Sir Anthony Eden: Full Circle*. London: Cassell, 1960.
Faroughy, Abbas. *The Bahrein Islands (750–1951); A Contribution to the Study of Power Politics in the Persian Gulf, an Historical, Economic, and Geographical Survey*. New York: Verry, Fisher, 1951.
Ferrier, R. W. *The History of the British Petroleum Company: The Developing Years, 1901–1932*. Cambridge: Cambridge University Press, 1982.
Freeth, Zahra. *Kuwait Was My Home*. London: Allen and Unwin, 1956.
Fromkin, David. *A Peace to End All Peace: Creating the Modern Middle East, 1914–1922*. London: A. Deutsch, 1989.
Gibran, Khalil. *The Prophet*. London: William Heinemann, 1926.
Glubb, John Bagot. *The Story of the Arab Legion*. London: Hodder and Stoughton, 1948.
———. *The Changing Scenes of Life: An Autobiography*. London: Quartet Books, 1983.
Goldschmidt, Arthur, Jr. *A Concise History of the Middle East*. 3rd ed. Boulder: Westview Press, 1988.
Graves, Philip. *The Life of Sir Percy Cox*. London: Thornton Butterworth, 1939.
Grutz, Jane Waldron. "Prelude to Discovery." *Aramco World* 50, no. 1 (January–February 1999): 31–34.
Haggard, H. Rider (Henry Rider, 1856–1926). *King Solomon's Mines*. London: Cassell, 1966. Reprinted as a Nelson Classic.
Hawley, Donald. *The Trucial States*. London: Allen and Unwin, 1970.
Hay, Rupert, Sir. *The Persian Gulf States*. Washington, DC: Middle East Institute, 1959.
Helms, Christine Moss. *The Cohesion of Saudi Arabia: Evolution of Political Identity*. London: Croom Helm, 1981.
Henry, J. D. *Baku: An Eventful History*. London: A. Constable, 1905.
Hewins, Ralph. *Mr. Five Per Cent: The Story of Calouste Gulbenkian*. New York: Rinehart, 1958.
———. *A Golden Dream: The Miracle of Kuwait*. London: W. H. Allen, 1963.
Hickey, Michael. *Gallipoli*. Melbourne: J. Murray, 1995.
Hitti, Philip K. *History of the Arabs from the Earliest Times to the Present*. 10th ed. London: Macmillan, 1970.
Hodgkin, E. C., ed. *Two Kings in Arabia: Letters from Jedda, 1923–5 and 1936–9*. Reading, UK: Ithaca Press 1993.
Holden, David, and Richard Johns. *The House of Saud*. London: Sidgwick and Jackson, 1982.
Hoover, Herbert. *Memoirs of Herbert Hoover: Years of Adventure, 1874–1820*. Philadelphia: Hollis Carter, 1952.
Hope, Stanton. *Arabian Adventurer: The Story of Haji Williamson*. London: R. Hale, 1951.
Howarth, David. *The Desert King*. London: William Collins, 1965.
Ingrams, William Harold. *The Yemen, Imams, Rulers and Revolutions*. London: John Murray, 1963.
Ireland, Philip Willard. *Iraq; A Study in Political Development*. New York: Russell and Russell, 1937.

James, Lawrence. *Raj: The Making and Unmaking of British India*. Boston: Little, Brown, 1997.

———. *The Rise and Fall of the British Empire*. 3rd ed. London: Abacus, 1998.

Jerrold, Douglas. *The Royal Naval Division: With an Introduction by the Right Hon. Winston S. Churchill*. London: Hutchinson, 1923.

Kelly, J. B. *Arabia, the Gulf and the West: A Critical View of the Arabs and Their Oil Policy*. London: Weidenfeld and Nicolson, 1980.

Kent, Marian. *Oil and Empire: British Policy and Mesopotamian Oil, 1900–1920*. London: Macmillan, 1976.

Klieman, Aaron S. *Foundations of British Policy in the Arab World: The Cairo Conference of 1921*. Baltimore: Johns Hopkins Press, 1970.

Kostiner, Joseph. "Britain and the Northern Frontier of the Saudi State, 1922–1925." In *The Great Powers in the Middle East, 1919–1939*. Edited by Uriel Dann. New York: Holmes and Meier, 1988.

Lacey, Robert. *The Kingdom*. London: Hutchinson, 1981.

Larson, Henrietta M., Evelyn H. Knowlton, and Charles S. Popple. *History of Standard Oil Company (New Jersey)*. Vol. 3, *New Horizons, 1927–1950*. New York: Harper, 1971.

Leatherdale, Clive. *Britain and Saudi Arabia, 1925–1939: The Imperial Oasis*. London: Frank Cass, 1983.

Lebkicher, Roy. "Aramco and World Oil." In *Arabian American Oil Company Handbook for American Employees* 1. New York: Russell F. Moore, 1950.

Lees, G. M. "The Physical Geography of South Eastern Arabia." Paper presented to the Royal Geographical Society London, January 16, 1928. *Geographical Journal* 71 (January–June 1928).

Levy, J. M. "US Company Wins Arabian Oil Grant." *New York Times*, August 8, 1939.

Liebesny, Herbert J. "International Relations of Arabia: The Dependent Areas." *Middle East Journal* 1 (1947): 148–68.

———. "Documents: British Jurisdiction in the States of the Persian Gulf." *Middle East Journal* 2 (1949): 330–32.

———. "Administration and Legal Development in Arabia, Aden Colony and Protectorate." *Middle East Journal* 9 (1955): 335–96.

———. "Administration and Legal Development in Arabia: the Persian Gulf Principalities." *Middle East Journal* 10 (1956): 33–42.

Longhurst, Henry. *Adventure in Oil: The Story of British Petroleum*. London: Sidgwick and Jackson, 1959.

Longrigg, Stephen Helmsley. *Oil in the Middle East: Its Discovery and Development*. London: Oxford University Press, 1954.

Mansfield, Peter. *The Arabs*. London: Penguin Books, 1982.

Marlowe, John. *The Persian Gulf in the Twentieth Century*. London: Cresset, 1962.

———. *Late Victorian: The Life of Sir Arnold Talbot Wilson*. London: Cresset, 1967.

Miller, Aaron David. *Search for Security: Saudi Arabian Oil and American Foreign Policy, 1939–1949*. Chapel Hill: University of North Carolina Press, 1980.

Mineau, Wayne. *The Go Devils*, London: Cassell, 1958.

Monroe, Elizabeth. *Philby of Arabia*. London: Faber and Faber, 1973.

———. *Britain's Moment in the Middle East, 1914–1971*. London: Chatto and Windus, 1981.

Moore, Frederick Lee, Jr. *Origin of American Oil Concessions in Bahrain, Kuwait and Saudi Arabia*. BA thesis, Department of History, Princeton University, 1948.

Moorehead, Alan. *Gallipoli*. London: H. Hamilton, 1956.

Mosley, Leonard. *Power Play: Oil in the Middle East*. New York: Penguin Books, 1973.

Mulligan, Bill. "Holmes: Negotiator Par Excellence." *Arabian Sun*. In-house magazine of Aramco Saudi Arabia, March 7, 1984.

Nash, George H. *The Life of Herbert Hoover: The Engineer, 1874–1914*. New York: Norton, 1973.

Odell, Peter R., and Luis Vallenilla. *The Pressures of Oil: A Strategy for Economic Revival*. New York: Harper and Row, 1978.

Owen, Wesley Edgar, ed. *Trek of the Oil Finders: A History of Exploration for Petroleum*. Semicentennial Commemorative Volume. Tulsa, OK: American Association of Petroleum Geologists, 1975.

Page, Bruce, David Leitch, and Phillip Knightley. *Philby, The Spy Who Betrayed a Generation*. London: Deutsch, 1968.

Pascoe, E. H. *Geological Notes on Mesopotamia with Special Reference to Occurrences of Petroleum*. Vol. 43, Memoirs of the Geological Survey of India, 1922. Australian National University Library, Document Search Service.

Philby, H. St. J. B. *Arabian Days, An Autobiography*. London: R. Hale, 1948.

———. *Arabian Jubilee*. London: Hale, 1952.

———. *Arabian Oil Ventures*. Washington, DC: Middle East Institute, 1964.

Philby, Kim. *My Silent War*. London: MacGibbon and Kee, 1968.

Pilgrim, Guy E. *The Geology of the Persian Gulf and the Adjoining Portions of Persia and Arabia*. Vol. 34, pt. 4. Memoir, 1908. Australian National University Library, Document Search Service: Geological Survey of India.

Pratt, Wallace E. "The Value of Business History in the Search for Oil." Paper presented to Harvard School of Business Administration. Filed in Owen Papers, American Heritage Center, 1960.

Redwood, Sir Boverton. *Petroleum: A Treatise*. 2nd ed. London: C. Griffin, 1906.

Rich, Claudius James. *Narrative of a residence in Koordistan, and on the site of ancient Ninevah, with journal of a voyage down the Tigris. Posthumous*, 1836.

Rihani, Ameen. *Ibn Sa'oud of Arabia: His People and His Land*. Delmar, NY: Caravan Books, 1983. First published 1928 by Constable & Co.

———. *Arabian Peak and Desert: Travels in Al Yaman*. London: Constable, 1930.

———. *Around the Coasts of Arabia*. Delmar, NY: Caravan Books, 1983. First published 1930 by Constable & Co.

Rumaihi, Muhammad G. *Bahrain: Social and Political Change Since the First World War*. London: Bowker, 1976.

———. *Beyond Oil: Unity and Development in the Gulf*. London: Al Saqi Books, 1986.

Saikal, Amin. *The Rise and Fall of the Shah, 1941–1979*. London: Angus and Robertson, 1980.

Sampson, Anthony. *The Seven Sisters: The Great Oil Companies and the World They Made*. London: Hodder and Stoughton, 1975.

Scholefield, Guy H. "A Tough Patient New Zealander." *Otago Daily Times*. n.d., 58–60.

———. "NZ Man Walked into Confidence of Sheik Oilmen." *Post*. n.d.

Scoville, Sheila A., ed. *Gazetteer of Arabia: A Geographical and Tribal History of the*

Arabian Peninsula, vols. 1–4. Graz, Austria: Akademische Druck Verlagsanstalt, 1979.

Shwadran, Benjamin. *The Middle East: Oil and the Great Powers*. London: Atlantic, 1956.

Simons, G. L. *Iraq: From Sumer to Saddam*. New York: St. Martin's Press, 1994.

Sparrow, Geoffrey, and J. N. Macbean-Roos. *On Four Fronts with the Royal Naval Division*. London: Hodder and Stoughton, 1918.

Stegner, Wallace. *Discovery! The Search for Arabian Oil*. Abridged for Aramco World. Beirut, Lebanon: Middle East Export Press, 1971.

Thesiger, Wilfred. *Arabian Sands*. London: Longmans Green, 1959.

———. *The Life of My Choice*. London: Collins, 1987.

Trench, Richard. "Windmills in the Sand." *Near East Business* 6, no. 6 (May 1981): 14–17.

———. "The King's Good Servant." *Near East Business* (September–October 1981): 16–19.

Troeller, Gary. *The Birth of Saudi Arabia: Britain and the Rise of the House of Sa'ud*. London: F. Cass, 1976.

Tugendhat, Christopher. *Oil: The Biggest Business*. New York: Putnam, 1968.

Tuson, Penelope. *The Records of the British Residency and Agencies in the Persian Gulf*. London: India Office Library and Records, 1979.

Twitchell, K. S., with Edward J. Jurji. *Saudi Arabia, with an Account of the Development of Its Natural Resources*. NJ: Princeton University Press, 1947.

Van der Meulen, D. *The Wells of Ibn Sa'ud*. London: Murray, 1957.

Van Ess, John. *The Spoken Arabic of Iraq*. 2nd ed. with rev. and additional vocabulary. London: Oxford University Press, 1917.

Wahba, Hafiz. *Arabian Days*. London: A. Barker, 1964.

Ward, Thomas E. *Negotiations for Oil Concessions in Bahrain, El Hasa (Saudi Arabia), the Neutral Zone, Qatar and Kuwait*. New York: privately published, 1965.

Whelan, John, ed. "Kuwait," "Bahrain," "Qatar," "Saudi Arabia," "Oman," and "UAE." In *MEED Practical Guide*. Edited by John Whelan. London: Middle East Economic Digest, 1980–1985.

White, Gerald T. *Formative Years in the Far West: A History of Standard Oil Company of California and Predecessors through 1919*. New York: Appleton-Century-Crofts, 1962.

Williams, Kenneth. *Ibn Sa'ud, The Puritan King of Arabia*. London: J. Cape, 1933.

Wilson, Arnold Talbot, Sir. *Mesopotamia, 1914–1917: A Personal and Historical Record*. London: Oxford University Press, 1930.

———. *Mesopotamia, 1917–1920: A Clash of Loyalties*. London: Oxford University Press, 1931.

Yergin, Daniel. *The Prize: Epic Quest for Oil, Money, and Power*. New York: Simon and Shuster, 1991.

INDEX

541